Coastal Environment and Water Quality

Coastal Environment and Water Quality

Proceedings of the AIH 25th Anniversary Meeting & International Conference "Challenges in Coastal Hydrology and Water Quality"

Edited by

Y. Jun Xu

Vijay P. Singh

Water Resources Publications, LLC WRP Highlands Ranch

For Information and Correspondence:

Water Resources Publications, LLC
P. O. Box 260026, Highlands Ranch, Colorado 80163-0026, USA

Coastal Environment and Water Quality

Proceedings of the AIH 25th Anniversary Meeting & International Conference "Challenges in Coastal Hydrology and Water Quality"

Edited by

Y. Jun Xu

Vijay P. Singh

ISBN-13: 978-1-887201-47-6
ISBN-10: 1-887201-47-5

U.S. Library of Congress Catalog Card Number: 2006924765

This publication is printed and bound in Chelsea, Michigan, United States of America.

The production of this book is supported by funds generously provided by

bp

PREFACE

Today, more than half of the world's population lives within 100 km of the coastline, and this number is expected to increase by 25% in the coming two decades. Population growth and urbanization have greatly altered coastal environments and the natural pathways of coastal water. With most of the water on land entering into the seas and oceans, man's activities in both coastal and inland watersheds have led to increased loadings of sediment, nutrients, toxic substances and pathogens with subsequent eutrophication, algal blooms, and hypoxia. Recent climate change has also led to a rise in sea level resulting in loss of coastal wetlands and saltwater intrusion. Unpredictable transformations in weather patterns that the world is currently witnessing are also being attributed to climate change. The world's coasts face increasingly serious environmental challenges, while comprehensive understanding and measurement of the complexities in coastal systems has yet to occur. Two recent devastating hurricanes, Katrina and Rita, that struck the coastal areas along the Gulf of Mexico - and more specifically the Louisiana and Mississippi coasts - in 2005, are a vivid reminder that man's relationship with nature is not only exceedingly fragile, but also highly unpredictable. The level of destruction from these storms is so widespread and staggering, that it is hard to say if these areas will ever be able to regain their previous vitality and prominence.

This conference, although more than two years in the making, is being held at a time when large parts of coastal Louisiana and Mississippi are still in various stages of recovery and while we continue to look for strategies to prevent large-scale impacts of extreme weather events, such as Katrina and Rita, in the world's coastal regions. This conference offers an opportunity to discuss a range of scientific and technical topics that are among the most critical issues in coastal regions. Some of these issues, such as wetland loss and restoration, climate change and sea level rise, and eutrophication and hypoxia in coastal waters, are of global significance. Located in the Northern Gulf of Mexico at the mouth of the Mississippi River, Louisiana is in a unique geographical location, ideal for attracting researchers, educators, practitioners, and industries working in coastal hydrology and water quality fields throughout the world to this conference. We hope this conference will help raise public and governmental awareness of a range of challenging environmental problems, such as erosion and land subsidence in coastal Louisiana.

Natural forces, such as floods, hurricanes, and tsunamis, can have a profound impact on the delicate balance of society and nature as these forces have the potential to be very destructive in our lives. On the other hand, we humans are not without blame, building and developing in coastal areas and thereby increasing our chances of confronting disasters head on. In retrospect, it is easy to argue what should have been done to help avoid this recent disaster, but it is the actions we take now that will play the most important role in the future. As Katrina and Rita have proven, we cannot conquer nature, but we can try and understand our limitations in controlling and interacting with our environment.

With this purpose in mind, the international conference on "Challenges in Coastal Hydrology and Water Quality" was organized. The objectives were to provide an international forum for the dissemination and exchange of the current art, science and technology on coastal hydrology, water quality and their relationships to environmental problems; promote interdisciplinary research on complex environmental systems in coastal regions from physical, biogeochemical, and socioeconomic perspectives; and foster interactions amongst climatologists, meteorologists, engineers, hydrologists, hydraulicians, geologists, hydrogeologists, environmentalists, ecologists, agricultural scientists, environmental and water resources planners, modelers, social scientists, economists, managers, and policy and decision makers under one roof.

We received a gratifying response to our call for papers. The number of abstracts received exceeded 120. Each abstract was carefully reviewed, and we accepted 81 full-length papers for publication in the conference proceedings. The subject matter of the conference was divided into 15 topics encompassing virtually all major aspects of coastal water and environment. Each topic comprises a number of contributed papers. These papers provide a natural blend to reflect a body of knowledge on each topic.

The 41 papers contained in this book, ***Coastal Environment and Water Quality***, represent one volume of the conference proceedings. The other volume is embodied in a companion book entitled, ***Coastal Hydrology and Processes***, which has been published simultaneously. Arrangement of the contributions in these two books under distinctive titles was a natural consequence of the diversity of the papers presented at the conference, and the topics covered therein. These books can be treated as independent themes, although considerable interconnectedness exists between them.

This book contains seven sections reflecting various aspects of coastal environment and water quality. Section 1 deals with coastal water pollution which has received considerable attention in recent decades. Section 2 looks at saltwater intrusion which is of fundamental importance for the management of coastal aquifers. Section 3 covers coastal water eutrophication. This subject takes on an added significance in light of growing hypoxia due to nutrient enrichment in the Gulf of Mexico. Section 4 deals with best management practices currently being employed for water quality management. Taken together, these first four sections constitute the building blocks for water quality modeling, which is the subject of section 5. Extending this discussion further, section 6 encompasses coastal ecology, wetlands and environment. In conclusion, section 7 presents some of the impacts of hurricanes Katrina and Rita, and discussion on recovery and rebuilding in the future. Lastly, we have developed a subject index based upon the authors' submitted keywords, which directs readers to the papers of interest for each topic.

We anticipate that researchers and practitioners in the field of coastal hydrology and water resources, environmental science and engineering, agricultural and forest resources, earth sciences, and watershed and range sciences, as well as those engaged in environmental resources planning and management will find this book to be of interest. Graduate students, as well as those wishing to conduct further research in coastal hydrology, environmental science and engineering, and coastal water resources planning, development and management, will also find the book to be of value.

The conference attracted a significant number of nationally and internationally well-known figures, who have long been at the forefront of coastal hydrology and environment education, research, development, management, and practice. Nearly 20 countries covering 5 continents, including a number of the major countries active in coastal water and environmental management throughout the world were represented. It is hoped that long-lasting, productive personal and professional associations and friendships will be developed as a result of this conference.

Y. Jun Xu, Conference Co-chair
Vijay P. Singh, Conference Co-chair
Baton Rouge, Louisiana, USA

ACKNOWLEDGMENTS

The editors wish to acknowledge the efforts of the hundreds of scientists, engineers, and practitioners who have contributed their research papers to this conference. We appreciate their coming to our state at a time when people in this part of the world are looking for support and encouragement. Numerous others have contributed to the conference in many ways, and the lack of space does not allow us to thank all of them by name here. This conference was made possible by the technical support of the paper presenters and keynote speakers. This book is the direct result of their collective efforts and contributions. The session chairpersons and co-chairpersons administered the sessions in a positive and professional manner. We owe our deep gratitude to all of these individuals and their organizations.

Our sincere appreciation goes out to all the members of the International Advisory Committee and the Local Organizing Committee, American Institute of Hydrology, Louisiana State University, and Louisiana State University Agricultural Center, Baton Rouge, U.S.A. Without their support and advice this conference would not have come to fruition.

The production of this book is supported by funds generously provided by BP America, for which we are grateful. We sincerely appreciate the great support from LSU alumnus Mr. Keith Seilhan, Asset Manager - NaKika, Deep Water Production BP America, and Ms. Ayana McIntosh-Lee, Director of Public Affairs of BP/Houston. LSU Foundation was instrumental in directing support and funds towards this conference, and we would especially like to thank Dr. Jeff Hale, Interim Vice President for Development, and Ms. Pinki Diwan, Director of Donor, Corporate & Foundation Programs, for their invaluable assistance.

With respect to their more direct contribution to the completion of this book, the editors wish to thank our graduate students who have helped with the preparation of papers for publication at various stages: Philip Saksa, Biao Zhong, April Mason, Adrienne Viosca, and Fugui Wang. We are also very appreciative of Water Resources Publications, LLC, for the professionalism and ethics of its staff. We especially thank Ms. Branka McLaughlin for her efforts in working towards the high quality completion of this book under an ambitious schedule. Her dedication to this project kept the pressure on to get the work done in a professional and timely manner.

The International Conference on Challenges in Coastal Hydrology and Water Quality was sponsored and co-sponsored by a number of organizations. These sponsors offered direct financial and/or inkind support to the conference. Their generosity towards supporting coastal water resources and environment research and education is gratefully acknowledged. The co-sponsors offered support by promoting the conference through their periodicals, newsletters, magazines, and internet websites. The following is a list of conference sponsors and co-sponsors:

Sponsors:

BP
American Institute of Hydrology
Louisiana State University
Louisiana State University Agricultural Center
Louisiana State University Foundation
The Shaw Group
URS Corporation
YSI Environmental
Baton Rouge Area Convention and Visitors Bureau
U.S. Army Corps of Engineers, New Orleans District

Co-Sponsors:

American Association for the Advancement of Science (AAAS)
American Geological Institute (AGI)
American Geophysical Union (AGU)
Association of State Wetland Managers, Inc. (ASWM)
Australian Government Emergency Management
Beijing Forestry University
Canadian Climate Impacts and Adaptation Research Network
Climatic Research Unit (CRU) at University of East Anglia, UK
Coastal Zone Research Network UK
Delaware Water Resources Center
Geological Society of America (GSA)
Great Lakes Regional Pollution Prevention Roundtable
GulfBase - Resource Database for Gulf of Mexico Research
Illinois-Indiana Sea Grant College Program
International Association of Hydrological Sciences (IAHS)
International Geosphere-Biosphere Programme (IGBP)
LaCoast
Lake Pontchartrain Basin Foundation
Louisiana Department of Natural Resources
Louisiana Department of Environmental Quality
Louisiana Department of Wildlife and Fisheries
Natural Hazards Center, University of Colorado at Boulder
Nordic Association for Hydrology (NHF)
Physical Sciences Information Gateway UK
Renewable Natural Resources Foundation
Southern Regional Water Program (SRWP)
U.S. Army Corps of Engineers, Engineer Research & Development Center
U.S. Department of Agriculture, Natural Resources Conversation Services Louisiana
U.S. Department of Agriculture, Forest Service, Southern Research Station
U.S. Department of Agriculture, Water Quality Information Center
U.S. Environmental Protection Agency, Mississippi River Basin & Gulf of Mexico Hypoxia
U.S. Geological Survey, Surface-Water Quality and Flow Modeling Interest Group
U.S. Water News
Virginia Water Resources Research Center
Water Conservation and Use in Agriculture InfoNET
Water Education Foundation
Water Quality Group Newsletter, North Carolina State University
Wisconsin Ground Water Association

TABLE OF CONTENTS

Section 1:

COASTAL WATER CONTAMINATION

1

HEAVY METALS AND HYDROCARBONS IN BEAUFORT LAGOON SEDIMENTS, NORTH ARCTIC ALASKA

Debasmita Misra[1], Sathy A. Naidu[2], John J. Kelley[2], M. Indira Venkatesan[3], and Franz J. Mueter[4]

[1] College of Engineering and Mines, P.O. Box 755800
University of Alaska Fairbanks, Fairbanks, AK 99775-5800, USA

[2]Institute of Marine Sciences, P.O. Box 757220
University of Alaska Fairbanks, Fairbanks, AK 99775-7220, USA

[3]Institute of Geophysics and Planetary Physics
Box 951567, 5863 Slichter
University of California, Los Angeles, CA 90095-1567, USA

[4]697 Fordham Drive, Fairbanks, AK 99709, USA

ABSTRACT

The arctic region, although most of which lies far away from major industries, could still be exposed to industrial contaminants. Anthropogenic discharges of chemicals and radioactive nuclides into arctic Russian margin have raised the possibility of transport of contaminants by ocean currents and atmosphere into the Alaskan seas. To understand the sources and levels of contaminants in the Beaufort Lagoon sediments, North Arctic Alaska, we have monitored the concentrations and historical changes of total Hg, Fe, Mn, Sn, Cu, Cr, Ni, V, Pb, Zn, Cd, Ba, and As and, saturated and polycyclic aromatic hydrocarbons. The study indicated that the sediments have remained generally uncontaminated in the last three decades, despite the increased petroleum-related industrial activities.

1.1. INTRODUCTION

During the past 30 years there has been increasing industrial growth in the North Slope of northern Arctic Alaska, relating to the activities of exploration and development of the onshore and offshore petroleum reserves. Concurrent to this growth increasing urbanization of the villages along the North Slope coast (e.g., Barrow, Kaktovik) has occurred. Additionally, there have been military-related [Distant Emergency Warning (DEW) line stations, rocket firing experiments] and enhanced municipal activities (sewage-treatment, power-generation, gravel mining). The petroleum-related developmental tasks include seismic surveys, drilling from onshore and offshore platforms, marine boat traffic, building of causeways, laying of submarine pipelines, among others. There is a concern that the anthropogenic activities can lead to contamination of the environment with metals and hydrocarbons, which could have a deleterious impact on marine biota. It has been theorized that polar food chains are relatively sensitive and susceptible to contaminants (Chapman and Riddle, 2005). Several environmental factors unique to the Arctic may specially contribute to higher levels of contaminant accumulation and biomagnification of toxic metals and organics (Naidu et al., 2001 and references therein).

Several investigations have been conducted since early 1970s to gather time-intervals data on metals and hydrocarbons in north Alaskan arctic nearshore sediments with an objective to monitor contaminants. Marine sediments were preferred as they represent a suitable environmental integrator and sink for particle-reactive contaminants (Forstner and Witmann, 1979; Chapman et al., 1998). Impacted sediments could be a major source of contaminants to the benthos (Lee and Wiberg, 2002; Mountouris et al., 2002) and other organisms, which have a close link with sediments (Naidu et al., 2001 and references therein). The earliest data on trace metals in the nearshore sediments of the Alaskan Beaufort Sea are those collected by Naidu and Mowatt (1975), Sweeney (1984), and Sweeney and Naidu (1989). Subsequent regional investigations on sediment trace metals were by Boehm et al. (1987), Crecelius et al. (1991), and Naidu et al. (2001, 2003, 2005). Additional site-specific studies were conducted to assess the environmental impact of discharge of metals entrained in drilling effluents (Snyder-Conn et al., 1997) and of activities near offshore drilling platform (Trefry et al., 2003). Investigations on hydrocarbons in sediments of the North Slope region appear in Shaw et al. (1979), Venkatesan and Kaplan (1982), Boehm et al. (1987), Steinhauer and Boehm (1992), and Naidu et al. (2001, 2003, 2005).

The objective of this paper is to synthesize the historical changes in the concentrations of a suite of trace metals and hydrocarbons in the sediments of Beaufort Lagoon (Fig. 1).

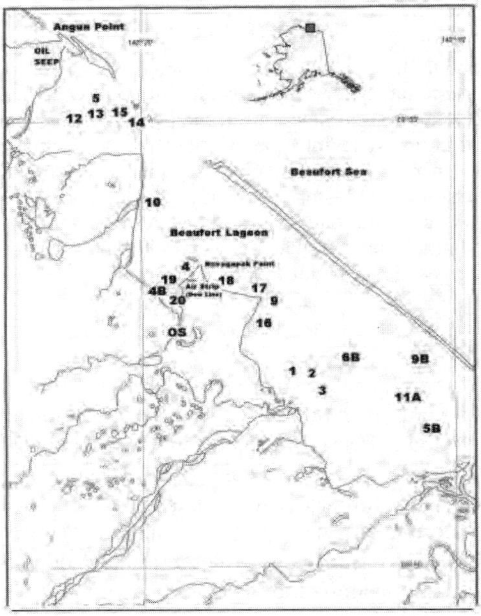

Fig. 1. Study area showing locations of sediment samples in the Beaufort Lagoon (Inset map of Alaska showing the location of the sampling sites).

This location is of particular interest for contaminant studies because it offers a range for comparing sediment chemistries between site specific sectors of the lagoon, which have been exposed to 1) a long-term natural oil seep, 2) past activities relating to a Dew Line military station on the lagoon shore and 3) sites that have presumably remained relatively pristine. The results of this study will provide base lines of trace metals and hydrocarbons for monitoring contaminants in the nearshore environment off the Arctic National Wildlife Refuge (ANWR), particularly of the refuge's 1.5-million acre coastal plain, which has potential petroleum reserves. The latter region is one of the most contentious areas for oil drilling and production because of possible impact on a variety of wildlife habitats there.

1.2. DATABASE: SOURCE, MATERIALS AND METHODS

The Beaufort Lagoon is located at the eastern margin of the north Alaskan Arctic coast and far from intense industrial activities (Fig.1). It consists of three contrasting environments: two of them, the Angun Point in the northwest and the Nuvagapak Point around a landing strip (Fig.1) have been exposed locally to long-term natural oil seeps and recent military activities, respectively. The latter activities were related to an abandoned DEW line station and an airstrip still active (Fig.1). The military operations may have exposed the lagoon adjacent to the station with refined petroleum products and metal contaminants. The rest of the lagoon presumably has remained pristine. It is anticipated that the chemistries of the sediments from the three diverse environments will be significantly different.

This paper draws essentially upon our data on trace metals and hydrocarbons in sediments included in several unpublished reports (Naidu et al., 2001, 2003, 2005). The concentrations of Fe, Mn, Sn, Cu, Cr, Ni, V, Pb, Zn, Cd, Ba, and As are in the mud fraction (<63 μm) of sediments, whereas the concentration of total Hg and hydrocarbons are in gross sediments. As the details in the analytical procedures, including the QA/QC, are included in the above reports, we describe them briefly in the following.

Mercury was analyzed by cold vapor atomic absorption spectrometry (Bloom, 1992; Bloom et al., 1999). For rest of the metals, an aliquot of the mud powder (<62 μm size) was dissolved in HF-HNO_3 mixture in a Teflon bomb and digested in a microwave oven. The resulting solution was analyzed for the 12 metals using either a Zeeman graphite furnace atomic absorption spectrometry, an ICP/MS, or by Excaliber automated hydride generation atomic fluorescence spectrometry. Hydrocarbon analysis for *n*-Alkanes, polycyclic aromatic hydrocarbons (PAHs), triterpanes and steranes were performed according to well-established methods (Venkatesan et al., 1987; Venkatesan, 1994). After solvent extraction and separation into individual compound classes, the alkanes were quantified by gas chromatograph (GC), using a flame ionization detector. Tricyclic di- and pentacyclic triterpenoids, steroids and PAHs were measured by GC/mass spectrometry. The contents of organic carbon (OC) and total nitrogen (N) and their stable isotopes ($\delta^{13}C$ and $\delta^{15}N$) were determined on carbonate-free muds, using a Thermo Finnigan Model Delta Plus XP isotope ratio mass spectrometer (Naidu et al., 2000). The values of $\delta^{13}C$ ($^{o}/_{oo}$) are referenced to the V-PDB standard and those of $\delta^{15}N$ to an air standard. Sediment grain size analysis was after Folk (1968).

Statistical Analysis consisted of determination of the correlation coefficients between trace metals, OC (organic carbon), nitrogen (N), and the silt and clay percents in the mud fraction, to deduce the geochemical partitioning of the metals in the mud. The time-intervals comparison was obtained between the concentrations of selected trace metals in sandy muds (>75% silt and clay, after Naidu, 1981) collected in 1977, and mud of 2003, using student 't' test. Cluster analysis (Euclidean distances, Ward's linkage) was undertaken to identify grouping, if any, of stations based on the trace metal compositions adjacent to presumptive point sources and the pristine lagoon. This was followed by principal component analysis of the trace metal data to correlate station group separation by cluster analysis with sediment variables. Surface trend analysis was conducted using trace metal data, granulometry and OC content to elucidate if there are any geographical gradient(s) in the individual sediment parameters and the metals, from presumptive point sources, for example the oil seeps and the abandoned DEW line station around Angun and Nuvagapak Points, respectively. BLO3-15 was excluded from the cluster, the principal component, and the surface trend analyses.

1.3. RESULTS

Table 1 includes for the mud of the individual samples the percents of gravel, sand, silt, clay, mud in gross sediments, and OC and N, OC/N ratios (weight to weight basis of OC

Table 1. Grain size, organic carbon (OC), nitrogen (N), OC/N, and stable isotopes of OC ($\delta^{13}C$ $^o/_{oo}$) and N ($\delta^{15}N$ $^o/_{oo}$) of Beaufort Lagoon sediments.

Sample	Gravel (%)	Sand (%)	Silt (%)	Clay (%)	Mud (%)	OC(%)	N (%)	OC/N	$\delta^{13}C$	$\delta^{15}N$
BL03-1	0.00	6.43	77.39	16.18	93.57	2.39	0.22	10.86	-26.94	1.88
BL03-2	0.00	42.28	36.63	21.09	57.72	2.33	0.20	11.60	-26.89	2.02
BL03-3	0.00	18.11	51.87	30.02	81.89	2.17	0.20	10.85	-26.36	0.83
BL03-4	0.00	95.55	0.09	4.36	4.45	1.72	0.17	10.11	-24.98	2.90
BL03-4B	92.87	2.54	1.92	2.66	4.59	3.29	0.32	10.28	-26.99	3.09
BLO3-5	NA	NA	NA	NA	NA	NA	NA	NA	NA	NA
BL03-5B	0.00	13.94	50.36	35.70	86.06	2.00	0.18	11.11	-26.24	1.60
BL03-6B	0.00	6.56	50.22	43.22	93.44	2.02	0.19	10.63	-26.49	2.05
BL03-9	0.28	33.60	42.13	23.98	66.12	3.96	0.36	11.00	-27.35	1.95
BL03-9B	0.00	27.86	48.44	23.69	72.14	2.27	0.19	11.94	-26.24	1.79
BL03-10	0.00	81.25	13.24	5.50	18.75	2.38	0.20	11.90	-27.14	3.41
BL03-11A	0.00	22.51	55.19	22.30	77.49	1.55	0.13	11.92	-27.26	1.07
BL03-12	0.00	19.01	58.97	22.02	80.99	2.04	0.17	12.00	-26.85	2.21
BL03-13	0.00	89.76	7.45	2.79	10.24	1.43	0.12	11.91	-26.85	2.07
BL03-14	0.00	77.97	9.18	12.85	22.03	1.66	0.15	11.07	-26.96	1.75
BL03-15	0.00	83.42	3.15	13.42	16.58	1.76	0.17	10.35	26.49	1.76
BL03-16	0.00	24.24	52.46	23.30	75.76	3.20	0.26	12.30	-27.53	1.80
BL03-17	0.12	97.88	0.05	1.95	1.99	2.57	0.30	8.56	-25.84	3.12
BL03-18	0.00	25.17	36.81	38.02	74.83	2.60	0.23	11.30	-26.85	2.83
BL03-19	28.60	63.62	1.34	6.45	7.78	4.05	0.38	10.66	-26.76	1.80
BL03-20	37.75	58.80	0.72	2.73	3.45	3.75	0.37	10.13	-27.21	2.23
OS	NA	NA	NA	NA	NA	NA	NA	NA	NA	NA

and N) and stable isotopes of OC and N ($\delta^{13}C$ and $\delta^{15}N$). In Table 2 are shown the statistics of the concentrations (on a dry weight basis) of 12 metals (V, Cr, Mn, Ni, Cu, Zn, As, Cd, Sn, Ba, Pb and Fe) in the mud fraction and of total Hg in the gross sediments on the 21 samples. The data of the concentrations of the 21 samples are available at a website due to space constraints in this paper (ftp://ftp.ims.uaf.edu/naidu/).

The correlation coefficients (r values) determined between the concentrations of silt, clay, OC, N and the 13 metals are shown in Table 3. The concentrations of the *n*-alkanes,

Table 2. Statistics of concentrations of metals (μg/g, except for Fe in %) in the mud fraction (<62 μm size) and total mercury (ng/g) in gross sediments of Beaufort Lagoon.

Statistics	V	Cr	Mn	Ni	Cu	Zn	As	Cd	Sn	Ba	Hg	Pb	Fe
A-Mean	120	73.0	608	39.0	37.0	94	14.0	0.265	2.00	505	57.0	17.0	2.9751
Std.AM	13	7.0	643	6.8	8.9	17	3.2	0.110	0.98	85	37.3	4.1	0.4004
G-Mean	119	73.0	472	39.0	36.0	92	14.0	0.246	2.00	497	41.0	17.0	2.9481
Std.GM	13	7.0	658	6.9	9.0	17	3.2	0.110	0.99	85	40.7	4.1	0.4014
% CV	11	9.7	139	17.7	25.2	19	23.2	46.400	49.40	17	99.9	24.3	14

A-Mean: Arithmetic Mean
Std.AM: Std.deviation of Arithmetic Mean
G-Mean: Geometric Mean
Std. GM: Std. Deviation of Geometric Mean
% CV: Percent Coefficient of Variation

Table 3. Correlation coefficients for chemical and physical parameters of muds from Beaufort Lagoon (N=19; significant correlations [p<.05] are shown in bold and insignificant correlations in gray).

	V	Cr	Mn	Ni	Cu	Zn	As	Cd	Sn	Ba	Pb	Fe	OC (%)	Silt (%)	Clay (%)
V	**1.00**														
Cr	**0.89**	**1.00**													
Mn	0.07	-0.07	**1.00**												
Ni	**0.85**	**0.71**	**0.48**	**1.00**											
Cu	**0.80**	**0.64**	0.39	**0.91**	**1.00**										
Zn	**0.84**	**0.78**	-0.16	**0.69**	**0.76**	**1.00**									
As	**0.72**	**0.52**	0.12	**0.55**	**0.58**	**0.68**	**1.00**								
Cd	0.41	**0.54**	0.30	**0.57**	0.45	0.45	0.31	**1.00**							
Sn	**0.48**	0.43	**0.74**	**0.78**	**0.61**	0.35	0.40	**0.78**	**1.00**						
Ba	**0.53**	**0.57**	-0.38	0.31	**0.47**	**0.70**	0.26	0.08	-0.12	**1.00**					
Pb	**0.70**	**0.50**	**0.80**	**0.89**	**0.92**	**0.58**	**0.68**	**0.63**	**0.86**	0.10	**1.00**				
Fe	**0.73**	**0.85**	0.28	**0.58**	**0.62**	**0.77**	**0.71**	**0.59**	**0.50**	**0.48**	**0.70**	**1.00**			
OC (%)	0.37	0.40	0.12	0.42	0.31	**0.62**	0.44	**0.66**	**0.46**	0.07	0.38	0.44	**1.00**		
Silt (%)	**-0.63**	-0.40	**-0.75**	**-0.81**	**-0.73**	**-0.51**	**-0.72**	**-0.73**	**-0.93**	0.19	**-0.88**	**-0.52**	-0.44	**1.00**	
Clay (%)	**0.64**	0.40	**0.74**	**0.82**	**0.75**	**0.55**	**0.72**	**0.72**	**0.93**	-0.17	**0.88**	**0.52**	**0.58**	**-1.00**	**1.00**

polyaromatic hydrocarbons, triterpanes and steranes in sediments are available at a website due to space constraints in this paper (ftp://ftp.ims.uaf.edu/naidu/).

1.4. DISCUSSION

The Beaufort Lagoon sediments consist of sandy muds to muddy sands, with rare gravel (Table 1). The exceptions are sediments BL03-4B, BL03-19 and BL03-20, which have significant or predominant amounts of gravel. The mud fractions of sediments generally contain relatively higher contents of silt than clay particles. In the gross sediments, which were taken for Hg analysis, there are wide differences in the contents of sand relative to mud.

The OC/N ratios, and the stable isotopes of carbon ($\delta^{13}C$) and nitrogen ($\delta^{15}N$) in the Beaufort Lagoon sediments (Table 1) indicate that all the samples consist predominantly of OC/organic matter derived from terrestrial plant sources, with minor input, if any, from marine phytodetritus or macrophyte. This interpretation is based in the light of the evidence gathered previously from very exhaustive investigations relating to the isotopic and OC/N signatures of the end-member sources of OC/organic matter in the coastal region of Alaskan and Canadian Beaufort Sea (Naidu et al., 2000; Macdonald et al., 2004).

Trace Metal Distribution Pattern and Geochemical Partitioning: The high coefficient of variations (%CV) about the mean for all the 13 metals (Table 2) obviously indicate that there are wide inter-sample variations in the concentrations of all the metals, with relatively low variability though for V, Cr and Fe. Some of the reasons for the inter-sample variability may be due to the differences between sediments in the granulometry, contents of organic carbon (a measure of organic matter), Fe and Mn. It is to be expected that greater the amounts of clay size particles, Mn and Fe, and/or of organic carbon (and by implication organic matter) in the mud fraction or gross sediments there will be relatively greater concentrations of metals. The significant ($p < .05$) positive correlations between all trace metals (Table 3), except Cr and Ba, and the clay content and Fe suggest the important role of granulometry and Fe oxide/hydroxide in determining the metal contents in Beaufort Lagoon mud. Additional detailed investigations on metal partitioning on sediments from Beaufort Lagoon based on an extraction technique, further substantiate the role of Fe oxy-hydroxides in sequestering metals in the Beaufort Lagoon sediments (Naidu et al., 1981). Likewise, organic carbon and Mn in the mud to some extent bond selected few metals (Zn, Cd , As and Sn by OC, and Ni, Sn and Pb by Mn) by forming organic-metal complexes. This interpretation is consistent with our earlier conclusions on partitioning of trace metal on lagoon sediments from a wide region off the North Slope coast (Naidu et al., 1981, 2001, 2003).

Univariate distributions and tests for normality: Histograms, box-and-whisker plots, and normal probability plots suggested that most metal concentrations were near-normally distributed, but there are a number of outliers for several of the metals. These are of concern in statistical modeling or hypothesis testing when normality is assumed. Outliers will have a strong influence on Pearson's correlations, cluster analyses and ordinations. Observed outliers include, Cd (2 extremely large values at stations BL03-19, and BL03-20), Cr (one extremely low value at station BL03-4), Mn (2 extremely large value at stations BL03-4 and BL03-17), Ni (unusually low value at station BL03-4, unusually high value at station BL03-17 and BL03-20), Pb (unusually large value at station BL03-17), Sn (unusually large values at stations BL03-17, BL03-19, and BL03-20), V (unusually low value at station

BL03-4). Based on outliers alone, several stations clearly stand out as unusual. These are BL03-4, which has unusually low concentrations of Cr, Cu, Fe, Ni, Pb, V, and Zn, and BL03-17, which is characterized by high concentrations of Cu, Mn, Ni, Pb, and Sn. Stations BL03-19 and/or BL03-20 have similarly high concentrations of some, but not all, of these metals (e.g., Sn, Ni). These two stations are easily identified as outliers in the multivariate analysis (cluster analysis, ordination) as follows.

Cluster analysis based on metal concentrations: A cluster analysis based on Euclidean distances clearly indicated the unusual metal concentrations at stations BL03-4 and BL03-17, which did not cluster with any of the other stations. These two stations had a very different sediment composition from most other stations with a very low percentage of silt in the mud fraction (< 3%). The remaining stations clustered into three groups. Only one of these groups (Group C, consisting of stations BL03-4B, BL03-19, and BL03-20) formed a geographically coherent cluster (Fig. 2). Stations in group C were characterized by a relatively low percentage of silt in the mud fraction, and a high percentage of organic

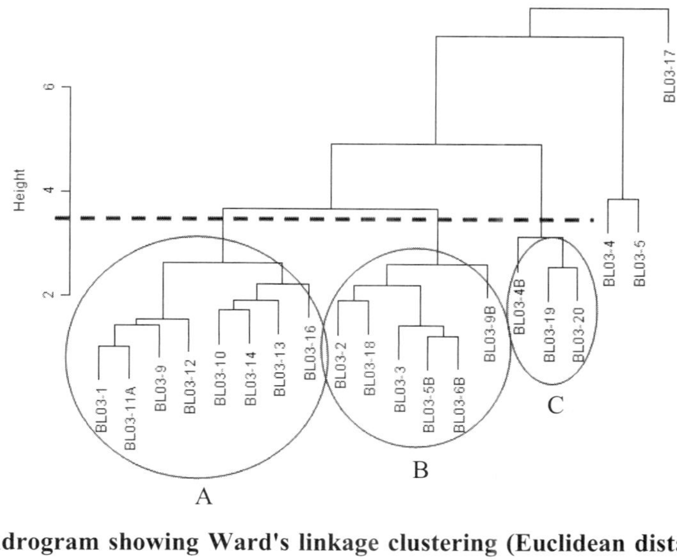

Fig. 2. Dendrogram showing Ward's linkage clustering (Euclidean distances) based on concentrations of 12 metals measured at each station.

matter. These stations contained relatively high concentrations of a number of metals such as Mn, Ni, Pb, Sn, & Cd. Cluster A included the four westernmost stations near Angun Point, as well as a number of stations in the southeastern part of the lagoon. Cluster B consisted of the remaining stations in the southeastern part of the lagoon and also included station 3, which was located immediately adjacent to the three cluster C stations. Cluster A and B were distinguished by differences in sediment composition with a higher percentage of clay in the mud fraction. Stations in cluster B had generally higher concentration of all metals, although the differences in means were small for Mn, Ni, Cd, and Cr. Figure 3 summarizes the clustering of stations based on metal concentrations.

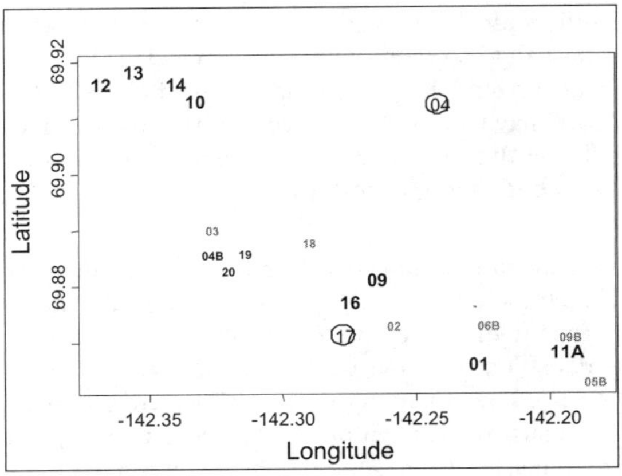

Fig. 3. Map showing relative locations of stations (coastline not shown) and group membership corresponding to clusters as shown in Figure 2. Stations BL03-4 and BL03-17 (circled) did not cluster with any other stations.

Principal components analysis based on metal concentrations: A Principal Components Analysis (PCA) suggests that much of the variability in the metal concentrations (78%) can be accounted for by only two Principal Components. The loadings for the first two PCs (Table 4) suggest that the PC 1 reflects the average concentration of all metals, with high concentrations corresponding to high values of PC 1. The PC 2 contrasts Mn, Cd, and Sn with Ba. Large values of the second PC correspond to high concentrations of Ba, while small values of PC 2 correspond to high concentrations of Mn, Cd, and Sn.

Table 4. Component loadings for the first two Principal Components.

Metal	Comp.1	Comp.2
V	0.338	0.120
Cr	0.311	0.131
Mn	0.116	-0.473
Ni	0.343	-0.106
Cu	0.335	0.010
Zn	0.316	0.260
As	0.269	0.074
Cd	0.242	-0.232
Sn	0.264	-0.387
Ba	0.162	0.434
Pb	0.338	-0.162
Fe	0.323	0.143

Correlations between PC 1/PC 2 and latitude / longitude / sediment composition: Table 5 presents correlations between PC 1 and PC 2 with latitude, longitude, silt (%) and OC (%). PC 1 is negatively correlated with latitude, implying that most of the metals (positive loadings on PC 1) decrease with latitude from southeast to northwest. PC 1 is also positively correlated with organic carbon content, reflecting a tendency for metal content to be higher at stations with high organic carbon content. This is also evident in generally positive correlations between metal concentrations and organic carbon. PC 2 is strongly correlated with the percentage of silt in the mud fraction, implying that Mn, Cd, Sn, and Pb (metals that are negatively correlated with PC 2, see above) decrease with silt content, while Zn and Ba increase with mud content (this is also obvious in direct correlations between metal concentrations).

Table 5. **Correlations between PC 1 and PC 2 with latitude, longitude, silt and organic carbon.**

Parameter	PC 1 (p-value)	PC 2 (p-value)
Latitude	-0.544 (0.0160)	-0.304 (0.2058)
Longitude	0.021 (0.9324)	0.328 (0.1709)
Silt	-0.274 (0.2572)	0.758 (0.0002)
OC	0.547 (0.0153)	-0.242 (0.3173)

The above correlations are strongly influenced by the outlying stations (BL03-4 and BL03-17). Therefore the correlations excluding those 2 stations were computed (Table 6). After exclusion of the 2 "outliers", PC 1 was most strongly correlated with the proportion of silt in the mud fraction and the percentage of organic carbon. This implies a strong decrease in the silt fraction and an increase in the organic carbon along the x-axis in Fig. 4 (Group C > Group B > Group A). In contrast, the PC 2 is positively correlated with the silt fraction.

Table 6. **Correlations between PC 1 and PC 2 with latitude, longitude, silt and organic carbon (excluding outliers BL03-4 and BL03-17).**

Parameter	PC 1 (p-value)	PC 2 (p-value)
Latitude	-0.428 (0.0862)	-0.473 (0.0551)
Longitude	0.147 (0.5742)	0.604 (0.0102)
Silt	-0.726 (0.0010)	0.516 (0.0339)
OC	0.516 (0.0341)	-0.390 (0.1218)

Multiple regressions of PC 1 and PC 2 with latitude, longitude, and sediment characteristics: The model that best explained variability in metal concentrations (as reflected in PC 1) was a regression of PC 1 on the silt fraction and latitude (best model chosen based on Akaike Information Criterion (AIC)). The regression (Table 7) captured 72% of the variability in PC 1 and also captures the strong decrease in PC 1 with the silt fraction and with latitude. While the silt fraction explained most of the overall variability, a

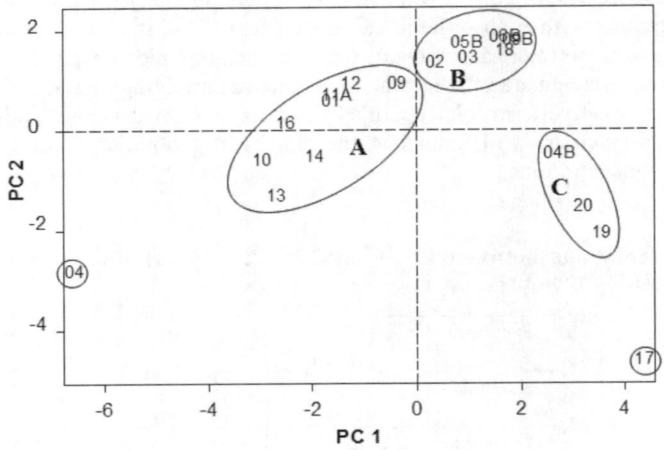

Fig. 4. Beaufort Lagoon stations in the space of the first two principal components. Groups identified by a cluster analysis are indicated.

Table 7. Analysis of Variance for the response of PC 1.

	Degree of Freedom	Sum of Squares	Mean Square	F Value	Pr (F)
Silt	1	39.30097	39.30097	26.30329	0.000153386
Latitude	1	14.34431	14.34431	9.60034	0.007856686
Residuals	14	20.91805	1.49415		

significant effect of latitude on PC 1 remained after accounting for the effects of silt. This effect reflected a southeast-northwest gradient in metal concentrations that primarily separates stations BL03-10, BL03-12, BL03-13, and BL03-14 in the northwestern part of the lagoon from the remaining stations. The geographic gradient was confirmed by simple linear regressions of individual metal concentrations on latitude/longitude (rotated axis to reflect SE-NW direction), which showed significant increases ($p < 0.05$) from NW to SE in Cu ($t = 2.72$, $p = 0.016$), Zn ($t = 2.23$, $p = 0.041$), and Ba ($t = 2.90$, $p = 0.011$). The best model explaining variability in PC 2 was a simple linear regression on mud content, which explained 77% of the variability in PC2. After accounting for the effects of mud, there was no significant effect of latitude, longitude or other sediment variables.

In summary, a PCA based on metal concentrations primarily separates stations with relatively high metal concentrations from stations with relatively low concentrations along PC 1 (56% of overall variability). PC 2 (23% of overall variability) contrasts stations that tend to be high in Mn, Cd, and Sn and low in Ba (Group C and stations BL03-4 and BL03-17) with stations that tend to show the opposite pattern (e.g. Group A). These differences in metal concentration reflect both geographical differences and differences in sediment characteristics. In general, there was a tendency for total metal concentrations (PC 1) to decrease with the silt fraction and with latitude from S to N and to increase with organic matter content.

While much of the variability in metal concentrations was explained by sediment characteristics, a significant geographical gradient remained even after accounting for the effects of sediment composition on PC 1. This gradient reflects differences in metal concentration between the "pristine" southeastern lagoon (higher concentrations of most metals) and the northwestern part of the lagoon that is not explained by differences in sediment composition. It seems clear, therefore, that there are no definite distributional patterns in the concentrations of trace metals in mud of Beaufort Lagoon, especially in context of the hypothesis that we had enunciated at the start of the project.

Time-interval comparison of the mean concentrations of metals: A time-intervals comparison is shown in Table 8 of the mean concentrations of selected six metals between 5 sandy mud (>75% silt + clay) and 21 mud samples, which were collected in 1977 (Naidu, 1981) and 2003 respectively from Beaufort Lagoon. A significant ($p < .05$) decrease in V and an increase in Mn and Cu are noted from 1977 to 2003, whereas no differences in the time-series mean concentrations in Cr, Ni and Zn are identified. The differences may be artifacts of the differences in the granulometry and organic carbon contents between the two sets of samples rather than metal pollution, as alluded to previously.

Table 8. Time-interval changes in the mean concentrations (μg/g) of trace metals in muddy sediments of the Beaufort Lagoon [significant differences ($p < 0.05$) in Bold].

YEAR	V	Cr	Mn	Ni	Cu	Zn
1977						
N=5						
X	**139**	69	**359**	48	**22**	81
SD	**17**	12	**69**	6	**4**	11
CV%	**12**	17	**19**	13	**18**	14
2003						
N=21						
X	**119**	73	**472**	39	**36**	92
SD	**13**	7	**659**	7	**9**	17
CV%	**11**	10	**139**	18	**25**	19

Hydrocarbon Studies: The discussion on the hydrocarbon data is focused on elucidating the sources of hydrocarbons in the Beaufort Lagoon sediments. Total n-alkane level varies from 0.1 to 34μg/g. The alkane gas chromatogram exhibits baseline separation of the components and is generally bimodal with the major maximum at n-C_{27}, n-C_{29} or n-C_{31}. Normal alkanes >C_{25} predominate in all the sediment samples with odd/even carbon alkanes ratio in the range from 3.5 to 8.1 characteristic of higher plants. The secondary maximum at n-C_{17} is typical of aquatic algae. The C_{20} and C_{21} olefins found are derived from plankton and bacteria. In contrast, the O.S. (Oil Seep) sample contains 100 to 1000 fold concentration of n-alkanes compared to the sediments. Very high amounts of n-C_{10} through n-C_{15} and significant levels of alkanes from n-C_{20} to n-C_{32} over a pronounced unresolved hump are present. The C_{20} and C_{21} planktonic olefins are absent in the O.S. The odd/even ratio is 1.3 at the high molecular weight end, typical of petroleum input. It appears that the O.S. may be degraded petroleum. The n-alkanes profile of the sediment samples is markedly different from that in the O.S. and is relatively pristine with no measurable petroleum contribution.

The levels of total PAH in the Beaufort Lagoon sediments range from 30 to 710 ng/g and are comparable to those in sediments collected from a wide nearshore region of the Alaskan Beaufort Sea (Naidu et al., 2001; Naidu et al., 2003). The PAH composition is dominated by the homologous series of phenanthrenes. The sum of parent PAH and their methyl homologs distribution follows generally in the order: phenanthrenes > naphthalenes ≥ chrysenes/triphenylenes > fluoranthenes/pyrenes.. The general dominance of parent and monomethylated PAHs over higher methylated homologs suggests the absence of significant petroleum input in the sediments.

The O.S. (oil seep) sample contains 208μg/g of total PAHs, about 1000 folds greater than the lagoon sediments, also consistent with the trend in *n*-alkanes content. Unlike in the sediments, naphthalenes are the most dominant PAHs in the O.S., and, similar to the dominance of low molecular weight C_{10}-C_{15} *n*-alkanes over high molecular weight analogs. Further, anomalously high value of the naphthalenes/phenanthrenes ratio in the O.S. is observed (7.62) in contrast to the very low range in the sediments (0.03-0.91). C_2-naphthalenes, C_2-phenanthrenes and C_3-chrysenes/triphenylenes are the most dominant methyl homologs, characteristic of petroleum. The molecular composition of PAHs in the Beaufort lagoon sediments and a comparison to that of the O.S. sample, imply absence of petroleum in the Beaufort Lagoon sediments, which is consistent with that noted above from the alkanes profiles.

Triterpanes are mostly biogenic in the sediments as reflected by the presence of five major components, i.e., 27(17β)-, 29ββ-, 29βα-, 30ββ- hopanes and diploptene. Thermally mature 29αβ- and 30αβ-hopanes are present in much smaller/trace amounts relative to biogenic hopanes. Extended hopanes with $\geq C_{31}$ are either not detected or present only in trace quantities and, only the R isomer is detected. Similar to *n*-alkanes, total triterpanes are at the μg/g level and about 1000 fold greater in the O.S. sample than in the lagoon sediment samples where they occur at the level of ng/g. Further, the O.S. contains a whole suite of thermally mature αβ-hopanes and none of the biogenic ββ- and βα-hopanes or the hopenes. Both S and R diastereomers of the extended hopanes (i.e., $\geq C_{31}$-hopanes) are present. This hopanes distribution in the O.S. is typical of petroleum/thermally mature sediments. The fingerprint of the triterpanes in sediments of the Beaufort Lagoon do not support any contribution from the O.S. Steranes are generally absent in the Beaufort Lagoon sediment samples. If present, they are only in small/trace amounts, usually about 10 times lower than triterpanes. Considering the extremely low levels of steranes in these samples, the input from petroleum may not be warranted. All steranes usually associated with natural weathered oil are present in the O.S. sample and it is about 1000 fold greater than in the lagoon sediments. The steranes profile of O.S. clearly indicates petroleum characteristics.

In summary, normal and cyclic alkanes distribution is characteristic of biogenic origin in all the sediment samples and is different from that of the O.S. (oil seep) sample, thus indicating very little or no petroleum input to the Beaufort Lagoon sediments.

1.5. CONCLUSIONS

Interpretations of the trace metal and hydrocarbon data suggests that there are no significant disparities in the metal concentrations between the regions exposed to long-term natural oil seep, past anthropogenic activities related to the DEW line military station and pristine environment of the Beaufort Lagoon. Comparison of the mean concentrations of metals in muddy sediments taken at 26 years interval (1977 to 2003) show significant decrease in V, increase in Mn and Cu, and similar values in Cr, Ni and Zn. Data on rest of the metals were unavailable for 1977. The differences mentioned above likely reflect the differences in granulometry and organic carbon contents between 1977 and 2003 samples. Correlation coefficient analysis suggests that clays and Fe presumably play an important role in metal scavenging, a conclusion consistent with our results on metal partitioning elsewhere in the Beaufort Sea near-shore.

The hydrocarbon components in gross sediments from the Beaufort Lagoon are biogenic and terrestrial with little (below detection levels) petroleum inputs. Their general composition is very similar to those found in several investigations on nearshore sediments from the Alaskan and adjacent Canadian inner shelf region of Beaufort Sea. The trace metal and hydrocarbon levels are either similar or below those reported for unpolluted near-shore regions of the world, which suggests that the Beaufort Lagoon sediments have remained uncontaminated in the analyzed trace metals and hydrocarbons.

ACKNOWLEDGEMENTS

This study was funded the Coastal Marine Institute/MMS, University of Alaska Fairbanks. Thanks are due to Zygmunt Kowalik for assistance in the sample collection and to Thomas Ooomen for his help in formatting this manuscript.

REFERENCES

Bloom, N. S., 1992. On the chemical form of mercury in edible fish and marine invertebrate tissue. Canadian Journal of Fisheries and Aquatic Sciences 49(5):1010–1017.

Bloom, N. S., Gill, G. A., Cappellino, S., Dobbs, C., McShea, L., Driscoll, C., Mason, R., and Rudd, J., 1999. Speciation and cycling of mercury in Lavaca Bay, Texas, sediments. Environmental Science & Technology 33(1): 7-13.

Boehm, P.D., Steinhauer, M., Crecelius, E., Neff, J., and Tuckfield, C., 1987. Beaufort Sea Monitoring Program: Analysis of trace metals and hydrocarbons from outer continental shelf (OCS) activities. Anchorage, AK, USDOI, MMS, Alaska OCS region. OCS Study MMS 87-0072. Final Report to MMS. 2 vols.

Chapman, P.M., Wang, F.Y., Janssen, C., Persoone, G., Allen, H.E., 1998. Ecotoxicology of metals in aquatic sediments: binding and release, bioavailability, risk assessment and remediation. Canadian Journal of Fisheries and Aquatic Sciences 55(10): 2221-2243.

Chapman, P. M., and Riddle, M. J., 2005. Polar marine toxicity – future research needs. Marine Pollution Bulletin 50(9): 905-908.

Crecelius, E.A., Trefry, J.H., Steinhauer, M.S., and Boehm, P.D., 1991. Trace metals in sediments from the inner continental shelf of the western Beaufort Sea. Environmental Geology and Water Sciences 18(1): 71–79.

Folk, R., 1968. Petrology of Sedimentary Rocks. Hemphills, Austin, Texas. Pp. 70.

Forstner, U., and Wittman, G.T. W., 1979. Metal Pollution in the Aquatic Environment. Springer-Verlag, Berlin. 486 pp.

Lee, H. J., and Wiberg, P. L., 2002. Character, fate, and biological effects of contaminated, effluent-affected sediment on the Palos Verdes margin, Southern California: an overview. Continental Shelf Research 22(6-7): 835-840.

Macdonald, R. W., Naidu, A. S., Yunker, M. B. and Gobeil, C. (2004). The Beaufort Sea: distribution, sources, fluxes and burial of organic carbon. In: R. Stein and R. W. Macdonald (Eds.), The Organic Carbon Cycle in the Arctic Ocean. Springer Verlag, Berlin/New York, 177-193.

Mountouris A., Voutsas, E., and Tassios, D., 2002. Bioconcentrations of heavy metals in aquatic environments: the importance of bioavailability. Marine Pollution Bulletin 44:1136-1141.

Naidu, A. S., 1981. Sources, transport pathways, depositional sites and dynamics of sediments in the lagoon and adjacent shallow marine region, northern Arctic Alaska. Annual Report to NOAA–OCSEAP Office, Boulder, Colorado. Pp.39.

Naidu, A. S., and Mowatt, T. C., 1975. Aspects of size distributions, mineralogy and geochemistry of deltaic and adjacent shallow marine sediments, North Arctic Alaska. In V. Alexander and 15 others, Environmental studies of an arctic estuarine system – Final Report, EPA-660/3-75-026, U. S. EPA,Corvallis, Or.143-205.

Naidu, A.S., Cooper, L.W., Finney, B.P., Macdonald, R.W., Alexander, C., and Semiletov, I.P., 2000. Organic carbon isotope ratios ($\delta^{13}C$) of Arctic Amerasian continental shelf sediments. *In* R. Stein [ed.], Circum-Arctic River Discharge and Its geological Record. International Journal of Earth Sciences 89(3): 522–532.

Naidu, A. S., Goering, J. J., Kelley, J. J., and Venkatesan, M. I., 2001. Historical changes in trace metals and hydrocarbons in the inner shelf sediments, Beaufort Sea: prior and subsequent to petroleum-related developments. Final Report to Coastal Marine Institute, Univ. Alaska Fairbanks, Fairbanks, AK. OCS Study MMS 2001-061. Pp.80.

Naidu, A. S., Kelley, J. J., Goering, J. J., and Venkatesan, M. I., 2003. Trace metals and hydrocarbons in sediments of Elson Lagoon (Barrow, Northwest Arctic Alaska) as related to Prudhoe Bay industrial region. Final Report submitted to Coastal Marine Inst., Univ. Alaska Fairbanks, Fairbanks, AK. Pp. 33.

Naidu, A. S., Kelley, J. J., Misra, D., and Venkatesan, M. I., 2005. Trace metals and hydrocarbons in sediments of the Beaufort Lagoon, Northeast Arctic Alaska, exposed to long-term natural oil seepage, recent anthropogenic acivities and pristine conditions. Final Report to Coastal Marine Institute, Univ. Alaska Fairbanks, Fairbanks, AK. In Press.

Shaw, D.G., McIntosh, D.J., and Smith, E.R., 1979. Arene and alkane hydrocarbons in nearshore Beaufort Sea sediments. Estuar. Coast. Mar. Sci. 9: 435–449.

Snyder-Conn, E., Garbarino, J. R., Hoffman, G. L., and Oelkers, A., 1997. Soluble trace elements and total mercury in arctic Alaskan snow. Arctic 50: 201–215.

Steinhauer, M.S., and Boehm, P.D., 1992. The composition and distribution of saturated and aromatic hydrocarbons in nearshore sediments, river sediments, and coastal peat of the Alaskan Beaufort Sea: Implications for detecting anthropogenic hydrocarbon inputs. Mar. Environ. Res. 33:223–253.

Sweeney, M.D., 1984. Heavy Metals in the Sediments of an Arctic Lagoon, Northern Alaska. M.S. Thesis, University of Alaska Fairbanks, 256 p.

Sweeney, M.D., and Naidu, A.S., 1989. Heavy metals in the sediments of the inner shelf of the Beaufort Sea, northern arctic Alaska. Marine Pollution Bulletin 20:140–143.

Trefry, J. H., Rember, R. D., Trocine, R. P. and Brown, J. S., 2003. Trace metals in sediments near offshore oil exploration and production sites in the Alaskan Arctic. Environmental Geol., 45: 149-160.

Venkatesan, M.I., and Kaplan, I.R., 1982. Distribution and transport of hydrocarbons in surface sediments of the Alaskan outer continental shelf. Geochim. Cosmochim. Acta 46:2135–2149.

Venkatesan, M. I., 1994. Historical trends in the deposition of organic pollutants in the Southern California Bight. Final Report to NOAA, UCLA. Los Angeles, 35 pp.

Venkatesan, M. I., Ruth, E., Steinberg, S., and Kaplan, I. R., 1987. Organic geochemistry of sediments from the continental margin off southern New England, USA – Part II. Lipids. Marine Chemistry 21: 267–299.

INVESTIGATING THE SOURCE(S) OF VINYL CHLORIDE IN A MUNICIPAL WELL WITHIN THE MINNEAPOLIS-ST. PAUL METROPOLITAN AREA, MINNESOTA

Lifeng Guo and Nile Fellows
Minnesota Pollution Control Agency
520 Lafayette Road North, St. Paul, MN 55155, USA

Piotr A. Rzepecki and Paul Putzier
STS Consultants, Ltd.
10900 73rd Avenue North, Suite 150, Maple Grove, MN 55369, USA

Scott Alexander
Department of Geology & Geophysics, University of Minnesota
310 Pillsbury Dr. SE, Minneapolis, MN 55455, USA

Stephen Robertson and James Lundy
Minnesota Department of Health
P.O Box 64975, St. Paul, MN 55164, USA

ABSTRACT

This paper presents a case study of contaminant migration through a complex, multi-aquifer system under heavy pumping conditions within the Minneapolis-St. Paul metropolitan area in Minnesota. Ground-water monitoring from multi-aquifers, borehole geophysics and discrete ground-water sampling were conducted to study vinyl chloride contamination detected in the City of Edina Municipal Well Number 7 (ED-7). The geochemical and tritium analysis revealed distinct age stratification of water within ED-7. The results of this investigation also revealed a potential tetrachloroethylene (PCE) release in the old commercial/industrial area two miles north of ED-7, near the Reilly Tar federal superfund site. The conceptual model based on those data suggests a contaminant migration through buried bedrock valley and old, multi-aquifer wells present in the area. Further investigation to test the conceptual model included a land use survey, continuous water level monitoring, and ground water modeling. These additional data, however indicate that the contaminants found in ED-7 may migrate from other directions including from the west.

2.1. INTRODUCTION

Ground water is the sole source of drinking water for the City of Edina, a western suburb of Minneapolis-St. Paul, Minnesota. Since the year 2000, cis-1,2-dichloroethene (cis-1,2-DCE), trans-1,1-dichloroethene (trans-1,1-DCE), trichloroethene (TCE), and vinyl chloride have been consistently detected in the Edina Wells 2, 7, and 15 during routine monitoring.

The concentrations of vinyl chloride detected in the City of Edina Municipal Well Number 7 (ED-7) have recently been above the federally mandated Maximum Contaminant Level (MCL), and concentrations of cis-1,2-DCE and vinyl chloride in this well appear to be increasing. The City discontinued the use of ED-7 in late 2003. There is concern that additional Edina municipal wells may be affected in the near future. In addition, the City is concerned about the possibility that private drinking water wells in the area may be adversely affected by these chemicals.

The Minnesota Pollution Control Agency (MPCA) and its contractor, STS Consultants, began field investigation in Spring 2004, with objectives of identifying the source(s) of chlorinated chemicals in the aquifers, and ultimately a remedy to the municipal well contamination. The 505 ft deep ED-7 is cased to 350 ft below land surface and open to the Prairie du Chien-Jordan aquifers.

2.2. GEOLOGY AND HYDROGEOLOGIC SETTING

The study area is covered by a complex sequence of Quaternary glacial drift, consisting of alternating Wisconsin Age glacial till and outwash deposits associated with both the Superior lobe and the Des Moines lobe (Wright, 1973). The stratigraphy of the unconsolidated materials is highly heterogenous and is complicated by the presence of buried bedrock valleys in the vicinity of the study area. The thickness of the glacial deposits (Drift) ranges from 80 to 190 feet, with the thickest deposits present in the bedrock valleys.

A sequence of generally horizontal, undeformed Paleozoic sedimentary rocks is present below the glacial deposits and forms a system of aquifers and confining layers. The Platteville Limestone is the uppermost bedrock formation in the area, is up to 35 ft thick, consists of thin to medium bedded, fractured dolomitic limestone and dolomite, and is absent in bedrock valleys. Secondary porosity features in the Platteville are the principal reason it behaves as an aquifer. The underlying Glenwood Shale is comprised of green fissile shale with argillaceous sandstone and scattered coquinoidal limestone beds. This confining layer is approximately 3 to 6 ft thick in the study area (ERT, 1983). Below the Glenwood is the St. Peter Sandstone, composed of fine- to medium-grained massive sandstone up to 200 ft thick (Bloomgren, 1985). The lower portion of the St. Peter (basal OSP) is considered a confining layer, consisting of 20 to 60 ft of siltstone and shale. The basal OSP separates the St. Peter aquifer from the Prairie du Chien-Jordan aquifer system which constitutes the most developed aquifer in the Twin Cities (Horn, 1983). Two separate bedrock units comprise the Prairie du Chien-Jordan aquifer system: the highly fractured thin-to thickly bedded, up to 170 ft thick dolomite (Prairie du Chien Group), and fine- to coarse- grained, up to 105 ft thick Jordan Sandstone (Runkel et al., 2003).

In the vicinity of the investigation area, ground water generally flows toward the east and south-east in both the Drift and bedrock aquifers (Lowry, 1989; Hult and Schoenberg, 1984; Stark and Hult, 1985). A down-ward vertical gradient exists between the upper Drift, Platteville and St. Peter aquifers, and the Prairie du Chien-Jordan aquifer, partly due to heavy pumping from the later.

2.3. INVESTIGATION METHODS

The investigators considered three hypotheses describing the likely origin of vinyl chloride contamination detected in ED-7:

1. A spill or near surface contaminant release of chlorinated chemicals might have occurred in the vicinity of the well. Contaminants could be carried down to the deep bedrock aquifer system by surface runoff and/or shallow aquifer then through the damaged casing or leaky annular space of ED-7 (Santi and et al., 2006);
2. The Prairie du Chien-Jordan aquifer, in which ED-7 is completed, may have been contaminated by a distant, up-gradient source. In this case, contaminants could have been transported through many multi-aquifer wells present in the area and/or through buried bedrock valleys where confining formations are missing, juxtaposing younger drift with older and generally more deeply buried bedrock aquifers;
3. The vinyl chloride detected in ED-7 is a result of chemical transformation processes, such as reductive dechlorination of chlorinated chemicals or solvents, such as PCE and/or TCE (Butler and Barker, 1996). The initial source(s) of contaminants detected in ED-7 would be TCE and/or PCE.

The study tasks and investigative methods were designed/selected and carried out to test thse hypotheses. The methods employed included borehole geophysics, continuous water level monitoring, discrete ground water chemistry sampling and ground water age dating from the municipal well itself. Ground water samples were also collected from the wells up-gradient of ED-7. These wells are open to the Prairie du Chien-Jordan and other, shallower aquifers, such as the St. Peter, Platteville and Drift aquifers. Ground water sampling was conducted in phases. First, sampled the Prairie du Chien-Jordan wells in close proximity to ED-7. Next, sampled the Prairie du Chien-Jordan wells that are successively further away in the up-gradient direction until the entire plume was delineated. Upon evaluation of the analytical results from the Prairie du Chien-Jordan aquifer, the wells in the shallower aquifers, St. Peter, Platteville and Drift, were sampled last.

2.3.1. Investigation of Edina Well 7

Borehole Geophysics

Borehole geophysical methods conducted at ED-7 included video, caliper, fluid temperature and electric resistivity logging. The geophysical equipment used included a normal resistivity probe (2PEA-1000), a temperature and fluid resistivity probe (2PFA, 2SFA,B, 2WQA-1000), a 3-arm caliper probe (2PCA-1000), and a gamma, spontaneous potential, single point resistance probe (2PGA-1000) (all probes from Mount Sopris Instrument Co., Inc., Golden, CO). Flow logging was performed using an Impeller Flowmeter probe (FLP-2492, Mount Sopris Instrument Co., Inc.). Separately, continuous measurements of specific conductance (SC), dissolved oxygen (DO), temperature and redox potential were conducted using a HydroLab (Hydrolab minisonde 3A) at Edina Well 7.

2.4. VERTICAL WATER QUALITY PROFILING BY DISCRETE INTERVAL GROUNDWATER SAMPLING

Ground-water samples were collected from ED-7 at four selected depths (360’, 400’, 450’ and 500’ below grade) on January 5, 2005. The 1-liter discrete sampler used was manufactured by the Century Geophysical Corp. (Tulsa, OK) (2FSA-1000). In the field ground-water pH, alkalinity, dissolved oxygen (DO), redox potential and specific conductance were measured. Collected samples were analyzed for cations (Ca, Mg, Na, K, Al, Si, Fe, Mn, and others) by ICP-MS and anions (SO_4^-, Cl^-, NO_3^-, Br^-, and others) by an ion chromatography in the laboratory. Alkalinity was determined through colorimetric titration method. Volatile organic compounds (VOC) were analyzed using the Minnesota Department of Health Method 465E. In addition, water samples were also collected for tritium analysis.

2.4.1. Monitoring of Multi-Aquifer System

Well Search and Selection

Various available databases were searched to identify candidate wells for VOC sampling. These databases included the Minnesota County Well Index (CWI), city utility records from the Cities of Edina and St. Louis Park, the Monitoring Inventory Database developed for the Reilly Tar and Chemicals Corporation Superfund site (Figure 1), and the Well Information Database maintained by the Minnesota Department of Health (MDH). A total of 81 municipal and private wells was sampled. These wells are within the limits of the Cities of Edina, Hopkins, Minneapolis and St. Louis Park and are completed in four different aquifers: 29 wells in the Drift aquifers, 20 wells in the Platteville, 10 wells in the St. Peter, and 22 wells in the Prairie du Chien-Jordan aquifer system.

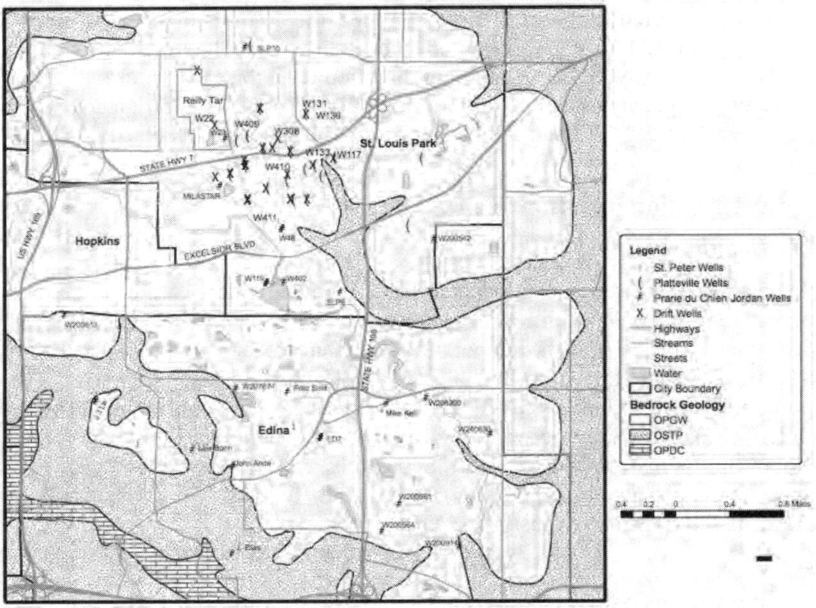

Fig. 1. Location of sampled wells in the Drift, Platteville, St. Peter, and Prairie du Chien-Jordan aquifers, and the bedrock geology in the vicinity of Edina. Not all wells are labeled, except those in the Prairie du Chien-Jordan aquifer.

Phased Approach

A phased approach for ground-water sampling was used to evaluate the extent of chlorinated contamination. In the initial phase, only the wells completed in the Prairie du Chien-Jordan aquifer were sampled, with ground water samples collected first from those wells near ED-7 and then from those located farther up-gradient. The second phase of investigation was guided by the results of the initial phase and included sampling of wells completed in the St. Peter, Platteville and Drift aquifers in the area of maximum contamination within the Prairie du Chien-Jordan aquifer. This sampling led to discovery of highly contaminated Platteville and Drift wells.

2.5. RESULTS

2.5.1. ED-7 Testing Results

Borehole Geophysics

Video logging provided visual evidence indicating that the casing and joints of ED-7 are intact. Good condition of the well casing diminishes the possibility of ground water getting into the well from potentially contaminated shallow aquifers. The caliper log demonstrated that the well diameter below the casing varies from 23 to 32 inches (Figure 2). Two densely fractured zones appear to be located at 380-400 ft and 460 ft below ground surface. Sections of the open hole containing fewer fractures than the fractured zones appear to correlate with high peaks of fluid electric resistivity values **(**Figure 2**).** Field parameters, such as specific conductance (SC) and chemistry data indicate the presence of two distinctly different water zones within the open-hole section of the well. The first zone, designated as the 'shallow zone', is within the upper portion of the Prairie du Chien Group, with a depth from the bottom of the well casing at 350 ft to approximately 400 ft. The second or the 'deep zone' is from 400 ft on to the bottom of the well. This deep zone is within the lower portion of the Prairie du Chien Group and the upper portion of the Jordan Sandstone.

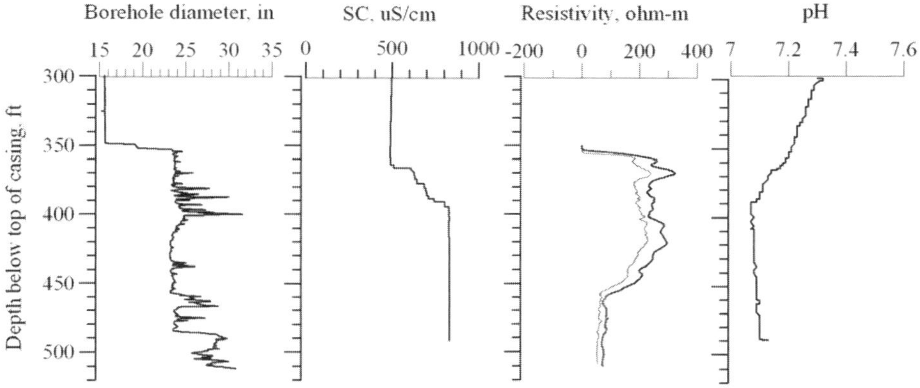

Fig. 2. Results of caliper, selected borehole geophysics and geochemical measurements from ED-7.

2.5.2. Ground-Water Chemistry and Tritium Analysis

Logging of the shallow zone (350 ft to 400 ft below ground surface) demonstrated a drastic increase with depth in the values of several parameters. SC increased from 500 uS/cm to 800 uS/cm, Cl from 1.08 mg/L to 60 mg/L, SO_4^{2-} from 5.19 mg/L to 35.4 mg/L, and Ca^{2+} from 70.9 mg/L to 102.3 mg/L. Vinyl chloride and other chlorinated chemicals were found at very low concentrations at the 350 ft depth. On the other hand, the deep zone (400 ft to 505 ft) was characterized by elevated and nearly homogeneous chemistry at and below 400 ft (Figure 3). In the 'deep zone' the SC was at 800 uS/cm, Cl^- from 62.0 to 64.4 mg/L, a uniform SO_4^{2-} concentration of 35.4 mg/L. Among the detected chlorinated compounds in the deep zone, vinyl chloride concentrations ranged between 3.3 to 3.5 μg/L and cis1,2-DCE concentrations varied in a tight range from 30.0 to 35.0 μg/L. The detected concentrations of TCE were between 2.2 μg/L and 2.3 μg/L (Figure 3). As no DO was detected, and dissolved iron of between 1.1 and 1.3 mg/L and detectable amounts of manganese were found in the ground water (Figure 3), indicating a reduced environment in the aquifer. The chlorinated chemicals detected, such as vinyl chloride and cis1,2-DCE could have been the result of reductive dechlorination of PCE or TCE.

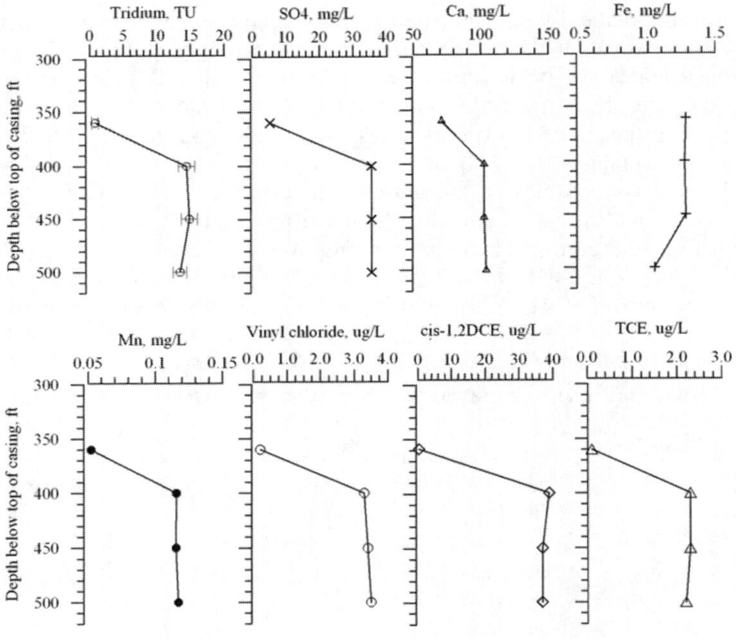

Fig. 3. Results of selected inorganic chemistry, VOCs and tritium analysis of discrete samples from ED-7.

Prior to the initiation of atmospheric nuclear testing in 1952, the natural tritium (^3H) concentration in precipitation water was in the range of a few tritium units (TU). Large amounts of tritium were introduced into the atmosphere by the above ground thermonuclear testing conducted in the mid-1960's in the northern hemisphere. With a half-life of 12.43

years, ^3H from the nuclear testing has been used as a tracer for waters that have recharged the ground water system since 1952. Due to various methodological and interpretative difficulties, the estimates of ground water age based on ^3H concentrations are in most cases qualitative in nature (Hendry, 1988; Clark and Fritz, 1997).

The concentrations of tritium measured in ED-7 samples varied significantly between the 'shallow' and 'deep' zones. The tritium concentration was 0.8+/-0.5 TU below the bottom of well casing at 350 ft, and changed from 14.4+/-1.2 TU at 400 ft, to 14.8+/-1.2 TU at 450 ft, and 13.4+/-1.0 TU at 500 ft below the ground surface (Figure 3). Water containing ^3H concentrations of less than 1.0 TU is considered Vintage water (before 1953), while water with ^3H concentrations between 8 and 20 TU is interpreted as Recent water, a mixture of pre-1953 and post-1953 water or after 1972 water (Alexander and Alexander, 1989). The uniform tritium concentrations in the 'deep zone' indicate the presence of relatively young water coming from a different source compared to the older water encountered in the 'shallow zone' of the well. A certain degree of mixing of waters between the two zones could be expected within the well. However, the vertical stratification observed in ED-7 is likely because the well was not pumped for several months prior to the sampling, allowing natural vertical gradients to equilibrate within the well.

2.5.3. Extent and Magnitude of Ground Water Contamination with Chlorinated Compounds

Prairie du Chien-Jordan Aquifer

A total of 29 wells were selected from the Prairie du Chien-Jordan aquifer for ground-water sampling. Most of these wells were municipal wells and ground-water monitoring wells located in the City of Edina and St. Louis Park. The results of VOC analysis of the ground-water samples collected from these wells are provided in Table 1.

Vinyl chloride was detected (above the detection limit of 0.2 μg/L) in 12 wells, all of them north, northeast and northwest west of ED-7. The highest concentration of vinyl chloride, 16 μg/L, was detected in the well W48, approximately 1.5 mile north-northwest of ED-7 (Figure 4). If one assumes that vinyl chloride is continuously present between ED-7 and W48, then the plume stretches over the distance of two miles north of ED-7. Cis-1,2-DCE and TCE were also detected in these wells and the plumes of these chemicals appear to extend over the area similar to vinyl chloride plume. The detected cis-1,2-DCE and TCE concentrations ranged from 105 μg/L to 1.9 μg/L and from 7.2 μg/L to 0.2 μg/L, respectively. No PCE was detected in any of the Prairie du Chien-Jordan wells.

2.5.4. St. Peter Aquifer

Only 9 wells were sampled in this investigation primarily due to lack of existing wells in the St. Peter aquifer in the area. These wells are located north of ED-7. Low levels of vinyl chloride were detected in two wells, 2.8 μg/L at W133 and 0.5 μg/L at W409 (Figure 1). Cis-1,2-DCE was detected in four wells at concentrations ranging from 5.0 μg/L at W133, 2.4 μg/L at W409, to 0.6 μg/L at W410 and 0.5 μg/L at W411 (Figure 1). TCE was detected in these four wells at concentrations between 0.4 μg/L and 3.2 μg/L. PCE was detected in two wells at concentrations of 1.7μg/L (W133) and 0.4 μg/L (W410).

Table 1. Chlorinated compounds detected in the Prairie du Chien-Jordan, Platteville and Drift wells.

Well Name	**Mike Kelly**	**Mike Kelly**	**W208399**	**ED7**	**ED13**	**W207674**	**Milastar Co.**	**W200542**	**SLP6**	**SLP10**	**W23**	**W48**	**W48**	**W119**	**W119**	**W119**
Unique No.	232315	232315	208399	206474	203613	207674	206454	200542	206457	206442	216050	216067	216067	216009	216009	216009
Sample Date	11/1/04	11/1/04	10/22/04	1/26/04	10/22/04	10/22/04	11/19/04	8/16/04	5/10/05	8/16/04	12/9/04	11/10/04	5/10/05	1/20/05	1/20/05	5/10/05
Aquifer	Prairie du Chien-Jordan															
cis-1,2-DCE	**16**	16	2.0	<0.2	0.8	7.1	47	<0.2	28	0.7	43	**100**	**124.0**	3.8	4.7	<0.2
trans-1,2-DCE	0.7	0.7	<0.1	<0.1	<0.1	0.4	2.4	<0.1	1.4	0.3	2.5	7.3	6.7	0.4	0.4	<0.1
PCE	<0.2	<0.2	<0.2	<0.2	<0.2	<0.2	<0.2	<0.2	<0.2	<0.2	<0.2	<0.2	<0.2	<0.2	<0.2	<0.2
TCE	1.2	1.1	0.2	<0.1	0.1	1.9	2.9	<0.1	7	<0.1	1.2	6.3	5	0.2	0.2	5
Vinyl Chloride	**2.0**	**2.1**	**0.4**	**3.9**	<0.2	**0.9**	**5.3**	**1.5**	**5.1**	<0.5	**4.8**	**16.0**	**15.0**	**0.3**	**0.4**	**13.0**

Well Name	W18	W20	P 62	W101	W131	W132	W143	W421	W421	W426	W431	W433	W434	W434	W437	W438
Unique No.	216046	216048	227948	149711	165586	165587	216051	434044	434044	439812	462935	462933	463012	463012	498917	498919
Sample Date	6/3/05	5/2/05	6/2/05	5/2/05	5/3/05	6/2/05	5/3/05	12/9/04	5/3/05	5/2/05	5/3/05	5/2/05	12/9/04	5/3/05	5/2/05	5/3/05
Aquifer	Platteville Formation															
cis-1,2-DCE	7.2	1	0.8	0.8	0.3	<0.2	14	**410**	**810.0**	<0.2	**89.0**	1	**290**	**1100.0**	**86**	**190.0**
trans-1,2-DCE	11	<0.1	0.2	1.8	<0.1	<0.1	4.7	**260**	**330.0**	0.1	41	3	43	63	2.5	18
PCE	<0.2	5.1	<0.2	<0.2	**12**	<0.2	**22.0**	**42.0**	**27.0**	<0.2	<0.2	<0.2	**1200.0**	**760.0**	**13000.0**	3.6
TCE	0.6	2.9	<0.1	<0.1	2.7	<0.1	**93.0**	**760.0**	**259.0**	<0.1	7.4	0.4	**900.0**	**680.0**	**2600.0**	**200.0**
Vinyl Chloride	**3.8**	**0.4**	**0.2**	**0.6**	<0.2	**0.7**	**2.4**	**100.0**	**150.0**	<0.2	9.1	**0.8**	**86.0**	**110.0**	**8.6**	**36.0**

Well Name	P 8	P112	P307	P308	P309	P310	P312	W-2	W117	W136	W136 - DUP	W420	W422	W427	W439 - USGS 135	W439
Unique No.	216117	216166	462926	462927	462928	462929	462932	216031	160031	165591	165591	434045	434043	439811	538134	538134
Sample Date	6/7/05	6/7/05	4/25/05	4/25/05	4/25/05	4/25/05	4/26/05	6/3/05	4/26/05	4/26/05	4/26/05	5/2/05	4/26/05	4/26/05	12/9/04	4/25/05
Aquifer	Drift Aquifer															
cis-1,2-DCE	**98**	3.5	0.6	**2100.0**	<0.2	0.5	1.9	0.2	**330.0**	**630.0**	**590.0**	28	4.7	4.1	2.3	0.6
trans-1,2-DCE	23	<0.1	<0.1	55	1.2	0.1	<0.1	<0.1	13	74	76	23	1.1	<0.1	1.0	0.3
PCE	<0.2	<0.2	<0.2	**560.0**	<0.2	<0.2	3.9	<0.2	**11.0**	**2800.0**	**2700.0**	<0.2	1.6	5	<0.2	<0.2
TCE	23	0.4	0.3	**420.0**	<0.1	<0.1	2.5	<0.1	5.3	**1900.0**	**1800.0**	1.7	0.9	0.6	0.1	0.2
Vinyl Chloride	**6.2**	**4.7**	<0.2	**130.0**	1.5	**2.1**	<0.2	<0.2	**160.0**	**62.0**	**63.0**	**12.0**	**1.9**	**6.1**	**1.9**	**1.2**

Note: (1) Only wells with the chemical concentrations above the respective detections limit are listed in the table;
(2) The concentrations above the respective Maximum Contaminant Level (MCL) are shown in bold.
(3) All concentrations are in ug/L.

2.5.5. Platteville Aquifer

20 wells completed in the Platteville aquifer were sampled. These wells are mainly located in the old commercial/industrial area along Highway 7 in St. Louis Park (Figure 4). Vinyl chloride, cis-1,2-DCE, TCE and PCE were detected in almost half of the sampled Platteville wells. High vinyl chloride concentrations of 150 μg/L and 110 μg/L were measured in wells W421 and W434, respectively. The highest detected concentrations of cis-1,2-DCE and TCE were 1,100 μg/L and 2,600 μg/L at W434 and W437, respectively. A very high concentration of PCE, 13,000 μg/L, was found in W437. Chlorinated compounds within the Platteville aquifer is relatively well defined, has an elongated shape with its long axis running through wells W421 in the west and W434, about 2600 feet east of W421 (Figure 4) Both of these wells serve as the Reilly Tar and Chemical Corporation Superfund Site gradient control wells to contain that site's PAH plume.

2.5.6. Drift Aquifer

As in the case of the Platteville wells, the Drift aquifer wells that were available for sampling are also concentrated north of ED-7, in the St. Louis Park area. The contamination hot-spots appear to be more dispersed within Drift deposits compared to the distribution observed in the Platteville Formation. This is likely due to the very complex structure of these drift deposits. The highest concentrations of vinyl chloride, cis-1,2-DCE, TCE and PCE detected in drift wells were 160 μg/L, 3,800 μg/L, 1,900 μg/L and 2,800 μg/L, respectively.

2.6. DISCUSSION AND CONCLUSIONS

Analysis of inorganic chemistry, VOCs and tritium of discrete samples from ED-7 indicates two distinctly different types of ground water in the Prairie du Chien-Jordan aquifer system. The low concentration of tritium in the Prairie du Chien aquifer just below the bottom of the ED-7 well casing shows an unusual pattern of ground water flow in the area. Older ground water is present on top of younger ground water. The shallower-older water was found to be uncontaminated while the deeper-younger water was observed to be contaminated. This pattern proves that the first hypothesis of a source in close proximity to ED-7 does not stand and should be rejected. The presence of young water in ED-7 points to the possibility that most of the production from the high capacity Prairie du Chien-Jordan wells in the area represents a fast moving ground water that is short-circuiting most of the existing shallow ground-water system. The deeper-younger water apparently travels along preferential pathways, like discrete fracture zones, buried bedrock valleys and, perhaps, multi-aquifer wells. The great production pressure may be inducing ground water flow toward the water supply wells from unexpected directions, not coinciding with the general gradients observed on a regional scale (Santi et al., 2006).

The presence of highly contaminated Platteville and Drift wells two miles north of ED-7 indicates that the second hypothesis of a distant, up-gradient source is likely to be true. The PCE concentration found at well W437 in the Platteville formation was near 6% of PCE solubility in water, indicating a potential existence of a DNAPL source in the Drift and/or the Platteville formation. In addition to PCE, vinyl chloride, cis-1,2-DCE and other chlorinated compounds were also found in the Platteville formation at high concentrations.

The likely pathways for this contaminated ground water or potential DNAPL to move to the Prairie du Chien-Jordan aquifer include unsealed multi-aquifer wells present in the area. It has been known that wells were completed in the past with an open hole through Platteville, St. Peter, and Prarie du Chien-Jordan aquifer system in the area. Even with the ongoing effort to seal off those wells, many of them still exist and may act as a conduit for contaminant transport from shallow to deeper aquifers. Ground-water withdrawals from the Prairie du Chien-Jordan aquifer by a cluster of Edina municipal wells located south of the suspected release of PCE could have accelerated the migration of vinyl chloride within the Prairie du Chien-Jordan aquifer.

Another type of a pathway may be buried bedrock valleys where Glenwood Shale and Basal St. Peter Formation (regionally serving as confining layers) are absent. There are two bedrock valleys located north and west of ED-7. In the bedrock valley to the north the drift aquifer is in direct contact with the St. Peter aquifer. The northern tip of the north bedrock valley coincides with the area of highly contaminated Platteville and Drift wells. However, the few boring logs available for this area indicate that this bedrock valley is filled predominantly with glacial till that may substitute the missing Glenwood Shale in its function as a confining layer. Also, St. Peter wells in that area are less contaminated than the deeper Prairie du Chien-Jordan wells. Those two factors seem to weaken the hypothesis of the Platteville and Drift sources along Highway 7 ('north source') as responsible for most of contamination detected in the Prairie du Chien aquifer and in ED-7.

Another possibility is that the contaminants detected in the Prairie du Chien-Jordan wells were introduced to this deep bedrock aquifer directly by discharges to the deep wells, or through accidental spills near such wells and leaking along the well casing (like near well W48). However, this hypothesis would be difficult to verify as such releases could have happened long time ago and intensive, year-around pumping from this aquifer would have 'smeared' the sources and distributed contaminants in all directions.

Chemical analysis of the discrete samples that were collected from ED-7 indicates the presence of anaerobic conditions in the 'deeper-younger' ground water that is carrying contaminants. Such conditions support reductive dechlorination of PCE and/or TCE into cis-1,2-DEC and vinyl chloride. Thus it is likely that vinyl chloride and cis-1,2-DCE detected in ED-7 originated from PCE and/or TCE source.

The preliminary results from ground water flow modeling using MODFLOW conducted for another related project shows a predominant eastern direction of ground-water flow in the Prairie du Chien-Jordan aquifer for the area. In the buried bedrock valley to the west of ED-7 both the Glenwood Shale and the entire St. Peter Sandstone were eroded, and the Drift aquifer is in direct contact with the Prairie du Chien-Jordan aquifer. Such geological configuration might provide an easy pathway for contaminants. A contaminant source to the west of ED-7 has yet to be found. However, a steady-state, or a 'roughly time-discretized' transient ground water model may not be able to capture the possible, short lived gradient reversals caused by a complicated pattern of pumping form a network of high capacity municipal wells present in the area. The limited continuous monitoring of water level data collected to date from the Meadowbrook Golf Course well (one and a half mile north of ED-7) afford a glimpse into a profoundly dynamic nature of this system. Summer-winter water level fluctuations measured in that well, which is located considerable distance from any production well, exceed 40 feet. It is possible that during some pumping configurations hydraulic gradients undergo drastic reversals. The recently initiated continuous water level monitoring in several wells will verify is such reversals take place.

The material presented in this paper illustrates the challenges in finding a contaminant source in a complex and dynamic aquifer system under heavy water production.

ACKNOWLEDGMENTS

We would like to thank Scott E. Anderson of the City of St. Louis Park, Roger Glanzer of the City of Edina, Bill Greg and Andrew Tarara of ENSR International for their cooperation and sharing with us variety of data that we used to develop this paper. There were also numerous other individuals who contributed to this project and we would like to acknowledge their efforts.

REFERENCES

Alexander, E.C. and Alexander, S.C., 1989. Residence times of Minnesota groundwaters, Journal of the Minnesota Academy of Science, Vol. 55(1): 48-52.

Bloomgen, B.A., 1985. Bedrock geology and topographic maps of the Minneapolis-St. Paul urban area, Minnesota, Minnesota Geological Survey, St. Paul, Minnesota.

Butler, B.J., and Barker, J.F., 1996. Chemical and microbiological transformations and degradation of chlorinated solvent compounds, In: Dense Chlorinated Solvents in Groundwater, J.F. Pankow and J.A. Cherry (Eds.), Waterloo Press, Portland, Oregon: 267-312.

Clark, I.D., and Fritz, P., 1997. Environmental Isotopes in Hydrogeology, CRC Press, Boca Raton, Florida, 33431, 352 p.

Hendry, M.J., 1988. Do isotopes have a place in ground-water studies. Ground Water, Vol. 26 (4): 410-415.

ERT, 1983. Recommended Plans for a Comprehensive Solution of the Polynuclear Aromatic Hydrocarbons Contamination Problem in the St. Louis Park Area, Environmental Research and Technology, Inc., Vol. II, Appendices A-F.

Horn, M.A., 1983. Ground-water use trends in the Twin Cities Metropolitan area, Minnesota, 1880-1980, U. S. Geological Survey, Water-Resources Investigations Report 83-4033, 37 p.

Hult, M.F. and M.E. Schoenberg, 1984. Preliminary Evaluation of Ground-Water Contamination by Coal-Tar Derivatives, St. Louis Park Area, Minnesota, U. S. Geological Survey Water-Supply Paper 2211, 53 p.

Lowry, G.L., 1989. A three-Dimensional Finite-Difference Ground-Water Flow Model of the Drift and Platteville Aquifers in the St. Louis Park Area, Minnesota, Masters of Science Thesis, University of Minnesota, Minneapolis, Minnesota. 187 p.

Runkel, A.C., Tipping, R.G., Alexander, F.C., Jr., Green, J.A., Mossler, J.H., and Alexander, S.C., 2003. Hydrogeology of the Paleozoic Bedrock in Southeastern Minnesota, Minnesota Geological Survey Report of Investigations 61, 105 p., 2 pls.

Santi, P.M., McCray, J.E., and Martens, J.L., 2006. Investigating cross-contamination of aquifers, Hydrogeology Journal 14(1-2): 51-68.

Stark, J.R., and Hult, M.F., 1985. Ground-Water Flow in the Prairie du Chien-Jordan Aquifer Related to Contamination by Coal-Tar Derivatives, St. Louis Park, Minnesota, U. S. Geological Survey Water-Resources Investigations Report 85-4087, 57 p.

Wright, H.E., Jr., 1973. Tunnel valleys, glacial surges, and subglacial hydrology of the Superior Lobe, Minnesota, Geological Society of America Memoir 136: 249-276.

3

THE OCCURRENCE AND FORMATION POTENTIAL OF TRIHALOMETHANES IN DRINKING WATER IN TULKARM DISTRICT, PALESTINE

Amer Marei[1], Amer Kanan[1], and Ahmed Nasser[2]
[1]Environmental Lab, Department of Applied Earth and Environmental Sciences
Al-Quds University-Jerusalem, POBox 20002, Bet Dagan 50250, Israel
[2]Inst. Of Soil, Water and Environmental Sciences, The Volcani Center,
ARO. P.O. Box 6, Bet Dagan 50250, Israel

ABSTRACT

An investigation into the occurrence of Trihalomethanes (THMs) (chloroform, bromodichlorometane, chlorodibromomethane, and bromoform) was conducted for the first time in Palestine. THMs, free chlorine and other water quality and operational parameters were monitored in 22 drinking water samples from points along the distribution system, between the chlorination points and the system extremities, which represented different variable traveling time from different user points (tap water). Results showed that THM concentrations vary significantly (from 0.7 μg /L to 52.7 μg /L) between chlorinated water samples as they leave the plant and at the system extremities. The effect of travel time in increasing THMs occurrence was noticed through the higher concentration of total THMs at the most distant points. Free chlorine measured at four chlorination points ranged from 0.13 to 0.43 mg/L which is lower than the Palestinian Water Authority (PWA) recommended dose (0.3-1 mg/L). At the user points free chlorine range was from 0.02 to 0.43 mg/L. This study also reports, for the first time, trihalomethanes formation potential (THMFP) of the drinking water in Palestine. THMFP was measured in three samples of three different ground waters affected by infiltration of wastewater. It was revealed that THMFP in the three tested samples was more than 100 μg /L (153.6, 290.7 and 1149.7 μg /L) after 7 days of chlorination and incubation at 25 °C and pH 7. Brominated THMs were dominant due to the relatively high existence of bromide (0.7 to 1.03 mg/L) in the raw ground waters.

3.1. INTRODUCTION

Disinfection of water using chlorine is a common practice in Palestine. Chlorine is applied to water in order to deactivate microorganisms and to ensure residual concentration in distribution systems to prevent microorganisms' regrowth. Chlorine is widely used as a disinfection agent because it is effective and inexpensive. In the 1970s the presence of disinfection by-products (DBPs) in drinking water was reported (Rook, 1974; Bellar et al., 1974; Symons et al., 1975). Chlorine reacts with natural organic compounds contained in water producing hundreds of DBPs such as Trihalomethanes (THMs), Haloacetic Acids (HAAs), Haloacetnitriles (HANs,) Haloketons (HKs), and others (Wingle, 2000). In recent years special interest has been evolved to the occurrence of DBPs in chlorinated drinking water. More attention has been paid to the concentration of trihalomethanes (THMs) which are the most commonly occurring group of DBPs. The four main THMs formed are chloroform ($CHCl_3$), bromodichloromethane ($CHCl_2Br$), dibromochloromethane ($CHClBr_2$) and bromoform ($CHBr_3$) (APHA, 1992; Golfinopoulos, 2000).

The main operational and quality parameters that influence the formation of THMs are chlorine dose, organic matter in water, water temperature, pH and travel time (residence time). Bromide present in raw water, from either natural or anthropogenic sources has an important effect on THMs speciation (Stuart et al., 2001).

Epidemiological studies have suggested that THMs are potential carcinogenic and there is an excess risk of bladder, rectal, liver cancer and exposure to DBPs, also an association between exposure to THMs and spontaneous abortion, growth retardation, and birth effects (Wingle, 2000).

Information about the concentration of THMs in Palestine is lacking. There is no information concerning THM occurrence in drinking water distribution systems and nothing is known about THMFP. The aim of this study is to measure the THMs in the chlorinated drinking water through the mains of the distribution system at the end users' homes and to investigate the THMFP in the raw waters. The result of this study will be valuable to the Palestinian Water Authority (PWA) and the Municipalities in defining the best technology for water disinfection and in creating the Palestinian THMs regulations.

3.2. MATERIALS AND METHODS

3.2.1. Utilities Studied

The water utilities taken as a case study are those supplying the population of Tulkarm and Anabta. Tulkarm and Anabta are two cities in Palestine located in the northwestern part of the West Bank. The two cities are 15 and 24 km east of the Mediterranean respectively (Fig. 1). Tulkarm and Anabta Municipalities provide drinking water for 90,000 and 10,000

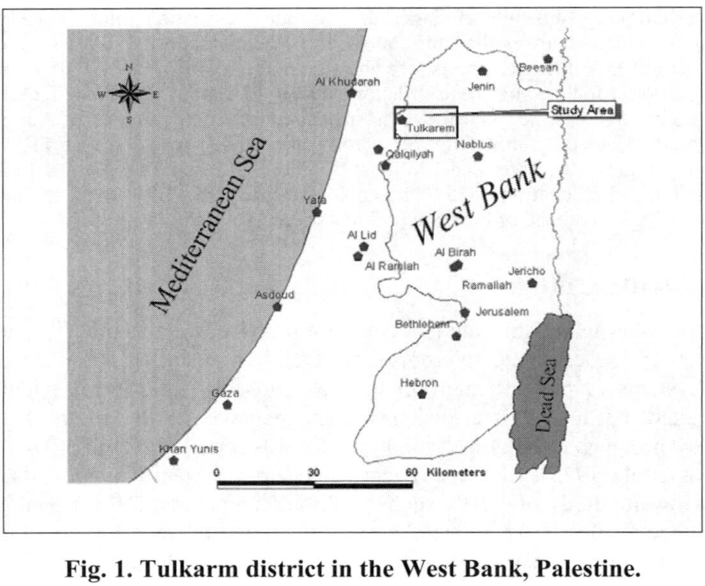

Fig. 1. Tulkarm district in the West Bank, Palestine.

inhabitants respectively. Ground water, the source of drinking water, is abstracted from wells of the Upper Cenomanian aquifer which consists from limestone and dolomite (Marei, 2003; Scarpa et al., 1998). Chlorination for the purpose of disinfection is the only treatment technique applied on the raw ground water in the two Municipalities before distributing to the end users' homes. The Municipal wells in Tulkarm and Anabta cities are close to Wade Zeimar, a raw wastewater stream running around the year (Fig. 2). Chemical analysis and Total and Feacal Coliform investigations show that ground water abstracted from these wells is contaminated (Scarpa et al., 1998; PWA, 2005).

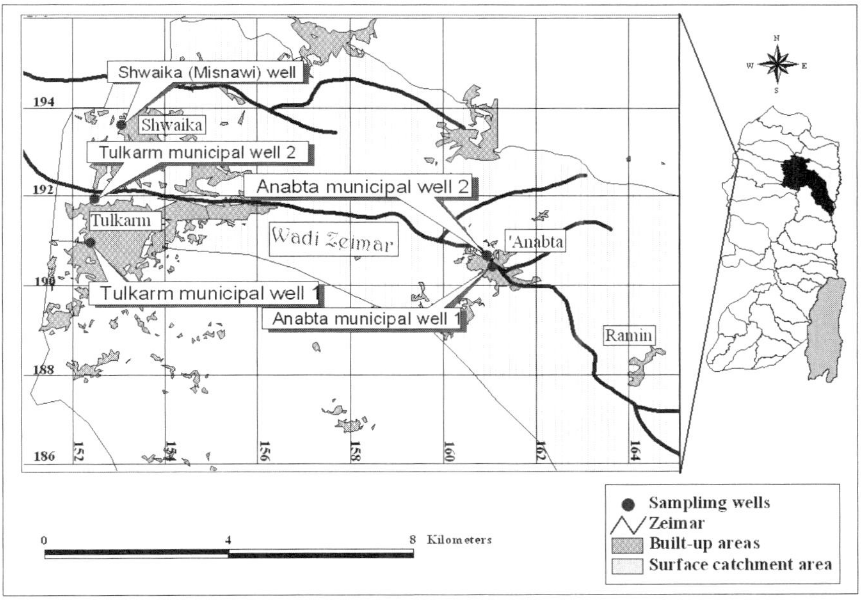

Fig. 2. Study Site.

3.2.2. Sampling and Analysis

To investigate the occurrence of THMs within the water distribution system of the concerned utilities, 22 water samples were collected at and after the chlorination points. Samples were taken at different points, from the chlorination points (Tulkarm Municipal wells 1 and 2, Shwaika (Misnawi) Municipal well, and Anabta Municipal wells 1 and 2) to measure the initial formation of THMs and from tap points with moderate and high traveling distances (contact time).

Samples for THM measurements were collected in 40 mL brown vials with Teflon lined screw caps from Supelco. The tap was allowed to run 3-5 minutes after which the vials were filled carefully just to over flowing, without passing air bubbles through the sample or trapping air bubbles in sealed vials. Before sampling Ascorbic Acid as a neat material (10 mg for 10 mL sample) was added to vials as a quenching material to prevent further THM formation. HCl (3 drops 1 N) was added when the vial was half filled to prevent dehydrohalogenation and biodegradation (Golfinpoulos, 2000).

Raw water samples from the prechlorination points were obtained for the purpose of TOC and other chemical analyses and for the THMFP experiment. 30 mL glass vials with Teflon lined caps were used for the collection of samples for TOC analysis. Polyethylene high density two liter bottles were used for collecting the samples used in the THMFP experiment and bromide analysis. The samples were analyzed directly or stored up to a maximum of 14 days at 4 ºC.

3.2.3. Trihalomethane Formation Potential Experiments

Laboratory chlorination experiments were carried out with raw water samples collected before disinfection from three wells, Tulkarm well 2, Anabta well 2 and Shwaika (Misnawi) well. One litter of each water sample was buffered with phosphate buffer in a one litter brown reagent bottle. The pH was adjusted to 7 with NaOH. The required dose of chlorine (5 mL of calcium hypochlorite solution has available chlorine ≤20%) was added to the samples. The chlorinated solutions were incubated at 25 ºC. From each solution for each desired reaction time, a sample was obtained using the same type of 40 mL vials used in tap water sampling and treated in the same way. The THMFP of the three samples was determined as total THM (TTHMs) formed during 7 days at 25 ºC, pH 7 and by maintaining free chlorine residual throughout the experiment (3.35-4.16 mg/L).

3.2.4. Analytical Procedures

Basic water quality parameters, such as water temperature, pH, EC, and free chlorine residuals were monitored at the time the THMs samples were collected. Temperature, pH and EC were determined using a multi WTW meter. Free residual chlorine measurements were conducted using a portable spectrophotometer by the DPD colorimetric method (HACH, 1997). Nonpurgeable Organic Carbon (NPOC) was determined in the water samples using N/C 2000 with an autosampler analyzer from Analytikajena after acidification the samples with 2 N HCl and then purging via nitrogen gas. Bromide was determined in the water samples by an ion selective electrode.

3.2.5. Trihalomethane Extraction and Analysis

Head-space gas chromatography with mass spectrometry detection was used to analyze THMs in the samples. The head-space method is suitable for the analysis of samples with high contents of volatiles without contamination risk (Deok-Hee Cho et al., 2003). It also has the advantages of being rapid with repeatability and minimum of solvent handling and waste production (Kuivinen Jorma and Hakan Johnsson, 1999; Nikolaou et al., 2002). Five milliliters of the water samples was transferred quantitatively into a 10 mL headspace vial which was immediately sealed with stainless steel crimp top cap equipped with PTFE-lined septa. Vials were statically incubated at 95 ºC for 20 minutes in a CombiPal autosampler (CTC Analytics AG, Switzerland). An aliquot (1000 µL) of the headspace was subsequently withdrawn and injected into the injector of a 6890N Agilant GC/MS at 300 µL/s. Separation was performed on a J&W-VRX column. The column oven temperature was held at 40 ºC for 4 min., ramped to 70 ºC at 2.0 ºC/min., and ramped to 150°C at 10°C/min. Injections were done in split mode (split ratio 2.0), with injector temperature at 220°C, transfer line temperature at 280°C and ion source temperature at 230°C.

The SIM method was developed for 4 compounds (chloroform, bromodichloromethane, dibromochloromethane, bromoform) following the USEPA method No. 501. Quantification ions and method validation were performed with external standard calibration. Calibration curves in the concentration range from 5 to 200 μg/L were prepared from standard solution (Supelco Cat. No 48746).

3.3. RESULTS AND DISCUSSION

Water quality parameters, of the ground water that affect the THM formation during chlorination, are summarized in Table 1. According to this table, the raw water of Tulkarm and Anabta municipal wells has relatively high bromide (Kampioti and Stephanou, 2002) and organic carbon content. The EC of the water is within the range of fresh water.

Results of free chlorine residuals determined in two sampling campaigns in the two distribution systems, Tulkarm and Anabta are presented in Table 2. Samples were taken at increasing distances from the chlorination points of each system. Active residual chlorine

Table 1. Water quality parameters.

Well	EC (us/cm)	pH	Temp ^0C	Br μg/L	TOC mg/L	Chlorine dose mg/L
Shwaika (Misnawi) Well	898	7.51	23.5	720	4.85	0.19
Anabta Municipality 1	1098	7.34	23.6	940	4.37	0.43
Anabta Municipality 2	864	7.29	23	780	3.16	0.41
Tulkarm Municipality 2	1070	7.62	23.9	820	*	*

* Not measured

Table 2. Free chlorine residual throughout the distribution system of Tulkarm and Anabta.

Sample Site	Free chlorine (mg/L) 19/05/2005	18/06/2005
Tap water/Tulkarm	0.43	*
Tap water/Tulkarm	0.14	*
Tap water/Tulkarm	0.08	*
Tap water/Tulkarm	0.06	0.03
Tap water/Tulkarm	0.02	0.03
Tap water/Tulkarm	0.02	0.02
Tap water/Shwaika	0.24	*
Tap water/Shwaika	0.18	*
Tap water/Anabta	0.16	0.21
Tap water/Anabta	0.07	0.01
Tap water/Anabta	0.05	0.04
Tap water/Anabta	0.03	0.03
Tap water/Anabta	0.02	0.14
Tap water/Ramin	0.08	0.11
Tap water/Ramin	0.07	0.13
Tap water/Ramin	0.03	0.03

* Not measured

concentrations decreased with increasing distance from the chlorination point. Free chlorine residuals are in the range of 0.13-0.43 mg/L at the chlorination points. Along the distribution system the free chlorine residuals are in the range of 0.01-0.43 mg/L having an average of 0.1 mg/L. These free chlorine residuals do not meet the WHO requirement for residual disinfectant concentration (≥0.5 mg/L) for effective disinfection (WHO, 2004).

THMs were detected in all water samples, after the chlorination point, collected from the treatment sites and from the distribution system of the two cities (Table 3). The concentration of THMs varied among the treatment plants (at the chlorination points) and throughout the sampling points at different distances in the distribution system of the drinking water in the two cities, Tulkarm and Anabta. TTHMs at the chlorination points at the site of ground water abstraction and chlorination are in the range of 0.7-23 μg/L. Throughout the distribution system TTHMs are in the range of 2.7-52.7 μg/L. The mean and the median of TTHMs measured throughout the distribution systems are 24.4 and 22.7 μg/L respectively. The variability of the TTHM concentrations can be attributed to unsteady operational parameters, such as chlorine doses and travel time; and to the ground water composition (bromide and organic material content) which is not constant in the different wells providing the net work in Tulkarm and Anabta.

Table 3. Free chlorine and THMs in the distribution systems of Tulkarm and Anabta. Samples are arranged according to their relative distance from the chlorination points.

point	Water type and source	Free chlorine mg/L	$CHCl_3$ μg/L	$CHCl_2Br$ μg/L	$CHClBr_2$ μg/L	$CHBr_3$ μg/L	TTHMs μg/L
1	Point of chlorination	0.13	0.2	0.1	0.2	0.2	0.7
7	Point of chlorination	19	0.1	0.2	0.1	0.3	0.7
2	Point of chlorination	0.41	0.2	0.2	0.3	0.1	0.8
6	Point of chlorination	0.38	0.8	0.3	0.3	0.3	1.7
3	Point of chlorination	0.43	12.9	0.3	3.1	6.7	23
14	Tap water/Tulkarm	0.07	0.1	0.3	0.3	2	2.7
15	Tap water/Tulkarm (Noor Shams)	0.24	0.1	0.5	0.2	2.3	3.1
11	Tap water/Tulkarm	0.16	0.7	0.2	0.9	6.5	8.3
9	Tap water/(Tulkarm) Shwaika	0.03	0.2	0.2	0.6	8.9	9.9
13	Tap water/Tulkarm	0.08	0.2	0.1	0.8	9.2	10.3
10	Tap water/Tulkarm	0.02	1.3	0.2	1.6	7.6	10.7
12	Tap water/Tulkarm	0.03	0.2	0.4	0.5	19.97	21.07
8	Tap water/ (Tulkarm) Shwaika	0.5	0.2	0.3	0.4	21.8	22.7
18	Tap water/Anabta	0.08	0.9	0.6	1.6	6.1	9.2
4	Tap water/Anabta	0.065	1.7	3. 3	10.3	16.1	31.4
17	Tap water/Anabta	0.14	20.1	0.4	3	8.5	32
22	Tap water/Anabta	0.43	20.2	1	4.3	7.8	33.3
16	Tap water/Anabta	0.18	3.1	1	8.5	34.8	47.4
5	Tap water/Anabta	0.07	1.1	2.4	9	29.7	42.2
19	Tap water/Ramin	0.02	0.2	0.4	3.2	38	41.8
20	Tap water/Ramin	0.02	0.1	0.3	5.5	36.9	42.8
21	Tap water/Ramin	0.06	0.2	0.7	6.8	45	52.7

The influence of the travel time on the THMs production throughout the distribution system of Anabta is noticed (Fig 3). At the site of chlorination (zero distance) the concentration of THMs formed are the lowest, but at the extremities of the distribution system an elevation of THMs is present.

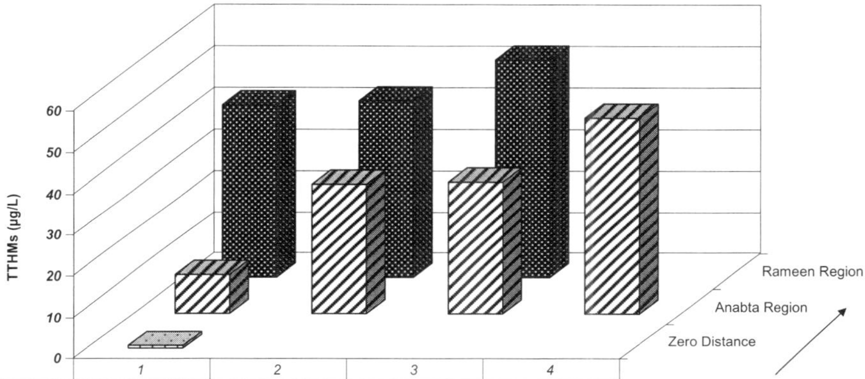

	1	2	3	4
Zero Distance	0.70			
Anabta Region	9.20	31.40	32	47.4
Rameen Region	41.8	42.8	52.7	

Fig. 3. Travel distance effect on THM occurrence in the distribution system of Anabta.

Brominated THMs compounds are prominent compared with chloroform. This relatively high level of brominated THMs is most certainly associated with the bromide level (720-940 μg/L) in the raw water. Bromide affects THM formation because chlorine oxidizes it producing hypobromous acid, which increases the formation of brominated THMs (Golfinopoulos, 2000; Hong Anh Duong et al., 2003). The sources of bromide in the study area are the natural source, mainly the sea origin rain water, and the waste water running around the year in Wade Zeimar crossing the catchment area of the ground water resource and from methyl bromide used in agriculture.

In all samples, the TTHMs measured in the drinking water distribution systems were lower than the maximum levels set by the USEPA, WHO and the EC.

The THMFP determined for Anabta Municipal well 2, Shwaika (Misnawi) well and Tulkarm Municipal well 2, are 153.6, 1149.7, 290.7 μg/L respectively (see Table 4). The high THMFP found for ground water used in drinking water corresponds to its high content of bromide and NOM represented by the NPOC values (see Table 1). Brominated THMs dominated in the formation potential test. The prominent THMs formed are bromoform and

Table 4. Trihalomethane formation potential (THMFP) for ground water samples from three different wells in Tulkarm and Anabta

Time (hour)	Anabta Municipal well 2	Shwaika (misnawi) well	Tulkarm Municipal well 2
		TTHMs (µg/L)	
0	3.2	24.7	65.3
2	12	*	73.5
6	31.4	114.8	*
8	31.1	*	103.3
22	63.7	*	115.9
28	68.7	342.2	*
52	96.9	*	185.9
76	115.9	655.4	*
100	148.3	*	212.7
168	153.6	1149.7	290.7

* Not measured

dibromochloromethane (Fig. 4). This domination of brominated THMs is the result of low chlorine dose with a relatively high bromide level that result in higher hypobromous acid producing more brominated THMs (Magazinovic et al., 2004).

The variation of the THMFP in the three ground water samples may be due to the differences in the breakpoint and peakpoint of chlorination for each water type (Hong Anh Duong, 2003) which was not practically verified in this study. The rate of THM formation in Anabta well 2 and Shwaika (Misnawi) samples is shown in Fig. 4. THMs were formed rapidly under the chlorine-saturated conditions used. After one day concentrations reached 30-40% and after 2 days reached 57-64% of the full measured THMFP (7 days).

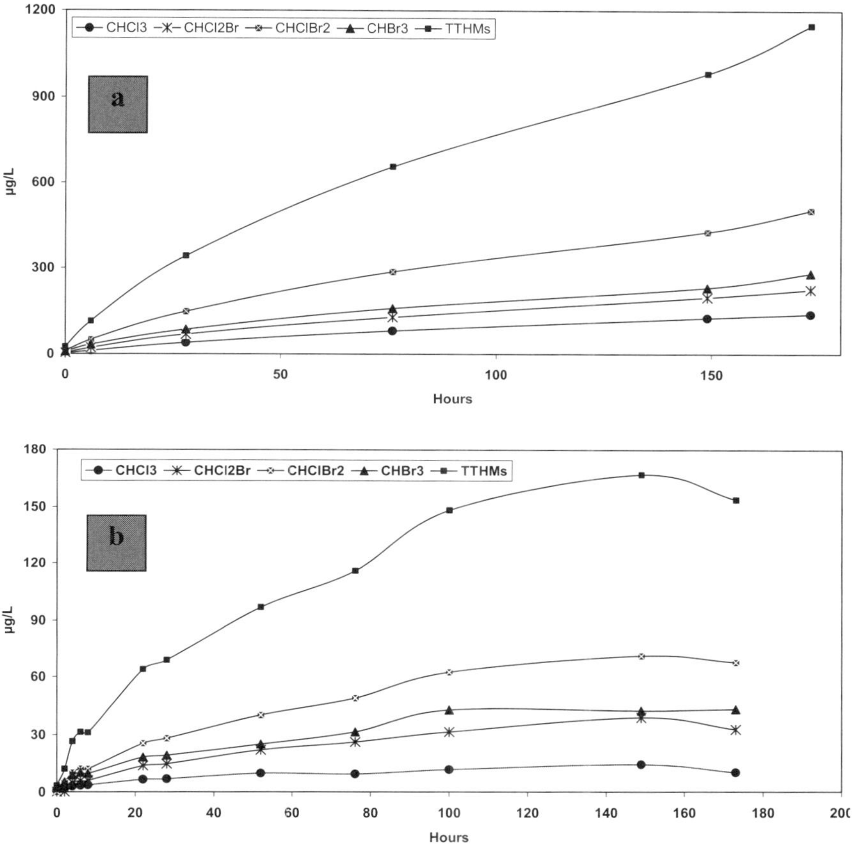

Fig. 4. The rate of THM formation and their speciation in (a) Anabta Municipal well 2 and (b) Shwaika (Misnawi) well.

3.4. CONCLUSION AND RECOMMENDATIONS

This study shows that the THMs concentrations occurred in Tulkarm and Anabta drinking water are below WHO, EC and USEPA drinking water standards. Free chlorine along the distribution system of Tulkarm and Anabta is much lower than the effective concentration. The residence time represented by the travel time along the distribution system increased the THM concentrations. THMFP results indicate that THMs may happen at levels more than the permissible level in the distribution system. Thus it is necessary to verify the operational parameters mainly the chlorination dose and the residence time to inhibit the levels of THMs exceed the required limits. Bromide and organic carbon in the raw water affected both the THMFP and THMs speciation. With regard to the low residual chlorine in the distribution systems in Tulkarm and Anabta, it is recommended to verify the efficiency of the existing monitored free chlorine and to estimate the required chlorination dose. It is recommended that a regular monitoring program of THMFP is instigated where ground water is affected by wastewater.

ACKNOWLEDGMENTS

The authors are grateful to the German Ministry of Research and Education (BMBF) for the funding of this study as part of project GAPI "Ground water Availability and the fate of Pesticide Input, Fundamentals of Human Health in the Middle East". We are indebted to the VOLCANI center for the cooperation and assisting the GC analysis.

REFERENCES:

APHA, 1992. Standard Methods for the Examination of Water and Wastewater, 18th ed. American Public Health Association, Washington DC.

Bellar, T.A., Lichtenberg, J.J., and Kroner, R.C., 1974. The occurrence of organohalides in chlorinated drinking waters. Journal of American Water Works Association 66: 702-706.

Deok-Hee Cho, Sung-Ho Kong, and Seong-Genun Oh, 2003. Analysis of trihalomethanes in drinking water using headspace- SPME technique with gas chromatography. Water Research 37: 402-408

Golfinopoulos Spyros K., 2000. The occurrence of trihalomethanes in the drinking water in Greece. Chemosphere 41: 1761-1767.

HACH, 1997. Water Analysis Handbook,3rd ed. Hach company, USA

Hong Anh Duong, Michael Berg, Minh Hang Hoang, Hung Viet Pham, Herve Gallard, Walter Giger, and Urs von Gunten, 2003. Trihalomethane formation by chlorination of ammonium-and bromide- containing ground water in water supplies of Hanoi, Vietnam. Water Research 37: 3242-3252

Kampioti Adamantia A., and Steephanou Euripides G., 2002. The impact of bromide on the formation of neutral and acidic disinfection by-products (DBPs) in Mediterranean chlorinated drinking water, Water Research 36, 2596-2606

Kuivinen Jorma, and Hakan Johnsson, 1999. Determination of trihalomethanes and some chlorinated solvents in drinking water by headspace technique with capillary column Gas-Chromatography. Water Research 33 No 5 pp 1201-1208

Magazinovic, R. S. , Nicholson B. C. , Mulcahy D. E. , and Davey D. E., 2004. Bromide levels in natural waters: its relationship to levels of both chloride and total dissolved solids and the implications for water treatment. Chemosphere 57: 329-335

Marei, A., 2003. Effects of wastewater on the groundwater Quality of Calcareous Aquifer: A case study of Wade Zeimar- West Bank Palestine – Unpublished-

Nikolaou Anastasia D., Themistokles D. Lekkas, Spyros K. Golfinopoulos, and Maria N. Kostopoulou, 2002. Application of different analytical methods for determination of volatile chlorination by-products in drinking water. Talanta 56 pp 717-726

PWA, 2005. Microbiological Water Quality Results, Palestinian Water Authority, Rammalah-Palestine.

Rook, J. J., 1974. Formation of halo forms during to chlorination of natural waters. Water Treat. Exam. 23: 234-243.

Scarpa D.J., Rabbo, A.A., Qannam, Z. S., and Abdul-Jaber, Q., 1998. Groundwater pollution in unconfined aquifer in the northern West Bank, Palestine. Hydrology in a Changing Environment Vol. II

Stuart, M.E., Gooddy, D. C., Kinniburgh, D. G., and Klinck, B. A., 2001. Trihalomethane formation potential: a tool for detecting non-specific organic ground water contamination. Urban Water 3: 173-184

Symons, J. M., Bellar, T. A., and Caroswell, J. K., et al., 1975. National organics reconnaissance survey for halogenated organics. Journal of American Water Works Association 67: 634-647.

WHO, 2004. Guidelines for Drinking-water Quality, third ed. Vol 1, Recommendations, Geneva.

Wingle, D. T., 2000. Safe drinking water: A public health challenge. Chronic diseases in Canada. 19, No.3.

4

NATURAL CHEMICAL TRACERS OF URBAN POLLUTION: CASE OF BORON EXAMPLE OF APPLICATION ON THE AREA OF ALGERIAN CHERIA

F. Baali[1], A.K. Rouabhia[2], N. Kherici[3], and L. Djabri[3]

[1] Maître assistant C.U de Tebessa Algerie, Étudiant en recherche libre Polytech'Lille

[2] Maître assistant C.U de Tebessa, Algerie

[3] Professeurs de L'Université de BADJI Mokthar SIDI Ammar 23000 Annaba, Algerie

ABSTRACT

Underground water exploitation from carbonated aquifers is currently the only source that caters to human, industrial and farming water use in the area of Tebessa, north east of Algeria. Chéria is the town located on the study plateau widely leveled by Eocene limestone and provides a real water tower in the area. In the recent years, the area has undergone a widespread agricultural and para-agricultural activity. Many hydrogeologic and hydrochemical studies in the area showed a high degree of contamination by nitrates. This result required deeper research work in order to identify the origins of this pollution. For this, the hypothesis that pollution is due to urban wastewater was researched, by means of boron as a natural chemical tracer and urban pollution marker. The results showed considerable boron concentrations and confirmed the hypothesis of contamination by the urban wastewater, drained off by Chéria Wadi, behaving as the main water route for natural drainage in the area.

4.1. INTRODUCTION

Between August 2000 and October 2004, several campaigns were undertaken for the study of water resources of the aquifer of Eocene limestone in the Chéria area. The study refers to a 720 km^2 area including 300 km^2 in the urban periphery of the town of Chéria which shelters more than 66,027 inhabitants. The area is characterized by a semi-arid climate, a high hydrous deficit as well as an intensive exploitation of subsoil waters. The average air temperature is about a few degrees Celsius and can reach negative values in winter, thus snow cover persists on the mount edge, particularly on the mount called Doukkane, for an average period of five to ten days per annum. The Chéria weather station reached an average annual precipitation of 304mm between 1983 and 1999. The maximum of monthly precipitation for the average of the same period is about 34mm in October, weakening the infiltration, knowing that precipitations are stormy and the temperature of this month can reach thirty degrees. The control network makes it possible to study the seasonal evolution of the water level (low and high waters) with 162 wells (Baali, 2000). The levels which were measured seem very variable (few centimeters with some meters). In order to control the composition of subsoil waters in the area, a sampling of 50 points was established.

4.2. GENERAL CHARACTERISTICS OF THE AREA

The plate of Chéria forms part of the catchment area of Chott-Melghir, located 65 km south-west of Tébessa (Fig. 1) from coarsely triangular whose edges consist of culminating mounts (Dj. Doukkane with 1550m and Metaguinaro 1712m). Its altitude varies in the

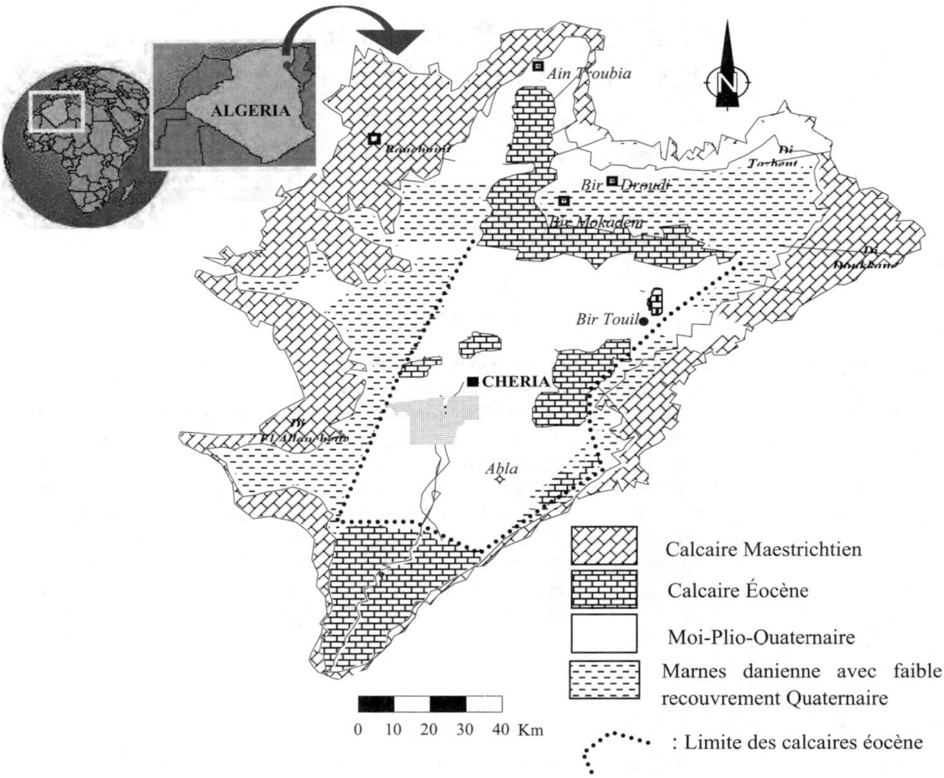

Fig. 1. Map of the situation and the geological Formations of the area of Chéria.

center between 1000 and 1200m. The town of Chéria constitutes the geographical, economic and administrative center. The work completed by Mr. LAMBERT, DUROZOY (1948) and J RICHARD (1973) described that the Quaternary downstream from Chéria, is presented by fluviatile gravels which can reach 20 m thick (Ain Babouche) whose surface is encrusted tufas whereas Mio-Pliocene is unmatched and transgressive to the Eocene limestone. This horizon contains an important aquifer, which is currently uninteresting, considering the dryness which prevailed during the 10 last years (Gaud, 1977). The Eocene is made of a hard limestone alternation, zoo gene, of chalk, marls more or less limestone. Silica separates there in the form of flint kidneys in the limestone bars and various phosphate levels are observed there. The maestrichtien, reached by the survey of Bir Droudj shows it of 220 m thickness and the marls of the campanien whose thickness can reach 250m.

The piézomètrie area observed on the various charts shows a flow which is generally made south towards north the discharge system of the catchments area, as well as a depression around the town of Chéria. (Figs. 2 and 3)

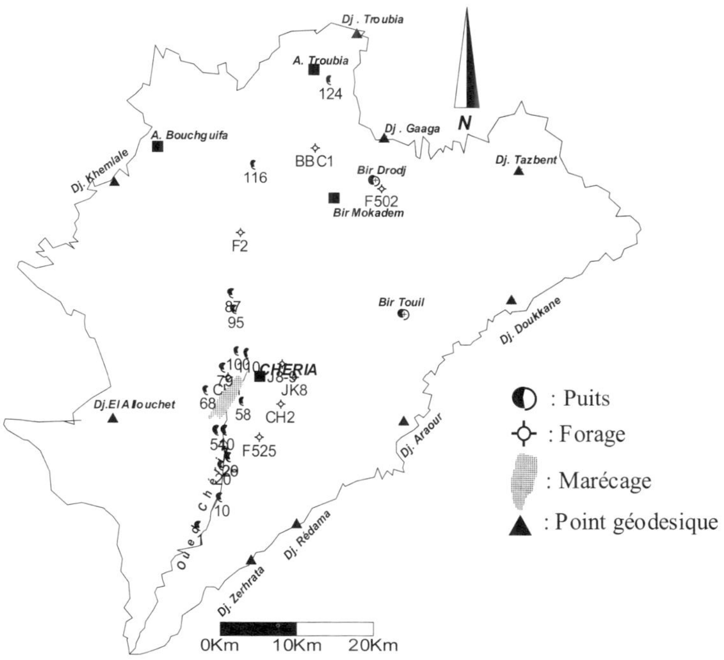

Fig. 2. Map of Cheria waters resource.

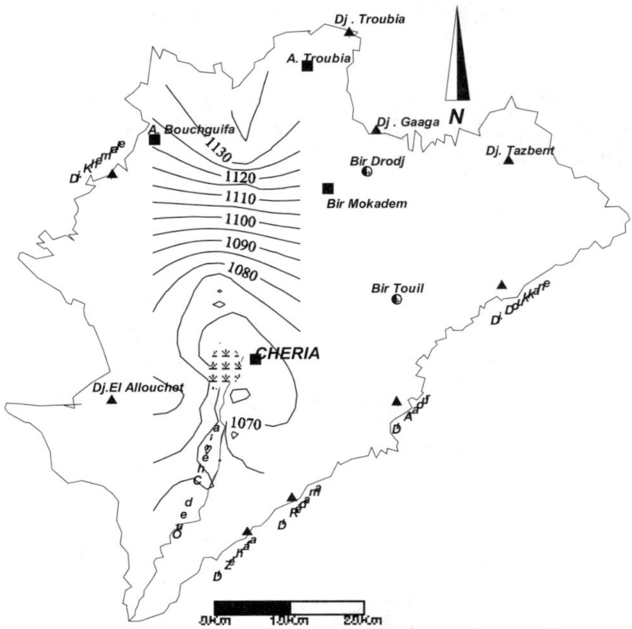

Fig.3. Piezometric map of the area of Chéria August 2000.

4.3. METHODS OF DETECTION

Many methods can be used for the detection and the proportioning of the boron of its compounds in solution in the water, among which one can quote:

* ***Spectrometry of emission by plasma (ICP-AES):*** The solution to be proportioned is introduced into the argon plasma gun in the form of aerosol which is atomized under the effect of the temperature (6000°C). The atoms emit then, before returning in their fundamental state, a radiation of which the spectrum characteristic comprises lines usable in qualitative analysis. This method of rather simple employment can have a threshold of detection about 20µg/l with a precision of 5µg/l approximately. It has the advantage of the proportioning of boron in subsoil waters. These performances were improved by the use of the horizontal plasma gun with flow (BRGM, 1974).

* ***Spectrophotometry:*** The solution is complexed by a reagent involving a colouring whose intensity is proportional to the concentration in element to proportion (Pain, 1995). The measurement of the absorption of a pencil of light wavelength given makes it possible to determine the concentration.

The cartography of the concentration of boron in the area of study allowed the highest concentrations of boron (the standard of WHO 0.3mg/l exceeds) are visualized in the northern area in the place says Abla, the area most touched by nitrate pollution is seen by its presence on two banks of the on-quoted wadi and its discharge system, which indicates consequently the participation the contamination of subsoil waters by urban water.

4.4. RESULTS AND DISCUSSION

The results of grand water chemical composition analysis shows that the concentration of nitrate in the southern area is more than 100 mg/l (Fig. 4). This result encouraged us into looking for the cause of the pollution which is harmful to the public health, especially to babies and contamination of the environment.

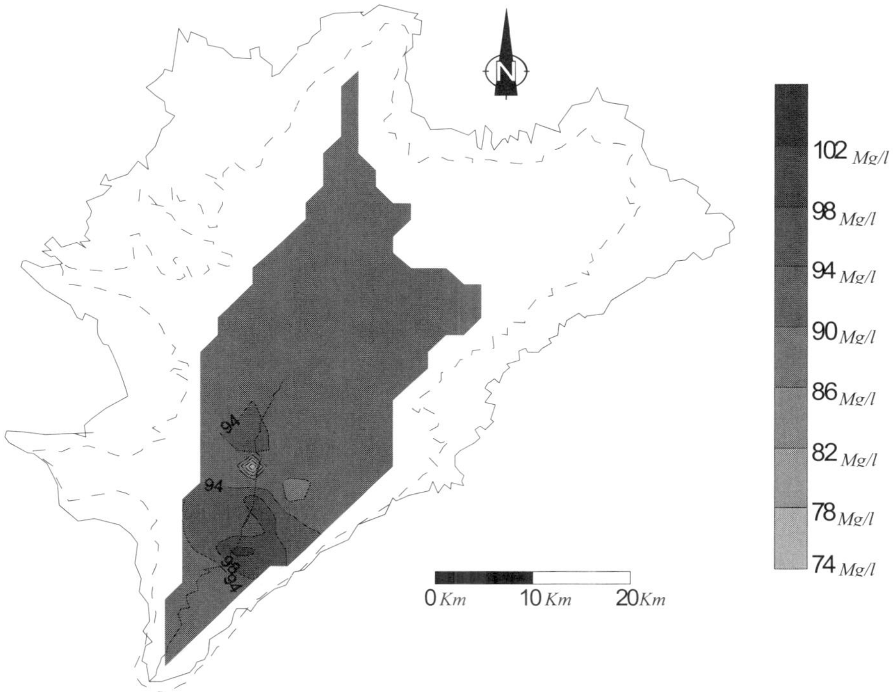

Fig. 4. Map of the nitrate concentrations Chéria Sept 2002.

The first prevision of pollution causes is from the wadi Of Cheria in the southern area, where the waste water and rainfall water were collected. The wadi pass through sandy soil which allowed to exchange the contaminant between the soil and the grand water (Hassil and Khaouni, 2003).

Until now, boron has not been an element having to answer to a criteria of water portability. It is not considered by the administration as one of the water quality monitoring parameters, especially in the third world countries.

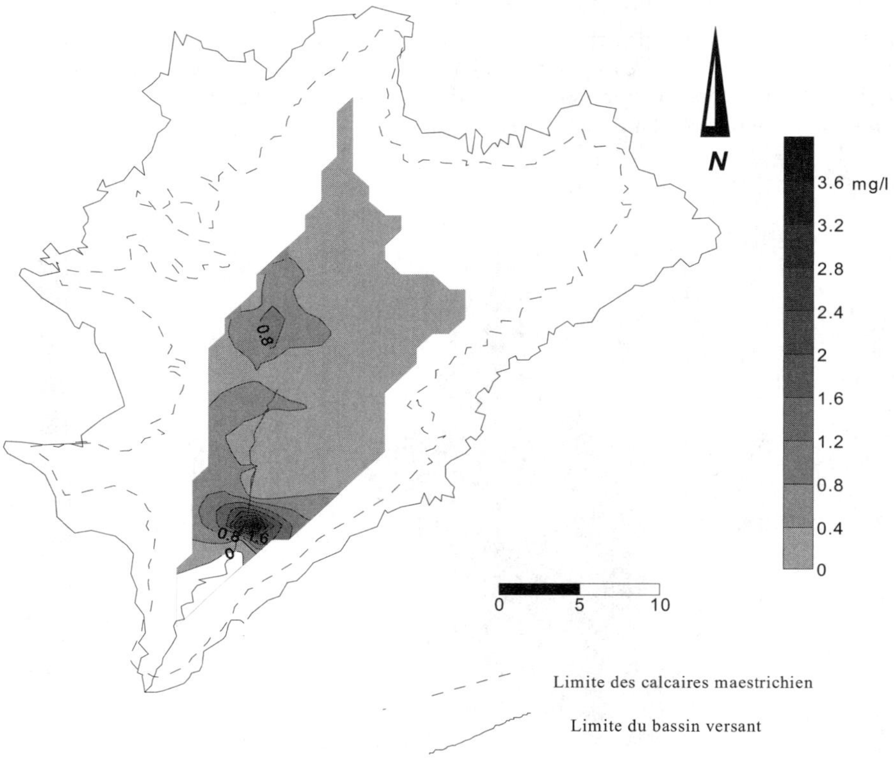

Fig. 5. Map of the Born concentrations Chéria, September 2002.

The absence of grand water monitoring and the previous studies in this subject, increase the difficulty of comparison and the statistical treatment of the data. Therefore, the results which are carried out by us in the domain of our research were for the objectives of differentiating between the origins of pollution: agriculture, breeding of animals and urban.

Born in the form of sodium perborate, it is a major component in domestic detergents. It is stable, easy and inexpensive. Therefore, it constitutes a good marker of pollution from domestic and urban origins (Almahallawi, 2005). Its presence in water is due to the large and harmful use of the detergents and it is directly injected in the natural environment.

Collected and analyzed water showed important concentrations in the wells indicating water pollution. The cartography carried out shows a distribution similar to that of nitrates characterized by strong concentrations of about 3.4 mg/l recorded in the southern part on the level of the wadi peripheral.

4.5. CONCLUSIONS

The use of boron detection shows the contamination contributed by urban water. It has the capacity to differentiate the origins of underground pollution to waters in the area, which will allow thereafter the proposal of the recommendation for rational solution of fighting against the latter. Thus and consequently, the wadi of Chéria represents a public danger that requires an immediate intervention from the state, at least to minimize pollution so that one can regard it as pollution free, and to protect the public health for the ordinary citizens that are subject to bouts of disease by hydrous transmission.

REFERENCES

Almahallawi, K., 2005. Modeling interactions of the land use, urbanization and hydrological factors for the analysis of groundwater quality in Mediterranean zone (example the Gaza Strip, Palestine). 147 – 181pp

Baali, F., 2000. Hydrochimic hydrogeologic Study of the karstic area of Algerian Chéria. memoir of magister University of. Annaba Algeria. 80pp

BRGM, 1974. Test of classification and cartography of the chemical quality of subsoil waters. Departement of. Hydrogéolgy. Orleans.

Gaud, B. 1977. Geological and hydrogeologic investigation of the plate of Chéria Tébessa. Internal report N°2. National Agency of water resource of Tébessa. 60pp

Hassil, N., and Khaouni, S., 2003. Natural tracers of urban pollution case of Boron. Memory of ingenir Tébessa University Centre, Algeria. 90pp

Pain, J.-M., 1995. Natural chemical tracers, markers of urban pollution: example of Boron. DEA. School of ingenir of Lille1Univrsity 31pp

5

WIND AND TIADAL EFFECTS ON CHEMICAL SPILL IN ST ANDREW BAY SYSTEM

Peter C. Chu and Patrice Pauly
Department of Oceanography, Naval Postgraduate School, Monterey, CA 93943, USA

Steven D. Haeger
Naval Oceanographic Office, Stennis Space Center, MS 39529, USA

Mathew Ward
Applied Science Associates, Inc., 70 Dean Knauss Drive, Narragansett, RI 02882, USA

ABSTRACT

A coupled hydrodynamic-chemical fate model is used to investigate mechanisms for chemical dispersion in the St Andrew Bay system. It is found that the time for the pollutants transporting outward the bay mainly relies on the winds and source location. If the application of the stochastic model somehow handles the wind factor, the release location must be shifted to other places in order to evaluate the relative weight of this factor. Because the flux originating from Gulf of Mexico predominantly flows westward, a release point located between St Andrew Bay and West Bay or even more inside West Bay is likely not to deeply impact East Bay. The pollution will only reach the end of East Bay after 15 days if the pollutant is spilled at point A, which only stands 8km away from the previous spot. As this chemical is not volatile, it does not evaporate and its mass is roughly conserved into the system until its natural decay acts. West Bay is much shallower than St. Andrew Bay, thus the small concentration decrease close to the Gulf entrance involves a large pollutant mass transfer into West Bay.

5.1. INTRODUCTION

St. Andrew Bay system is located in the western part of Florida State close to Pensacola (Fig. 1). It is a long, narrow and rather shallow (water column extends from 4 to 10 m) bay oriented in the northwestern direction. Deer Point Lake located at the northern edge of North Bay, provides the major freshwater inflow into the estuary, along with a number of smaller creeks. Two major passes, East Pass and West Pass, have provided surface water connections via its central part (St Andrew Bay) with the Gulf of Mexico. West Pass was artificially cut in 1934 in Shell Island as the primary navigation channel to the Gulf, while most exchange between the estuary and the Gulf had historically occurred through East Pass. East Pass was recently filled in (1998), however, by the movement of shoreline sediments. Dredging was conducted for reopening the pass in 2001. None of these channel

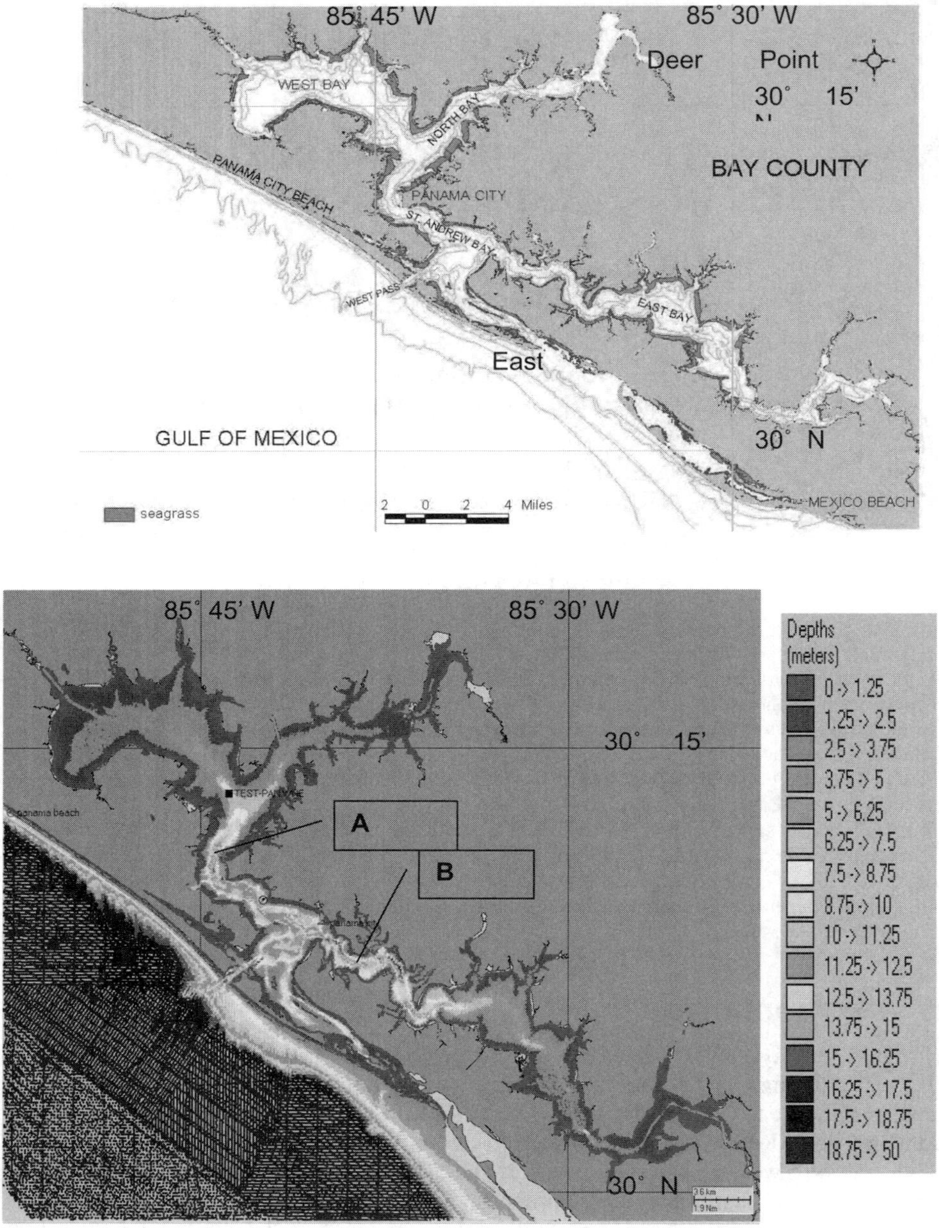

Fig. 1. St. Andrew Bay system geography (upper panel) and bathymetry (lower panel). Note that two points "A" and "B" will be used.

widths exceed 1 km. St Andrew Bay system also links the Intracoastal Waterway on both the very end of the East Bay and the West Bay. The waterway depth is constantly maintained at 4m which causes some water exchanges to occur at these locations.

Drainages into these bays are small and mainly due to groundwater seepage. For example, it is suggested that almost two thirds of the creeks which flow into Deer Point Lake are supposed to be spring fed (Musgrove et al., 1965). Nevertheless their influence can be crucial when modeling the whole system. In terms of estuarine classification, the four basins are generally positive, i.e., drainage inflow exceeds evaporation (Pritchard, 1952) except in St. Andrew bay where neutral conditions, due to its connection with the Gulf, are found (Ichyie and Jones, 1961). The bottom contours are part of the WQMAP geographic information system (GIS) layer and have been provided by NOAA. They depict a 5.2m average depth for St Andrew bay, the mean depth respectively being 2.1, 1.7 and 2m for East, North and West bays.

The shoreline landscape of St Andrew Bay system is known for its high biodiversity, seagrass beds, and its importance as a nursery for coastal fishery resources. Due to its role in fishery production, recreation, and related tourism, it is a key element in the economy of Bay County. Chemical pollutants threaten public health and the environment. In this study, a coupled hydrodynamic-hydrochemical fate model is used to identify the pollutant dispersion patterns. The hydrodynamic sub-model is driven by tides and winds and predicts the velocity field. The hydrochemical model is driven by the velocity field and predicts the chemical spill.

5.2. BACKGROUND

5.2.1. Hydrographic Features

Temperature is rather uniform within the water column through out the bay system with very small vertical variations even in summer (the depth is too shallow for the thermocline to really develop). But its seasonal variability is evident. A five year-study of surface temperature from both inside and outside the bay, conducted using NOAA time series data, shows that the mean temperature varies between 15° C in winter and 30° C in summer. The comparison of these two 5-year-long time series (between 2000 and 2004) portrays a strong zonal sensibility and the difference ranges from 0 during winter time up to 2.5°C during summer (Fig. 2). The rather shallow bay allows the water column to heat up very quickly; the maximum difference is reached no later than March and lasts almost 9 months.

Salinity, however, presents some zonal features. Close to the Gulf of Mexico connection, the salinity is almost constant within the water column (gradient of 0.17 psu/m) and reaches its highest value (34.3 psu). Its value can drop down to 9 psu at surface nearby with a gradient of 3.3 psu/m (Ichyie, 1961; Blumberg and Kim, 2000). Nethertheless, the 24 h variation of salinity does not correlate closely with variations of the tidal current, specifically away from the bay entrance. The salinity follows the net flows caused by wind driven currents and freshwater inflow events. At last, the hydrographic study conducted by Ichyie proved the water column stability being time independent and mainly subjected to salinity vertical variations (Sverdrup et al, 1942).

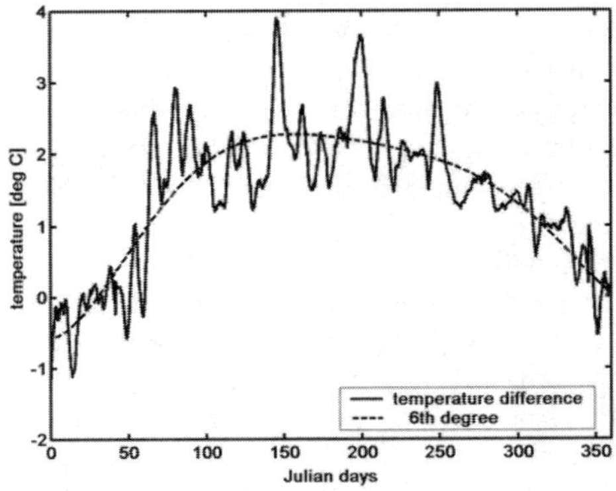

Fig. 2. Temperature difference between St Andrew Bay and Gulf inner shelf.

5.2.2. Tides

The Panama City Beach (30° 12.8'N – 85° 52.7'W) and the St Andrew Bay (30° 09.1'N – 85° 40.0'W) data study shows that according to the form ratio (ratio between diurnal -K1+O1- and semidiurnal -M2+S2-), the tidal constituents are diurnal (ratio greater than 3). High and low water times study show the tidal waves first enter the bay through East Pass (5 minutes difference between East and West Pass) which correlates the general motion of tidal waves. The maximum current is found in the West Pass and can exceed 0.5m/s. It then decreases as low as 0.15m/s at the western end of West Bay. Most of the time, ebb currents duration is longer than flood current duration, mainly due to the effect of drainage, source of a quasi permanent outward current in the upper layer if any (Pritchard, 1952). This process causes "tidal pumping" phenomenon (Fischer et al, 1979).

Compound tides (for ex. M4 ...), often important in shallow water and hence on inner shelves, are generated by the nonlinear interactions of the primary constituents. The long-period tidal constituents are not resolvable in the month-long series, and contributes little (1.1 cm) to the tidal elevation. It is crucial to understand any seasonal effects which may occur on tides. Tidal results from a yearly sea-level time series were compared with those from a monthly subset of the sea-level time series to examine the seasonal effects. Additionally, this provides insight into the effects of unresolved tidal constituents contained in shorter time series. Overall, none of the major diurnal and semidiurnal tidal constituents differed more than 1 cm between the year- and month-long series except K1 constituent which varies significantly. The five main constituents have been computed though 12 successive month-long time series collected in 2004 outside St Andrew Bay and are associated with a 95% confidence interval (Fig. 3). This general trend is consistent with the observations of the last 5 years and denotes the occurrence of a seasonal dependence of both K1 and S2 constituents.

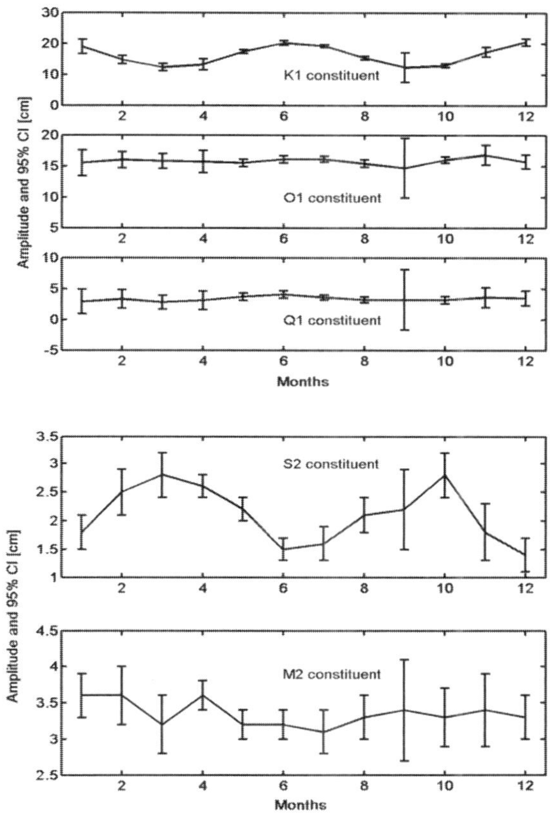

Fig. 3. Seasonal variability of tidal constituents at Panama City Beach.

5.2.3. Fresh Water Supply

Freshwater discharge in the Gulf of Mexico totals approximately 1110 km^3 per year and is dominated by the Mississippi and Atchafalaya rivers, which contribute 55% of this discharge (Solis and Powell, 1999). The extent of the impact of this discharge, both vertically and horizontally, is variable and modulated by discharge rate and processes that disperse the plume, primarily wind stress. The river water accumulates in the inner shelf, forming a low salinity band that supports a persistent baroclinic flow to the south (Atkinson et al., 1983). The mid shelf zone (21-40 m isobaths) is vertically homogeneous during the fall (September, October, November) and winter (December, January, February), due to enhanced wind mixing and decreased runoff. This region undergoes a transition to a vertically stratified state in the spring (March, April, May) and summer (June, July, August) as wind mixing decreases and runoff increases (Atkinson et al., 1983). In addition to runoff, groundwater sources have been shown to be important in the SE region of United States and prevail in the St Andrew bay system.

5.2.4. Alongshore Current and Littoral Transport

The main current affecting the surf zone is the alongshore current created by waves breaking at an angle to the shore. The magnitude of the alongshore current depends on the breaking wave characteristics, breaking angle and local bottom and shore configurations. The alongshore currents are responsible for sand transport along the coast (Chu et al., 2005a). For the study area, the net littoral transport is generally westward as the predominant waves are from the east and is estimated to be about 65,000 cubic yards per year. This estimate seems to be in agreement with the field evidence as indicated by the lack of strong erosion and accretion at the west and east side of the gulf entrance to the St. Andrew Bay (U. S. Army Corps of Engineers, 1971). However, the accumulation of hurricane-induced waves during the 1990's is likely to be responsible for East Pass closure that happened in 1998. Van de Kreeke (1990) showed that a two-inlet system could not be stable for a one-bay system. Even without erosion evidence, one of the inlets has to close. Moreover, Jain et. al (2004) showed that both inlets were actually unstable and hence will request continuous dredging maintenance unless St Andrew bay is enlarged.

All beaches are experiencing erosion, the more severe locating on the northern edge of main entrance channel with a rate of 2.44 m/yr. This process is associated with the construction of St Andrews Inlet in 1934. Similar rates occur along Shell Island and vary from 0.53m/yr in its western part to 1.95 m/yr close to the secondary channel entrance. The lack of construction on Shell Island does not make the eroding process as critical. However, bypass placements of sand have been helping in the stabilization of the shoreline.

These changes are mainly due to extreme weather system hitting the area. Hurricane Eloise in September 1975 caused a 100,000 m^3 beach-dune erosion. Hurricane Opal in October 1995 impacted the area by dredging almost 100,000 m^3 from both beaches and dunes (Bureau of Beaches and Coastal Systems Division). Finally, during hurricane Ivan event in September 2004, beaches alongside Bay County suffered a vertical loss of sand between 1.2 and 1.5 m (Bureau of Beaches and Coastal Systems Division, 2004).

5.2.5. Winds

Located at average latitude of 30º10'N, the St Andrew Bay system is under the influence of rather weak winds (Table 1). The climatology of wind forcing in the Gulf of Mexico exhibits a pattern of seasonal variation. During the fall and winter months (mid-September to mid-February), winds are primarily from the north. In the late spring and summer (from March to July), the northern Gulf of Mexico is dominated by tropical weather with winds mainly southerly. During summer, the influence of the subtropical high (Bermuda High) increases as the frontal zone between subtropical and mid-latitude air masses moves north and out of the Gulf. Weaker pressure gradients and, hence, calmer winds associated with high pressure produce less vigorous wind stress forcing of oceanic or, in particular, shelf

Table 1. Monthly mean wind direction and speed (m/s).

	Jan	Feb	Mar	Apr	May	Jun	Jul	Aug	Sep	Oct	Nov	Dec
Dir.	N	N	SSE	S	S	WSW	WSW	E	ENE	N	N	N
Speed	3.5	3.5	4	3.5	3	3	3	2.5	3	3	3	3.5

circulation. During summer, warm fronts move generally from south to north. The average data collected throughout the last 60 years show very uniform winds.

5.3. HYDRODYNAMIC MODEL

5.3.1. Model Description

The numerical hydrodynamic model implemented for St. Andrew Bay system is a depth-averaged, boundary fitted tidal and residual circulation model known as WQMAP (Muin and Spaulding, 1996; 1997) developed at the Applied Science Associates Inc.. The numerical techniques incorporated in the model are well documented, thus only a summary of the model characteristics is presented. WQMAP is an integrated hydrodynamic and water quality modeling system designed for use within coastal and fresh water environments. This commercial off-the-shelf program was developed by Applied Science Associates, Inc. out of Narragansett, Rhode Island. WQMAP consists of three basic components: a boundary-fitted coordinate grid creation module, a three-dimensional hydrodynamics model, and a water quality or pollutant transport model. These models are executed on a boundary fitted grid system. They can also be operated on any orthogonal curvilinear grid or a rectangular grid, which are special cases of the boundary fitted grid. The model is configured to run in a vertically averaged (barotropic) mode or as a fully three-dimensional (baroclinic) mode. Several assumptions are made in the model formulation, including the hydrostatic (shallow water) approximation, the Boussinesq approximation, and incompressibility. In this study, the 2D version is used.

Most striking feature of WQMAP is its hybrid orthogonal curvilinear-terrain following coordinate system. Let (φ, λ, z) be the latitude, longitude, and height, and (ξ, η, σ) be a hybrid coordinate system with a generalized orthogonal curvilinear coordinate system (ξ, η) in the horizontal and terrain-following σ-coordinate in the vertical. The metric coefficients connecting (ϕ, λ) to (ξ, η) are defined by

$$g_{11} = (\frac{\partial \lambda}{\partial \xi})^2 \cos^2 \phi + (\frac{\partial \varphi}{\partial \xi})^2, \quad (1)$$

$$g_{22} = (\frac{\partial \lambda}{\partial \eta})^2 \cos^2 + (\frac{\partial \varphi}{\partial \eta})^2 . \quad (2)$$

The coefficient g_{11} is the metric tensor in ξ-direction and the coefficient g_{22} metric tensor in η-direction. These tensors permit the model to transform the user defined boundary fitted grid to a numerical grid employed for spatial discretization utilized in an Arakawa *C* Grid.

Let (ζ, *H*) be the surface elevation and bathymetry. $D = H + \zeta$, is the total water depth. The σ - and z-coordinates are connected by

$$\sigma = \frac{z + H}{\zeta + H}, \quad (3)$$

which makes $\sigma = 1$ for the ocean surface and $\sigma = 0$ for the ocean bottom.

The 2D WQMAP represents a depth-averaged shallow-water system (similar to Wang et al., 1998). Let (*U*, *V*) be the vertically averaged velocity components in (ξ, η) directions. The momentum equations for (*U*, *V*) are given by

$$\frac{\partial UD}{\partial t} + \frac{1}{\sqrt{g_{11}g_{22}}}\left[\frac{\partial\left(U^2 D\sqrt{g_{22}}\right)}{\partial \xi} + \frac{\partial\left(UVD\sqrt{g_{11}}\right)}{\partial \eta} + UVD\frac{\partial\left(\sqrt{g_{11}}\right)}{\partial \eta} - V^2\frac{\partial\left(\sqrt{g_{22}}\right)}{\partial \xi}\right] - fDV$$
$$= -\frac{gD}{R\sqrt{g_{11}}}\left[\frac{\partial \zeta}{\partial \xi} + \frac{D}{\rho_0}\int_{-1}^{0}\int_{\sigma}^{0}\left(\frac{\partial \rho}{\partial \xi} - \frac{\sigma}{D}\frac{\partial D}{\partial \xi}\frac{\partial \rho}{\partial \sigma}\right)d\sigma\right] + \frac{1}{\rho_0}\left(\tau_\xi^w - \tau_\xi^b\right) + A_h D\nabla^2 U, \tag{4}$$

$$\frac{\partial VD}{\partial t} + \frac{1}{\sqrt{g_{11}g_{22}}}\left[\frac{\partial\left(UVD\sqrt{g_{22}}\right)}{\partial \xi} + \frac{\partial\left(V^2 D\sqrt{g_{11}}\right)}{\partial \eta} + UVD\frac{\partial\left(\sqrt{g_{22}}\right)}{\partial \xi} - U^2\frac{\partial\left(\sqrt{g_{11}}\right)}{\partial \eta}\right] + fDV$$
$$= -\frac{gD}{R\sqrt{g_{22}}}\left[\frac{\partial \zeta}{\partial \eta} + \frac{D}{\rho_0}\int_{-1}^{0}\int_{\sigma}^{0}\left(\frac{\partial \rho}{\partial \eta} - \frac{\sigma}{D}\frac{\partial D}{\partial \eta}\frac{\partial \rho}{\partial \sigma}\right)d\sigma\right] + \frac{1}{\rho_0}\left(\tau_\eta^w - \tau_\eta^b\right) + A_h D\nabla^2 V. \tag{5}$$

The continuity equation is represented by

$$R\sqrt{g_{11}g_{22}}\frac{\partial \zeta}{\partial t} + \frac{\partial\left(UD\sqrt{g_{22}}\right)}{\partial \xi} + \frac{\partial\left(VD\sqrt{g_{11}}\right)}{\partial \eta} = 0. \tag{6}$$

Here, *R* is the earth radius; ρ_0 (= 1025 kg m^{-3}) is the characteristic density for the seawater; *f* is the Coriolis parameter; *g* is the acceleration due to gravity; A_h is the horizontal eddy viscosity; $(\tau_\xi^w, \tau_\eta^w)$ are the wind stress; and $(\tau_\xi^b, \tau_\eta^b)$ are the bottom stress. As with any depth-averaged model, it is implicitly assumed that velocity and density are nearly constant over the water column. However, horizontal density gradients are treated explicitly in the momentum equations. As we mentioned in Section 2 that the freshwater flow and surface winds in the bay are low. The currents in St. Andrew Bay system are predominately produced by tides. Thus, the horizontal density gradient can be neglected in short-term prediction.

5.3.2. Model Configuration

The results from Blumberg and Kim (2000) being considered, in this particular study, we ran the model with a rectangular grid in its baroclinic mode. The number of layers was set to 10, every layer representing one tenth of the averaged water column depth within every cell. The model is rather sensitive to bathymetry changes occurring between two nearby cells. To avoid it to blow up, smoothing passes were to be applied on top of the grid averaging and each cell was assigned a minimum 2m-depth to avoid large computational flux increase. This process led to significant depth changes over the whole bay. As the model conserves the total mass, the resulting speeds computed in the different cases correspond to the smoothed depths. A careful attention must so be taken if these values have to be used for operational applications. A similar model popularly used in the coastal oceanographic community is the Princeton Ocean Model (POM). Chu et al. (2001, 2005b) show the capability of POM for littoral prediction.

Since the temperature is well mixed through the water column, we considered it as being constant in order to ease the computational effort. The modal choice is obvious and will not be discussed; the choice of a rectangular grid was driven by computational time restriction. For satisfying the CFL condition, the time step had to be set to 0.1 minute, and hence each run was requesting 17 hours.

5.3.3. Lateral Boundary Conditions

Overall, there are 15 sources of freshwater that are used as part of the model forcing (Fig. 4). The major source of freshwater to the bay is Deer Point Lake. Daily-based flow rates were calculated using measured weir heights at the dam itself. The long-term average flow is 27m^3/s (Musgrove et al., 1965). Freshwater also enters the bay at many other locations. Unfortunately, these sources are largely ungauged even through the flows are often substantial, especially during times of intense precipitation. Table 2 represents the volume transport estimated by Ichyie (2000). Nevertheless these data are not as crucial, and thereby as accurate as they would have been for mass transfer computation. The effect of their fluctuation is, in this study, what is sought.

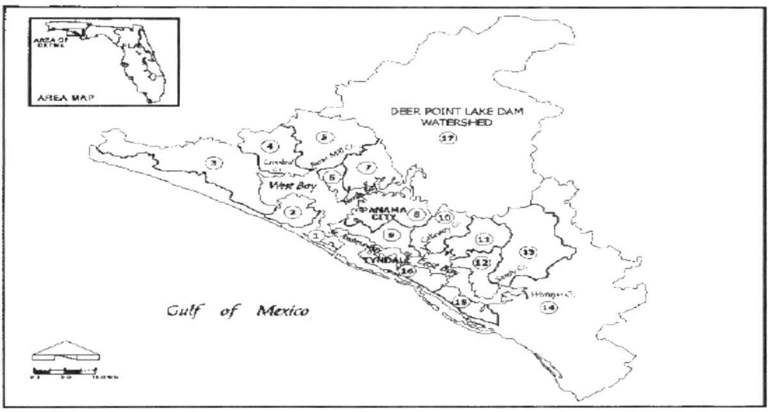

Fig. 4. St Andrew Bay drainage sub-basins (30° N - 85° 30' W).

Table 2. Calculated Mean volume transport calculated by Ichyie (2000).

Sub-basin	1	2	3	4	5	6	7	8	9	10	11	12	13	14	15	16	17
m^3/s	0.5	0.8	-	1.3	1.4	0.4	1.4	1.7	1.0	1.2	1.2	1.0	3.8	-	0.7	0.9	38.1

Despite the seasonal fluctuation of temperature, we ran the different configuration setting a constant 21°C temperature value. As the temperature is uniform over the water column, its influence on density distribution is negligible. Forcing the fresh water temperature to the same value has probably introduced the larger error. The different boundaries -West Bay, Gulf entrances and East Bay- were given salinity values -22, 35, 20 psu respectively- in agreement with measurements collected by Blumberg and Kim (2000).

Finally, the rivers cells were forced with pure fresh water which remains questionable as water enters the system through seepage.

5.3.4. Wind Forcing

The wind files were also collected at Panama City Beach and have been applied unmodified and uniformly over the whole domain. The station provides data hourly in standard weather conditions and increases the sampling to every 6 minutes when extreme weather does occur. WQMAP automatically manages the sampling differences.

5.3.5. Tidal Forcing

Data were collected at Panama City Beach where NOAA's National Ocean Service maintains a station providing among others water elevation in meters above MLLW (Mean Lower Low Water) every 6 (or 60) minutes. As we considered the flow entered the bay through two passes the evolution of each tidal constituent was examined as they propagate from East Pass onto West Pass on the continental shelf. The distance separating the two channels is 11.5 km. As a result, the phase shifts were of the order of the degree (1.1° for diurnal constituents and 2.5° for semi-diurnal constituents) and corresponded to a 5 minute-time lag. Moreover, the slight difference on the bathymetry in front of both entrances did not involve a large magnitude difference (0.1% for semi-diurnal constituents - K1 or O1- up to 2% for shallow water constituent – M4 – which only contributes little (1.1cm). At the end, these modifications were ignored. At last, even if the time series used at both locations were issued from Panama City Beach, 30km north of East Pass, we would not have had to correct them because the results from these series had not to be compared with any measured data set. However, NOAA/NOS does not maintain gauge at the three open boundary locations and therefore the elevation time series were got using Wxtide32 software (also called Tides and Currents) which is a free tool providing pure tidal elevation all around the world. Table 3 represents from left to right the Fourier decomposition of elevation gauge time series collected at NOAA station on Panama City Beach, the tidal constituents used by NOAA for predicting tides at Panama City Beach and the Fourier decomposition of tide predictions from Wxtide32 software. Numbers in parenthesis represents the 95% confidence interval for the computed values.

Table 3. NOAA tidal table for Panama City beach.

	Collected data study		NOAA prediction data		Tide and current software	
	Amplitude (cm)	Phase (deg)	Amplitude (cm)	Phase (deg)	Amplitude (cm)	Phase (deg)
K1	15.89 (0.3)	296.2 (1)	14.5	286.7	15.53 (0.1)	313.9 (1)
O1	15.84 (0.3)	284.8 (1)	14.1	284.5	15.95 (0.1)	302.3 (1)
Q1	3.4 (0.3)	270.7 (5.5)	3.1	273.4	3.5 (0.1)	295.0 (4)
M2	3.4 (0.1)	287.4 (1.5)	3.4	277.1	2.3 (0.1)	329.5 (1)
S2	2.0 (0.1)	303.1 (2.5)	2.0	274.5	0.7 (0.1)	325.2 (3)
Form Ratio	5.90		5.30		10.50	

The major difference between the data from Tides and Current software and NOAA data lies in the semi-diurnal constituents which seem to be underweighted. However the resulting error remains low, being considered the low amplitudes of both M2 and S2. A comparison of both NOAA gauge and Wxtide32 software time series (Fig. 5a) shows that differences are larger during spring tides. Finally, as NOAA could not provide time series for the lateral open boundaries, they were forced with Tides and Currents data. To be consistent, we also decided to force the Gulf open boundaries with time series provided by this software.

The residual elevation (Fig. 5b) composed from storm surge, wind-induced waves and wave setup was obtained from the NOAA gauge measurements after filtering the tidal elevation out. The times series depicts a rather uniform shape, except during the hurricane event in September. The residuals provide an average 7.7 cm setup, its standard deviation being 13.6 cm. It is to notice that the hurricane event does not impact much the statistics. When disregarding it, the mean drops to 7.3 cm and the standard deviation to 11.8 cm. These residuals were put on top of the tidal elevation when forcing the last run Gulf open boundaries.

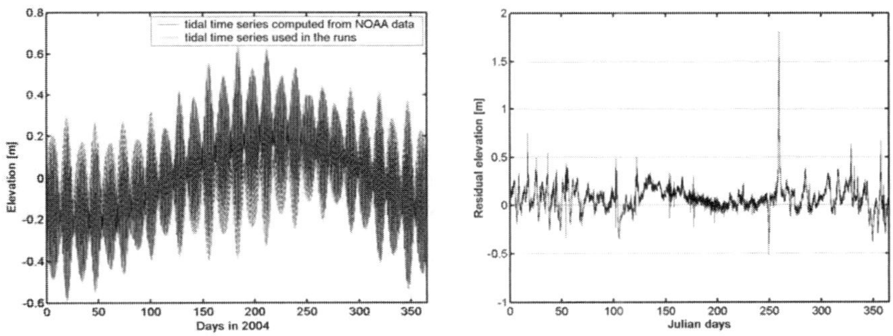

(a) (b)

Fig. 5. Time series of surface elevation: (a) comparison between NOAA (solid curve) and Wxtide32 datasets (dashed curve), and (b) residual from storm surge, wind-induced waves and wave setup.

5.4. CHEMICAL SPILL MODEL

5.4.1. Model Description

A chemical spill model, called CHEMMAP and developed at developed at the Applied Science Associates Inc., is used to predict the trajectory and fate of floating, sinking, evaporating, soluble and insoluble chemicals and product mixtures. It estimates the distribution of chemical elements (as mass and concentrations) on the surface, in the water column and in the sediments. The model is 3D, separately tracking surface slicks, entrained droplets or particles of pure chemical, chemical adsorbed to suspended particulates, and dissolved chemical (McCay and Isaji, 2002). The CHEMMAP model is used to predict the propagation of chemicals.

CHEMMAP can be either run as a certain scenario with tidal and wind forcing or in stochastic mode to estimate the probable distribution and concentrations resulting from hypothetical spills. In this study, CHEMMAP was used to predict the chemical spill in San

Diego Bay. Therefore, the physical characteristics (such as velocity, temperature, salinity) of this tidally dominated bay are simulated using WQMAP. It incorporates a number of model components including simulation of the initial release for surface and subsurface spills, slick spreading, transport of floating, dissolved and particulate materials, evaporation and volatilization, dissolution and adsorption, sedimentation and degradation. It uses physical and chemical properties such as density, viscosity, vapor pressure, surface tension, water solubility, environmental degradation rates, and adsorbed/dissolved partitioning coefficients. The Stoke's Law is used to calculate vertical velocities. Furthermore, its approach towards propagation is Langrangian. The outputs of the model include the trajectories, and concentrations. More specifically, it is possible to see the swept area by a floating chemical, as well as the total, absorbed, dissolved and particulate concentration in both the water column and the sediments. The most important is that it is then possible to determine the range of distances and directions of the contamination caused from the spill at a particular location. The database encloses all the physical/chemical properties required for simulating transport and fate of the spilled material. CHEMAP uses either the Chemical Abstract System (CAS) registry number or the UN number to reference each chemical. The chemical state can be defined as solid, liquid or gas, dissolved or not in aqueous solution or in solvent. As chemical properties do vary with temperature, there all referred to their value at 25°C.

5.4.2. Model Initialization

The model is initialized for the spilled mass at the location and depth of the release. The state and solubility are the primary determining factors for the initialization algorithm. If the chemical is highly soluble in water and is either a pure chemical (e.g., the benzene scenario) or dissolved in water (e.g., the methanol scenario), the chemical mass is initialized in the water column in the dissolved state and in a user-defined initial volume. For insoluble or semi-soluble gases released underwater (e.g., the naphthalene gas scenario), the spilled mass is initialized in the water column at the release depth in a user-defined plume volume, as bubbles. The median particle size is characterized by a user-defined diameter (McCay and Isaji, 2002).

For the state where the chemical of interest is both adsorbed to particles and dissolved in the water phase of the bulk liquid (e.g. our ammonia liquefied gas scenario), dissolved mass is also initialized in the initial plume volume. The mass of chemical spilled is corrected from the bulk spill volume using the appropriate density and concentration data from the database (McCay and Isaji, 2002).

Chemical mass is transported in three-dimensional space and time, by surface wind drift, other currents, and vertical movement in accordance with buoyancy and dispersion. The model simulates adsorption onto suspended sediment, resulting in sedimentation of material. Stoke's Law is used to compute the vertical velocity of pure chemical particles or suspended sediment with adsorbed chemical. If rise or settling velocity overcomes turbulent mixing, the particles are assumed to float or settle to the bottom. Settled particles may later re-suspend (assumed to occur above 20 cm/s current speed). Wind-driven current (drift) in the surface water layer (down to 5m) is calculated within the fates model, based on hourly wind speed and direction data. Surface wind drift of oil has been observed in the field to be 1-6% of wind speed in the direction of 0-30 degrees to the right (in the northern hemisphere) of the down-wind direction (Youssef and Spaulding, 1993). The user may also specify the wind drift speed and angle (McCay and Isaji, 2002).

CHEMMAP simulates degradation, volatilization, evaporation, dissolution, entrainment and spreading. More specifically, spreading is simulated using the Fay algorithm (Fay, 1971). Entrainment is modeled as for oil (Delvigne and Sweeney, 1988). Surface floating chemicals interaction with shorelines is simulated based on the algorithms developed for oil spills. The dissolution rate of pure chemicals is a function of solubility using a first order constant rate equation. The dissolved chemical in the water column is assumed to adsorb to particulate matter in accordance with the equilibrium partitioning theory (DiToro et al., 1991). Evaporation is modeled following the theory that the rate of mass flux to the atmosphere increases with vapor pressure, temperature, wind speed and surface area (Mackay and Matsugu, 1973). Volatilization from the water column is calculated from the chemical's vapor pressure and solubility (Lyman et al., 1982). Degradation is estimated assuming a constant rate of "decay" specific to the environment where the mass exists (i.e., atmosphere, water column or sediment).

The spilled chemical is modeled using the Lagrangian approach. At each time step, phase transfer rates (evaporation, dissolution, volatilization, and entrainment) are calculated and a proportionate percentage of the spillets are transferred to the new phase (McCay and Isaji, 2002).

The currents in the surface water layer (surface drift currents) are calculated with hourly wind field. The surface drift currents have the magnitude ranging between 1 and 6% of the wind speed and are in a direction 0-30° to the right (in northern hemisphere) of the down-wind direction. In this study, the horizontal turbulent diffusion coefficient ranges between 1 and $10 m^2/s$ and its value was set to $1\ m^2/s$ in the different cases. The vertical turbulent diffusion coefficient ranges from 10^{-5} (for stratified ocean) to $10^{-3}\ m^2/s$ (for well mixed ocean). The chemical is modeled using a Lagrangian approach in which spillets are tracked in both space and time. All the phase transfer rates are computed at each time step. This type of model is frequently used to trace back harmful substances to their sources and constitutes a valuable tool in the identification of environmental polluters.

5.5. CHEMICAL POLLUTANT- ETHYLENE GLYCOL

The choice of the chemicals used in this study is based on chemical/physical properties and toxicity data and are contained in a database compiled from published literature sources. Since several properties vary with temperature, the chemical data are for an initial temperature of 25°C. The model corrects these parameters to the ambient temperature for the spill incident. The algorithms for changing viscosity and vapor pressure to ambient temperature. For pure chemical processes, the increase per 10°C is assumed to be a factor of 2. For biological processes (e.g., degradation rates), the increase in rate per increase of 10°C is assumed to be a factor of 3 (McCay and Isaji, 2002). In this paper, an chemical pollutant, ethylene glycol, is selected as an example for the study.

Ethylene glycol ($OH\text{-}CH_2\text{-}CH_2\text{-}OH$) is a non volatile, soluble sinker, its density being $1.1132 \times 10^3\ kg/m^3$. It is referenced as CAS 107-21-1 or UN 8027. Ethylene glycol is a clear, colorless, slightly syrupy liquid at room temperature. It is odorless but has a sweet taste. It is used to make antifreeze and de-icing solutions for cars, airplanes, and boats; to make polyester compounds; and as solvent in the paint and plastics industries. Ethylene glycol is also an ingredient in photographic developing solutions, hydraulic brake fluids and in inks used in stamp pads, ballpoint pens, and print shops.

With eating or drinking very large amounts, ethylene glycol can result in death, while large amounts can result in nausea, convulsions, slurred speech, disorientation, and heart and kidney problems. Female animals that ate large amounts of ethylene glycol had babies with birth defects, while male animals had reduced sperm counts. However, these effects were seen at very high levels and would not be expected in people exposed to lower levels at hazardous waste sites. Ethylene glycol affects the body's chemistry by increasing the amount of acid, resulting in metabolic problems. Similar to ethylene glycol, propylene glycol increases the amount of acid in the body. However, larger amounts of propylene glycol are needed to cause this effect. Its primary hazard is the threat to the environment. Immediate steps should be taken to limit its spread to the environment. Since it is a liquid it can easily penetrate the soil and contaminate groundwater and nearby streams. The EPA has set a drinking water guideline for ethylene glycol of 7,000 μg in a liter of water for an adult. The American Conference of Governmental Industrial Hygienists (ACGIH) recommends a maximum level of 127 milligrams of ethylene glycol per cubic meter of air (127 mg/m^3) for a 15-minute exposure.

5.6. CHEMICAL SPILL PATTERNS

In all cases, 10 tons of the aforementioned chemical constituent is released within 10 hours with an original plume thickness of 0.5m. The choice of the date did not respond to any particular need. The release locations, however, are selected in order to identify the tidal pumping effect and minimize the vertical mixing process.

5.6.1. Control Run

The chemicals are released at the mouth of the bay (30° 08' 45.5"N, 85° 40' 46.8"W) 12 am on June 1st. The observed wind vector field is shown in Fig. 6a. We analyze differences between surface and bottom release. As ethylene glycol is not volatile, the evaporation process is weak. Because of the tidal pumping, the chemical dilutes far away from its source in both West Bay and East Bay directions (Fig. 6b). Even if this location is closer to East Bay, it essentially diffuses westward accordingly with the mean winds (from 155° at 4 m/s) whereas flow tends to push the water eastward. The maximum dissolved concentration rarely exceeds 1mg/m^3 beyond 85°35' toward East Bay and occurs between 4 and 10 days after the release. The same peak reaches a position as far as 8.5km south west of West Pass on the inner shelf 4 days after the release. A maximum concentration of 32000 mg/m^3 is obtained 1 hour after the release started (Fig. 7a). Figure 7b shows that the ethylene glycol is immediately mixed into the water column. Its decreasing rate is 0.3024 per day at 25°C and after 3 weeks ethylene glycol has disappeared from the bay.

It is however difficult to correlate the signal between two different locations as the plume goes back and forth with tides and as the concentration depends upon the depth at each location. As it propagates westward from the release point to the end of West Bay, the shape of the signal varies significantly as shown in Fig. 8. We can also notice that, the shape smoothes with time (only the largest peak can be tracked) but when the depth is shallow enough, we detect again higher peaks which were present nearby the release location. It is clear from these plots that the water depth influences the dissolved concentration between the last two locations (the depth decreases from 10 to 2m). The concentration naturally decreases as we move away from the source. Fig. 8d shows the

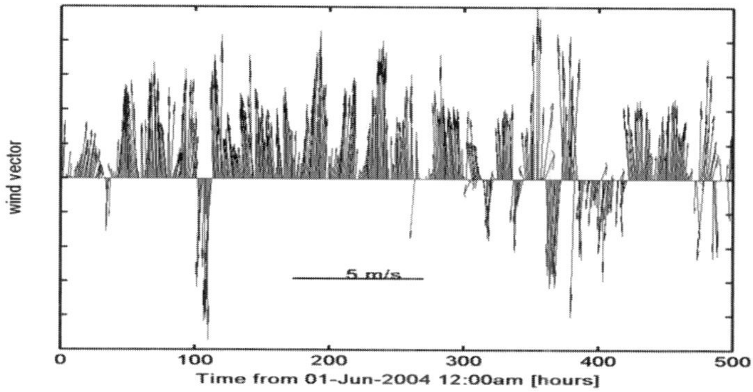

(a)

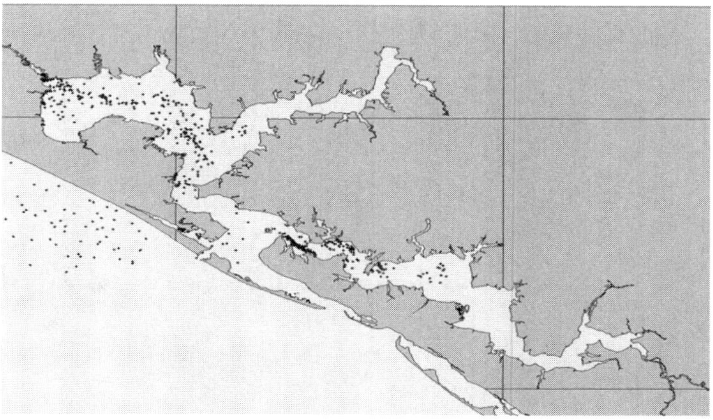

(b)

Fig. 6. **(a) Wind field over the domain between the 01-Jun-2004 12:00am and the 21-Jun-2004 11:00 pm, and (b) spillet distribution at 3 weeks after releasing from the bay mouth.**

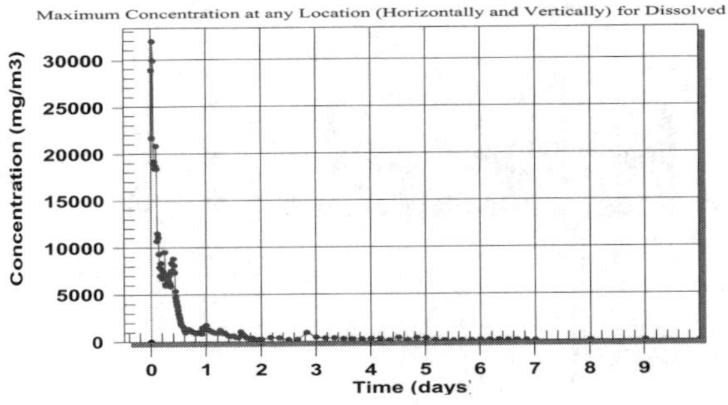

(a)

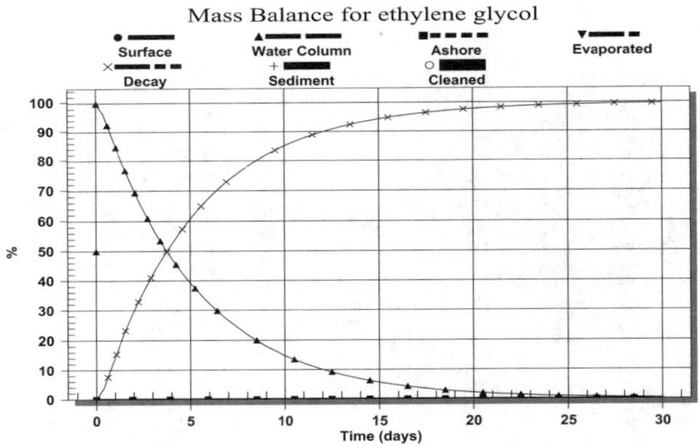

(b)

Fig. 7. (a) Temporally varying maximum concentration, and (b) mass balance for dissolved ethylene glycol at the west pass of the inner shelf.

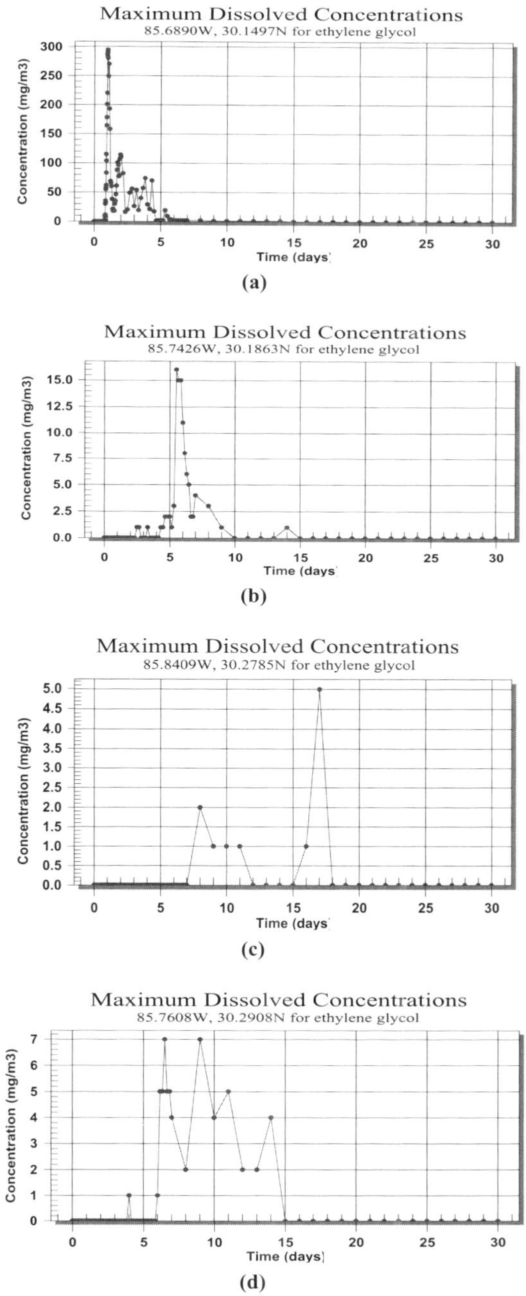

Fig. 8. Temporally varying maximum concentration at (a) St Andrew Bay, (b) location-A, (c) West Bay open boundary, and (d) northern West Bay.

enhancement of the effects a semi enclosed small creek can produce. With southeastern winds, the chemical piles up in the northern part of West Bay where it remains and accumulates until being diluted into West Bay by changing wind conditions. We can therefore find lumps of pollutant in very small areas which can more severely impact the environment.

5.6.2. Effect of Winds

If the tidal influence in the spreading of the plume can be viewed during the simulation, the impact of the wind is less obvious. Two cases are designed to show the wind effect: (1) reversing wind direction and (2) eliminating wind. When the simulation was done without wind, most of the pollution remained contained between points A and B defined in the previous chapters during the first two weeks (the 15th of June corresponds to the highest tidal range after the release). After that period, a large part of the chemical diffuses into West Bay (Fig. 9a). As the pollutant transits within the mean flow in West Bay, it seems to pile up in different places in East Bay. This does not reflect the truth but only the situation at that particular snapshot. When the wind field is reversed, it is not surprising that the pollutants mainly diffuse eastward (the average wind now blows from WNW at 4 m/s) and outward the bay system (Fig. 9b).

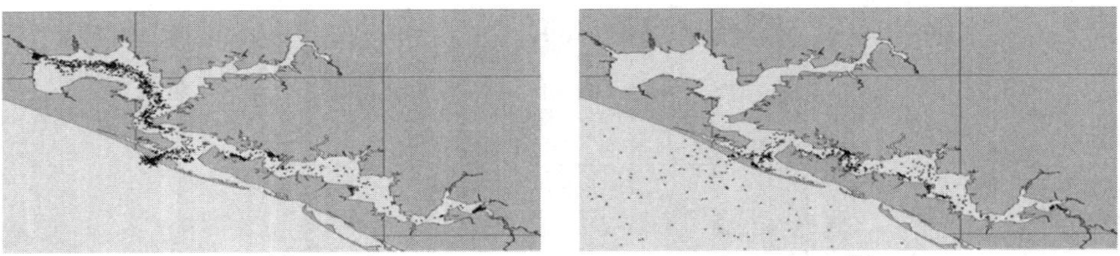

(a) (b)

Fig. 9. Spillet distribution at 3 weeks after releasing from the bay mouth with (a) no wind, and (b) wind reversal.

5.6.3. Effect of Release Time

The stochastic model is used to feature with 50 randomized dates of release taken between the 1st of June and 31st of August. Fig. 10a represents the run number of the worst case scenario for each cell (each run is featured with different release date, then different wind and current data but same location and same amount of chemical). The worst case scenario is defined for giving the maximum value encountered at each cell during all the runs. To each tone of grey correspond 5 consecutive runs (even if they are not related to each other).

Figure 10b shows a severe case with a maximum dissolved concentration and two major features. First, the pollutant predominantly diffuses westward which correlates the averaged wind encountered during this period of the year. Second, the pollutant tends to concentrate in shallow motionless waters on the edges of West and East Bay. Note that the two patches of lower concentration in St Andrew Bay correspond to the deepest locations of the bay.

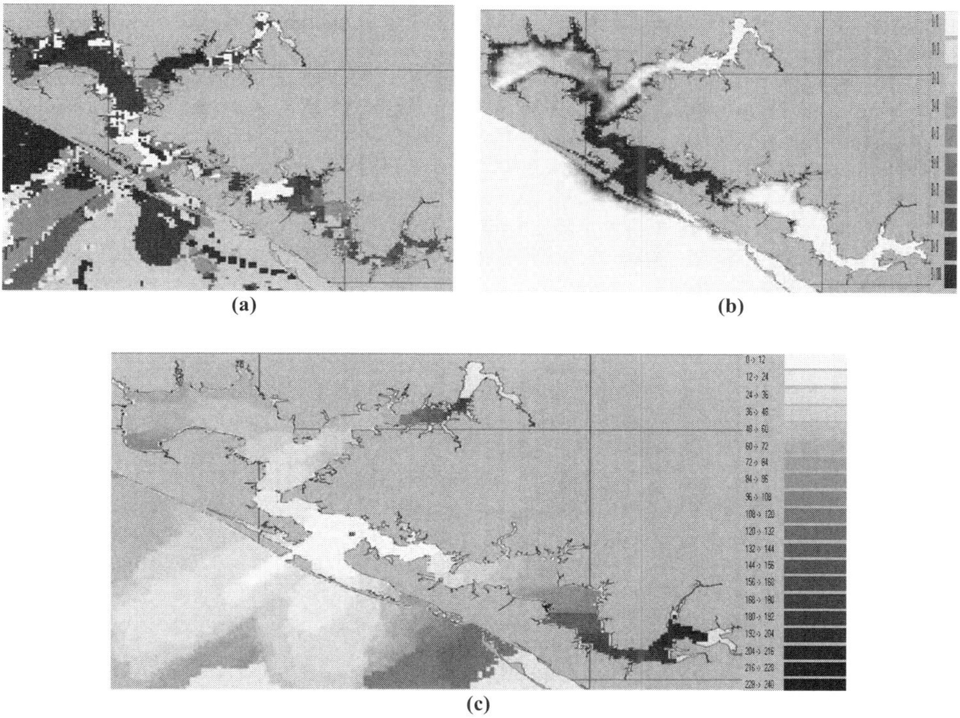

Fig.10. Worst case scenario: (a) run numbers, (b) maximum dissolved concentration (mg/m^3), and (c) minimum time (hours) to exceed threshold.

Therefore, defining the probability a threshold to be overshoot at each grid cell is important. This threshold represents a short term exposure limit (STEL) or an immediately dangerous for life or health (IDLH) limit. The model provides such a probability being given in percentage of runs during which this threshold has been exceeded. If we only consider the possibility of the bay system to be polluted (threshold is given a 0-value), then this percentage is 100% for the whole bay except in the northern part of North Bay and the eastern one of East Bay where probability never drops under 50%. The knowledge of such a probability is vital for deciding the deployment of antipollution devices. But of even furthest importance is the time given before this value can be reached. Fig. 10c then represents the minimum time requested for exceeding the threshold at each grid cell. Hence, for that particular release location, all St Andrew Bay will be polluted within 12 hours, within 24 hours the pollution will extend from point A to the first third of East Bay. It will last over 3 days for the pollutant to reach West Bay open boundary and 10 days for East Bay's.

5.6.4. Effect of Spill Location

It is obvious that the time for the pollutants transporting outward the bay mainly relies on the winds and source location. If the application of the stochastic model somehow handles the wind factor, the release location must be shifted to other places in order to evaluate the relative weight of this factor. We then applied the same concept to 4 different sites, point A, point B, the center of East Bay and the center of West bay which were supposed to depict different pollutant propagation features accordingly with the hydrodynamic model results.

Because the flux originating from Gulf of Mexico predominantly flows westward, a release point located between St Andrew Bay and West Bay or even more inside West Bay is likely not to deeply impact East Bay. The pollution will only reach the end of East Bay after 15 days if the pollutant is spilled at point A, which only stands 8km away from the previous spot. As this chemical is not volatile, it does not evaporate and its mass is roughly conserved into the system until its natural decay acts. West Bay is much shallower than St Andrew Bay, thus the small concentration decrease close to the Gulf entrance involves a large pollutant mass transfer into West Bay (Fig. 11).

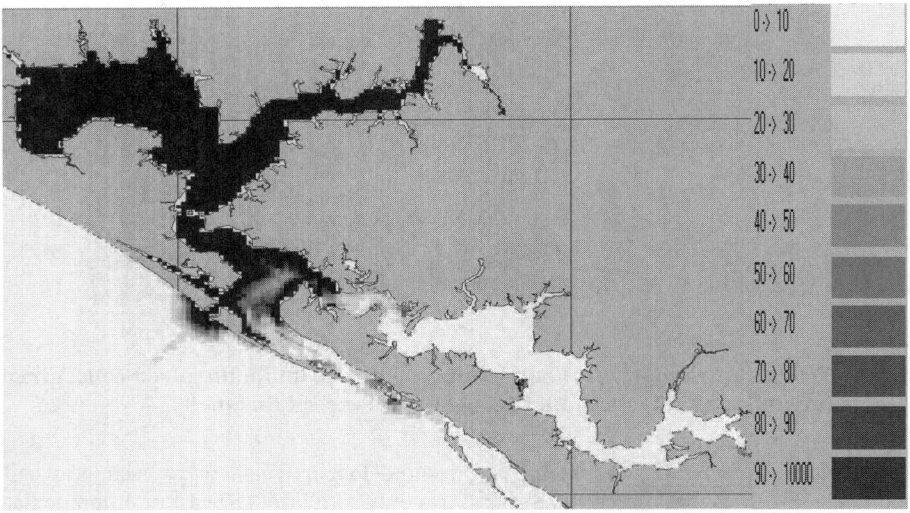

Fig. 11. Worst case maximum dissolved concentration (mg/m^3) with spill released at point A.

5.7. CONCLUSIONS

Because the chemical data embedded provided substances which were either immediately dissolved into water after their release, like ethylene glycol, or into the sediment, like tetraethyl lead, or even too volatile for remaining into the water, we could not study the differences implied by releases at surface and bottom. We plainly proved the furthest importance of the wind in the pollution drift, particularly in St Andrew Bay system where the shallowness gives an outstanding weight to the wind in driving the flow. Because the water flows inward at Gulf entrances and splits asymmetrically mostly towards West Bay, the dependence of the chemical dispersion on the release point is strong and East Bay is hardly impacted unless the pollution takes source in the bay itself or when winds are blowing eastward. The chemical clearly propagates westward during the simulation with a concentration generally damping away from the source except in very shallow zones where some accumulation can occur with favorable winds. These accumulations are finally responsible of secondary pollution events. It is, at last, obvious that these results were biased, again, by the bathymetry smoothing process. However, it is reasonable to conclude that, as East Bay presents a rather uniform depth, this bias impact was not so crucial in this area and also that most of the pollution drift relied on the wind. Furthermore, the description of different salt diffusion processes during ebb and flood tides besides the observation that spring tides were globally causing a general freshening of the system by pumping fresh water while neap tides let salt to diffuse from the saltier St Andrew Bay towards both East Bay and particularly West Bay constituted two central resulting outcomes. The tidal impact study finally described an estuarine circulation with imbalanced ebb and flood periods.

ACKNOWLEDGMENTS

This work was funded by the Naval Oceanographic Office, the Office of Naval Research, and the Naval Postgraduate School.

REFERENCES

Atkinson,L.P., Lee, T.N., Blanton,J.O., and Chandler, W.S., 1983. Climatology of the southeastern United States continental shelf waters. Journal of Geophysical Research 88: 4705-4718.

Blumberg Alan F. and B. N. Kim, 2000. Flow balances in St. Andrew Bay revealed through hydrodynamic simulations. Estuaries 23(1): 21–33.

Chu, P.C., S. H. Lu, and Y. C. Chen, 2001. Evaluation of the Princeton Ocean Model using the South China Sea Monsoon Experiment (SCSMEX) Data. Journal of Atmospheric and Oceanic Technology 18: 1521-1539.

Chu, P.C., L.M. Ivanov, and O.M. Melnichenko, 2005a. Fall-winter current reversals on the Texas-Louisiana continental shelf. Journal of Physical Oceanography, 35, 902-910.

Chu, P.C., C.-L. Fang,and C.S. Kim, 2005b. Japan/East Sea model predictability. Continental Shelf Research, 25, 2107-2121.

Fischer H.B., List, E.J., Koh, R.C.Y., Imberger, J., and Brooks, N.H., 1979. Mixing in Inland and Coastal Waters. Academic Press, 483 pp, San Diego.

Jain, M., Mehta, A. J., van de Kreeke, J., and Dombrowski, M. R., 2004. Observations on the stability St. Andrew Bay inlets in Florida. Journal of Coastal Research 20 (3): 913-919.

McCay, F. D.P., and Isaji, T., 2004. Evaluation of the consequences of chemical spills using modeling: chemicals used in deepwater oil and gas operations. Environmental Modeling & Software 19(7-8): 629-644.

Muin, M., and Spaulding, M. L., 1996. Two-dimensional boundary fitted circulation model in spherical coordinates. Journal of Hydraulic Engineering 122 (9): 512-520.

Muin, M., and Spaulding, M. L., 1997. Three-dimensional boundary fitted circulation model. Journal of Hydraulic Engineering 123(1): 2-12.

Musgrove, R. H., Foster, J. B., and Toler, L. G., 1965. Water resources of the Ecofina Creek basin area in northwestern Florida. Report of Investigation No. 41. Florida Geological Survey. United States Geological Survey, Florida.

Pritchard, D. W., 1952. Estuarine hydrography. Advances in geophysics I. New York: Academic Press, pp. 243-280.

Roberts,H.H., McBride,R.A. and Coleman,J.M. (1999). Outer shelf and slope geology of the Gulf of Mexico: An overview. The Gulf of Mexico Large Marine Ecosystem (Kumpf,H., Steidinger,K. and Sherman,K., Eds.), pp. 93-112. Blackwell Science, Malden, MA.

Solis, R.S., and Powell, G.L., 1999. Hydrography, mixing characteristics, and residence times of Gulf of Mexico estuaries. Biogeochemistry of Gulf of Mexico Estuaries (Bianchi,T.S., Pennock,J.S. and Twilley,R.R., Eds.), pp. 29-61. John Wiley & Sons, Inc., New York.

Spaulding, M. L., Mendelsohn, D., and Swanson, J. C., 1999. WQMAP: An integrated three-dimensional hydrodynamic and water quality model system for estuarine and coastal applications. Marine Technology Society Journal, invited paper, Special issue on state of the art in ocean and coastal modeling 33(3): 38-54.

U. S. Army Corps of Engineers. National Shoreline Study - Regional Inventory Report. South Atlantic Division, Atlanta, Georgia, August 1971.

Weisberg, R.H., Z. Li, and Muller-Karger, F.E., 2001. West Florida shelf response to local wind forcing: April 1998. Journal of Geophysical Research 106: 31239-31262.

Wiseman,W.J.,Jr., and Sturges,W., 1999. Physical oceanography of the Gulf of Mexico: Processes that regulate its biology. The Gulf of Mexico Large Marine Ecosystem (Kumpf,H., Steidinger,K. and Sherman,K., Eds.), pp. 77-92. Blackwell Science, Malden, MA.

6

TWO CHEMICAL SPILL PATTERNS IN TIDALLY DOMINATED SAN DIEGO BAY

Peter C. Chu and Kleanthis Kyriakidis
Department of Oceanography, Naval Postgraduate School, Monterey, CA 93943, USA

Steven D. Haeger
Naval Oceanographic Office, Stennis Space Center, MS 39529, USA

Mathew Ward
Applied Science Associates, Inc., 70 Dean Knauss Drive, Narragansett, RI 02882, USA

ABSTRACT

A coupled hydrodynamic-chemical spill model is used to investigate the chemical spill in the San Diego Bay. The hydrodynamic model shows that the San Diego Bay is tidally dominated. The chemical spill model shows the existence of two different patterns of chemical spill for pollution (methanol, benzene, liquefied ammonia, etc.) released at 0.5 m depth in the north (32°43'N, 117°13.05' W) and the south bays (32°39'N, 117°07.92' W). For the north-bay release, the chemical spill in the whole basin with a fast speed of spill in the northern part (12 hours) and a slow speed of spill in the southern part (20 days) with very small concentration. For the south-bay release, the chemical pollutants are kept in the southern part. Very few pollutants reach 32°41'N parallel (the boundary between the north and south bays).

6.1. INTRODUCTION

San Diego Bay (Figure 1) is located near the west coast of southern California. It is a relatively small basin (43-57 km^2) nearly 25 km long and 1-4 km wide. It shapes a flipped Γ-type and extends to the north to the city of San Diego and to the south to Coronado Island and Silver Strand, with a northwest to southeast orientation. It is a natural harbor sheltered by overlapping peninsulas (in west Point Loma and in east Coronado). San Diego, being the last southwest major city neighboring Mexico, is a very important port, especially for the U.S. Navy, since it hosts the headquarters of the 11th U.S. Naval District and a large portion of the American Fleet. The topography is not homogeneous (Figure 2), and the average depth is of 6.5 m (measured from the mean sea level). The northern/outer part of the bay is narrower (1-2 km wide) and deeper (reaching depth of 15 m) and the southern/inner part is wider (2-4 km wide) and shallower (depth less than 5 m). Since San Diego is a semi-closed bay, it exchanges with the Pacific Ocean only through a single channel at the mouth. Near the mouth of the bay, the north-south channel is about 1.2 km wide, bounded by Point Loma to the west and Zuniga jetty to the east with depths between 7 and 15 m (Chadwick and Largier, 1999a). The western side of the channel is shallower than the east side.

The shoreline landscape of San Diego Bay is spotted with highly polluting shipbuilding and ship repair facilities. Navy and civilian ship operations including recreational boating are other sources of pollution in the San Diego Bay. These toxins threaten public health and the environment. Investigation of the chemical dispersion of floating chemicals such as methanol, benzene and ammonia is very important for the water quality control. Besides, in the current threat environment, a chemical attack in a big city hosting a large portion of the U.S. Naval bases (such as San Diego) is anything but impossible. To evaluate the chemical treat, we first need to identify the chemical spill pattern for floating chemicals (e.g., methanol, benzene and ammonia), sinking chemicals (e.g., chlorobenzene and trichloroethylene) and dispersing chemicals to the air (e.g., naphthalene). In this study, a coupled hydrodynamic-hydrochemical fate model is used to identify the pollutant dispersion patterns. The hydrodynamic sub-model is driven by tides and winds and predicts the velocity field. The hydrochemical model is driven by the velocity field and predicts the chemical spill.

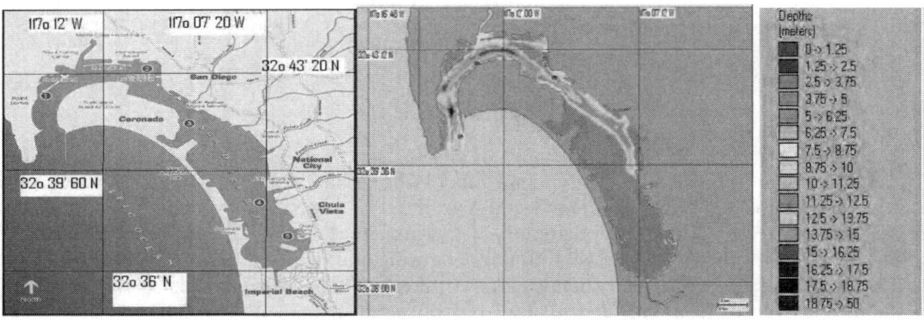

(a) (b)

Fig. 1. San Diego Bay (a) main geographical locations in (from: http://sdbay.sdsc.edu/html), and (b) bathymetry.

6.2. BACKGROUND

6.2.1. Tidal Basin

Since San Diego is a semi-closed bay, it exchanges with the Pacific Ocean only through a single channel at the mouth. Near the mouth of the bay, the north-south channel is about 1.2 km wide, bounded by Point Loma to the west and Zuniga jetty to the east with depths between 7 and 15 m (Chadwick and Largier, 1999a). The western side of the channel is shallower than the east side. The topography is not homogeneous (Figure 2), and the average depth is of 6.5 m (measured from the mean sea level). The northern/outer part of the bay is narrower (1-2 km wide) and deeper (reaching depth of 15 m) and the southern/inner part is wider (2-4 km wide) and shallower (depth less than 5 m).

San Diego Bay is a perfect example of a tidal basin connected to the ocean by an inlet with an artificial jetty (Zuniga) built to control beach erosion. The Zuniga jetty extends

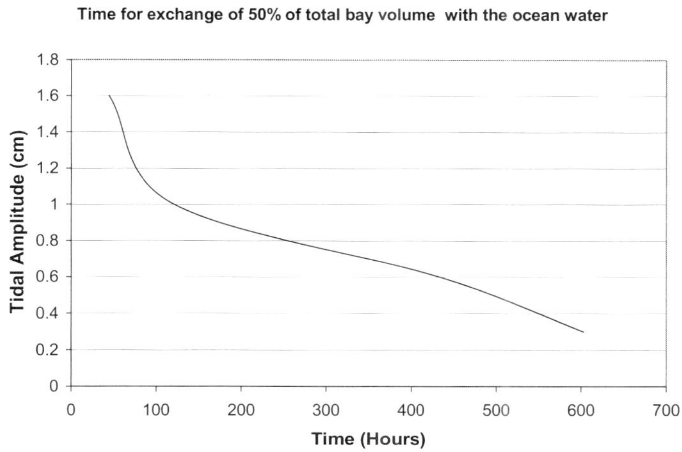

Fig. 2. Time for exchange of 50% of total bay volume with the ocean water (from the website: http://sdbay.sdsc.edu/html/modeling2.html).

almost one mile offshore Zuniga Point and most of it is not clearly visible at high water (Figure 3). Obviously, the bay has been intensively engineered to accommodate shipping activities. Ninety percent of all available marsh lands and fifty percent of all available inter-tidal lands have been reclaimed and dredging activities within the bay have been equally extensive (Peeling, 1975; Wang et al., 1998). Kelp forests extend approximately 2 km south of Point Loma and along its western side. They are quite thick and they create seasonal dumping of currents to about one-third their values outside (Jackson and Winant, 1983).

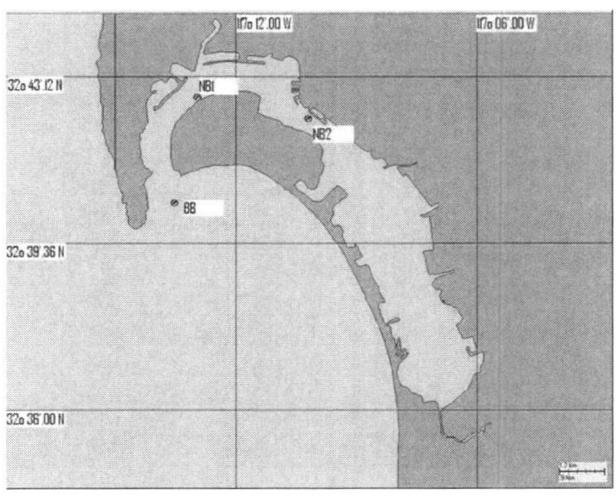

Fig. 3. Location of the ADCP stations deployed by SPAWAR in June to August 1993. Note that Station-bb is at the mouth of the San Diego Bay.

Currents in San Diego Bay are predominately produced by tides (Wang et al., 1998). This tidal exchange between the ocean and the bay is a result of a phenomenon called "tidal pumping" (Fischer et al., 1979). The "pumping" of water is caused due to the flow difference between the ebb and the flood flows.

Being located at mid-latitude, tides and currents within the San Diego Bay are dominated by a mixed diurnal-semidiurnal component (Peeling, 1975). The tidal range from mean lower-level water (MLLW) to mean higher-high water (MHHW) is 1.7 m with extreme tidal ranges close to 3 m (Chadwick and Largier, 1999a). Typical tidal current speeds range between 0.3-0.5 m/s near the inlet and 0.1-0.2 m/s in the southern region of the bay (Wang et al., 1998). The phase propagation suggests that the tides behave almost as standing waves with typical lags between the mouth and the back portion of the bay of 10 min and a slight increase in tidal amplitude in the inner bay compared to the outer bay. The overall tidal prism for the bay is 5.5×10^7 m^3 and the tidal excursion is larger than the mouth with a value of 4.4 km (Chadwick and Largier, 1999b). Offshore the Bay, the average current speed is 0.1 m/s. The currents are equator ward in all seasons and ninety percent of their measurements range between 1 and 0.25 m/s (Jackson and Winant, 1983).

The form ratio (ratio between diurnal -K1+O1- and semidiurnal -M2+S2-) shows that the tidal constituents are mixed (Table 1). Although there are 21 harmonic tidal constituents, it is possible to approach the solution by using four or five (including N2).

Table 1. The mixed diurnal-semidiurnal nature of San Diego Bay tides. (From: National Ocean Service (NOS) accepted harmonic constants for station number 9410170 in San Diego, CA Latitude: 32° 42.8' N Longitude: 117°10.4' W).

	Name	Amplitude (m)	Epoch (degrees)
1	M2	0.576	148.9
2	S2	0.233	145.9
3	N2	0.136	128.7
4	K1	0.352	210.5
5	O1	0223	195.6
6	NU2	0.027	134.3
7	MU2	0.010	109.7
8	2N2	0.018	108.7
9	OO1	0.010	225.4
10	LAM2	0.004	147.5
11	M1	0.011	194.2
12	J1	0.018	217.9
13	SSA	0.017	272.7
14	SA	0.063	182.0
15	RHO	0.008	189.2
16	Q1	0.041	188.7
17	T2	0.014	145.9
18	2Q1	0.006	180.7
19	P1	0.109	208.8
20	L2	0.013	121.7
21	K2	0.065	139.3

Since San Diego Bay has a relatively narrow mouth and a large portion of shallow water, the percentage of total bay water exchanged during a tidal cycle can be quite significant. At mean low low water, the bay has a surface area of 4.3×10^7 m^2 and a volume of 2.795×10^8 m^3. At the mouth of the bay, the tidal prism, or volumetric flux passing a cross-section during a single flooding cycle, can reach 40% of the mean volume of the bay during a strong flood. For an average tidal cycle, the tidal prism at the mouth is about 13% of the mean total bay volume. During a single tidal cycle, water from the front portion of the bay mixes with open ocean water and exchanges with the bay water that existed on the previous cycle. This open ocean water exchanges further into the bay on successive cycles (Figure 2).

6.2.2. Barotropic Basin

San Diego Bay is a barotropic basin. This can be verified by data from three ADCPs (Figure 3) deployed by SPAWAR in 1993. The first was a broad band one (bb) and was deployed in position (32°42'25.8"N, 117°13'30.6" W) from June 22 until July 23, 1993. The second was a narrowband one (nb1) and was deployed in position (32°42'43.98"N, 117°12'55.68"W) from June 22 until August 26, 1993. The last was a narrowband one (nb2) and was deployed in position (32° 42' 17.22"N, 117°10' 8.88"W) from June 23 until August 27, 1993. By checking the ADCP data inside the bay (nb1, nb2), San Diego Bay was concluded to be vertically well mixed. This can be confirmed from time series of (u, v) components at station-nb2 in three different depths from the ADCP measurements. For nb2, the correlation coefficient between the surface and bottom is 91.89% for the u-component and 94.71% for the v-component 96.32%. Between the surface and bottom, the phase matches very well and so does the trend in the respective amplitudes. In terms of actual amplitude matching, it is possible to optically verify the good match. The differences observed cannot contradict that San Diego Bay is well mixed.

6.2.3. Hydrographic Features

Freshwater inflow to the bay is minimal, comes less from the Otay River and mainly from the Sweetwater River located to the southern part of the Bay and occurs only during winter storms. Both rivers are regulated by storage reservoirs. The San Diego River has been diverted by the U. S. Army Corps of Engineers since 1875 and no longer empties into San Diego Bay.

San Diego Bay can be regarded as a vertically well-mixed estuary (Wang et al., 1998). In the south bay, the currents are much smaller than in the north and the model performance cannot be well documented. The average water temperature in San Diego Bay is 21°C and ranging between 14°C and 26°C. The average temperature during the summer (late June to late August) is 23°C. Using data from five different stations of the Bay Wide Water Quality Monitoring Program of the Port of San Diego, the conclusion was that there is zonal separation as regards the water temperature. In the northern part (off the Shelter Island – SI/1, number 1 in Figure 1a), the range is between 14°C and 19.5°C during the year and 17.5°C to 19.5°C during the summer. At the Laurel Street Anchorage (LSA/2, number 2 in Figure 1a), it reaches 21°C, with a range of 14°C to 21°C during the year but only 19°C to 21°C during the summer. Further south at the Bay Bridge Anchorage (BBA/3, number 3 in Figure 1a) the range is from 15°C to 22°C and again in the summer is from 20°C to 22°C. At the Sweet Water Channel (SWC/4, number 4 in Figure 1a) the range is from 14°C to 25°C with 22°C to 25°C in the summer. In Chula Vista Marina (CVM/5, number 5 in Figure 1a), it reaches a maximum of 26°C and a minimum of 23°C in the summer. As expected, the north part of the bay is colder than the south and in the summer the water is much warmer than in the winter. The salinity varies from 32.5 to 37.5 ppt and the average is 35 ppt. The zonal separation is similar to the temperature. In LSA and SI, the range is from 33 to 35.5 ppt. In BBA, it reaches 36 ppt. In SWC, the range is from 32 to 37 ppt. In CVM, it is from 34 to 37.5 ppt. Again, detailed analysis shows that in the summer, salinity increases with an average of 1 ppt because of the zero precipitation. Therefore, 36 ppt is used as the standard salinity during the summertime.

6.2.4. Atmospheric Forcing

The winds have a very small effect on the currents because of their intensity and the geography of the Bay. Both mean westerly winds in the afternoon and mean easterly winds in the morning and evening are less than 5 m/s. According to NOAA's weather description, there are practically no storms for San Diego during June, July and August. Wind forcing is always less significant than tidal forcing. The shallow waters make instrument deployment problematic. With small currents, the angular momentum of the instruments induces direction errors. Hence, the model is more difficult to validate in these areas. Annual precipitation is about 0.26 m (Woodward and Clyde, 1996) and occurs mostly in winter. Therefore, in terms of estuarine classification, San Diego Bay is generally positive, i.e., drainage inflow exceeds evaporation (Pritchard, 1952). However, during the summer, the precipitation is usually zero. The evaporation rate, about 0.16 m, exceeds precipitation (Peeling, 1975) and a "reversed estuary" phenomenon is observed (Defant, 1961). In general, the low inflow of fresh water provokes very small buoyancy forcing; hence the density-driven circulation is driven by seasonal heating and evaporation. Note that for precipitation, in June, the rainfall is negligible averaging only 0.17 cm, in July 0.5 cm and in August 0.25 cm. Therefore, small surface water mass flux (mostly in winter) and wind forcing for the San Diego Bay are ignored. This study area is a small basin with circulation driven by tidal flow (Fagherazzi et al., 2003).

6.2.5. Water Quality Monitoring

In 1960, an earthquake with a Richter scale of 9 in Chile caused the biggest sudden rise in sea level ever recorded in the San Diego area of 1.07 m at the Scripps pier. There is a natural protection due to the 160 km wide continental shelf of San Diego. There is a fault off San Diego Bay, but it is inactive. These are the reasons why from the 15 locally generated tsunamis in California since 1812, only two have occurred in Southern California, and only one in San Diego, dating back to 1862.

Several mineral resource extraction activities are now occurring or have occurred in the recent past in San Diego Bay. Among them, the most important are the production of sodium chloride (salt), bromine and other chemicals from sea water, magnesium, magnesium compounds and brine. There is widespread toxicity in San Diego Bay sediments attributable to copper, zinc, mercury, polycyclic aromatic hydrocarbons, polychlorinated biphenyls (PCBs) and chlordane. No single chemical or chemical group has a dominant role in contributing to the identified toxicity. Contributions of trace metals from vessel activities have long been suspected as a potentially large source to San Diego Bay. Actually, Shelter Island Yacht Basin, a semi-enclosed boat harbor, has been added to the State's list of impaired water bodies (the 303d list). These contributions arise from specially formulated paints, impregnated with biocides, and applied to boat hulls to retard the growth of fouling organisms such as barnacles.

6.3. HYDRODYNAMIC MODEL

6.3.1. Model Description

The numerical hydrodynamic model implemented for San Diego Bay is a depth-averaged, boundary fitted tidal and residual circulation model known as WQMAP (Muin and Spaulding, 1996; 1997) developed at the Applied Science Associates Inc.. The numerical techniques incorporated in the model are well documented, thus only a summary of the model characteristics is presented. WQMAP is an integrated hydrodynamic and water quality modeling system designed for use within coastal and fresh water environments. This commercial off-the-shelf program was developed by Applied Science Associates, Inc. out of Narragansett, Rhode Island. WQMAP consists of three basic components: a boundary-fitted coordinate grid creation module, a three-dimensional hydrodynamics model, and a water quality or pollutant transport model. These models are executed on a boundary fitted grid system. They can also be operated on any orthogonal curvilinear grid or a rectangular grid, which are special cases of the boundary fitted grid. The model is configured to run in a vertically averaged (barotropic) mode or as a fully three-dimensional (baroclinic) mode. Several assumptions are made in the model formulation, including the hydrostatic (shallow water) approximation, the Boussinesq approximation, and incompressibility. In this study, the 2D version is used.

Most striking feature of WQMAP is its hybrid orthogonal curvilinear-terrain following coordinate system. Let (φ, λ, z) be the latitude, longitude, and height, and (ξ, η, σ) be a hybrid coordinate system with a generalized orthogonal curvilinear coordinate system (ξ, η) in the horizontal and terrain-following σ-coordinate in the vertical. The metric coefficients connecting (ϕ, λ) to (ξ, η) are defined by

$$g_{11} = (\frac{\partial \lambda}{\partial \xi})^2 \cos^2 \phi + (\frac{\partial \varphi}{\partial \xi})^2, \tag{1}$$

$$g_{22} = (\frac{\partial \lambda}{\partial \eta})^2 \cos^2 + (\frac{\partial \varphi}{\partial \eta})^2. \tag{2}$$

The coefficient g_{11} is the metric tensor in ξ-direction and the coefficient g_{22} metric tensor in η-direction. These tensors permit the model to transform the user defined boundary fitted grid to a numerical grid employed for spatial discretization utilized in an Arakawa *C* Grid.

Let (ζ, *H*) be the surface elevation and bathymetry. $D = H + \zeta$, is the total water depth. The σ- and z-coordinates are connected by

$$\sigma = \frac{z + H}{\zeta + H}, \tag{3}$$

which makes $\sigma = 1$ for the ocean surface and $\sigma = 0$ for the ocean bottom.

The 2D WQMAP represents a depth-averaged shallow-water system (similar to Wang et al., 1998). Let (U, V) be the vertically averaged velocity components in (ξ, η) directions. The momentum equations for (U, V) are given by

$$\frac{\partial UD}{\partial t}+\frac{1}{\sqrt{g_{11}g_{22}}}\left[\frac{\partial\left(U^2D\sqrt{g_{22}}\right)}{\partial\xi}+\frac{\partial\left(UVD\sqrt{g_{11}}\right)}{\partial\eta}+UVD\frac{\partial\left(\sqrt{g_{11}}\right)}{\partial\eta}-V^2\frac{\partial\left(\sqrt{g_{22}}\right)}{\partial\xi}\right]-fDV$$
$$=-\frac{gD}{R\sqrt{g_{11}}}\left[\frac{\partial\zeta}{\partial\xi}+\frac{D}{\rho_0}\int_{-1}^{0}\int_{\sigma}^{0}\left(\frac{\partial\rho}{\partial\xi}-\frac{\sigma}{D}\frac{\partial D}{\partial\xi}\frac{\partial\rho}{\partial\sigma}\right)d\sigma\right]+\frac{1}{\rho_0}\left(\tau_\xi^w-\tau_\xi^b\right)+A_hD\nabla^2U, \tag{4}$$

$$\frac{\partial VD}{\partial t}+\frac{1}{\sqrt{g_{11}g_{22}}}\left[\frac{\partial\left(UVD\sqrt{g_{22}}\right)}{\partial\xi}+\frac{\partial\left(V^2D\sqrt{g_{11}}\right)}{\partial\eta}+UVD\frac{\partial\left(\sqrt{g_{22}}\right)}{\partial\xi}-U^2\frac{\partial\left(\sqrt{g_{11}}\right)}{\partial\eta}\right]+fDV$$
$$=-\frac{gD}{R\sqrt{g_{22}}}\left[\frac{\partial\zeta}{\partial\eta}+\frac{D}{\rho_0}\int_{-1}^{0}\int_{\sigma}^{0}\left(\frac{\partial\rho}{\partial\eta}-\frac{\sigma}{D}\frac{\partial D}{\partial\eta}\frac{\partial\rho}{\partial\sigma}\right)d\sigma\right]+\frac{1}{\rho_0}\left(\tau_\eta^w-\tau_\eta^b\right)+A_hD\nabla^2V. \tag{5}$$

The continuity equation is represented by

$$R\sqrt{g_{11}g_{22}}\frac{\partial\zeta}{\partial t}+\frac{\partial\left(UD\sqrt{g_{22}}\right)}{\partial\xi}+\frac{\partial\left(VD\sqrt{g_{11}}\right)}{\partial\eta}=0. \tag{6}$$

Here, R is the earth radius; ρ_0(= 1025 kg m^{-3}) is the characteristic density for the seawater; f is the Coriolis parameter; g is the acceleration due to gravity; A_h is the horizontal eddy viscosity; (τ_ξ^w, τ_η^w) are the wind stress; and (τ_ξ^b, τ_η^b) are the bottom stress. As with any depth-averaged model, it is implicitly assumed that velocity and density are nearly constant over the water column. However, horizontal density gradients are treated explicitly in the momentum equations. As we mentioned in Section 2 that the freshwater flow and surface winds in the bay are low. The currents in San Diego Bay are predominately produced by tides. Thus, the horizontal density gradient can be neglected in short-term prediction. A similar model popularly used in the coastal oceanographic community is the Princeton Ocean Model (POM). Chu et al. (2001, 2005) show the capability of POM for littoral prediction.

6.3.2. Model Implementation

WQMAP for San Diego Bay covers an area of 43 km^2. Different from Wang et al. (1998), the model domain is only The computational mesh has 150×200 (30,000) grid nodes with an average horizontal resolution of 40 m. Model bathymetry is determined from depth sounding data provided by NOAA and supplemented by data from published navigation charts. Recently Navy conducted bathymetry surveys show that the water depths in regions near the bay entrance are significantly deeper than the water depths shown on the NOAA navigation chart (Wang et al., 1998). The most up-to-date bathymetry data are used in the model.

Water surface elevation and velocity are set to zero, and temperature and salinity are assigned as the characteristic values for San Diego Bay (16°C, 34 ppt) at all grid points. The model is allowed to "spin up" from quiescent initial condition for one day before any model results are used for analysis. A six-minute time step is chosen for time step. At this time step the CFL condition is satisfied. Besides, the model parameters are given as follows: the wind drag coefficient (0.0014), the bottom drag coefficient (0.003), the vertical viscosity (0.005 m^2s^{-1}), the vertical diffusivity (0.001 m^2s^{-1}), and the horizontal diffusivity (1.0 m^2s^{-1}).

6.3.3. Tidal Forcing

Temporally varying sea surface elevation (or tidal harmonic constituents) along the open boundary (entrance of San Diego Bay) is taken as the model forcing function. Such data are available at the NOAA Center for Operational Oceanographic Products and Services website. In order to verify the model performance using the ADCP data shown in Figure 3, the elevation data with six-minute interval are archived from time 0000 on 22 June 1993 to 2354 on 27 August 1993 for San Diego Bay entrance, in accordance with NOAA San Diego Station number 9410170, located at (32°42'48"N, 117°10'24"W).

6.3.4. Model Verification

Statistical analysis shows that a good correlation exists between most of the compared signals in both phase and amplitude (correlation coefficient above 90% in all cases). For nb1, the u speed between the data and the model has a correlation coefficient of 91.87% and can be verified. The observational u-velocity ranges between -51.8 and 44.5 cm/s and the modeled u-velocity changes between -46.9 and 40.8 cm/s (Figure 4). The difference between the observational and modeled mean u-velocity is 0.49 cm/s. Furthermore, the root mean square error (between model and observation) is 9.02 cm/s. For nb1, the v speed between the data and the model has a correlation coefficient of 91.66% and can be verified optically. The observational v-velocity ranges between -31.6 and 29.6 cm/s and the modeled v-velocity changes between -37.0 and 32.0 cm/s. The difference between the observational and modeled mean v-velocity is -0.65 cm/s. The root mean square error of v-velocity is 6.83 cm/s.

For nb2, the u speed between data and model has a correlation coefficient of 92.60% and can be verified optically. The observational u-velocity ranges between -42.8 and 32.8 cm/s and the modeled u-velocity changes between -40.2 and 34.3 cm/s. The difference between the observational and modeled mean u-velocity is 1.0862 cm/s. Furthermore, the root mean square error (between model and observation) is 6.7356 cm/s. For nb1, the v speed between the data and the model has a correlation coefficient of 92.60 % and can be verified optically. The observational v-velocity ranges between -34 and 42.7 cm/s and the modeled v-velocity changes between -20.4 and 23.5 cm/s. The difference between the observational and modeled mean v-velocity is -1.2971. The root mean square error of v-velocity is 8.5035.

Comparison of the current speed between observation and the model is acceptable. As regards the point south of Coronado Island (bb), when filtering the data to keep only the part that results from the tidal influence, an improved correlation between the model and observation occur. Again, a band-pass butterworth filter is first applied to filter out all

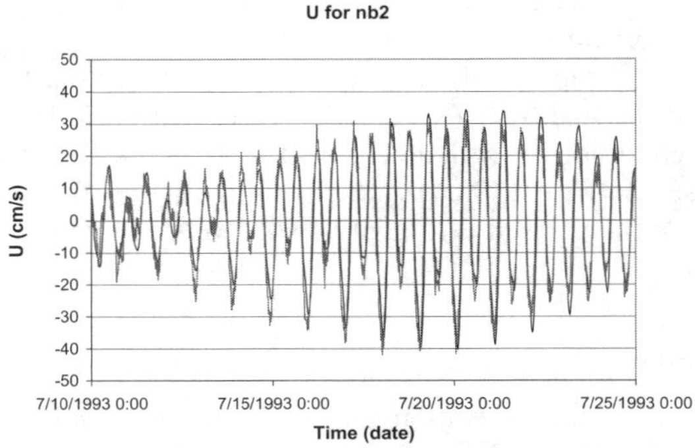

(a)

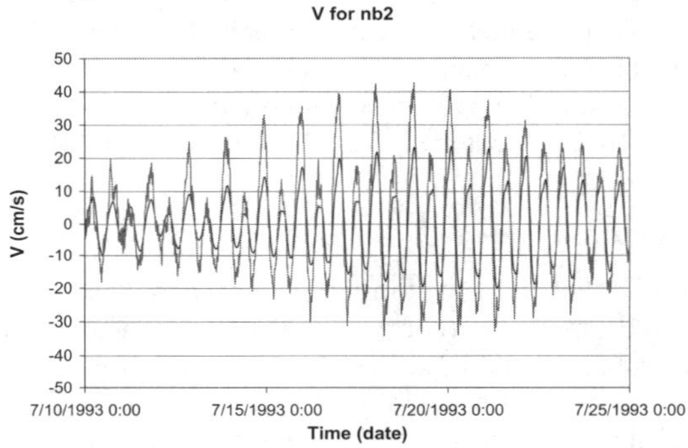

(b)

Fig. 4. Model (solid curve) and (ADCP) data (dashed curve) comparison for station-nb1: (a) u-component, and (b) v-component.

frequencies either smaller than 0.5 cycles per day or greater than 2.5 cycles per day. The correlation coefficient is 85.09% for u-velocity and 78.49% for v-velocity. Optically, the amplitude difference between the model and observation improves as well. Thus, it is possible to conclude that by filtering the results of the non-tidal influence, a 2-D model can be used in the entrance of San Diego Bay with relatively better results. The middle and upper levels contribute much more to the vertical average u component of bb than the lower (to bottom) depths and indicate its behavior much better. Actually the correlation coefficient of the average u component and the surface u is 91.63% and between the average u component and the middle depth u is even higher (92.14%). Moreover, the middle depth represents the u component better since it has also a smaller, relative mean error (RME), root mean square error (RMSE), and the error coefficient of variation (ecv),

$$\text{RME} = \frac{\sum_{i=1}^{n}(X_i^{\text{mod}} - X_i^{obs})^2}{\sum_{i=1}^{n}\left|X_i^{\text{mod}}\right|^2}, \quad \text{RMSE} = \sqrt{\frac{1}{n}\sum_{i=1}^{n}(X_i^{\text{mod}} - X_i^{obs})^2} \quad ECV = \frac{RMSE}{\sqrt{\frac{1}{n}\sum_{i}\left|X_i^{\text{mod}}\right|^2}}, \tag{7}$$

where $\left(X_i^{mod}, X_i^{obs}\right)$ are the modeled and observational data.

The bottom u component and the average u have a correlation coefficient of only 62.22%. For the v component, the correlation between the average and the v component in all depths is high: For bottom, the correlation coefficient is 96.24%, for surface 99.26%, and again for middle depth, it is the best with a coefficient of 99.43% and better rme, rmse and ecv than the other cases. In order to evaluate the model further, the water elevation in position nb2 is calculated and despite the fact that exact verified data is not available, an attempt will be made to compare it in two different ways. First, Table 2 describes all results after harmonic decomposition.

Table 2. Harmonic decomposition of the modeled results.

TIDE	FREQ	AMPL	AMP.ERR	PHASE	PH ERR	SNR
MSF	0.0028219	0.0061	0.014	266.81	160.24	0.2
*2Q1	0.0357064	0.0067	0.003	337.90	21.94	6.8
*Q1	0.0372185	0.0364	0.002	76.81	4.66	2.4e+002
*O1	0.0387307	0.1952	0.003	125.14	0.86	5e+003
*NO1	0.0402686	0.0096	0.003	19.17	16.04	13
*K1	0.0417807	0.3773	0.002	60.54	0.44	2.5e+004
J1	0.0432929	0.0026	0.002	97.99	69.16	1.2
*OO1	0.0448308	0.0157	0.002	129.23	9.23	40
*UPS1	0.0463430	0.0047	0.003	289.53	32.85	2.7
*N2	0.0789992	0.1226	0.014	203.96	7.74	75
*M2	0.0805114	0.5804	0.015	270.27	1.36	1.6e+003
*S2	0.0833333	0.2144	0.013	267.04	3.93	2.6e+002
ETA2	0.0850736	0.0077	0.011	7.45	98.73	0.48
*MO3	0.1192421	0.0042	0.001	258.54	22.76	8.5
*M3	0.1207671	0.0021	0.001	172.85	40.42	2.4
*MK3	0.1222921	0.0085	0.001	219.46	10.21	33
*SK3	0.1251141	0.0026	0.001	208.56	32.29	3.7
*MN4	0.1595106	0.0039	0.002	15.82	21.09	6.1
*M4	0.1610228	0.0107	0.001	75.84	8.11	71
*MS4	0.1638447	0.0074	0.002	71.22	11.13	23
S4	0.1666667	0.0014	0.001	66.29	51.91	1.2
*2MK5	0.2028035	0.0037	0.001	185.30	24.66	6.6
2SK5	0.2084474	0.0003	0.001	258.37	225.54	0.057
2MN6	0.2400221	0.0029	0.002	355.66	52.92	1.5
*M6	0.2415342	0.0059	0.002	52.23	22.32	6.5
*2MS6	0.2443561	0.0080	0.003	72.33	18.29	10
2SM6	0.2471781	0.0019	0.002	83.37	75.50	0.65
*3MK7	0.2833149	0.0042	0.002	108.25	31.28	3.4
*M8	0.3220456	0.0007	0.000	295.35	30.49	3.3

To compare the model results with a previous study (Wang et al., 1998), the four main tidal constituents (K1, O1, S2, M2) at downtown (further north and deeper than the nb2 position) are used. Amplitude difference (model minus observation) is 3.83 cm for M_2 (but only 1.08 cm, if compared to that of south San Diego), 3.73 cm for K_1, -2.19 cm for O_1, and 1.1 cm for S_2. Phase difference is 1.71° for M_2, 25.94° for K_1, 45.33° for O_1, and 5.41° for

S_2. Differences in bathymetry and analysis can justify such small discrepancies, especially since there is no certainty about the position for which the data was collected.

The best choice for a valid comparison is to compare the model results at nb2 with the NOAA San Diego station at the pier. The data is gathered too close to land, and therefore, cannot be exactly compared. However, it can be compared with a point very near the actual San Diego Station position. Amplitude difference (model minus observation) is 2.51 cm for M_2, 0.94 cm for K_1, 0.84 cm for O_1, and 0.71 cm for S_2. Phase difference is 0.75° for M_2, 26.08° for K_1, 29.58° for O_1, and 48.96° for S_2. Again, it can be stated that differences in bathymetry and analysis can justify such small discrepancies, especially since there is no precision about the position for which the data was compared. The results from this more accurate comparison are even better than the previous one.

Overall, the model results are reasonably good, especially taking into account that the comparison between data and model is not at exactly the same position and the proximity of the ADCPs to the shore. If finer grid and more accurate bathymetry are used, the model results may be further improved.

6.4. CHEMICAL SPILL MODEL

6.4.1. Model Description

A chemical spill model, called CHEMMAP and developed at developed at the Applied Science Associates Inc., is used to predict the trajectory and fate of floating, sinking, evaporating, soluble and insoluble chemicals and product mixtures. It estimates the distribution of chemical elements (as mass and concentrations) on the surface, in the water column and in the sediments. The model is 3D, separately tracking surface slicks, entrained droplets or particles of pure chemical, chemical adsorbed to suspended particulates, and dissolved chemical (McCay and Isaji, 2002). The CHEMMAP model is used to predict the propagation of chemicals.

CHEMMAP can be either run as a certain scenario with tidal and wind forcing or in stochastic mode to estimate the probable distribution and concentrations resulting from hypothetical spills. In this study, CHEMMAP was used to predict the chemical spill in San Diego Bay. Therefore, the physical characteristics (such as velocity, temperature, salinity) of this tidally dominated bay are simulated using WQMAP. It incorporates a number of model components including simulation of the initial release for surface and subsurface spills, slick spreading, transport of floating, dissolved and particulate materials, evaporation and volatilization, dissolution and adsorption, sedimentation and degradation. It uses physical and chemical properties such as density, viscosity, vapor pressure, surface tension, water solubility, environmental degradation rates, and adsorbed/dissolved partitioning coefficients. The Stoke's Law is used to calculate vertical velocities. Furthermore, its approach towards propagation is Langrangian. The outputs of the model include the trajectories, and concentrations. More specifically, it is possible to see the swept area by a floating chemical, as well as the total, absorbed, dissolved and particulate concentration in both the water column and the sediments. The most important is that it is then possible to determine the range of distances and directions of the contamination caused from the spill at a particular location.

6.4.2. Model Initialization

The model is initialized for the spilled mass at the location and depth of the release. The state and solubility are the primary determining factors for the initialization algorithm. If the chemical is highly soluble in water and is either a pure chemical (e.g., the benzene scenario) or dissolved in water (e.g., the methanol scenario), the chemical mass is initialized in the water column in the dissolved state and in a user-defined initial volume. For insoluble or semi-soluble gases released underwater (e.g., the naphthalene gas scenario), the spilled mass is initialized in the water column at the release depth in a user-defined plume volume, as bubbles. The median particle size is characterized by a user-defined diameter (McCay and Isaji, 2002).

For the state where the chemical of interest is both adsorbed to particles and dissolved in the water phase of the bulk liquid (e.g. our ammonia liquefied gas scenario), dissolved mass is also initialized in the initial plume volume. The mass of chemical spilled is corrected from the bulk spill volume using the appropriate density and concentration data from the database (McCay and Isaji, 2002).

Chemical mass is transported in three-dimensional space and time, by surface wind drift, other currents, and vertical movement in accordance with buoyancy and dispersion. The model simulates adsorption onto suspended sediment, resulting in sedimentation of material. Stoke's Law is used to compute the vertical velocity of pure chemical particles or suspended sediment with adsorbed chemical. If rise or settling velocity overcomes turbulent mixing, the particles are assumed to float or settle to the bottom. Settled particles may later re-suspend (assumed to occur above 20 cm/s current speed). Wind-driven current (drift) in the surface water layer (down to 5m) is calculated within the fates model, based on hourly wind speed and direction data. Surface wind drift of oil has been observed in the field to be 1-6% of wind speed in the direction of 0-30 degrees to the right (in the northern hemisphere) of the down-wind direction (Youssef and Spaulding, 1993). The user may also specify the wind drift speed and angle (McCay and Isaji, 2002).

CHEMMAP simulates degradation, volatilization, evaporation, dissolution, entrainment and spreading. More specifically, spreading is simulated using the Fay algorithm (Fay, 1971). Entrainment is modeled as for oil (Delvigne and Sweeney, 1988). Surface floating chemicals interaction with shorelines is simulated based on the algorithms developed for oil spills (French et al., 1999). The dissolution rate of pure chemicals is a function of solubility using a first order constant rate equation. The dissolved chemical in the water column is assumed to adsorb to particulate matter in accordance with the equilibrium partitioning theory (DiToro et al., 1991). Evaporation is modeled following the theory that the rate of mass flux to the atmosphere increases with vapor pressure, temperature, wind speed and surface area (Mackay and Matsugu, 1973). Volatilization from the water column is calculated from the chemical's vapor pressure and solubility (Lyman et al., 1982). Degradation is estimated assuming a constant rate of "decay" specific to the environment where the mass exists (i.e., atmosphere, water column or sediment).

The spilled chemical is modeled using the Lagrangian approach. At each time step, phase transfer rates (evaporation, dissolution, volatilization, and entrainment) are calculated and a proportionate percentage of the spillets are transferred to the new phase (McCay and Isaji, 2002).

6.5. CHEMICAL POLLUTANTS

The choice of the chemicals used in this study is based on chemical/physical properties and toxicity data and are contained in a database compiled from published literature sources, mainly French et al. (1996) and Mackay et al. (1992a, b, c, d). Since several properties vary with temperature, the chemical data are for an initial temperature of 25°C. The model corrects these parameters to the ambient temperature for the spill incident. The algorithms for changing viscosity and vapor pressure to ambient temperature are taken from French et al. (1996), who developed regression using the data in Gambill (1959). For pure chemical processes, the increase per 10°C is assumed to be a factor of 2. For biological processes (e.g., degradation rates), the increase in rate per increase of 10°C is assumed to be a factor of 3 (McCay and Isaji, 2002). Table 3 shows the comparison of the chemicals selected for this study.

Table 3. Comparison of chemicals used in the chemical spill scenarios.

	Methanol	Benzene	Ammonia	Chlorobenzene	TCE	Naphthalene (gas)
Floatation	Floater	Floater	Floater	Sinker	Sinker	Sinker/ Air dispersed
Solubility	High	High	High	Normal	High	Semi
Volatility	High	High	High	Semi	Semi	None
Absorption	Dissolves	Moderate	Slight	Moderate	Moderate	Moderate
Flammability	High	High		High		High
Water/Air rapid interaction	No	No		No		No

6.5.1. Floating Chemicals

Methanol (CH3OH) is a colorless fairly volatile liquid, belonging to the aliphatic alcohols chemical type. It is originally distilled from wood, but currently, it is synthetically produced from carbon oxides and hydrogen. It has a faintly sweet pungent odor like that of ethyl alcohol. It has a flash point 12.222°C and its density is 791 kg/m^3 (at 25°C). Its vapors are slightly heavier than air and may travel some distance to a source of ignition and flash back. Any accumulation of vapors in confined spaces, such as buildings or sewers, may explode if ignited, so it is very dangerous. It is used to make chemicals, to remove water from automotive and aviation fuels, as a solvent for paints and plastics, as an alternative motor fuel and as an ingredient in a wide variety of products, so it is not easy control. Actually, the most recent inventory is estimated to be 1,125 metric tons in the U.S. alone.

Methanol reacts violently with acetyl bromide. Mixtures with concentrated sulfuric acid and concentrated hydrogen peroxide can cause explosions. It reacts with hypochlorous acid either in water solution or mixed water/carbon tetrachloride solution to give methyl hypochlorite, which decomposes in the cold and may explode on exposure to sunlight or heat. It gives the same product with chlorine and can react explosively with isocyanates under basic conditions. The presence of an inert solvent mitigates this reaction. A violent exothermic reaction occurred between methyl alcohol and bromine in a mixing cylinder. A flask of anhydrous lead perchlorate dissolved in methanol exploded when it was disturbed.

Hence, methanol is high flammable, is a floater, is highly volatile, is highly soluble and remains dissolved. Furthermore, is a toxic element, and therefore, very dangerous. Its immediately dangerous to life or health indicator (IDLH) is very high (6,000 ppm), its short term exposure limit (STEL) is 250 ppm and its odor threshold is 4.2 ppm. Its degradation rate is 0.097835 (%/day) in air and sediments and 0.3024 (%/day) in water.

Even though it is not "acutely" toxic, inhalation can cause cough, dizziness, headache, nausea, weakness and visual disturbance. Ingestion can be even worse causing abdominal pain, convulsions, shortness of breath, unconsciousness and vomiting. However, what is more important is its ecotoxicity. For certain fish (such as red drums), shrimps, mussels and snails, it can be lethal. Benzene (C6H6) is the second chemical to be used. It is also toxic, causing the same symptoms as methanol, and can be fatal to many species in the eco system including all the aforementioned ones as well as oysters, clams, trout, salmon, catfish and goldfish. As regards its toxicity, benzene is a confirmed carcinogen, develops and reproduces toxins, and therefore, is extremely dangerous. The water maximum contaminant level is 0.000005 kg/m^3. Its immediately dangerous to life or health indicator (IDLH) is not as high as methanol (500 ppm), its short term exposure limit (STEL) is 2.5 ppm and its odor threshold is 34 ppm for detection and 97 ppm for recognition. Its degradation rate is 0.97835 (%/day) in air, 0.097835 (%/day) in water and 0.0097835 (%/day) in sediments.

For chemical characteristics, benzene is high flammable, floater, highly volatile, highly soluble, and moderately absorbable to particles. It is a clear colorless liquid with a petroleum-like odor. It is less dense than water (0.877 g/cm^3) and slightly soluble in water. Its vapors are heavier than air. Benzene reacts vigorously with alkyl chloride or other alkyl halides even at -70°C in the presence of ethyl aluminum dichloride or ethyl aluminum sesquichloride and explosions have been reported. It ignites in contact with powdered chromic anhydride. It is incompatible with oxidizing agents such as nitric acid. Mixtures with bromine trifluoride, bromine pentafluoride, iodine pentafluoride, iodine heptafluoride and other interhalogens can ignite upon heating.

The ammonia (NH3) liquefied gas is clear and colorless and has a strong odor. It is shipped as a liquid under its own vapor pressure. Its density in the liquid form is 12.8825 kg/m^3. Contact with the unconfined liquid can cause frostbite. Gas generally regarded as nonflammable but does burn within certain vapor concentration limits and with strong ignition. Fire hazard increases in the presence of oil or other combustible materials. Although gas is lighter than air, vapors from a leak initially hug the ground. Prolonged exposure of containers to fire or heat may cause violent rupturing and rocketing. Long-term inhalation of low concentrations of the vapors or short-term inhalation of high concentrations has adverse health effects. It is used as a fertilizer and refrigerant, and in the manufacture of other chemicals

Ammonia is a floater, highly volatile, highly soluble and slightly absorbable to particles, and reacts exothermically with all acids. Violent reactions are possible. It also readily combines with silver oxide or mercury to form compounds that explode on contact with halogens. When in contact with chlorates, it forms explosive ammonium. As for toxicity, its immediately dangerous to life or health indicator (IDLH) is relatively small (300 ppm), its short term exposure limit (STEL) is 35 ppm and its odor threshold is 0.019

ppm. Its degradation rate is 0.1586 in both air and water. Contact with ammonia could cause skin and eye burns and inhalation some burning sensation, cough, shortness of breath and sore throat. For the ecosystem, its slight toxicity can be lethal to shrimp, prawns, salmon, trout and catfish.

6.5.2. Sinking Chemicals

Chlorobenzene (C6H5Cl) is a sinking chemical. It is a colorless to clear, yellowish liquid with a sweet almond-like odor. It is insoluble in water and a little denser than water (1,107 kg/m^3). Its vapors are heavier than air. It is used to make pesticides, dyes, and other chemicals Chlorobenzene undergoes a sometimes explosive reaction with powdered sodium or phosphorus trichloride and sodium. It may react violently with dimethyl sulfoxide. It reacts vigorously with oxidizing agents. It attacks some forms of plastic, rubber and coatings and it forms a shock sensitive solvated salt with silver perchlorate. Chlorobenzene is, therefore, a sinker, semi-volatile, soluble, highly flammable and moderately absorbable to particles. For toxicity, its immediately dangerous to life or health indicator (IDLH) is quite big (1,000 ppm), and its odor threshold is 1.3 ppm. Its degradation rate is 0.09784 (%/day) in Air, 0.0097835 (%/day) in Water and 0.00098 (%/day) in Sediments. Inhalation of chlorobenzene can cause drowsiness, headache, nausea and unconsciousness. Ingestion causes abdominal pain. As regards eco-toxicity, it can be lethal to prawns, trout and goldfish.

Another sinker is trichloroethylene (C2HCl3). It is a toxic sinker with similar results as chlorobenzene when in contact with humans. Furthermore, it is a proven carcinogen. It is a clear colorless volatile liquid having a chloroform-like odor. It is denser than water, slightly soluble in water and is non-combustible. It is used as a solvent, fumigant, in the manufacture of other chemicals, and for many other uses. It has been determined experimentally that mixtures of finely divided barium metal and a number of halogenated hydrocarbons possess an explosive capability. Specifically, impact sensitivity tests have shown that granular barium in contact with monofluorotrichloromethane, trichlorotrifluoroethane, carbon tetrachloride, trichloroethylene, or tetrachloroethylene can detonate. It has been determined experimentally that a mixture of beryllium powder with carbon tetrachloride or with trichloroethylene will flash or spark on heavy impact. A mixture of powdered magnesium with trichloroethylene or with carbon tetrachloride will also flash or spark under heavy impact. Thus, trichloroethylene is a sinker, semi-volatile, highly soluble and moderately absorbable to particles. As regards toxicity, its immediately dangerous to life or health indicator (IDLH) is the same as chlorobenzene (1,000 ppm), its short term exposure limit (STEL) is 100 ppm and its odor threshold is 82 ppm. Its degradation rate is 0.09784 (%/day) in air, 0.03024 (%/day) in water and 0.003024 (%/day) in sediments. Its dangers for the eco-system include death to toads, trout and flagfish.

6.5.3. Gaseous Chemicals

Naphthalene (C10H8) is a gas chemical. It is a dark liquid mixture, with much different qualities than all the previous chemicals. It is insoluble in water and denser than water. A part disperses in the atmosphere and another sinks in water. Contact with naphthalene may cause irritation to skin, eyes, and mucous membranes. It can cause confusion, headache, sweating, nausea, vomiting and jaundice when inhaled. It is toxic by ingestion and can cause abdominal pain, convulsions, diarrhea, dizziness and unconsciousness. Its toxicity is moderate and immediately dangerous to life or health indicator (IDLH) is 250 ppm, its short term exposure limit (STEL) is 15 ppm and its odor threshold is 0.038 ppm. It is a known carcinogen, and therefore, dangerous to humans. A mixture containing naphthalene may react vigorously with strong oxidizing agents. It can also react exothermically with bases and with diazo compounds. Naphthalene reacts violently with chromic anhydride. Hence, naphthalene disperses in the atmosphere (but can be also a sinker), is not volatile, is semi-to non-soluble, moderately absorbable to particles, is highly flammable and does not react rapidly with either water or air. Its degradation rate is 0.97835 (%/day) in air, 0.09784 (%/day) in water and 0.00302 (%/day) in sediments. Its dangers for the eco-system include the death of toads, crabs, shrimp, cod, salmon, trout and oysters, being one of the most dangerous enemies of natural underwater life.

6.6. CHEMICAL SPILL PATTERNS

The coupled WQMAP-CHEMMAP is used to investigate the chemical spill patterns for floating, sinking, gaseous chemicals. Since the WQMAP is integrated for the period from 0000 on 22 June 1993 to 2354 on 27 August 1993 for San Diego Bay, the following scenarios were suggested: A small boat drops one barrel of chemical (methanol, …) in less than 12 minutes on midnight July 4, 1993 (Independence Day) at (1) northern San Diego Bay (32°43'N, 117°13.05' W) (Point 2 in Figure 1a) , and (2) southern San Diego Bay (32°39'N, 117°07.92' W) (Point 4 in Figure 1a). The release depth is 1 m and the initial plum thickness is 0.5 m. Two distinct spill patterns are found for all the chemicals. Here, spill of methanol is presented for illustration.

6.6.1. Pollutants Released at North San Diego Bay

The chemical spill pattern is described as follows. In 3 hours, the methanol is in San Diego port (Figure 5a) and in 10 hours it is spread all over the North San Diego Bay. However, the south part of the Bay is contaminated much later. After two days, there are no pollutant particles south of 32°40'N (Figure 5b). After 3 days the northern part is heavily impacted but after 9 days, there are still no pollutant particles south of 32°39'N. The methanol reaches the south end of the Bay only after 20 days (Figure 5c), but its concentration in the water column can be neglected. Figure 6 shows the swept area after 2 days and 32 days. In such a case, it can be concluded that there is plenty of time to take protective measures for the southern part of the Bay where the results of such an incident would be minimal.

Furthermore, after five entire days, one third of the methanol is still in the water column (Figure 7). Note that it takes almost 12 days for the concentration in the water column to reach 10% and 15 days for the decayed methanol to reach a level of 80%. Moreover, the end-state is the contamination not only of the San Diego Bay but also a considerable part of the sea outside the Bay. The scenario is repeated by increasing the amount of methanol, but nothing changes fundamentally. The mass balance curves and the area contaminated remain the same.

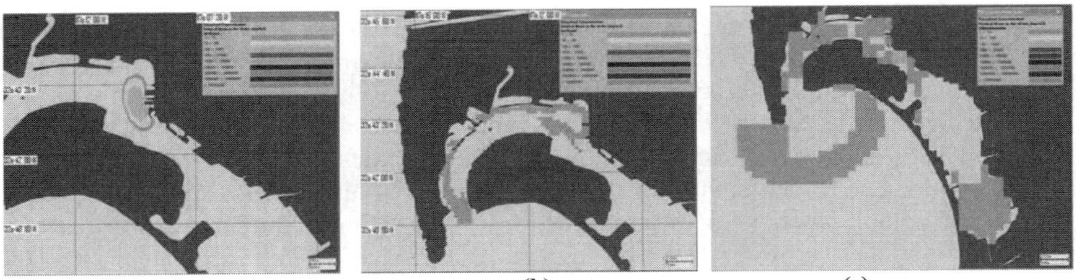

(a) (b) (c)

Fig. 5. Dissolved concentration in San Diego after (a) 3 hours, (b) 2 days, and (c) 20 days after methanol dropped in North San Diego Bay.

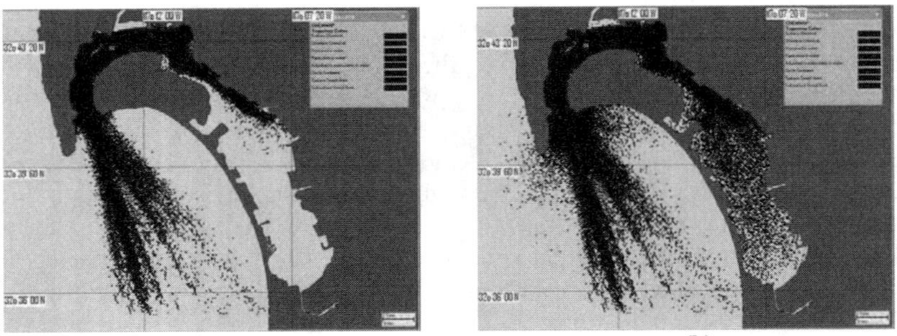

(a) (b)

Fig. 6. Swept area after (a) 2 days and (b) 32 days for methanol dropped in North San Diego Bay.

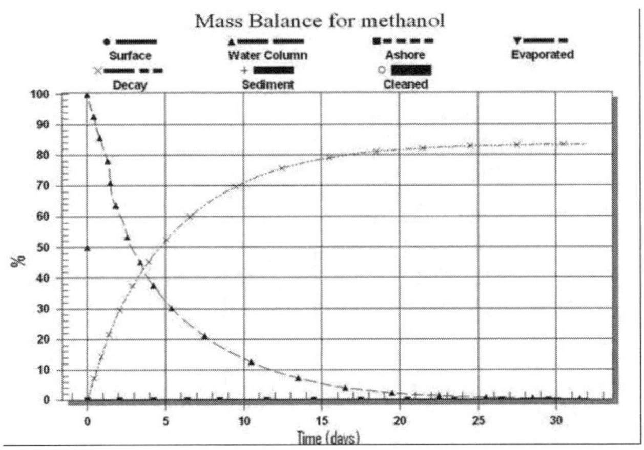

Fig. 7. Mass balance for methanol dropped in North San Diego Bay.

6.6.2. Pollutants Released at South San Diego Bay

The chemical spill pattern is described as follows. In 13 hours, the methanol reaches the central San Diego Bay (Figure 8). But, very few pollutants reach 32°41'N parallel. Figure 9 shows the swept area after 32 days. It is crucial for protective measures to highlight this fact because a chemical attack in the South San Diego Bay will have minimal effects, or at least much less considerable than an attack (or accident) in the north part of the bay. Figure 10 shows a similar but different result as regards the mass balance curves. Thus, the decayed methanol reaches 80% in only nine days, mainly due to the inert nature of methanol in combination to the shallow bathymetry of the southern part of the Bay. It is important to single out that in the first case (methanol spill over in the north), the dissolved concentration disappears after only 15 days, but in the second case (south), it needs 29 days. It is noted that the ecological catastrophe that can be caused with a relatively big amount of methanol spill over is very considerable, especially if the spill over is in the north. It can also be harmful to humans.

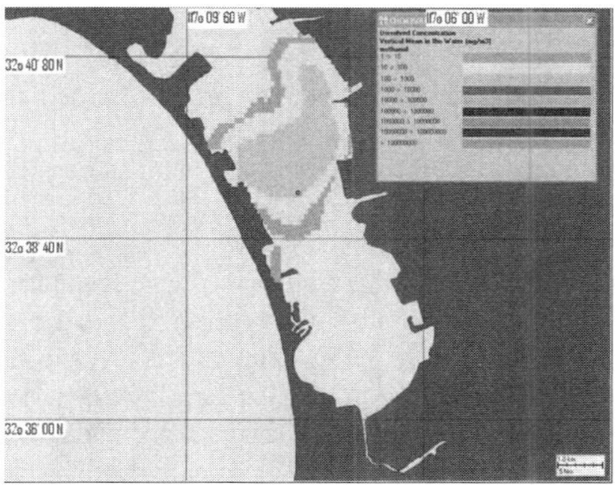

Fig. 8. Dissolved concentration in San Diego after 13 hours after methanol dropped in South San Diego Bay.

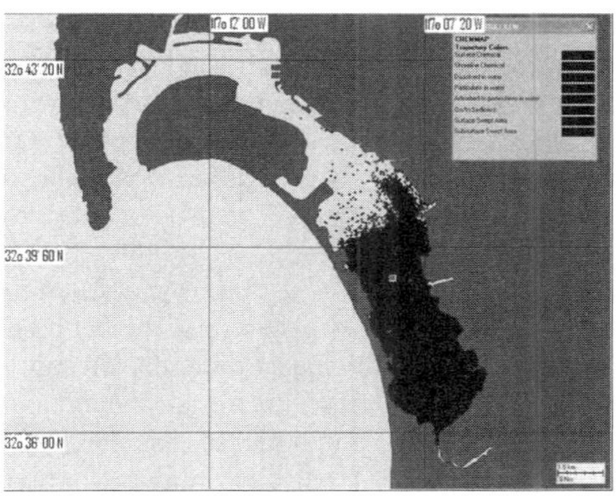

Fig. 9. Swept area after 32 days for methanol dropped in South San Diego Bay.

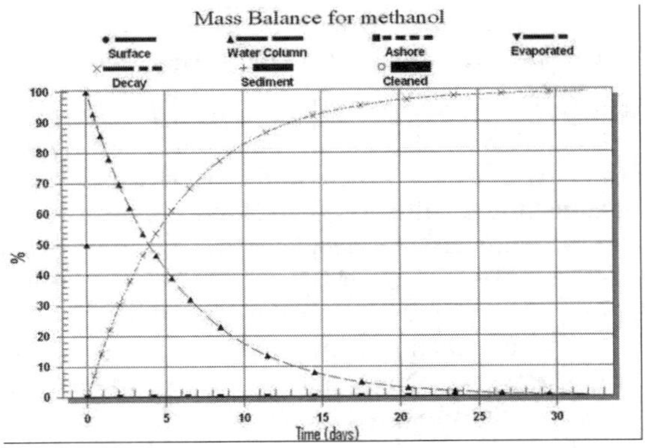

Fig. 10. Mass balance for methanol dropped in South San Diego Bay.

6.7. CONCLUSIONS

This study shows the vulnerability of a semi-enclosed tidal basin in a possible chemical attack or accident, with the aforementioned particular results for San Diego Bay. In order to summarize these results, it should be repeated that in a case of a chemical attack or accident, first the sensitive eco-system would be severely damaged, no matter the nature of the event and the location. If the chemical were a sinker, the results would be more catastrophic than if it were a floater. Since the water exchange with the Pacific Ocean occurs only through a narrow entrance, the water would be contaminated for long time.

Two regimes of the chemical dispersion were found in this thesis. The first was the case of an attack/accident in the North San Diego Bay. In that case the entire Bay would be contaminated. In 3 hours the chemical would reach San Diego port and city, in 12 hours the entire northern part of the Bay would be affected and in 2-5 days the south part of the bay would be contaminated as well. The rest of the Bay would be reached much later. The second regime was an attack/accident in the South San Diego Bay. In such case, the incident would have minimal effects on the city and the shores of Coronado Island (located in the north part of the bay) and none outside the Bay. On the other hand, when the spill occurs in the southern part of the Bay, a larger percentage of the chemical remains in the water column and for longer period of time, which makes it more "effective", which in a case of a chemical attack means lethal.

For the aforementioned reasons, the propagation model shows that the northern part of the Bay is more likely to be a target because it would affect the city, and it would reach, even slightly, the South San Diego Bay and would spread outside the Bay as well. In general, results concerning San Diego Bay can also be applied to studies in other semi-closed, barotropic, no-wind driven circulation basins.

As regards recommendations for future research, it should be mentioned that the use of more accurate bathymetry and of a finer grid would give better results in a similar case. Moreover, the use of more recent ADCP measurements, during a longer period of time would further improve the results and verify the overall conclusions. It would be helpful if the ADCPs used in the future were located in a bigger distance from the shore.

A more detailed comparison of 3D vs. 2D model is encouraged, as well as its application for drift and for instantaneous current prediction. Last but not least, as regards chemical propagation, a classified research with data unavailable to foreigners about real chemical threats (e.g. anthrax) should be conducted.

ACKNOWLEDGMENTS

This work was funded by the Naval Oceanographic Office, the Office of Naval Research, and the Naval Postgraduate School.

REFERENCES

Chadwick, D. B., and Largier, J. L., 1999a. Tidal Exchange at the Bay-Ocean Boundary. Journal of Geophysical Research 104 (C12): 29,901-29,924.

Chadwick, D. B., and Largier, J. L., 1999b. The Influence of Tidal Range on the Exchange Between San Diego Bay and the Ocean. Journal of Geophysical Research 104 (C12): 29, 885-29, 899.

Cheng R. T., Casulli, V. T., and Gartner, J. W., 1983. Tidal, Residual, Inter-tidal Mudflat (TRIM) Model and its Applications to San Francisco Bay, California. Estuarine, Coastal and Shelf Science 36: 235-280.

Chu, P.C., Lu, S. H., and Chen, Y. C., 2001. Evaluation of the Princeton Ocean Model using the South China Sea Monsoon Experiment (SCSMEX) Data. Journal of Atmospheric and Oceanic Technology 18: 1521-1539.

Chu, P.C., C.-L. Fang,and C.S. Kim, 2005b. Japan/East Sea model predictability. Continental Shelf Research, 25, 2107-2121.

Defant, A, 1961. Physical Oceanography, New York, Pergamon Press.

Delvigne, G. A. L., and Sweeney, C. E., 1988. Natural Dispersion of Oil. Oil and Chemical Pollution 4: 281-310.

DiToro, D. M., Zarba, C. S., Hansen, D. J., Berry, W. J., Swartz, R. C., Cowan, C. E., Pavlou, S. P., Allen, H. E., Thomas, N. A., and Paquin, P. R., 1991. Annual Review: Technical Basis for Establishing Sediment Quality Criteria for Nonionic Organic Chemicals Using Equilibrium Partitioning. Environmental Toxicology and Chemistry 10: 1541-1583.

Fagherazzi Sergio, Patricia, L. Wiberg and Alan D. Howard, 2003. Tidal Flow in a Small Basin. Journal of Geophysical Research 108(C3): 16-1 to 16-10.

Fay, J. A., 1971. Physical Processes in the Spread of Oil on a Water Surface. Proceedings at Joint Conference and Control of Oil Spills, Washington, D.C., June 15-17, 1971.

Fischer H. B., List, E. J., Koh, R. C. Y., Imberger, J., and Brooks, N. H., 1979. Mixing in Inland and Coastal Waters. Academic Press, p. 483, San Diego.

French, D., Reed, M., Jayko, K., Feng, S., Rines, H., Pavignano, S., Isaji, T., Puckett, S., Keller, A., French III, F. W., Gifford, D., McCue, J., Brown, G., MacDonald, E., Quirk, J., Natzke, S., Bishop, R., Welsh, M., Phillips, M., and Ingram, B.S., 1996. The CERCLA Type A Natural Resource Damage Assessment Model for Coastal and Marine Environments (NRDAM/CME). Technical Documentation, Vol. I -VI, Final Report, submitted to the Office of Environmental Policy and Compliance, U.S. Dept. of the Interior, Washington, D.C., Contract No. 14-0001-91-C-11. April 1996.

French McCay, D. P., and Isaji, T., 2002. Evaluation of the Consequences of Chemical Spills Using Modeling: Chemicals Used in Deepwater Oil and Gas Operations. International Marine Environmental Seminar 2001, Journal of Marine Systems, Special Issue 2002.

Gambill, W. R., 1959. How to calculate liquid viscosity without experimental data. Chemical Engineering, p. 127.

Jackson J. A. and C. D. Winant, 1983. Effect of a Kelp Forrest on Coastal Currents. Continental Shelf Research 2(1): 75-80.

Lyman, C. J., Reehl, W. F., Rosenblatt, D. H., 1982. Handbook of Chemical Property Estimation Methods. McGraw-Hill Book Co., New York.

Mackay, D., and Matsugu, R. S., 1973. Evaporation Rates of Liquid Hydrocarbon Spills on Land and Water. The Canadian Journal of Chemical Engineering 51: 434-439.

Mackay, D., Shiu, W. Y., and Ma, K. C., 1992a. Monoaromatic Hydrocarbons, Chlorobenzenes, and PCBs. Illustrated Handbook of Physical-Chemical Properties and Environmental Fate for Organic Chemicals, Vol. I. Lewis Publishers, Chelsea, Michigan, 668p.

Mackay, D., Shiu, W. Y., and Ma, K. C., 1992b. Polynuclear Aromatic Hydrocarbons, Polychlorinated Dioxins, and Dibenzofurans. Illustrated Handbook of Physical-Chemical Properties and Environmental Fate for Organic Chemicals, Vol. II. Lewis Publishers, Chelsea, Michigan, 566p.

Mackay, D., Shiu, W. Y., and Ma, K. C., 1992c. Volatile Organic Chemicals. Illustrated Handbook of Physical-Chemical Properties and Environmental Fate for Organic Chemicals, Vol. III. Lewis Publishers, Chelsea, Michigan, 885p.

Mackay, D., Shiu, W. Y., and Ma, K. C., 1992d. Oxygen, Nitrogen, and Sulfur Containing Compounds. Illustrated Handbook of Physical-Chemical Properties and Environmental Fate for Organic Chemicals, Volume IV. Lewis Publishers, Chelsea, Michigan, 930p.

Madala, R. V., and Piaczek, S. A., 1977. A Semi-Implicit Numerical Model for Baroclinic Oceans. Journal of Computational Physics 23: 167-168.

Muin, M., and Spaulding, M. L., 1996. Two-Dimensional Boundary Fitted Circulation Model in Spherical Coordinates. Journal of Hydraulic Engineering 122 (9): 512-520.

Muin, M., and Spaulding, M. L., 1997. Three-Dimensional Boundary Fitted Circulation Model. Journal of Hydraulic Engineering 123 (1): 2-12.

Peeling, T. J., 1975. A Proximate Biological Survey of San Diego Bay, California. Naval Undersea RD Center, San Diego, California, Technical Report No. TP389.

Pritchard, D. W., 1952. Estuarine Hydrography. Advances in Geophysics, Academic Press, pp. 342-280.

Spaulding, M.L., Mendelsohn, D.L., and Swanson, J.C., 1999. WQMAP: An Integrated Three-Dimensional Hydrodynamic and Water Quality Model System for Estuarine and Coastal Applications. Marine Technology Society Journal 33 (3): 38-54.

Swanson, J. C., and Ward, M., 2000. Improving Coastal Model Prediction through Data Assimilation. Proceedings of the 6th International Conference on Estuarine and Coastal Modeling, November 3-5, 1999, New Orleans, 10-11.

Wang, P. F., Cheng, R. T., Richter, K., Gross, E. S., Sutton, D., and Gartner, J. W., 1998. Modeling Tidal Hydrodynamics of San Diego Bay, California. Journal of the American Water Resources Association 34 (5): 1123-1140.

Woodward-Clyde, 1996. 1995-1996 City of San Diego and Co-Permittee NPDES Stormwater Monitoring Program Report.

Youssef, M. and Spaulding, M. L., 1993. Drift Current Under the Action of Wind and Waves. Proceedings of the 16th Arctic and Marine Oil Spill Program Technical Seminar, Calgary, Alberta, Canada, pp. 587-615.

Section 2:

SALTWATER INTRUSION

7

EFFECT OF POROSITY ON UNSTEADY SEAWATER INTRUSION IN CONFINED COASTAL AQUIFERS

Mohsen M. Sherif
Department of Civil and Environmental Engineering, College of Engineering
UAE University, Al Ain, United Arab Emirates

Anvar Kacimov
Department of Soils, Water and Agricultural Engineering, College of Agriculture
Sultan Qaboos University, Muscat, Sultanate of Oman

ABSTRACT

Ground water resources have been overexploited especially over the last three decades to meet the increasing demands for domestic, agriculture and industrial development. Many of the productive aquifers around the globe are located in coastal areas and are thus subjected to seawater intrusion phenomenon. Due to the direct hydraulic contact between the freshwater in the aquifers and the saline water of oceans and seas, the seawater intrudes the costal aquifers causing a tangible degradation in the ground water quality. The degree of intrusion depends on several geometric, geological, hydrogeological and transport parameters as well as pumping and recharge events. In this paper, SUTRA-GUI, is employed to study the effect of porosity on seawater phenomenon under unsteady state conditions. Previous studies related to the sensitivity and parametric analyses related to seawater intrusion problems were mainly conducted under steady state conditions. Therefore, the effect of porosity has not been fully understood. Based on the results of the numerical simulations, it is concluded that the porosity has a major effect on the degree of seawater intrusion under transient conditions.

7.1. INTRODUCTION

Seawater intrusion phenomena are of main concern in almost all coastal aquifers around the globe. The problem is more severe in arid and semi-arid regions where the ground water constitutes the main freshwater resource. A 3% mixing of seawater with the freshwater in a coastal aquifer would render the freshwater resource unsuitable for human consumption. A 5% mixing of seawater with the freshwater in an aquifer is enough to make the aquifer unsuitable for any use. Therefore, ground water resources in coastal aquifers should be carefully studied to maintain the dynamic balance between the fresh and saline water bodies.

The management of the ground water resources in costal aquifers is of prime concern to ensure the sustainable development of the costal regions. These aquifers should be carefully studied to assess the current seawater intrusion level and predict any further intrusion in the future. The critical (sustainable) long-term pumping from these aquifers should be accurately determined. The understanding of the effect of the different factors that affect the seawater intrusion is therefore vital in any ground water management plan.

The shape and degree of the seawater intrusion in a coastal aquifer depend on several factors. Some of these factors are natural and cannot be controlled while others are manmade and could, thus, be managed. These factors, include among others, the type of the costal aquifer (confined, phreatic, leaky, or multi-layer) and its geology and geometry, water table and/or piezometric head, seawater concentration and density, natural rate of flow, capacity and duration of water withdrawal or recharge, rainfall intensities and frequencies, evaporation rates, physical and geometric characteristics of the porous media, geometric and hydraulic boundaries, tidal effects, variations in barometric pressure, earth tides, earthquakes and other vibrational effects, water wave actions, and chemical changes. The depth of the aquifer at the seaside through which the seawater intrudes inland and pumping and recharge rates and locations are the most critical factors to be considered, Sherif and Singh (2002).

Several investigations have been conducted to assess the effect of the different parameters on the seawater intrusion process. Researchers distinguished between three cases of solute or pollutant transport: (1) when the effects of molecular diffusion and randomness of the flow pattern are neglected, (2) when the effects of the molecular diffusion are dominate and (3) when the randomness of the flow pattern is of equal importance to molecular diffusion in the mixing process. Pinder and Cooper (1970) investigated the movement of seawater intrusion in homogeneous confined aquifers with a constant freshwater flux at the land side and a constant saline water head at the seaside. Anand et al. (1980) elaborated the effect of the finite element mesh size on the numerical convergence. Anand and Pandit (1982) and Pandit and Anand (1984) studied the transient conditions of seawater intrusion for a wide range of field parameters and the effect of the selected time step was also investigated. The movement of saline water front was found to be more rapid at the early stage of simulation as compared to the stage prior to the steady state conditions. Huyakorn et al. (1987) employed the Picard sequential solution algorithm to simulate the seawater intrusion problem in single and multiple coastal aquifers. Sherif et al., (1988) developed a two dimensional finite element model (2D-FED) to simulate the seawater intrusion problem in confined and leaky aquifers under steady state conditions. Sherif et al., (1990) conducted a sensitivity analysis and studied the effects of some hydraulic and geometric parameters on the extent of seawater intrusion and equipotential lines using 2D-FED. The effect of the element size was also investigated. Sherif and Singh (2001) studied the effect of global warming and the associated seawater rise on the seawater intrusion phenomenon.

Most of the parametric studies were conducted under steady state conditions and thus the effect of porosity was not investigated. This paper is devoted to the study of the effect of porosity on seawater intrusion under unsteady state conditions. To that end, SUTRA-Argus-One, a USGS model, has been employed and the effect of porosity was evaluated.

7.2. SEAWATER INTRUSION MECHANISM

The transport of salt ions from saline water bodies into freshwater aquifers is governed by advection, mechanical dispersion and diffusion processes. Advection is the transport of salt in the direction of flow with the average flow velocity. Therefore, the advection depends on the both the magnitude and direction of the velocity vectors in ground water systems. Mechanical mixing depends on the pore size, velocity distribution in pores and the length of the different paths of flow. Mechanical mixing depends on the isotropy and homogeneity of the porous media. Diffusion is a process in which salt ions move from areas of high concentrations to areas of lower concentrations. More discussions on the transport processes in porous media are provided in Bear (1979).

Consider the idealized leaky system as shown in Figure 1. The aquifer is recharged by freshwater entering from the landward boundary and by leakage through the upper clay cap (aquitard) in parts where the free water table is higher than the aquifer piezometric head (before the point of the intersection, hinge point, between the free water table and the aquifer piezometric head). In some cases, the free water table may be higher than the piezometric head all over the upper clay cap, therefore the system will be recharged by the freshwater through out the entire upper boundary as well as the land ward boundary. An unconfined (phreatic) aquifer will be recharged by freshwater through infiltration of precipitation. A confined aquifer will be recharged only by freshwater from the boundary at the land side. The upper boundary will be impermeable.

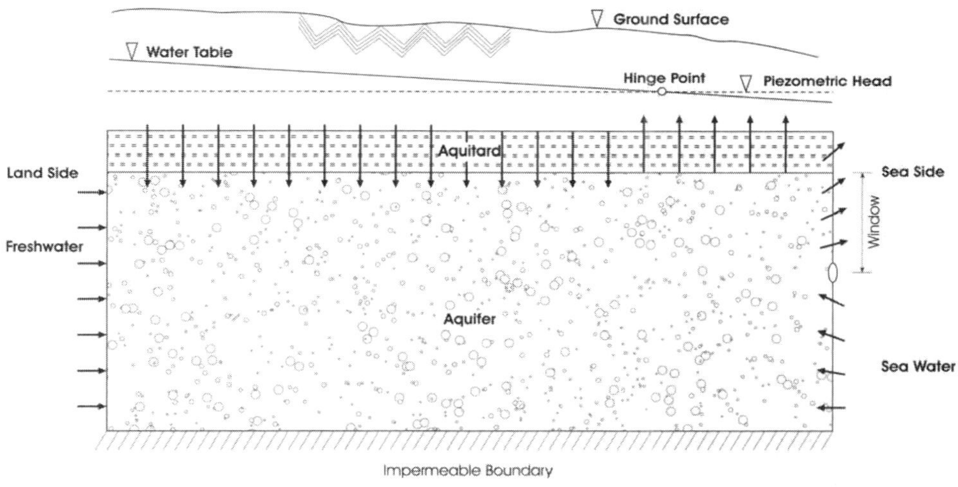

Fig. 1. Mechanism of seawater intrusion in coastal aquifers.

At the seaward boundary, there will be an influx of seawater into the system which, because of its greater density, will migrate to the bottom of the aquifer and displace the freshwater. Upward leakage of mixed water will also take place through the aquitard in parts where free water table is lower than the piezometric head (near the seaside). The rest of the mixed water will find its way out of the system through the window at the seaside. This discharge through the window and the upward flux through the aquitard cause a loss of salt from the system which is replenished from the new seawater moving in from the seaward boundary. The bottom boundary is an impermeable one. The flux of water and salt ions through this boundary is therefore equal to zero.

If the boundary conditions remain constant, a state of dynamic equilibrium will eventually be attained by the system. At equilibrium, the total fluid mass entering at both ends of the aquifer plus the leakage influx, will be balanced by the upward leakage through the aquitard plus the flux out through the window in the sea side. Likewise, the salt mass entering at the seaward will be balanced by the salt mass swept out from the system with the mixed water through the aquitard and the window at the seaward boundary. Again if the aquifer is confined, then all the water flux and salt ion exchange will be through the land and sea side boundaries.

7.3. GOVERNING EQUATIONS

The governing equations describing the dispersion zone and the flow pattern in coastal aquifers subjected to saltwater intrusion under the unsteady state conditions are (Bear and Veruijt, 1987):

1. The general Darcy equation for ground water flow,

$$q = -\frac{k}{\mu}(\nabla p + \rho g \nabla z) \tag{1}$$

where q is the specific discharge vector (LT^{-1}), k is the permeability tensor (L^2), μ is the dynamic viscosity ($ML^{-1}T^{-1}$), p is pressure ($ML^{-1}T^2$), ρ is the fluid density (ML^{-3}), g is the gravitational acceleration (LT^2) and z is a space coordinate (L). Substitution of

$$\psi = \frac{p}{\rho_f g} + z,\ K = \frac{k \rho_f g}{\mu}\ and\ \rho_r = \frac{\rho}{\rho_f} - 1$$

into the general Darcy equation, then equation (1) can be written as,

$$q = -K(\nabla \psi + \rho_r \nabla z) \tag{2}$$

where K is the hydraulic conductivity tensor (LT^{-1}), ψ is the equivalent hydraulic head (L), and ρ_r is the relative density (dimensionless).

2. The basic fluid continuity equation or the mass balance equation for the fluid which can be written as

$$\frac{\partial n\rho}{\partial t} = -\nabla \rho q + R\rho^* + P\rho \tag{3}$$

where, n is the effective porosity (dimensionless), R and P are the recharge and pumping rates per unit volume of aquifer medium, respectively (T^{-1}), and ρ^* is the density of the recharged water (ML^{-3}).

3. The hydrodynamic dispersion equation or the mass balance equation for the salt ions can be written as,

$$\frac{\partial nC}{\partial t} = -\nabla(qC - nD\nabla C) + RC^* - PC \tag{4}$$

where C is the solute concentration in mg/l. In the above equation, the effect of adsorption on dispersion process and solute transport is neglected. For two-dimensional vertical cross-section problems, the components of the hydrodynamic dispersion tensor D can be calculated as follows (Bear, 1979):

$$D_{xx} = \alpha_L \frac{V_x^2}{|V|} + \alpha_T \frac{V_z^2}{|V|} + D^*$$

$$D_{zz} = \alpha_T \frac{V_x^2}{|V|} + \alpha_L \frac{V_z^2}{|V|} + D^*$$

$$D_{xz} = D_{zx} = (\alpha_L - \alpha_T) \frac{V_x V_z}{|V|} \tag{5}$$

4. A constitutive equation relating fluid density to solute concentration, which can be expressed as:

$$\rho = \rho_f + a(C - C_f) \tag{6}$$

where C_f is the freshwater (reference) concentration (ML^{-3}), and a is a known constant (dimensionless) which can be calculated as

$$a = \frac{\rho_s - \rho_f}{C_s - C_f} \tag{7}$$

where ρ_s is the seawater density (ML^{-3}), and C_s is the seawater concentration (ML^{-3}). A linear relationship between the density and concentration is assumed in equation (6). Baxter and Wallace (1916) developed an empirical relation which relates the salt concentration to fluid density as

$$\rho = \rho_f + (1 - E)\rho_s \tag{8}$$

where E is a constant (dimensionless) and has a value of 0.3 for concentrations as high as seawater.

Examination of the main equations (2, 3, 4 and 6) reveals that there are four unknowns (ψ, V, C and ρ) in four equations. However, these equations can be combined together into two nonlinear partial differential equations in only two variables, namely, the hydraulic head ψ and the concentration C.

7.4. STUDY DOMAIN AND SELECTED SIMULATION RUNS

An idealized confined aquifer system with dimensions of 2000 m x 400 m was considered. At the land side, the piezometric head was set as 2 m above mean seawater level. At the sea boundary it was set equal to zero with reference to the same datum. A freshwater concentration was considered at the land side and seawater concentration was considered at the seaside. The equivalent hydraulic head was calculated at the seaside in order to account for the higher density of the seawater.

The default parameters were set as, porosity, n = 0.2, hydraulic conductivities in the x and z directions, K_{xx} and K_{zz} = 10 m/d, and longitudinal and lateral dispersivities α_l and α_t = 1.0 m. Therefore, the porous medium was considered isotropic for both the hydraulic conductivity and dispersivity. The simulation was conducted without any pumping. This run was designated as the basic run to which all other parameters were compared. In the next runs the porosity, n was assigned as 0.05, 0.1, 0.15, 0.25, 0.3, 0.35 and 0.4. All other parameters were kept unchanged. A freshwater concentration was considered as an initial condition in the study domain.

SUTRA (Voss, 1984; Winston and Voss, 2004), a USGS finite element model for saturated-unsaturated, fluid density dependent ground water flow with energy transport or chemically reactive single species solute transport, was used to simulate the unsteady seawater intrusion under different porosities.

7.5. RESULTS AND DISCUSSIONS

Seawater intrusion and transport processes in ground water systems may require a few decades to develop. Under natural conditions, seawater and other transport processes may never reach the steady state conditions. Recharge and discharge events depend on rainfall patterns and ground water demands that are highly variable over time and can not be considered uniform. In addition boundary conditions, such as seawater levels, may also change if the simulations are to be conducted over a large time scale.

To provide a base for comparison and assess the effect of porosity on seawater intrusion, the simulation was conducted for a total duration of 20 years. The resulted equiconcentration lines, equipotential lines and velocity vectors were presented after 2, 10 and 20 years.

Figures 2a-e present the results of the basic run in which the porosity was set as 0.2. After 2 years (Figure 2a) equiconcentration line 0.5 (freshwater) intruded inland to a distance of 450 m, while equiconcentration line 31.5 (almost seawater) intruded to a distance of 360 m measured along the bottom boundary from the sea side. The width of the dispersion zone is 90 m measured along the bottom boundary. Figure 2b and c represent the resulted equiconcentration lines after simulation times of 10 and 20 years, respectively. After 20 years of simulation, equiconcentration lines 0.5 and 31.5 intruded inland to about 1428 m and 1217 m, respectively, as shown in Figure 2c and the width of the dispersion zone was 211 m. The equipotential lines and velocity vectors revealed a strong cyclic flow near the sea boundary, Figure 2d and e. The seawater (of higher density) intrudes the aquifer from the bottom of the sea boundary and rotates back to the sea after mixing with the fresh water in the dispersion zone.

In the first run, where the porosity was set as 0.05, the seawater intrusion was much more severe, Figure 3a and b. The seawater reached the land side boundary after a simulation time of about 8.5 years and the entire domain was fully occupied by seawater after a simulation time of 11 years. Reducing the porosity prevented the freshwater to flow with sufficient quantities to the sea side and hence the seawater migrated vigorously into the aquifer.

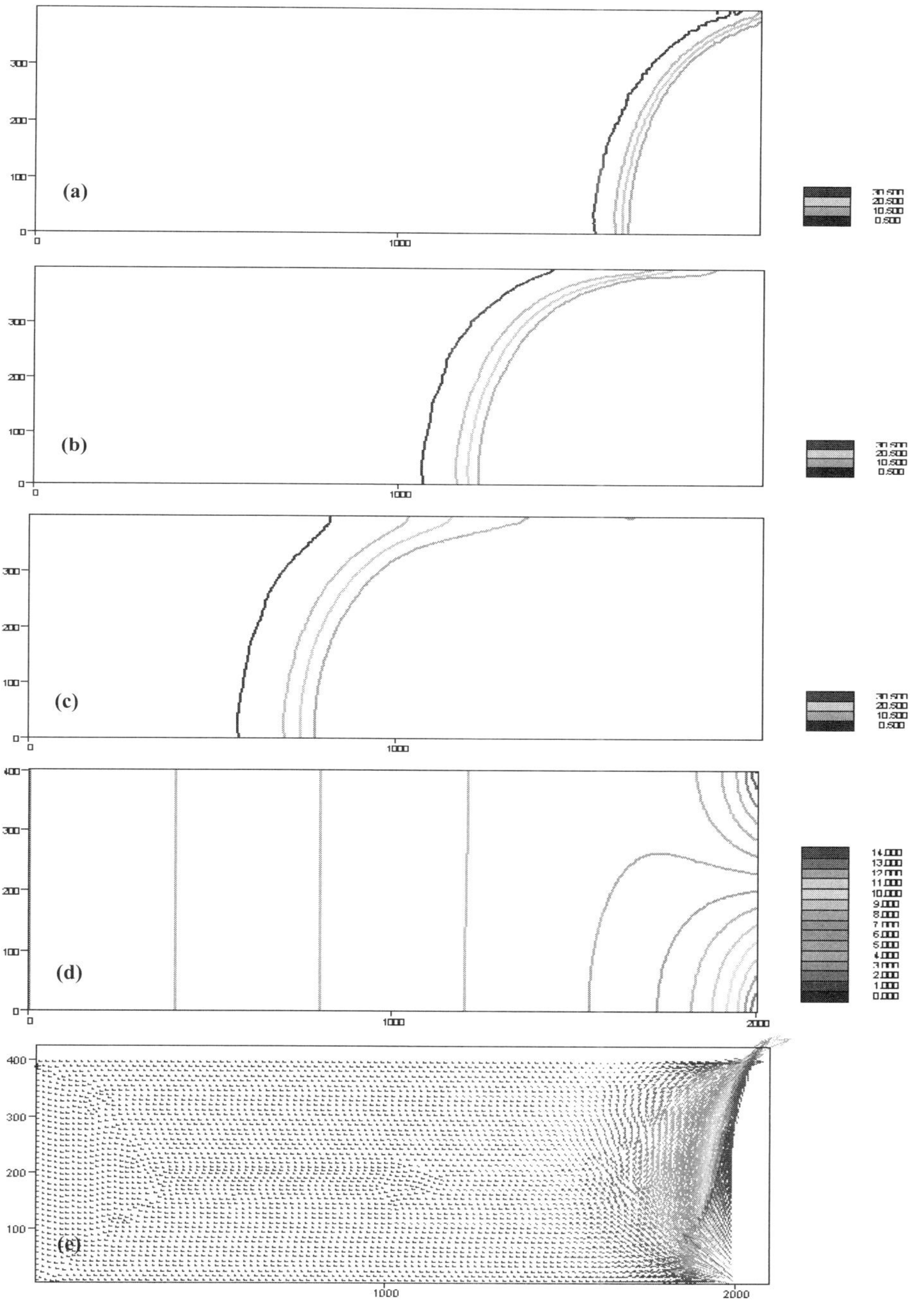

Fig. 2. Results of the basic run, n=0.2. (a) equiconcentration lines after 2 years; (b) equiconcentration lines after 10 years; (c) equi-concentration lines after 20 years; (d) equi-potential lines after 20 years; (e) velocity vectors after 20 years.

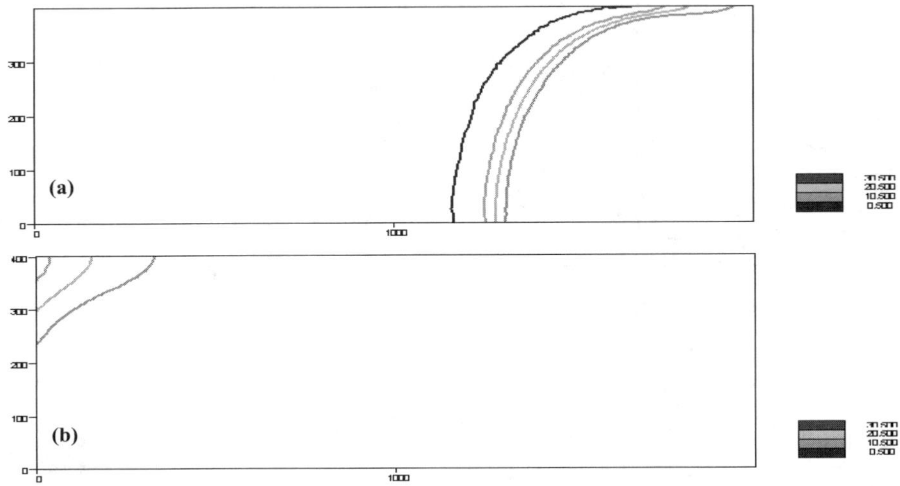

Fig. 3. Results of Run, n = 0.05. (a) equiconcentration lines after 2 years; (b) equiconcentration lines after 10 years.

In run 2, the porosity was set as 0.1 and all other parameters were kept unchanged. The resulted equiconcentration lines for simulation times of 2, 10 and 20 years are presented in Figures 4a, b and c. Again, the seawater reached the land side boundary after about 17 years of simulation. However, the seawater intrusion after 2 and 10 years was less as compared to that of the first case.

In run 3, the porosity was increased to 0.15 and all the other parameters remained constant. The results are presented in Figures 5a, b and c. After 20 years of simulation, equiconcentration lines 0.5 and 31.5 moved inland to a distance of 1751 m and 1523 m measured along the bottom boundary from the seaside. In runs 4 and 5, the porosity was increased to 0.25 and 0.35, respectively. Figures 6a and b present the resulted concentration distribution after 20 years of simulation.

The results of the runs were consistent and indicated that the porosity of the porous medium has a significant effect on the migration of seawater intrusion under the unsteady state conditions. In isotropic systems, aquifers of higher porosity will be subject to less seawater intrusion assuming that other parameters are unchanged. Increasing the porosity allowed more freshwater to move from the land side toward the sea boundary and reduce the seawater intrusion. It should also be noted that the width of the dispersion zone is relatively big. Thus, the sharp interface approach may not provide a complete understanding of the seawater intrusion. The effect of porosity may also need to be investigated under the sharp interface approach.

It should also be noted that the coastal aquifers will eventually be subjected to the same seawater intrusion degree if the system is allowed to reach the steady state conditions. However, the time required to achieve the steady state conditions differs significantly based of the porosity. On the other hand it is practically impossible to maintain the uniform pumping and/or recharge from rainfall events and/or irrigation practices to allow for the steady state conditions to be achieved. Thus seawater intrusion simulations should be conducted under transient conditions.

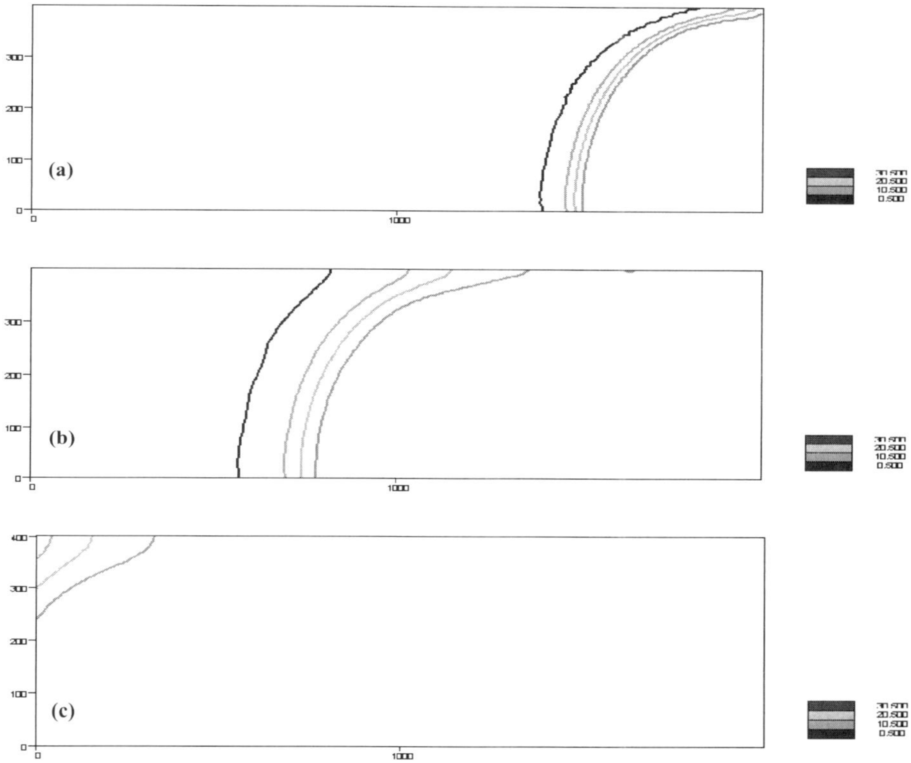

Fig. 4. Results of run 2, n =0.1, (a) Equicencentration lines after 2 years; (b) equiconcentration lines after 6 years; (c) equi-concentration lines after 20 years.

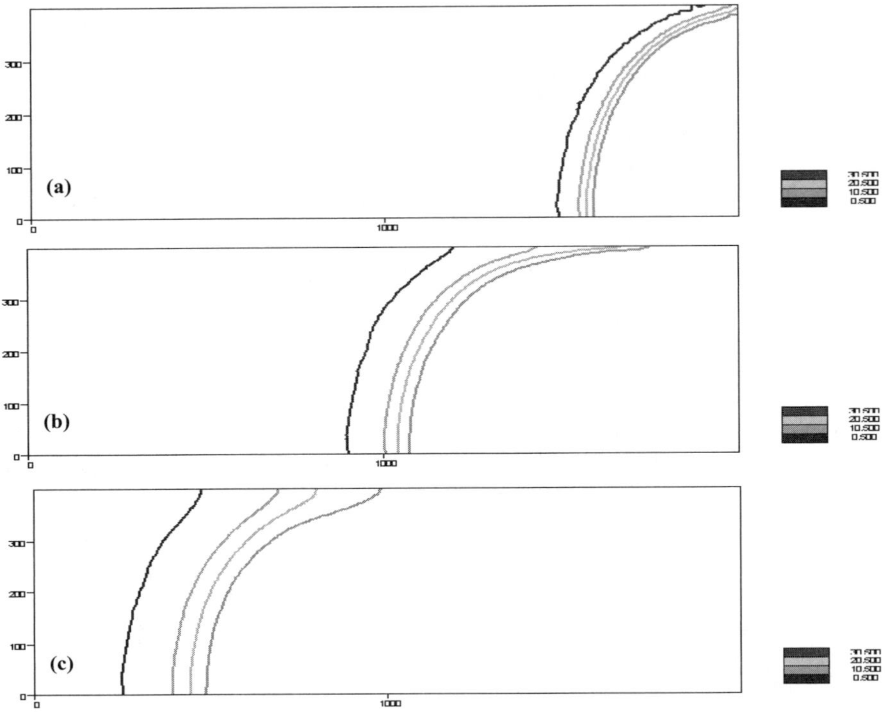

Fig. 5. **Results of run 3, n=0.15. (a) equiconcentration lines after 2 years; (b) equiconcentration lines after 10 years; (c) equiconcentration lines after 20 years.**

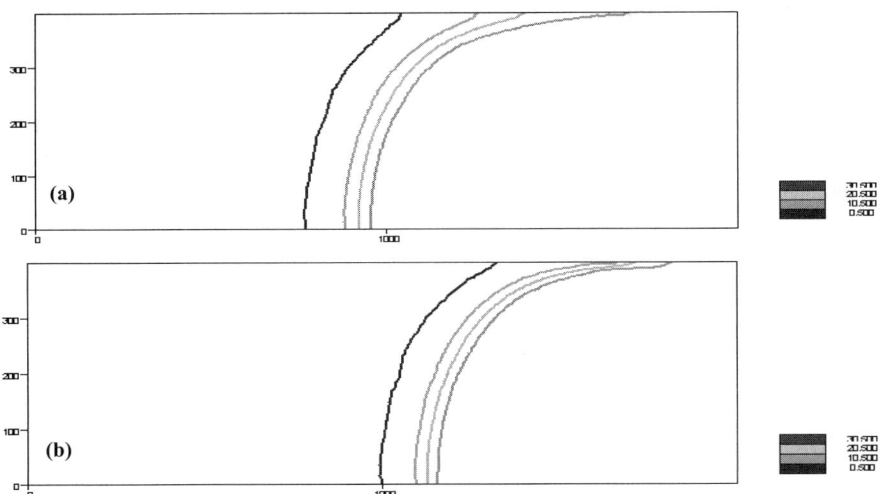

Fig. 6. **Resulted equiconcentration lines after 20 years of simulations, (a) n = 0.25; (b) n = 0.35.**

7.6. CONCLUSIONS

Numerical simulation of seawater intrusion phenomenon in coastal ground water systems should be conducted under unsteady state conditions. The time scale of the seawater intrusion process is quite large and could be in the order of several centuries. In order to achieve the steady state conditions, all boundaries, pumping and recharge events should remain constant during this period. This situation is practically impossible as both the natural recharge from rainfall events and the pumping from the aquifers to meet the demands are variable. However, steady state regimes are important to identify the ultimate seawater intrusion for a given sent of conditions.

SUTRA-Argus-One was used to simulate the seawater intrusion process in confined coastal aquifers and assess the effect of porosity on the degree of seawater intrusion under the unsteady state conditions. A fixed head boundary was employed at the landside and the hydraulic head at seaside was adjusted to account for the higher density of the seawater. All the aquifer parameters were kept unchanged except the porosity. Several simulation runs were conducted and the equiconcentration lines were presented for the different values of the porosity. High values of porosity would allow more freshwater to travel from the land side to the sea boundary and hence reduce the rate of seawater intrusion.

ACKNOWLEDGEMENT

This study is conducted within the framework of a project entitled "Modeling of Ground water in two Selected Coastal Aquifers of UAE and Oman as a Precursor for Water Resources Management." The project is jointly funded by the United Arab Emirates University and Sultan Qaboos University.

REFERENCES

Anand, S.C., and Pandit, A., 1982. Finite element solution of coupled groundwater flow and transport equations under transient conditions including the effect of selected time step size. Proc. 4th International Conference on Finite Elements in Water Resources, Univ. of Hannover.

Anand, S.C., and Pandit, A, and Sill, B.L., 1980. Some considerations in finite elements solutions to coupled groundwater flow and convective-dispersion equation. Proc. 3rd International Conference on Finite Elements in Water Resources, University of Mississippi, Oxford, Miss.

Baxter, G.P., and Wallace, C.C., 1916. Changes in volume upon solution in water of the halogen salts of alkali metals. J. Amer. Chem. Soc. 38: 70-104.

Bear, J., 1979. Hydraulics of Groundwater. McGraw Hill, New York, N.Y.

Bear, J., and Veruijt, A., 1987. Modeling Groundwater Flow and Pollution. Reidel Publishing Company, Dordrecht, Holland.

Huyakorn, P.S., Andersen, P.F., Mercer, J.W., and White, H.O., 1987. Saltwater intrusion in aquifers: Development and testing of a three-dimensional finite element model. Water Resources Research 23(2): 293-312.

Pandit, A., and Anand, S.C., 1984. Groundwater flow and mass transport by finite elements-a parametric study. Proceedings of the 5th International Conference on Finite Elements in Water Resources, University of Vermont.

Pinder, G.F., and Cooper, H.H., 1970. A numerical technique for calculating the transient position of the saltwater front. Water Resources Research 6: 875-882.

Sherif, M. M., Singh, V.P., and Amer, A.M., 1988. A Two Dimensional Finite Element Model for Dispersion (2D-FED) in Coastal Aquifers. Journal of Hydrology 103: 11-36.

Sherif, M.M., Singh V.P., and Amer, A.M., 1990. A Note on Saltwater Intrusion in Coastal Aquifers. Water Resources Management 4: 123-134.

Sherif, M.M., and Singh, V.P., 1999. Effect of Climate Changes on Seawater Intrusion in Coastal Aquifers. Hydrological Processes 13(8): 1277-1287.

Sherif, M.M., and V.P., Singh, 2002. Effect of Groundwater Pumping on Seawater Intrusion in Coastal Aquifers, Journal of Agriculture Sciences 7(2): 61-67.

Voss, C.I., 1984. SUTRA- A finite element simulation model for saturated-unsaturated fluid density dependent groundwater flow with energy transport or chemically-reactive single-species solute transport. U.S. Geol. Surv. Water Resour. Invest. Rep., 84-4269, 409 pp.

Winston, R. B., and Voss, C.I., 2004, SutraGUI: A Graphical User Interface for SUTRA, A Model for Ground-Water Flow with Solute or Energy Transport: U.S. Geological Survey Open-File Report 03-285, 114 p.

8

MECHANISM OF GROUNDWATER RECHARGE AND ITS SIGNIFICANCE TO GROUNDWATER QUALITY IN A WELL DEVELOPED QUATERNARY AQUIFER SYSTEM OF KERALA, SOUTH INDIA

Lal Thompson[1] and G.V. Rajmanickam[2]

[1]Kerala Highway Research Institute, Kariyavattom, Thiruvananthapuram-695 581, Kerala, India
[2]Department of Disaster Management, SASTRA (Deemed) University
Tanjavur-613 402, Tamilnadu, India

ABSTRACT

During the last five decades, the state Kerala had taken up a number of water resource development schemes for various purposes like irrigation, hydro-power generation, domestic and industrial water supply and salinity exclusion. But most of these projects have evolved in an isolated manner without giving due consideration to future developments. This oversight view at the planning stage has been responsible for a variety of groundwater problems witnessed in the coastal regions today. Over the years, various anthropogenic practices along the coast also seem to have affected the natural movement of groundwater considerably, and the modern groundwater recharge is no more the same. The present study carried out in the quaternary aquifer system of Alleppey district reveals that the shallow and deeper groundwater zones are distinctly different. The analysis of the data indicates that the quality of groundwater is directly related to the modern groundwater recharge and nature of present local environment. It is also revealed that the shallow layers up to 15mbgl are more sensitive to the rapid changes in the new hyrogeological environment. The geochemical data shows the presence of various hydro-geochemical types viz. $Ca\text{-}HCO^3$, Na-Cl, $Na\text{-}HCO^3$ and mixed in which the predominance of $Ca\text{-}HCO^3$ is noticed in the upper layers. The electrical conductivity appears to be increasing in the shallow groundwater regime during the post monsoon period. However at deeper levels, an anomalous higher set of values with a decreasing trend is noticed at the end of the recharge season where the fluctuation of groundwater is extremely low. Trend analysis of hydrograph depicts multiple recharge sources to the shallow aquifers, in which horizontal and vertical recharge accounts high.

8.1. INTRODUCTION

The Alleppey district of Kerala, is one of the thickly populated coastal districts in India with a population density of 1496/Sq.km. It is a low-lying coastal strip of land with a maximum width of 35 Km in the center and has a general elevation ranging from 0-6m above msl. However vast expanse of land in the eastern side is lying below msl, waterlogged almost throughout the year. It is also one of the most important physiographic units consisting of beach dunes, ancient beach ridges, coastal alluvial plains, river terraces,

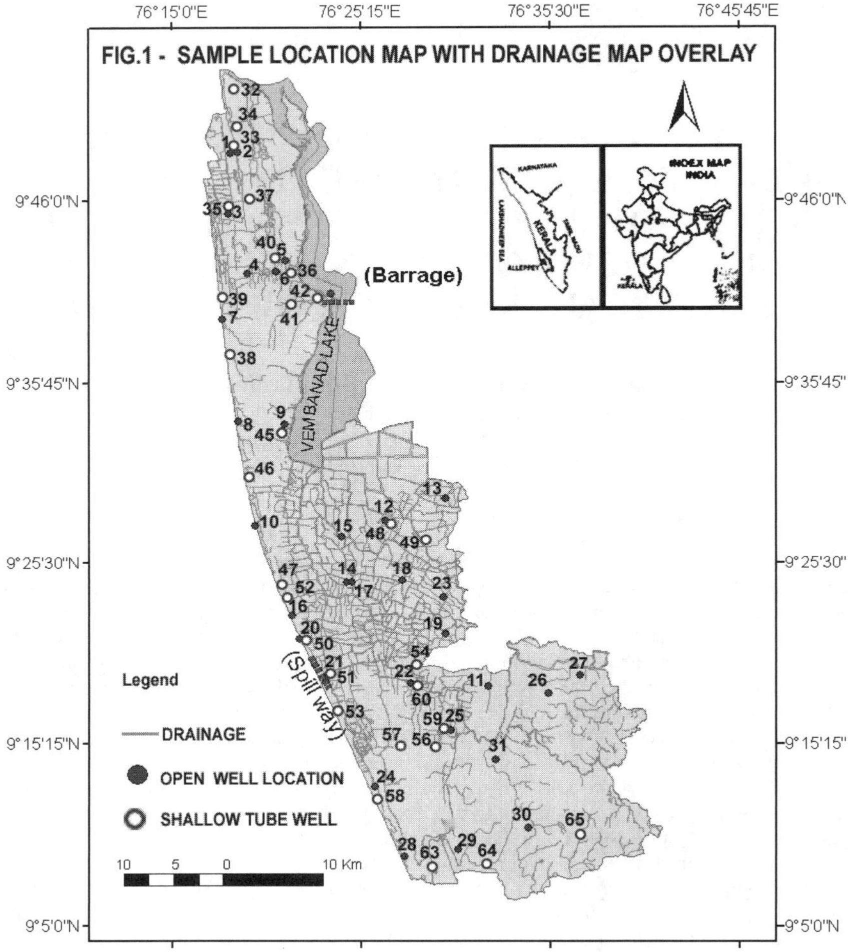

Fig. 1. Sample location map with drainage map overlay.

marshes and lagoons covering a total geographic area of 1414 Sq.km (Fig. 1). Unscientific exploitation of groundwater and poor management practices already raises many serious issues as the growing demand of water for industrial, agricultural and drinking purpose has increased many folds in the recent years. Along with the galloping population and the use of land for various constructional purposes creates barrier to the natural groundwater movement and did reduce the groundwater recharge considerably elsewhere. The impact of urbanization is an irreversible process, which would be vulnerable to the local environment. Therefore, it is necessary to understand the behavior of groundwater by proper scientific assessment to reduce the adverse impact. Hence the study has been taken up to decipher the present nature of groundwater recharge and groundwater quality, which would be of high significance to extract base line information for planning and longer sustainability of coastal groundwater resources.

8.2. HYDROGEOLOGIC SETUP

The quaternary formation comprised of unconsolidated alluvium of recent age, occurs as layers and intercalations of beach sand, sand, clay with peat beds, alluvium and soil bed. It forms the shallow coastal aquifer system. The groundwater occurs in the unconfined condition (0 to 15mbgl) to semi-confined condition (15 to 30mbgl) on an average scale (Thompson, 2004). The shallow quaternary aquifer system is demarcated from the underlying tertiary sediments by a semi-permeable layer of laterite having an average thickness of 20m, which acts as an unconformity. The area is recharged by five eastern rivers viz. Achenkoil, Pamba, Meenachail, Manimala and Muvattupuzha that debouch into the southern side of the Vembanad estuary and its environs. The drainage system is generally dendritic to sub-parallel, with a gentle slope towards north. The area receives an average annual rainfall of 3100mm of which 69% is received during the SW monsoon, 16% during NE monsoon and the remaining as summer showers (Anon, 2002). It is remarkable to note that none of the rivers flow directly into sea rather takes a natural linear coarse along the estuary in the northern direction connected to the sea by tidal channels. There is an artificial barrage constructed at Tannermukam in the northern side across the estuary to prevent the entry of seawater into the low-lying areas during summer. There is also a spillway constructed at Thottapally in the west central portion of the district that helps monsoon flood discharge during the rainy season and remain closed during rest of the year. This situation promotes growth of weeds in such areas and is responsible for a number of water related issues including the trapping of pesticides, effluents from the industries, acute crop failure and general ecological degradation, which is compounding year after year. The gentle slope in the terrain and urbanization also contributes to the slow movement of groundwater. The coastal aquifer is exposed to natural tidal activity along the interconnected tidal channels to a distance of about 2km inland from the coastline, making the system more complex.

8.3. MATERIALS AND METHODS

A total number 31 open wells (1-5mbgl) and 34 shallow tube wells (15-30mbgl) had been selected for the present study. These wells were selected in order to gather maximum possible information on the water quality and water levels from two different depths and also during two extreme seasons. Samples were collected during the month of May and December 2000 representing pre and post monsoon periods respectively. Monthly water level observations were made in association with State Groundwater Department for a period of five years from Jan 1999 to July 2004.

8.4. RESULTS AND DISCUSSIONS

The data on the water quality during the pre and post monsoon periods has been provided in Table 1. It is used for the classification of groundwater and to understand its general behavior. The water level data has been used to determine the groundwater level trends and also to understand the water table fluctuation during pre and post monsoon periods. The results of the investigation are discussed hereunder.

8.5. TRILINEAR DIAGRAM

The trilinear diagram of Piper has been very useful in bringing out the chemical relationship of water samples for the classification into various hydrogeochemical types (Walton, 1970). The chemical data of the pre-monsoon and post monsoon periods were subjected to graphical treatment by plotting them in a Piper (1944) trilinear diagram (Fig. 2 and 3).

Distribution of the groundwater samples in the diamond-shaped field of the Piper diagram reveals that majority of the open well samples during the pre and post monsoon period fall in the Area 5. This field demonstrates the excess presence of alkaline earth and weak acids. Based on the dominant cations and anions, these are of Ca-HCO^3 type. However few samples tend to fall in the Area 9 of the field, where no cation-anion pair exceeds 50% indicating mixed type of water. The predominance of the said type of water throughout the region irrespective of seasons is an indication of active groundwater

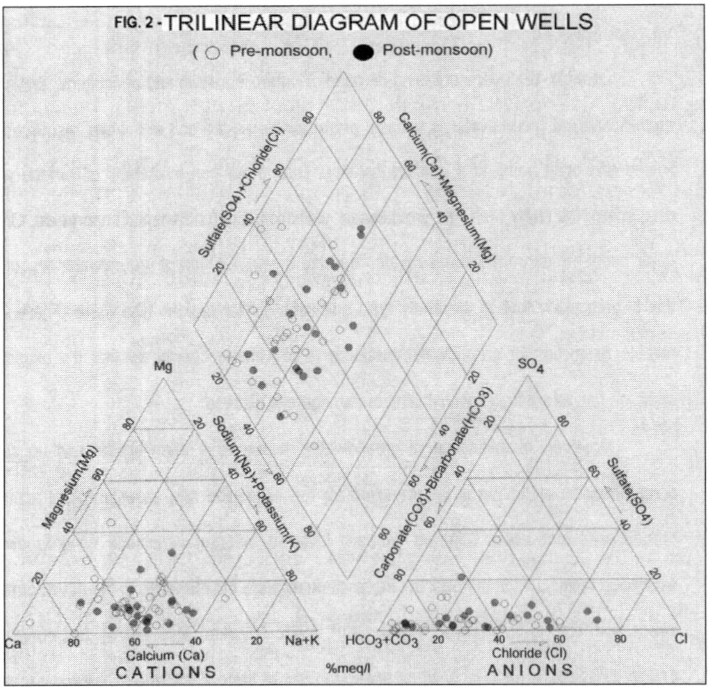

Fig. 2. Trilinear diagram of open wells.

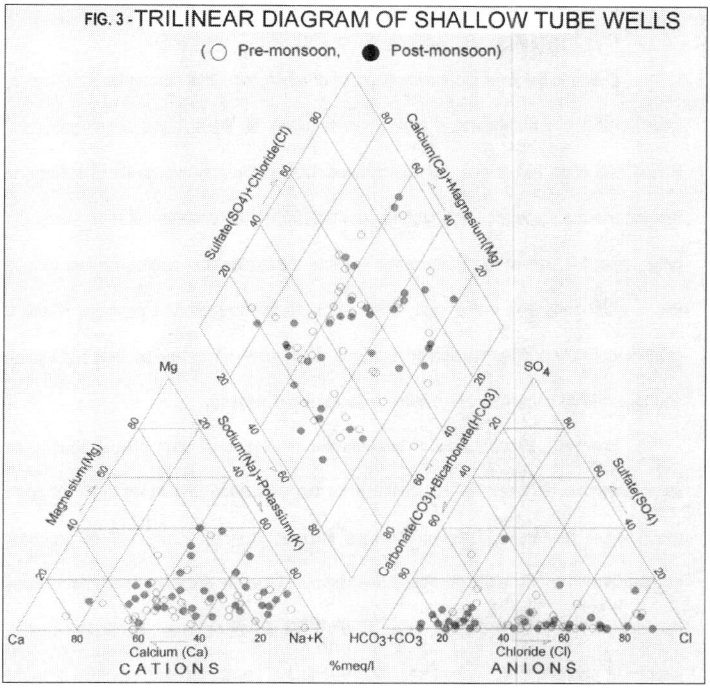

Fig. 3. Trilinear diagram of shallow tube wells.

recharge in the top sandy layers. The terrain has a very low relief and gentle gradient and subject to the construction of earthen bunds for irrigation and embankments for roads. This in fact reduced groundwater runoff and was expected to promote favorable conditions for vertical recharge that keeps the top aquifer layers rather fresh.

There was a distinct change noticed in the grouping of shallow tube well samples. They are distributed in the Area 9, 5 and 7 respectively, irrespective of seasons. Close observations of the samples reveal that the chemical properties of the samples in the northern side are dominated by alkalis and strong acids and are of Na-Cl type. The samples distributed in the central and eastern regions are of Mixed and Na-HCO_3 types as majority of samples fall in the Area 9 and 5 of the field. However in the south and southeastern side, they fall in the Area 7 and 9 indicating the predominance of Na-Cl and Na-HCO_3 types. In the deeper zones, long residence time of groundwater promotes base exchange in clays replacing Ca^{2+} by Na^+. Moreover the base exchange of Ca^{2+} by Na^+ in the clay decreases the permeability of soil drastically (Singal and Gupta,1999). The presence of clay lenses and semi-confined nature of aquifer system in the bottom creates a different type of environment that promotes slow movement of groundwater. Thus it may be assumed that the quality of groundwater in the area depends on the nature of aquifer material and local hydrogeological environment.

Table 1. Chemical parameters of water samples during pre monsoon (May) and post monsoon (Dec) periods.

Well No	Month	EC (µmhos/cm)	Na (mg/l)	K (mg/l)	Ca (mg/l)	Mg (mg/l)	Cl (mg/l)	CO3 (mg/l)	HCO3 (mg/l)	SO4 (mg/l)
1	May	697.00	90.00	22.00	116.00	6.10	110.00	12.00	84.00	32.00
	Dec	640.00	82.20	32.70	85.00	4.16	78.00	17.10	87.00	30.00
2	May	836.00	85.00	3.50	66.00	39.70	207.00	0.00	64.00	15.00
	Dec	390.00	49.30	1.90	48.10	5.10	82.83	1.43	61.42	12.00
3	May	632.00	75.00	3.50	117.00	12.30	112.00	0.00	72.00	2.50
	Dec	370.00	3.30	3.10	118.10	3.48	64.39	8.57	74.27	1.00
4	May	ND	ND	ND	ND	ND	ND	ND	ND	ND
	Dec	273.00	24.00	2.00	40.00	4.00	9.00	0.00	38.00	11.00
5	May	455.00	60.10	11.00	86.00	6.10	72.00	10.00	103.00	3.00
	Dec	180.00	3.20	2.70	18.10	10.46	21.27	12.86	43.57	5.00
6	May	886.00	41.00	4.80	167.00	17.20	109.00	0.00	170.00	6.00
	Dec	740.00	80.30	5.10	135.24	17.89	160.49	4.20	87.12	12.00
7	May	95.00	10.00	2.10	15.00	5.00	9.00	0.00	19.00	3.45
	Dec	150.00	4.50	1.40	28.57	8.13	23.00	0.00	22.85	3.20
8	May	318.00	45.00	8.20	41.00	1.80	28.00	0.00	63.00	8.50
	Dec	410.00	43.10	7.10	67.60	11.62	34.39	3.57	82.84	9.00
9	May	180.00	21.00	3.50	31.00	1.20	40.00	0.00	14.00	6.00
	Dec	210.00	54.20	1.50	17.91	6.13	12.49	0.00	56.78	0.00
10	May	240.00	34.00	9.00	25.00	3.50	30.00	0.00	26.00	5.00
	Dec	580.00	68.50	4.00	77.43	13.79	34.39	4.91	161.01	4.00
11	May	ND	ND	ND	ND	ND	ND	ND	ND	ND
	Dec	217.00	22.00	3.00	40.00	5.00	15.00	6.00	52.00	6.00
12	May	ND	ND	ND	ND	ND	ND	ND	ND	ND
	Dec	530.00	74.00	17.00	60.00	7.30	12.00	12.00	128.00	21.00
13	May	477.00	72.00	8.00	60.00	11.00	42.00	0.00	111.00	5.00
	Dec	ND	ND	ND	ND	ND	ND	ND	ND	ND
14	May	ND	ND	ND	ND	ND	ND	ND	ND	ND
	Dec	390.00	24.33	3.10	72.33	13.01	8.65	3.73	121.01	0.00
15	May	ND	ND	ND	ND	ND	ND	ND	ND	ND
	Dec	280.00	11.90	6.90	53.34	11.18	5.38	0.00	84.91	0.00
16	May	128.00	23.00	7.00	10.00	3.70	7.00	0.00	32.00	3.50
	Dec	183.00	30.00	4.00	25.00	4.00	8.00	0.00	62.00	3.00
17	May	378.00	32.00	21.00	65.00	6.10	11.00	0.00	113.00	6.00
	Dec	418.00	38.00	17.00	87.00	7.00	38.00	7.00	108.00	6.00
18	May	669.00	72.00	33.00	101.00	1.20	60.00	0.00	158.00	5.00
	Dec	490.00	65.00	4.00	61.09	13.42	25.74	5.74	155.42	4.00
19	May	388.00	28.00	6.30	80.00	6.10	17.00	0.00	118.00	3.00
	Dec	370.00	24.00	5.00	72.00	9.00	8.00	9.00	110.00	2.00
20	May	ND	ND	ND	ND	ND	ND	ND	ND	ND
	Dec	309.00	33.00	5.00	65.00	5.00	8.00	8.00	99.00	3.00
21	May	377.00	34.00	6.00	71.00	7.20	12.00	0.00	118.00	6.00
	Dec	470.00	49.50	3.20	63.66	6.67	5.17	8.64	131.05	7.00
22	May	547.00	72.00	9.00	102.00	12.00	120.00	0.00	79.00	11.00

Well No	Month	EC (µmhos/cm)	Na (mg/l)	K (mg/l)	Ca (mg/l)	Mg (mg/l)	Cl (mg/l)	CO3 (mg/l)	HCO3 (mg/l)	SO4 (mg/l)
	Dec	550.00	77.80	2.60	79.52	9.49	64.96	4.64	129.14	8.00
23	May	444.00	64.00	5.00	73.00	5.00	27.00	0.00	82.00	45.00
	Dec	566.00	55.00	31.00	65.00	19.00	37.00	0.00	62.00	83.00
24	May	396.00	63.20	7.00	75.00	3.33	36.00	0.00	113.00	12.00
	Dec	290.00	24.30	3.10	43.94	3.55	26.91	0.00	51.18	5.00
25	May	818.00	162.00	5.50	96.00	12.00	113.00	0.00	125.00	21.00
	Dec	460.00	63.80	1.40	60.34	18.26	50.49	2.48	104.52	0.00
26	May	ND	ND	ND	ND	ND	ND	ND	ND	ND
	Dec	263.00	26.00	6.00	35.00	2.00	41.00	0.00	32.00	2.00
27	May	ND	ND	ND	ND	ND	ND	ND	ND	ND
	Dec	125.00	14.00	4.00	15.00	5.00	19.00	0.00	20.00	2.00
28	May	484.00	47.00	7.00	77.00	15.00	47.00	0.00	67.00	21.00
	Dec	320.00	32.10	5.00	52.27	3.55	34.20		66.10	6.00
29	May	560.00	42.10	18.30	97.12	2.02	36.49	3.73	125.42	7.00
	Dec	262.00	13.00	7.00	65.00	6.00	13.00	0.00	64.00	12.00
30	May	ND	ND	ND	ND	ND	ND	ND	ND	ND
	Dec	195.00	17.00	1.00	60.00	0.00	11.00	0.00	40.00	12.00
31	May	ND	ND	ND	ND	ND	ND	ND	ND	ND
	Dec	215.00	30.00	1.00	30.00	1.00	26.00	0.00	34.00	0.00
32	May	10500.00	1975.00	332.00	515.00	425.00	2597.00	22.00	760.00	ND
	Dec	4637.00	1125.00	147.00	258.00	189.00	1384.00	40.00	170.00	144.00
33	May	1318.00	140.00	51.50	224.00	6.10	269.00	0.00	139.00	11.00
	Dec	380.00	74.60	21.80	24.00	9.30	64.39	0.00	64.28	8.00
34	May	7406.00	1275.00	412.00	325.00	198.00	2072.00	50.00	380.00	32.00
	Dec	3480.00	1050.00	53.00	65.00	51.00	691.00	80.00	292.00	4.00
35	May	933.00	242.00	13.20	72.00	13.50	100.00	24.00	196.00	1.20
	Dec	540.00	120.90	12.80	35.70	9.75	53.22	12.85	135.70	0.80
36	May	4770.00	1280.00	222.00	141.00	107.00	1274.00	0.00	413.00	38.00
	Dec	1300.00	256.00	33.60	95.63	17.00	125.00	8.55	285.69	31.00
37	May	5953.00	1254.00	516.00	250.00	59.00	1267.00	0.00	463.00	112.00
	Dec	2938.00	475.00	213.00	222.00	60.00	504.00	0.00	220.00	114.00
38	May	848.00	114.00	12.00	147.00	33.20	208.00	0.00	110.00	2.00
	Dec	770.00	112.20	5.70	122.86	56.93	181.56	7.42	100.00	3.00
39	May	5663.00	1200.00	385.00	195.00	148.00	1117.00	0.00	728.00	52.00
	Dec	3347.00	775.00	115.00	206.00	163.00	945.00	0.00	350.00	6.00
40	May	5087.00	1114.00	288.00	160.00	317.00	1332.00	19.00	423.00	12.00
	Dec	810.00	86.00	8.63	130.00	5.00	120.00	14.00	70.00	16.00
41	May	235.00	33.00	2.00	35.00	3.70	38.00	0.00	42.00	0.30
	Dec	170.00	22.50	2.20	28.57	2.43	31.27	0.00	24.36	1.00
42	May	1791.00	335.00	45.00	85.00	155.00	286.00	22.00	245.00	56.00
	Dec	1271.00	287.00	29.00	110.00	12.00	191.00	23.00	172.00	21.00
43	May	822.00	220.00	21.00	109.00	9.80	192.00	15.00	147.00	1.00
	Dec	580.00	108.50	4.90	117.20	16.62	125.30	2.85	115.00	3.00
44	May	968.00	136.00	35.00	145.00	16.00	162.00	20.00	145.00	13.00
	Dec	782.00	167.00	24.00	125.00	13.00	136.00	14.00	156.00	7.00
45	May	478.00	92.00	11.00	61.00	8.00	94.00	0.00	60.00	8.00

Well No	Month	EC (µmhos/cm)	Na (mg/l)	K (mg/l)	Ca (mg/l)	Mg (mg/l)	Cl (mg/l)	CO3 (mg/l)	HCO3 (mg/l)	SO4 (mg/l)
	Dec	210.00	502.00	3.40	24.22	1.26	35.96	0.00	37.46	4.00
46	May	8512.00	1986.00	112.00	290.00	124.36	2406.00	45.00	225.00	ND
	Dec	4085.00	1181.00	70.00	280.00	26.00	1248.00	32.00	350.00	7.00
47	May	740.00	124.00	15.00	95.00	37.00	65.00	30.00	140.00	12.00
	Dec	514.00	95.00	8.00	80.00	5.00	31.00	40.00	116.00	10.00
48	May	1292.00	316.00	51.00	105.00	25.60	62.00	52.00	355.00	15.00
	Dec	920.00	121.00	23.00	85.00	38.00	169.00	68.00	80.00	3.00
49	May	4727.00	1245.00	84.00	60.00	115.00	867.00	84.00	335.00	12.00
	Dec	2221.00	585.00	47.00	65.00	67.00	355.00	52.00	334.00	2.00
50	May	366.00	42.00	9.00	60.00	9.70	32.00	0.00	100.00	6.00
	Dec	272.00	34.00	7.00	40.00	6.00	26.00	0.00	66.00	4.00
51	May	240.00	36.00	9.00	30.00	7.30	27.00	0.00	59.00	12.00
	Dec	230.00	34.00	3.00	55.00	2.00	23.00	0.00	50.00	21.00
52	May	376.00	80.00	18.00	45.00	13.50	20.00	0.00	120.00	0.91
	Dec	160.00	34.50	2.10	17.00	3.58	17.57	0.00	38.59	4.00
53	May	1988.00	602.00	102.00	112.00	43.00	218.00	60.00	470.00	72.00
	Dec	1469.00	347.00	53.00	65.00	35.00	132.00	28.00	302.00	61.00
54	May	1373.00	88.00	18.00	331.00	32.00	94.00	19.00	355.00	15.00
	Dec	580.00	45.30	12.60	132.20	12.66	44.21	23.73	124.91	20.00
55	May	1144.00	213.00	32.00	192.00	35.40	285.00	35.00	185.00	12.00
	Dec	1339.00	206.00	28.00	235.00	44.00	297.00	23.00	206.00	9.00
56	May	385.00	84.00	12.00	52.00	33.00	22.00	15.00	100.00	12.00
	Dec	220.00	23.00	3.00	35.00	0.00	15.00	2.00	30.00	8.00
57	May	4260.00	712.00	62.00	320.00	395.00	1117.00	15.00	105.00	302.00
	Dec	3186.00	610.00	40.00	216.00	277.00	1097.00	28.00	112.00	113.00
58	May	898.00	96.00	12.00	154.00	31.00	102.00	26.00	187.00	11.00
	Dec	570.00	80.02	4.40	116.25	6.29	62.08	21.18	139.66	9.00
59	May	791.00	90.00	18.00	146.00	19.60	69.00	0.00	184.00	4.00
	Dec	430.00	70.00	23.80	78.48	11.97	60.74	0.00	113.44	1.00
60	May	386.00	72.00	6.00	45.00	3.70	22.00	0.00	88.00	14.00
	Dec	320.00	54.00	5.00	37.00	1.00	18.00	0.00	80.00	6.00
61	May	393.00	72.00	8.00	35.00	8.50	32.00	0.00	100.00	ND
	Dec	132.00	23.00	2.00	20.00	2.00	21.00	0.00	18.00	4.00
62	May	463.00	89.00	16.00	70.00	7.30	42.00	0.00	65.00	90.00
	Dec	262.00	39.00	6.00	45.00	4.00	15.00	0.00	48.00	41.00
63	May	467.00	65.00	2.00	105.00	11.00	42.00	20.00	105.00	4.00
	Dec	382.00	23.00	0.00	120.00	5.00	36.00	0.00	118.00	1.00
64	May	375.00	61.00	3.00	35.00	27.00	102.00	0.00	13.00	8.00
	Dec	156.00	21.00	1.00	25.00	6.00	33.00	0.00	12.00	3.00
65	May	212.00	52.00	1.00	20.00	6.10	36.00	0.00	20.00	17.00
	Dec	176.00	34.00	1.00	25.00	4.00	27.00	0.00	22.00	11.00

ND Not determined, Well No 1-31 - Open wells, Well No.32-65 Shallow tube wells

8.6. ELECTRICAL CONDUCTIVITY

Electrical conductivity indicates total concentration of ionic (salt) soluble constituents and is closely related to the sum of cations and anions. Chemically pure water has very low conductivity, but with the presence of dissociated ions in solution, it becomes conducting. The EC values of open wells are in the range of 195 to 836 μmhos.cm^{-1} with an average of 200-300μmhos.cm^{-1}, though there are areas where the values exceed that limit. Generally, recharge water dilutes the chemical concentrations and the EC values are low, but the values tend to increase during non-recharging periods due to the effect of evaporation. But in the study area, the isocone map of open wells during the post monsoon period indicates higher concentration of ions than the pre monsoon period (Fig.4 and 5). This might be due to the drying up of clayey aquifer material above the water table during high water level condition in the summer months. That leads to oxidation, which would increase the solubility of minerals by the infiltrating water during the recharge season.

The terrain was subjected to several anthropogenic practices and this might have impaired the aquifer properties of upper layers due to compaction up to certain depths bgl. but it provides favorable pockets for vertical recharge as the natural surface runoff in many places has been modified by earthen bunds, road developments, and other constructional

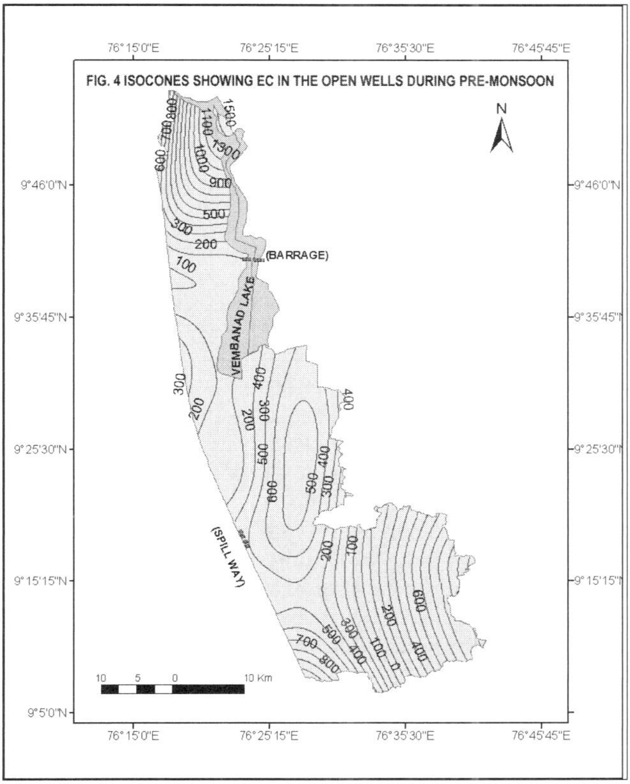

Fig. 4. Isocones showing EC in the open wells during pre-monsoon.

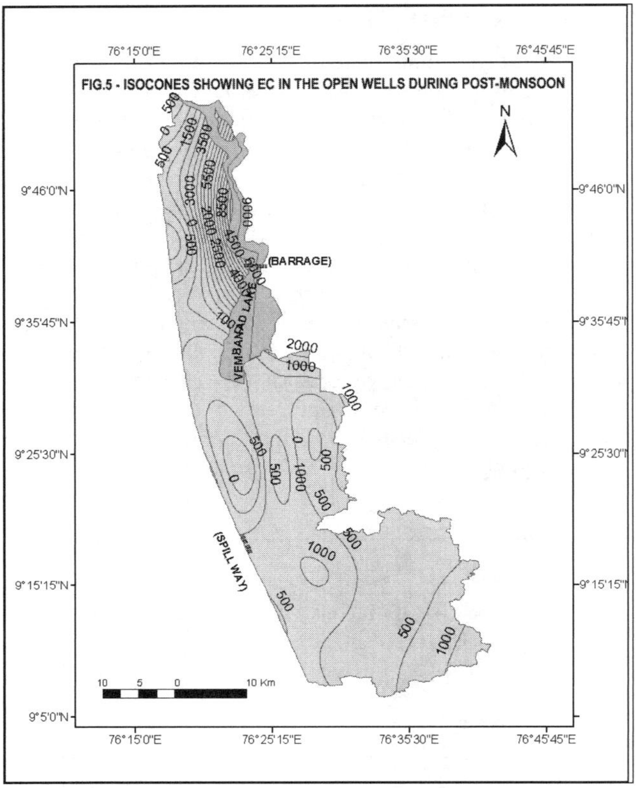

Fig. 5. Isocones showing EC in the open wells during post-monsoon.

practices. This might be the reason for the higher EC values in the open wells during the post monsoon period. Such kind of mechanism responsible for the increase in EC during the recharge season has been reported by Ballukrya and Ravi (1999) and Subba Rao et al. (2002) from the study of shallow aquifers in the Chennai and Guntur area respectively.

The groundwater derived from the perennial sources in the east and southeastern side move towards north along the Vembanad estuary during the monsoon period (June-Nov) when the barrage is in the open condition. During the summer months (Dec-May) when the barrage is kept closed to prevent the entry of brackish water, the flow of fresh water from the natural drainage system is arrested by the barrage. This builds up pressure down south of the barrage that takes a shift towards the western peripheral regions flushing the aquifer system through and within the porous formation upto an average depth of 15bgl by horizontal movement. This mechanism is cyclic and has been noticed especially during the summer months and is regulated by the periodic opening and closing of the barrage. This is another major mechanism through which the recharge of groundwater takes place. The influence of horizontal recharge of groundwater through rainwater harvesting in the coastal areas under similar conditions has been already reported by Rajendran et al. (2002).

The shallow tube wells on the other hand depict a set of higher EC values ranging from 145 to 10,550 with an average of 500-1000 µmhos.cm^{-1}. The values are much higher as compared to the open wells (Fig.6 and 7). The EC in this zone seems to decrease with recharge due to gradual percolation of fresh water from the surface layers and other perennial sources reaching underground. Here the groundwater is in the semi-confined condition that permits only partial mixing. The anomalous higher set of values might be due to the combined effect of seasonal accumulation of dissociated ions aided by slow groundwater movement under reduced natural flushing. In such conditions the ions tend to increase within the aquifer system and part of the same derived from the overburden might be trapped in the clay lenses and laterite formation. Hence the groundwater present in the deeper layers is highly mineralized as inferred from the isocon maps.

Thus the movement of groundwater in the upper layers is mainly through and within the porous formation up to a 15mbgl because beyond that depth the groundwater is in the semi-confined condition, where free movement is not possible due to hydrostatic pressure. The deeper zone further permits longer residence time of groundwater within the formation. It may thus be inferred that the hydro-geochemical behavior of the top and the bottom layers are distinctly different even though they belong to the same aquifer system.

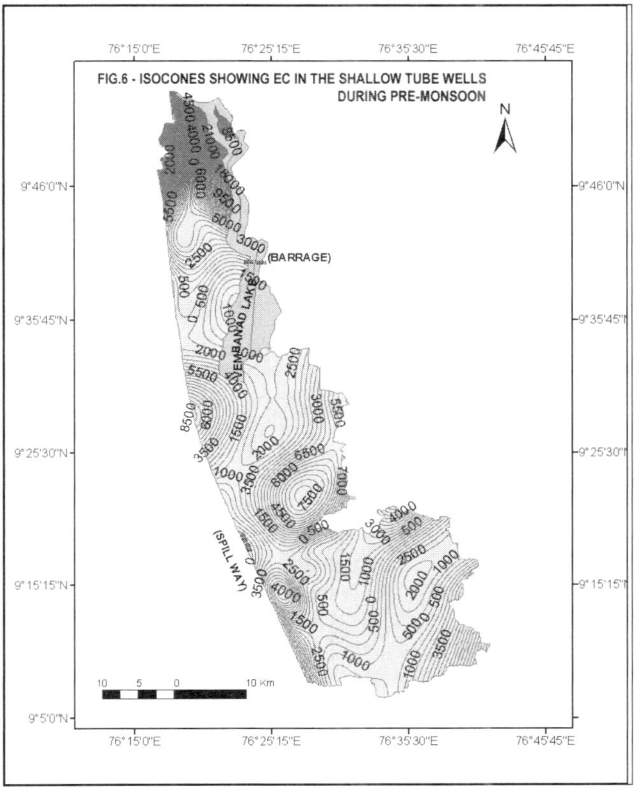

Fig. 6. Isocones showing EC in the shallow tube wells during pre-monsoon.

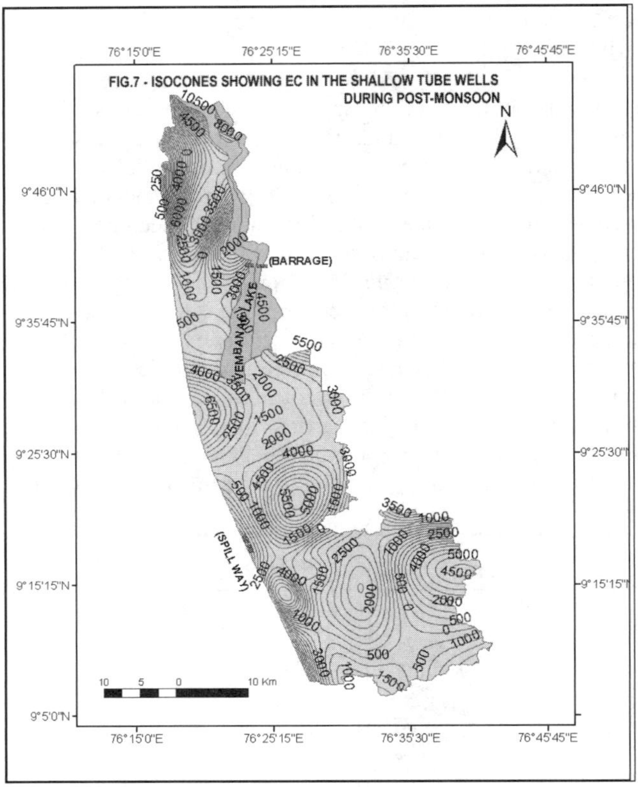

Fig. 7. Isocones showing EC in the shallow tube wells during post-monsoon.

8.6.1. WATER TABLE AND PIEZOMETRIC CONTOURS

The water table contour maps of open wells during the pre monsoon and post monsoon period are significantly different (Fig. 8 and 9). The water table during the pre monsoon period is deep and contours exhibit a broad trough especially in the south and center portions of the district. But the contours during the post monsoon period are shallow but indicate a different pattern in the spatial distribution. This might be due to the gradual change achieved by the water table in response to the horizontal recharge through the seasonal regulation of artificial barrage. The piezometric surface of shallow tube wells during the post and pre monsoon does not show much difference in the spatial distributional pattern, which might be attributed to the slow groundwater recharge as compared to the open wells (Fig. 10 and 11).

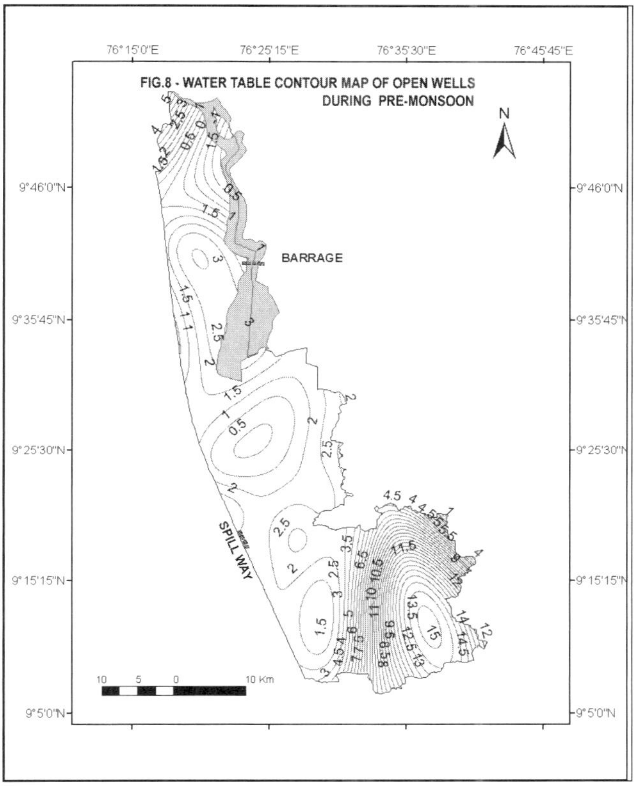

Fig. 8. Water table contour map of open wells during pre-monsoon.

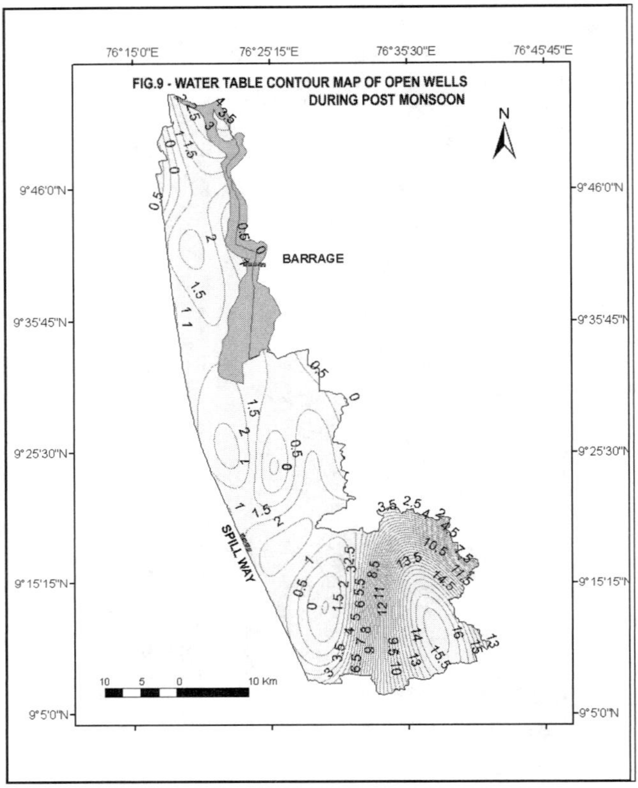

Fig. 9. Water table contour map of open wells during post-monsoon.

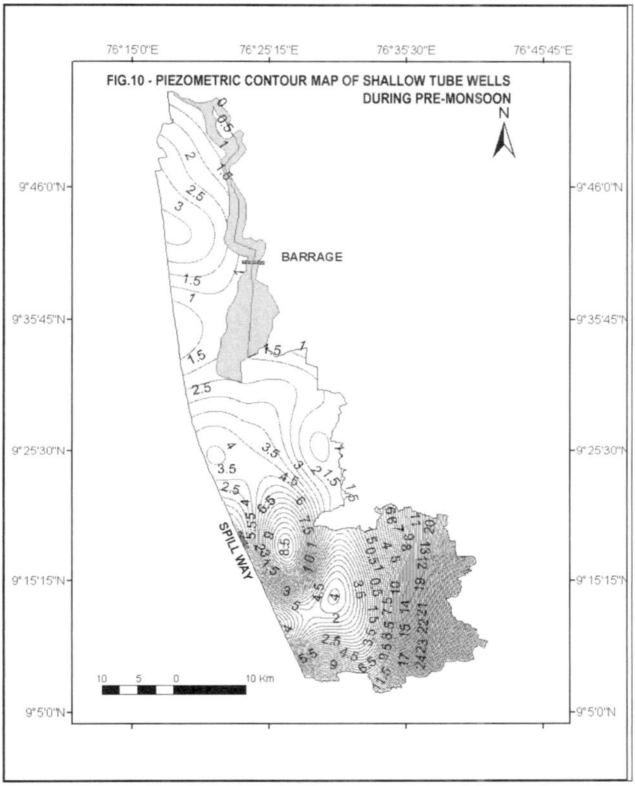

Fig. 10. Piezometric contour map of shallow tube wells during pre-monsoon.

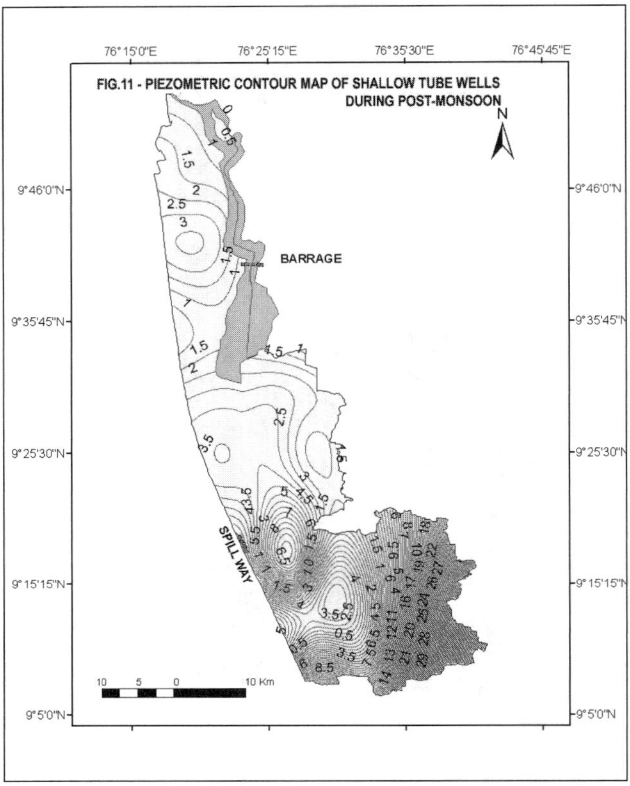

Fig. 11. Piezometric contour map of shallow tube wells during post-monsoon.

8.7. TREND ANALISIS OF WATER LEVEL

To monitor the response of water level to groundwater recharge, trend analysis has been carried out for a period of five years. The method of least square for best fitting line was adopted to find out the slope of the trend line. Trend line will give the indication of the general behavior of water levels by taking into account the influence of different seasonal cycles. Based on the significance of the different geographic localities, the hydrographs analyzed for the trend of water level points to three prominent trends by the open and tube wells in the area viz decreasing, increasing and steady. The open wells exhibit a trend in the range of –0.051 to 0.287. The northern region depicts almost a steady trend with a positive slope of 0.028 indicating that the equilibrium of groundwater in the northern side is very delicate (Fig. 12). The southern side on the other hand, depicts an increasing trend with a slope of 0.238, which might have resulted for the constant recharge from the perennial rivers (Fig. 13). The trend analysis on the eastern side is also increasing whereas in the west central portion along the coast, it is decreasing with a negative trend slope of –0.051 (Fig. 14 and 15). The hydrograph of open wells exhibit typical sharp rise and fall in the peak and

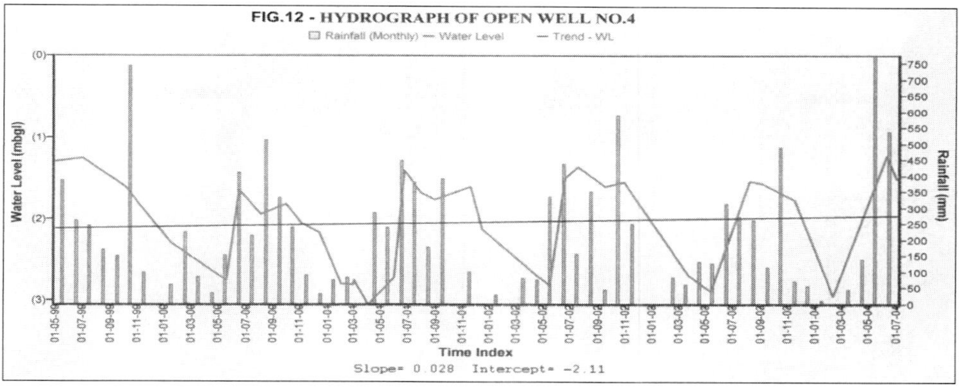

Fig. 12. Hydrograph of open well No. 4.

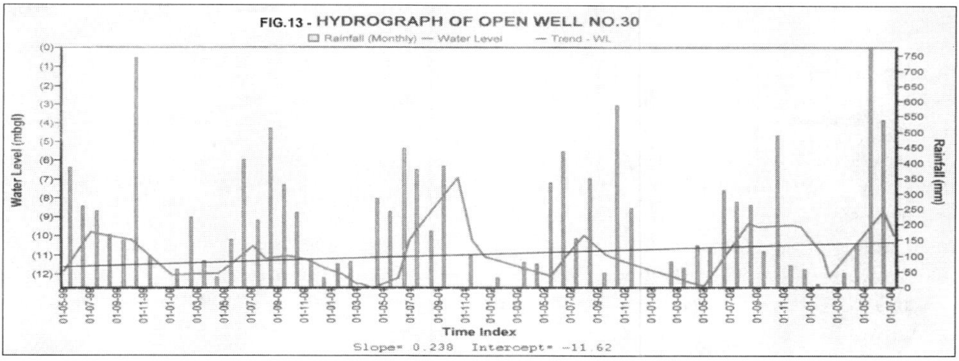

Fig. 13. Hydrograph of open well No. 30.

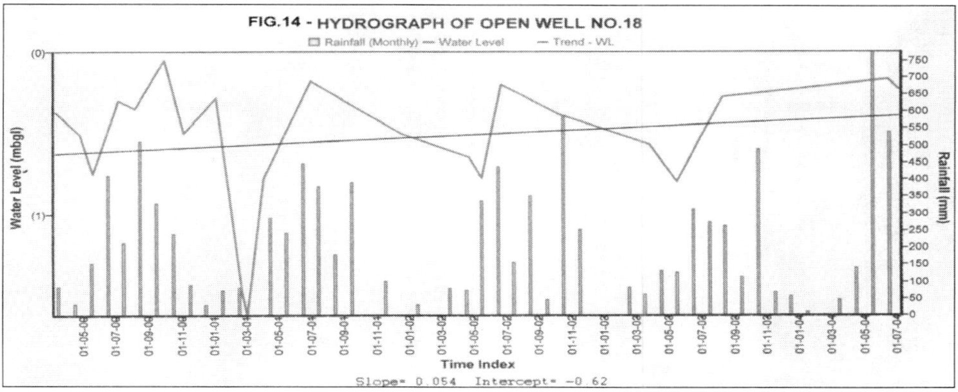

Fig. 14. Hydrograph of open well No. 18.

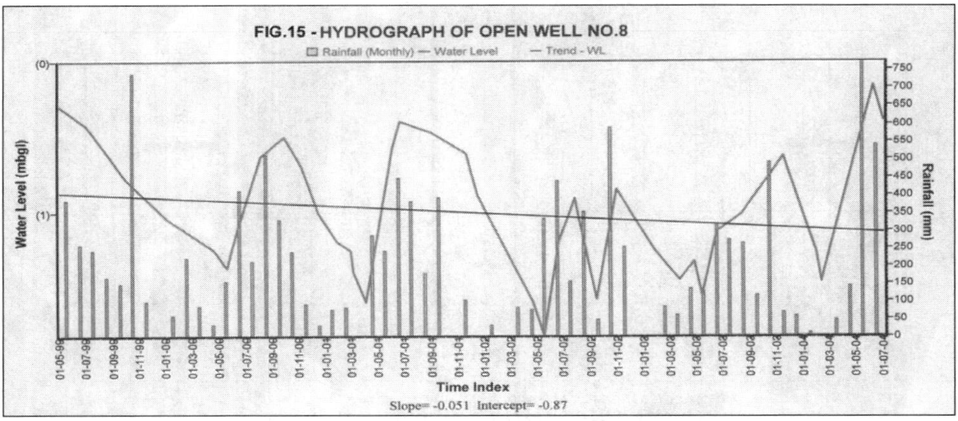

Fig. 15. Hydrograph of open well No. 8.

limbs throughout the area. This may be attributed to the response of dominant vertical recharge and other anthropogenic practices as explained earlier. The distribution pattern of water levels of open wells over different seasons is in good agreement with the corresponding rainfall which further supports the recharge mechanism.

The shallow tube wells indicate a trend in the range of –0.035 to 0.585. The north and southern side exhibits a steady positive trend with slopes corresponding to 0.043 and 0.009 respectively (Fig. 16 and 17). Unlike the shallow zones, the direct recharge seems to have insignificant effect in the southern side. However eastern side reflects a highly increasing trend of 0.0420, which might be due to the presence of undulations and slopping nature of the terrain that exerts enormous pressure by the perennial recharge from the east peripheral regions (Fig. 18). The laterite, which is seen at the bottom of the river basins, extends

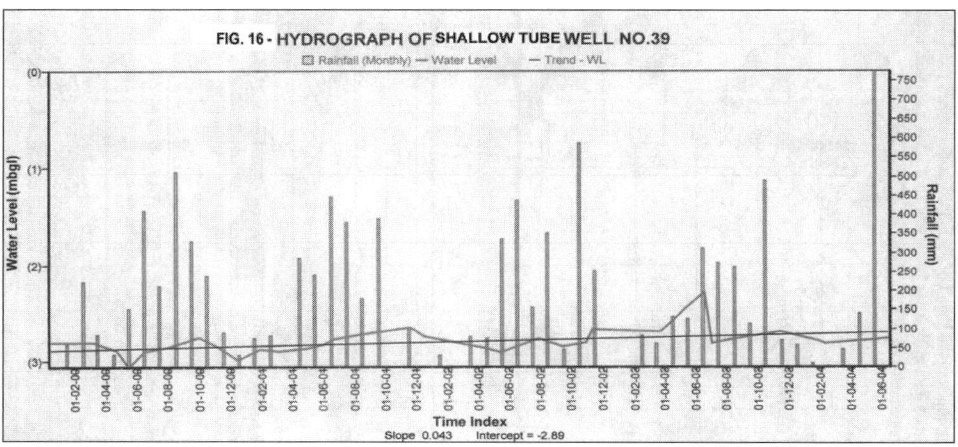

Fig. 16. Hydrograph of shallow tube well No. 39.

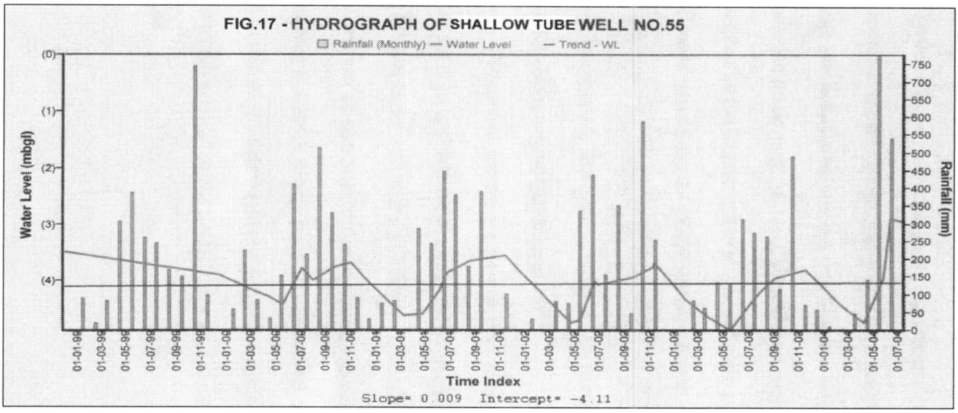

Fig. 17. Hydrograph of shallow tube well No. 55.

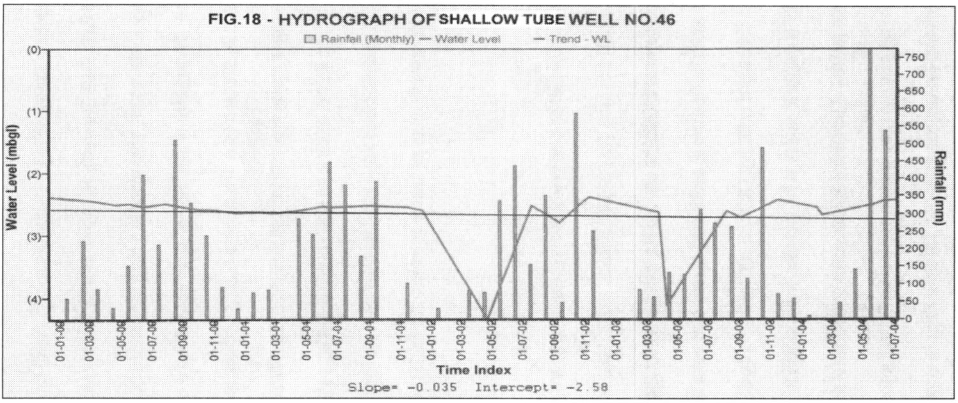

Fig. 18. Hydrograph of shallow tube well No. 46.

laterally over the tertiary sediments and acts as a favorable structure for the free movement of groundwater into the quaternary aquifer system. Here also a decreasing trend with a negative trend slope is noticed in the west central region indicating excess withdrawal and lack of replenishment (Fig. 19). Innumerous ice factories situated in this sector consumes enormous quantity of water that might have lead to this alarming situation which is to be immediately taken care off for longer sustainability of the aquifer system. The shallow tube wells exhibit flat topped peaks and gentle recession limbs clearly suggesting a slow response to the groundwater recharge and is not in good agreement with the corresponding rainfall. It may thus be inferred that the response of the groundwater to the modern groundwater recharge with respect to upper layers are fast where as the bottom layers are rather slow.

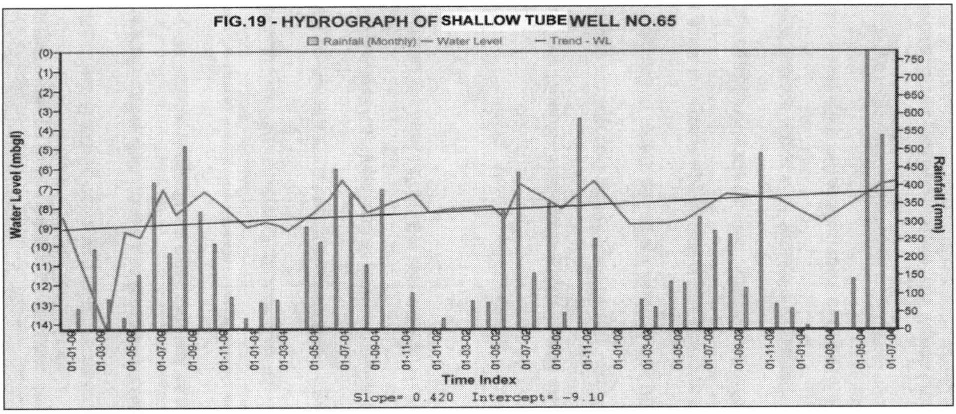

Fig. 19. Hydrograph of shallow tube well No. 65.

8.8. CONCLUSIONS

Based on the study, it may be concluded that the groundwater quality and behavior of the shallow and deeper layers of the same aquifer system are distinctly different. The trilinear diagram reveals that the shallow groundwater in the area has more affinity towards Ca-HCO_3 type. The geochemistry of groundwater beyond 15mbgl is distinctly different in which Ca-HCO_3, Na-Cl, Na-HCO_3 and Mixed types are noticed. The Na-Cl type is noticed in the northern side but the central portion is dominated by mixed and Na-HCO_3 type, whereas the south and southwestern side is dominated by Na-Cl and Na-HCO_3 types respectively. The concentrations of dissolved constituents in the bottom layers are extremely high as compared to the top layers. The frequent groundwater recharge is reflected by the presence of low dissociated ions in the upper layers. The higher concentrations of ions in the deeper groundwater regime might be due to the stagnant unconfined condition prevailing at depths. The change in the land use practices and the regulation of artificial barrage promotes movement of groundwater through vertical and horizontal recharge flushing the upper groundwater regime through and within the porous formation up to a depth of 15mbgl. The deeper groundwater regime exhibits a very slow response to the groundwater recharge as compared to the upper layers. The trend analysis of open wells exhibit typical sharp rise and fall in the peak and limbs throughout the area which holds in good agreement with the corresponding rainfall and this further supports the dominant mechanism of recharge in the upper layers. The tube wells on the other hand display gentle flat-topped peaks and a gentle recession limb, which is attributed to the slow response of groundwater recharge and is not in good agreement with the corresponding rainfall. This clearly demonstrates that the quality of groundwater in the quaternary aquifer system is directly controlled by the modern groundwater recharge and present local environment.

REFERENCES

Anon, 2002. Land Resources of Kerala. Pub., Kerala State Land Use Board, Thiruvananthapuram, 379p.

Ballukrya, P.N, and Ravi, R., 1999. Characterisation of groundwater in the unconfined aquifers of Chennai city, India, P-1, Hydrogeochemistry. Joul.Geol.Soc.India 54: 1-11.

Piper, A.M., 1944. A graphic procedure in the geochemical interpretation of water analysis. Trans. Amer. Geophys. Union, v.25.

Rajendran, K., Ajeeb, P., and Thompson, L., 2002. Implications of rainwater harvesting in the coastal aquifer - A case study pp135-143, In Proc., Int. conference on water resources and management in arid regions, March 2002, Kuwait.

Singal, B.B.S, and Gupta, P.P., 1999. Applied Hydrogeology of Fractured Rocks, Kluwer, Acd. Publ. Lonon.223p.

Subba Rao, N., Prakasha Rao, J., John Devadas, Srinivasa Rao, K.V., Krishna. C and Nagamalleswar Rao,K.V., 2002.

Hydrogeochemistry and Groundwater Quality in a Developing Urban Environment of a Semi-Arid Region, Guntur, A.P, Joul.Geo.Soc. India 59: 159-166

Walton, W.C., 1970. Groundwater Resources Evaluation. McGraw-Hill Book Co., New York.

9

SOME HYDROGEOLOGICAL ASPECTS OF THE SHALLOW BEACHROCK (CALC-ARENITE) AQUIFERS ALONG MAHARASHTRA COAST, INDIA

S. S. Thigale[1], R. A. Duraiswami[1,2], and Bhavana Umrikar[1]
[1]Department of Geology, University of Pune, Pune 411 007, India
[2] Groundwater Surveys & Development Agency
Bhujal Bhavan, Shivajinagar, Pune 411 005, India

ABSTRACT

Sandy spits are quite common along the Konkan coast of Maharashtra, India. Sand and beachrock (calc-arenite) constitute the shallow aquifer system within these spits. The lateral and vertical occurrence of beachrock within the sandy spits along the coast is fairly continuous and uniform (up to 8 m). Lithosections at several localities generally consist of a calcareous conglomerate at the base, coarse calc-arenite in the middle and fine-grained calc-arenite at the top. Field studies, primary sedimentary structures, petrography and stain tests reveal that the calcareous conglomerate and the coarse calc-arenite were deposited in the breaker zone and intertidal zones respectively, while the fine grained calc-arenite, in the upper parts of most sections, represent beach dune environment. Groundwater chemistry data from dug-wells were used to measure the vertical variation in salinity within the shallow beachrock aquifers. Wide variation in the salinity of groundwater along the coast is landform specific and can be attributed to the seasonal water level fluctuations, surface and sub-surface ingress of seawater and excessive pumping. Lateral and / or upconing process of saltwater intrusions cause intermingling of fresh-saline water that results in changes in groundwater composition leading to base-exchange reactions that alter the relative concentration of ions. Degassing of carbon dioxide within carbonate-rich sands causes precipitation of carbonates that consolidates sands into beachrock. Natural and human activities further enhance the mixing process that ultimately affects the storativity and transmissivity of the beachrock aquifers.

9.1. INTRODUCTION

Migration of saltwater into fresh water aquifers is known as sea water intrusion (Atkinson et al., 1986). Aquifers that are in hydraulic connection with the sea are susceptible to flow of saltwater from the sea due to lowering of water table and development of hydraulic gradient during excessive pumping. Thus, coastal areas world over are being subjected to increased anthropogenic pressures leading to wide spread saltwater intrusion. This has prompted several recent studies leading to the understanding of the dynamics of saltwater intrusion and suggestions regarding management of groundwater in the coastal areas

(Cartwirght, 2004; Tran, 2004; Hong, 2005). The mainland coastline of India extends for a length of 6100 km and is susceptible to varying degrees of saltwater intrusion (Das, 2004). Of this the 720 km long coastline of Maharashtra, popularly known as the Konkan coast, includes four districts viz. Thane, Raigad, Ratnagiri and Sindhudurg, in addition to the metropolis of Mumbai (Fig. 1). The coast is characterized by alternating rocky basaltic headlands and sandy beaches. As many as 54 creeks of big and small rivers present along the coast are subjected to tidal influence up to 40 km inland (Patil, 1998) and is responsible for more than 45,465 hectares of saline soils along the coast (Anonymous, 1990).

The rural population along the Konkan coast is dependent upon groundwater as the main source of drinking water, especially during the critical summer months. Population explosion and rapid industrialisation along the coast have already increased the demand for fresh water and has lead to the deterioration of the hydrogeological regimes in the region. In a recent review of coastal aquifers and seawater ingress in India, barring the occurrence of salinity in Greater Mumbai, the problem of salinity in the coastal aquifers of Maharashtra have received little attention (Das, *op cit*). In spite of the fact that aquifers along the coast are under severe stress, systematic efforts to quantify the hydrogeology and aquifer characteristic in the region are lacking. The present paper is an attempt to bridge this gap by undertaking detailed studies along the Konkan coast. The paper also provides insight in to some aspects like framework of beachrock aquifers, water-rock interaction and the role of groundwater chemistry and performance of aquifers vis-à-vis cementation processes.

9.2. STUDY AREA

The study area is included in Survey of India toposheets no. 47B/14 to 16, 47 F/4 and G/2 extends from Thal (72°52'18"; 18°41'12") in the north to Guhagarh (73°11'02"; 17°29'20") in the south (Fig. 1). The coast experiences sub tropical climate, where the mean maximum temperature is 35.2° C and the mean minimum temperature is 22.4° C. The area receives rainfall from the south-west monsoon between June and October that ranges from 2000 to 3000 mm. Relative humidity seldom remains below 55% and generally lies between 65 to 75%, but averages over 80% during the monsoon.

The area is characterized by a variety of depositional and erosion landforms common to the shore zone (Thigale et. al., 1998). Sandy beaches in the study area occur either along the shoreline between headlands or are present on the seaward margins of spits. Beaches that are developed along the spits are almost straight, while those occurring between headlands are crescent shaped with the concave side facing the sea. The beaches vary in length from less than a kilometer to as long as 10 km and their widths vary from 40 m to 200 m. Sandy beaches are invariably backed by 2 to 3 rows of sand dunes.

Spits constituted of beach rock with one-end attached to the mainland are developed at several places, but those found at Akshi-Nagaon, Aravi, Dive Agar, Srivardhan and Guhagarh need special mention. The spits are 300 m to 500 m wide, up to 6 km in length and attain elevations from 3 m to 6 m. These are largely stabilised and support luxuriant vegetation. Mud flats, tidal marshes and mangrove swamps are developed along creeks, tidal inlets and inner sides of spits all along the coast. Profile across a typical beach where the beach is backed by younger and older dunes and followed by mudflats or a tidal marsh is depicted in Fig. 2.

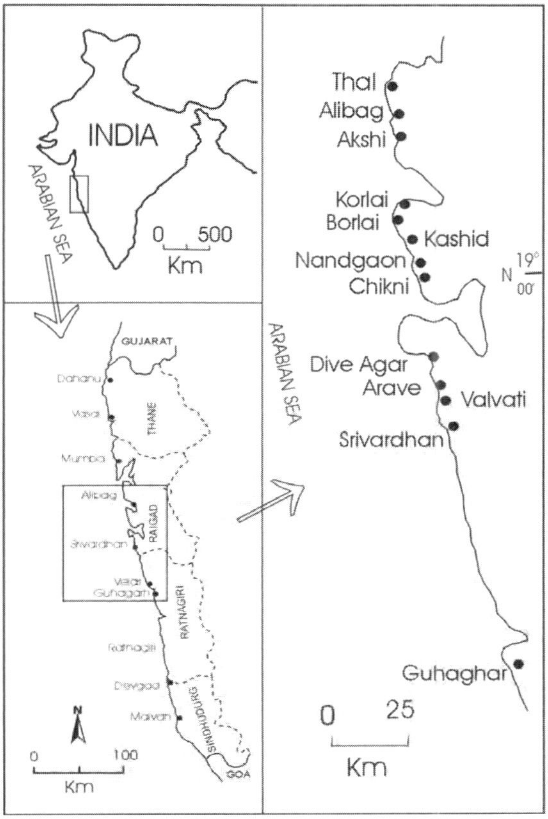

Fig. 1. Map of the Maharashtra, India with inset showing sampling locations.

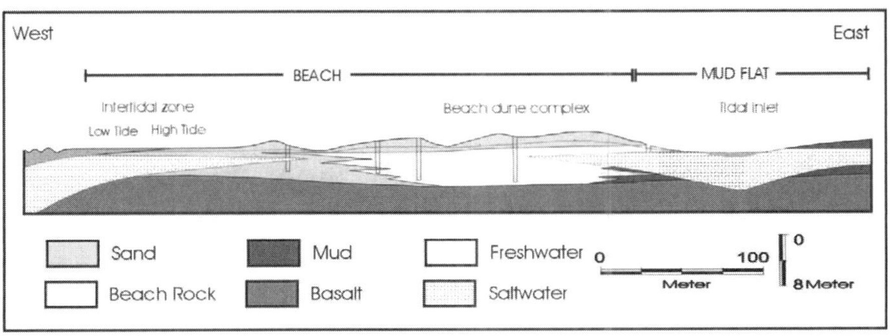

Fig. 2. Schematic diagram across the coast showing coastal land forms and seawater intrusion.

9.3. GEOLOGY OF THE BEACHROCK

Beachrocks, also described as littoral concrete (Ahmed, 1972), raised beaches (Powar et al., 1978) or calcareous grit (Mathur and Naidu, 1932) are developed all along the coast in discontinuous patches. They extend laterally up to a maximum of 500 m inland and are usually concealed under sand and/or soil. They show low angle seaward dips (6°-8°) and rest unconformably on the basaltic flows. The beachrocks are also deposited along mud flats. The surfacial outcrops of the beachrock measure about 2 m thick, but can extend up to 8 m below surface (Fig. 2). Lithosections from different localities are illustrated in (Fig. 3). Primary sedimentary structures such as parallel-, cross-,wave-ripple- and current ripple laminations are present implying deposition in the intertidal zone. (Soman et al., 1996). A number of unconsolidated sand layers and distinct ilmenite- and shell-fragment layers are present at different horizons. Thick layers of fine-grained beachrock is present at the upper portion of most sections and represent wind blown beach sand. At Akshi, Dive Agar, Aravi and Srivardhan lenses and thin intercalations of mud are also present.

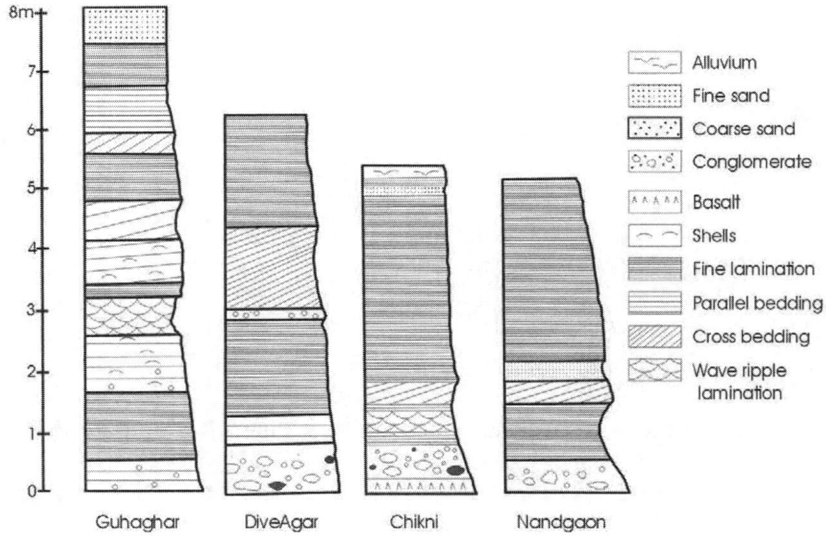

Fig. 3. Lithosections of beachrocks from different locations in the study area.

In thin sections, beachrocks show presence of bioclasts and terrigeneous clastics cemented together by microsparite (Fig. 4a). Fragments of foraminifers, bivalves, gastropods and echinoids represent the bioclasts, and basalt, laterites, grains of pyroxenes, plagioclase and agates represent terrigeneous clastics. In sections around Thal and Akshi a large amount of fine-grained quartz, pelloids and small foraminifera are cemented by microspar (Fig. 4b). Fine sand sized quartz grains are distributed throughout (fig. 4c). Based on the texture and mineralogy the beach rocks are referred to here as calc-arenite (Soman et al., op cit). Petrographic observations indicate that the interclast pore spaces are generally lined with layers of micritic cements cemented by microsparite (Fig. 4d). Some pore spaces, however, show deposition of silica.

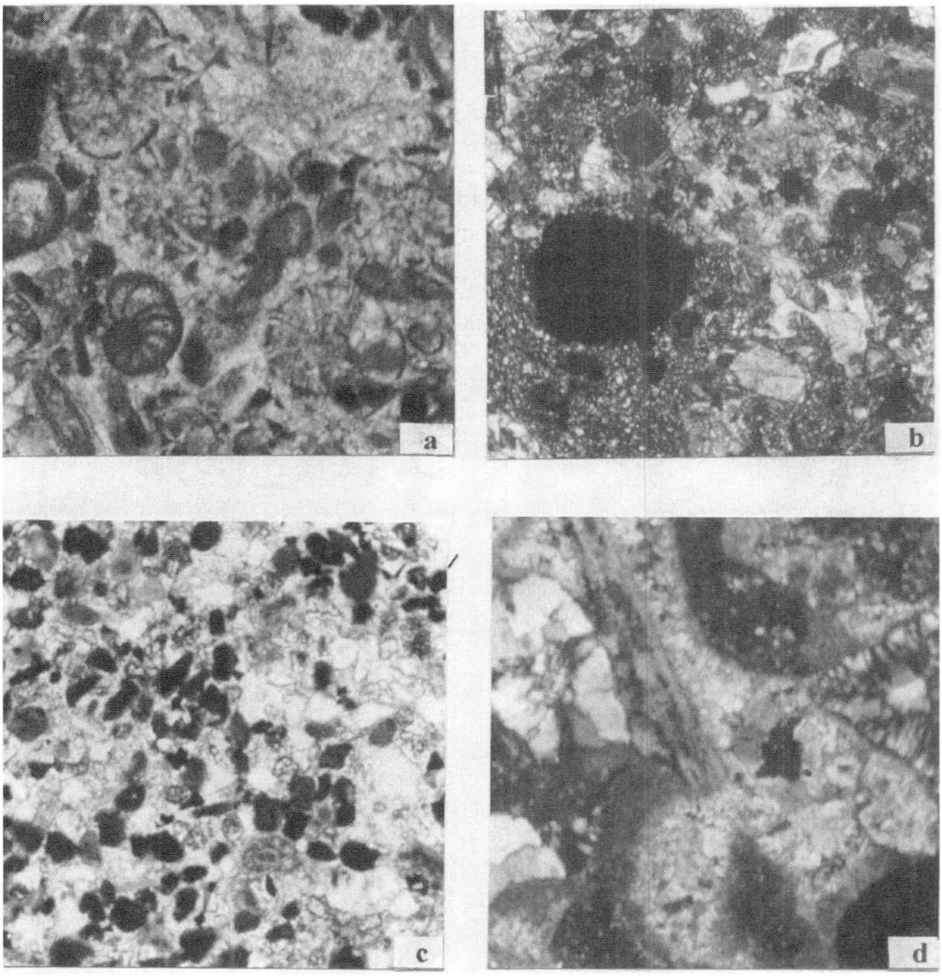

Fig. 4. Photomicrographs of beachrocks showing (a) bioclasts surrounded by microsparite, (b) laterite fragments and pyroxenes that constitute the terrigeneous clastics (c) fine grained terrigeneous clastics, quartz and bioclasts cemented by microsparite (d) small pore-void that is lined by spary calcite that coarsens towards the centre (Scale 1cm- 1μm).

Eighteen beachrock samples from different locations representing different horizons were selected for stain test to distinguish various types of carbonate minerals. Standard staining procedures for calcite, ferron calcite and dolomite (Dickson, 1965) and aragonite (Tucker, 1988) were followed. All the samples show presence of nonferroan calcite but dark traces of aragonite were also seen in lower conglomeratic horizons from Chikni and Nandgaon. Similarly, thirty beachrock samples from twelve localities were subjected to

Insoluble Residue Analysis (Ireland, 1971). The data is incorporated in Table 1. Carbonate percentage > 75%, between 50 to 70% and < 50% were obtained in seventeen, ten and three samples respectively. Examination of residue under binocular microscope shows dominance of lithic fragments of basalt and laterite, quartz, agate and opague minerals. Systematic increase in the insoluble residue percentage from top to bottom within the individual lithosections is noticed at Borlai, Chikni, Nandgaon and Guhagarh. However, at Akshi and Velas the reverse is true. There appears to be no systematic variation in the insoluble residue percentage in the sections at Dive Agar. Such variations in trends can be attributed to the variation in grain size, amount of bioclasts, digenetic changes and degree of cementation.

Table 1. Insoluble Residue Percentage of beachrocks.

Locality	Sample No.	Insoluble (%)	Carbonate (%)
Thal	Th1	09.29	90.71
Akshi	Ak1	31.62	68.38
	Ak2	17.27	82.73
	Ak3	10.80	89.20
Nagaon	N7	19.90	80.10
Korlai	Kor1	52.48	47.52
	Kor3	71.43	28.57
Borlai	Borl	44.48	55.52
	Bor3	74.37	25.63
Kashid	Kasl	17.94	82.06
	Kas3	29.41	70.59
Chikni	Chkl	13 .45	86.55
	Chk3	29.13	70.87
Nandgaon	Ndl	09.29	90.71
	Nd2	34.34	65.66
	Nd3	43.85	56.15
Velas	Vn1	21.24	78.76
	Vn2b	41.84	56.16
	Dla	15.55	84.46
Dive Agar	D1b	24.70	75.30
	D3	05.14	94.86
	D5	28.36	71.64
	D6	23.26	76.74
	D15	15.06	84.94
	Dl1	30.10	69.98
	D21	22.32	77.68
Aravi	Ar1	07.47	92.53
Valvati	Val1	24.35	75.65
Guhagarh	Gh1	34.31	65.69
	Gh3	05.62	94.38
	Gh4	16.38	83.62

9.4. HYDROGEOLOGY OF BEACHROCK

The data of 180 dug wells have been used to understand the hydrogeological framework of beachrocks. The seasonal fluctuation in waterlevels was recorded for winter and summer seasons between 1994 and 1997 and subsequently in the year 2001 to 2003 (Umrikar, 2004). Besides, 25 representative wells were selected to record rise in water levels due to tidal ingress during new moon days. Eleven dugwells (Thigale et al., 1998) and 25 dugwells from Guhagar (Umrikar, 2004) were also subjected to pumping test for evaluating various aquifer parameters (Table 2). The electrical conductivity values for different depths were measured during drawdown tests for obtaining information of fresh-saline water interface.

Table 2. Aquifer Performance Test (APT) results of beachrocks from select localities.

Well. No.	Location	Carbonate (%)	Specific capacity 'C' (Lpm/m)	Storativity 'S'	Transmissivity 'T' (m^2 /day)
N7	Nagaon	80.10	96	3.33×10^{-3}	34.2
D3	Dive Agar	94.86	34	0.1×10^{-3}	17.9
D15	Dive Agar	84.94	36	0.2×10^{-2}	27.3
D5	Dive Agar	71.64	326	7.91×10^{-3}	133.9
D11	Dive Agar	69.98	456	$0.96\ 10^{-3}$	156.8
W9*	Guhagarh	94.38	152	4.95×10^{-3}	28.4
W17*	Guhagarh	94.38	54	10.06×10^{-3}	24.2
W23*	Guhagarh	83.62	183	17.70×10^{-3}	31.3
Gh1	Guhagarh	65.69	125	5.82×10^{-3}	54.7
Gh3	Guhagarh	94.38	27	2.06×10^{-3}	21.2
Gh4	Guhagarh	83.62	50	1.47×10^{-3}	44.8

* Carbonate percentage from the nearest sampled well.

Beach rock intercalated sand constitute the shallow aquifer system within the beach and littoral terrace, and margins of mud flat (Fig. 2). The dug wells vary widely in shape and size and are dependent upon the location of well vis-à-vis geomorphic environment and the aquifer type. The wells are either circular or rectangular in shape. A wide variation in depth of dug wells from 2.7 to 9.6 m is noticed.

Water level fluctuations: In coastal area, water level in wells can fluctuate due to seasonal variations and seawater ingress. Seasonal variation in the groundwater is landform specific and ranges from 2.30 to 4.10 m in sandy beaches and spits and from 1.85 to 3.20 meters in the mudflat environment (Thigale et al., 1998). The rise in water level in wells located along mudflats is dependent upon the spread of seawater during high tide. The geomorphic setup and daily diurnal fluctuations in height of tides under the influence of moon control this rise. The aquifer type and mud intercalations are additional components that govern the amount of rise. The maximum rise of 1.7 m was observed in Guhagarh area on new moon day whereas minimum of 0.40 m was recorded in Akshi. The distance from the creek or tidal inlet also affects the rise of water level, the wells in close proximity show maximum rise up to 1.5 m while those located away show 0.30 m rise. The amount of subsurface seawater ingress during high tide along the sandy beaches or spit varies considerably and decreases from the high tide mark to inland.

Fresh-seawater/saline interface: Being two immiscible fluids, fresh- and saltwater develop a sharp contact but convectional dispersion and molecular diffusion cause mixing of the two and form a transitional zone of varying degree of salinity. Since water column in wells is much more in winters than in summer season. It means that since the head of the fresh water is higher than that of the saline water in winter months. Freshwater, therefore, is continuously discharged into the sea restricting the movement of seawater to enter inland. The decrease in the head of the freshwater due to lowering of water table in summer or due to excessive pumping, however, allows seawater to migrate inland. This migration of interface was observed to be occurring either by lateral or upconing process.

When electrical conductivity data obtained during pumping test were plotted as a function of depth, the graphs (Fig. 5) show straight-line relationship until the fresh-seawater interface is reached (Duraiswami et al., 2000). The interface is indicated by a sharp bend in the straight line due to sudden increase in electrical conductivity values (T 25, T 26, K0, D12). The interfaces with gentle bend (N2, R23a, R23b and B35) are diffuse and indicate mixing of fresh and seawaters.

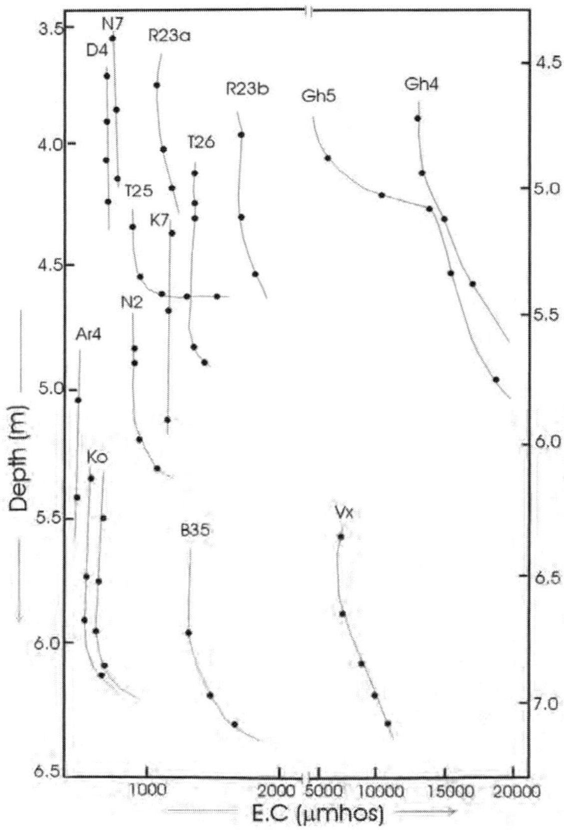

Fig. 5. Plots of depth vs. electrical conductivity values obtained during APT.

Analysis of Aquifer Performance Test (APT) test data following methods suggested by Slichter (1906) and Cooper and Jacob (1946) show wide variation in aquifer parameters (Table 2). The specific capacity, storativity and transmissivity values obtained for the beachrock aquifers range from 18 to 456 LPM/m, 0.2×10^{-2} to 2.40×10^{-2} and 28 to 1010 m^2/day respectively (Duraiswami et al., 1997; Thigale et al., 1998; Umrikar, 2004). The variation in values obtained in the present study can be attributed to several factors such as thickness of the beachrock aquifers, degree of cementation, location of wells vis-à-vis specific geomorphic environment, presence of clay/mud intercalations and influence of tides during pumping. The average carbonate percentages for samples from dug wells where Aquifer Performance Test were conducted were plotted against the aquifer parameters like specific capacity and transmissivity (Fig. 6). The parameters are negatively correlated indicating the possibility of reduction of these parameters with increasing carbonate percentage and cementation.

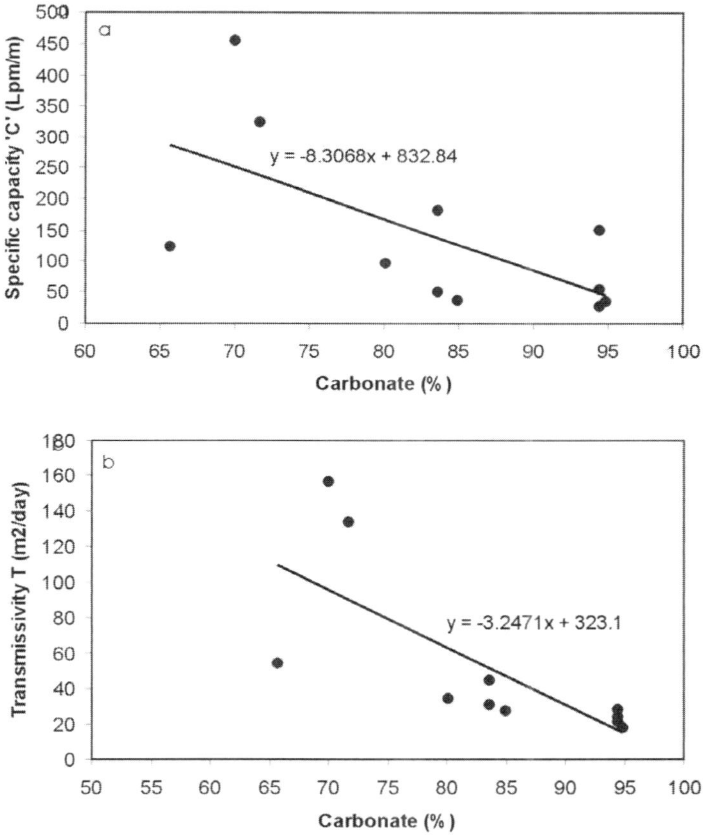

Fig. 6. Negative correlation between carbonate percentage a) Specific capacity, and b) Transmissivity in beachrock aquifers indicating strong control of cementation in reducing porosity.

9.5. WATER CHEMISTRY

Groundwater samples were collected from 100 wells for winter and summer season between 1995-1997 and 2001-2003, and analysed following standard procedures (APHA,

AWWA and APCF, 1975). The compositions of groundwater from within the beachrock aquifers are given in Table 3 and plotted in salinity hazard diagram (Fig. 7). It can be seen from the data that there is a wide variation in the salinity of groundwaters from beachrock aquifers. A systematic spatial and temporal variation in the hydrochemical facies (Thigale et al., 1996; 2004) is seen across most beaches/spits. The pH of groundwaters from the coast varies from 6.1 to 8.9 and the electrical conductivity (E.C.) varies from 410 to 19740 μmhos/cm, depending upon the geomorphic environment. Similarly, variations in alkalinity (205-1986 mg/l) and hardness (70-3228 mg/l) are also reported. The seasonal variation in the pH of the groundwater in the coastal aquifers controls the precipitation of calcium carbonate in the pore space. The Groundwater in coastal aquifers dominantly fall in Na+K>Ca+Mg; $Cl+SO_4>HCO_3+CO_3$ facies (Thigale, 2004). The variation in the chemistry can be attributed to the nature of saltwater intrusion in the coastal aquifers. Addition of potash and other organic fertilisers to the coconut and beetle nut plantations have led to the increase in the phosphate and nitrate content of the groundwaters leading to algal blooms in some dug wells tapping the beachrock aquifers. Domestic sewerage and waste has also contributed to the deterioration in the quality of groundwater (Thigale et al., 1998).

Table 3. Compositional range of groundwaters from beachrock aquifers.

Parameters	Season	Environment	
		Beach/lithoral terrace	Creek / Mudflat
pH	Summer	6.0 to 8.5	6.4 to 7.8
	Winter	7.6 to 8.8	7.8 to 8.9
Ca	Summer	8.0 to 3697.4	25.3 to 817.6
	Winter	4.7 to 1123.5	35.3 to 216.4
Mg	Summer	1 to 1229.0	9.8 to 1330.3
	Winter	1 to 426.4	1.5 to 189.7
Na	Summer	29 to 8000	17.0 to 440
	Winter	24 to 746	27 to 390
K	Summer	3 to 200	1 to 60
	Winter	1 to 49	1 to 32
Alkalinity	Summer	1 to 828	1 to 205
	Winter	118 to 1160	298 to 1986
Cl	Summer	21.3 to 17000	25.6 to 624.8
	Winter	24.9 to 2016.4	30.5 to 781.0
SO_4	Summer	7.12 to 519.5	6.68 to 249.4
	Winter	1.1 to 393	3.8 to 264.3
PO_4	Summer	1 to 280.7	0.4 to 214.2
	Winter	1 to 17.4	0 to 210.7
Reactive silica	Summer	19.83 to 59.8	5.8 to 64.9
	Winter	6.6 to 57.9	4.1 to 92.4

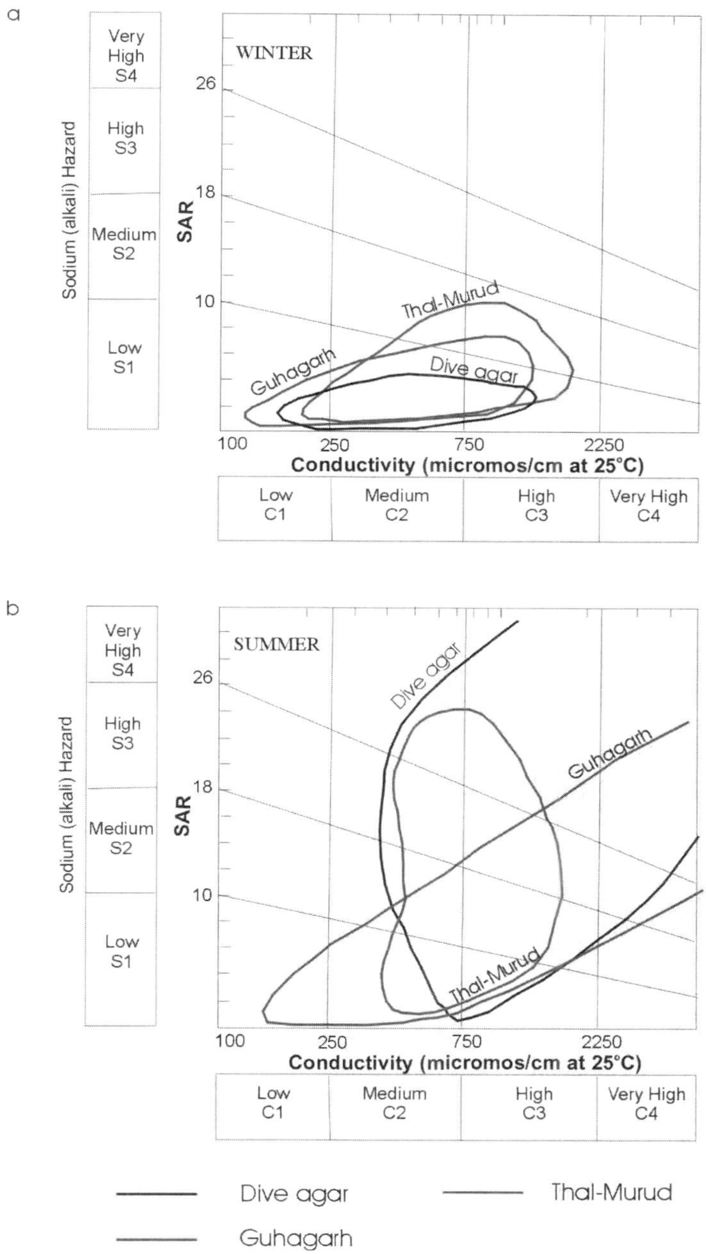

Fig. 7. Envelops showing variation in groundwater chemistry during a) winter and b) summer season in the salinity hazard diagram (USAD, 1974).

9.6. PRECIPITATION OF BEACHROCK CEMENT

The precipitation of carbonate cement in Recent tropical and sub-tropical beach sediments is wide spread and is known to occur near the water table. It is also known to be rapid and beaches are reported to get consolidated within months (Hanor, 1978). The origin of cement has been a topic of considerable debate and is attributed to various processes such as mixing of fresh and seawater, evaporation and biological activity. Groundwater flow is discharged in to the sea under a hydraulic gradient at and just below sea level. Tidal fluctuations cause the water table to oscillate vertically as well as laterally. This fluctuation is responsible for a good deal of mixing of groundwater and seawater causing precipitation of carbonates. Seasonal fluctuations further enhance the process of cementation where the precipitation takes place within the zone of fluctuation. However, the process does not account for the precipitation outside the zone of fluctuation i.e. above and below the water table.

Carbonates can be precipitated due to evaporation of seawater but such a process appears less extensive. Hanor (1978) rightly argues that continued evaporation of beach groundwater will produce a denser fluid, which should reflux seaward at depth rather than precipitate carbonates at the water table or in the inter tidal zone. Porous sedimentary sections along beaches can get saturated with carbon dioxide produced biologically by plants living in the soil zone or by algae in the beach waters. The precipitation of calcium carbonate around root moulds and within rhizoliths especially in the upper horizons of most sections lends support to this contention. Bacterial decay of organic matter washed down into the beach groundwater or present in mud found at the base of many sections may release carbon dioxide that diffuses upwards and enhances precipitation of carbonates. Although the biological contribution of carbon dioxide to the system cannot be entirely neglected the process cannot account for the complete cementation of sands to beachrocks.

In processes such as mixing of waters of different compositions or biological activity one factor is common i.e. the release of carbon dioxide. In fact, precipitation of beachrock cements due to carbon dioxide degassing from carbonate-saturated groundwater was postulated way back in the beginning of the nineteenth century (Field, 1919) and have been amply demonstrated later on by Thorstenson et al., (1972) and Hanor (1978). According to Hanor (1978), factors that could cause degassing include increase in the dispersion coefficient of dissolved carbon dioxide due to tidal oscillation, tidal pumping across sediment-atmosphere interface, decrease in the average depth of the water table and increase in the porosity from back beach clayey sediments to pure forebeach sands. Since degassing of carbon dioxide from both carbonate saturated groundwater and seawater can cause precipitation of carbonate cements, it is evident that most beach sands would be eventually consolidated. This implies that with time the storativity and transmissivity of beach aquifers would be lowered considerably. On the contrary, it is noticed that sands rich in carbonates are seldom consolidated. This can be explained by the fact that lowering of porosity due to cementation inhibits upward diffusion of carbon dioxide thereby decreasing the ability of the sediment to degas. The process thus appears to be self-limiting and accounts for the unconsolidated sands within and at the top of most beachrock sections. In the present study, interfaces of fresh-sea water were observed in the depth range of 4.90 to 6.60 meters, reason good enough to believe that the cements in beachrock were dominantly precipitated due to mixing of fresh and saline waters.

9.7. CONCLUSIONS

The shallow coastal aquifers along the Konkan coast of Maharashtra faces a serious threat of over exploitation due to demand for water for horticulture and rapid urbanisation. Excessive pumping caused due to indiscriminate drilling has lowered water tables and is responsible for wide spread mixing of fresh groundwater with seawater. This has not only led to saltwater intrusion within coastal aquifers but is also responsible for artificial human induced degassing of carbon dioxide. Moreover, pumping also interferes with the air-water balance of the pore spaces and causes precipitation of carbonates. All this reduces the porosity of the sediments and affects the storativity and transmissivity of the beachrock aquifers. Eutrification of groundwaters from beachrock aquifers due to use of fertilizers and domestic runoff aids plant growth in the soil zone as well as in wells thereby adding biological carbon dioxide to the system.

Since coastal beachrock aquifers are susceptible to saltwater ingress, it becomes imperative to understand the hydrogeological dynamics of such aquifers and to suggest suitable preventive and remedial measures. Although previous attempts have concentrated on effective remedial measures, proper understanding of the hydrogeological and hydrogeochemical characteristics of the beachrock aquifers may add to the scientific management of coastal aquifers. It is suggested that attempts to reduce saltwater ingress should involve enhancement of the hydraulic head of seaward flowing groundwater by suitable recharge techniques. Further, systematic withdrawal of groundwater by reducing pumping hours and raising of foot-valve above the fresh-saline water interface in wells that tap saltwater not only help in better management of beachrock aquifers but will also reduce the residence time of groundwater in spit environment. The low residence time of groundwater lowers the possibility of carbon dioxide degassing and precipitation of carbonate cements. This will ensure a higher storativity and transmissivity of the aquifers and provide good quality drinking water to the inhabitants.

ACKNOWLEDGEMENTS

The authors wish to acknowledge the financial assistance received from Department of Science and Technology, Government of India. Dr. D. C. Meshram is thanked for his assistance in the field.

REFERENCES

Ahmed, E., 1972. Coastal geomorphology of India. Orient Longman Ltd., New Delhi, 222p.

Anonymous, 1990. Report of the committee appointed to examine the present status of Khar Land Development, Government of Maharashtra.

APHA, A WW A and APCF, 1975. Standard methods for examination of water and waste waters. Am. Public Health Assoc. Inc. 14th Ed. New York, pp 1193.

Atkinson, S. F., Miller, G. D., Curry, D. S., and Lee, S. B., 1986. Saltwater intrusion: status and potential in the contiguous United States. Lewis Publishers Inc., Michigan, 1-390.

Cartwirght, N., 2004. Groundwater Dynamics and the Salinity Structure in Sandy Beaches. Unpubl. Ph.D. Thesis, University of Queensland, Australia.

Cooper, Jr., H. H., and Jacob, C. E., 1946. A generalized graphical method for evaluating formation constant and summat1zing well field history. Trans. Am. Geophy. Union 27(4): 525-534.

Das, S., 2004. Coastal aquifers and sea water ingress in India. Jour. Geol. Soc. Ind., v. 63: 567-568.

Dickson, J. A. D., 1965. A modified staining technique for carbonates in thin sections. Nature, v. 205, pp. 587.

Duraiswami, R. A., 1997. Aquifer characteristics of beachrocks along Coastal Maharashtra. Abst. XIV Convention of Indian Association of Sedimentologists, Chennai.

Duraiswami, R. A., Thigale, S. S., Pawar, N. J., and Meshram, D. C., 2000. Coastal environment of Maharashtra: Issues vis-à-vis development. In: A. M. Pathan and S. S. Thigale (eds.) Contribution to Environmental Geosciences. Prof. K. B. Powar Commemorative Volume, Aravali Books International, New Delhi. 59-73.

Filed, R. M., 1919. Investigations regarding calcium carbonate oozes at Tortugas and the beachrock at Loggerhead Key. Carnegie Intl. Wash., Yearbook, v. 18: 197-198.

Hanor, J. S., 1978. Precipitation of beachrock cements: mixing or marine and meteoric waters vs CO_2- degassing. Sed. Petrology 48: 489-501.

Hong, S.-H., 2005. Optimization Model for development and management of groundwater in coastal areas. Unpubl. Ph.D. Thesis, Dong-A University, Korea.

Ireland, H. A., 1971. Insoluble Residues. In: Procedures in Sedimentary Petrology. (Ed.) R. E. Carver, John Wiley and Sons, New York, 49-68.

Karlekar, S., 1993. The coastal dunes and the dune building plants of Kalbadevi spit bar. In: Coastal Geomorphology of Konkan, (Ed.) S. Karlekar. Aparna Publ. Pune, 175-180.

Mathur, K. K., and Naidu, P. R. L., 1932. Volcanic activity on the coastal tract of Bombay, Salsette and Bassein. Malviya Commem. Vol., B. H. U., 787-805.

Patil, D. N., 1998. Geomorphology and tectonics of the West Coast of India between Palghar and Thal, Maharashtra. Unpubl. Ph. D. thesis, University of Pune.

Powar, K. B., Sukhtankar, R. K., Patil, D. N., and Sawant, P. T., 1978. Geomorphology and Tectonics of West Coast of India. Tech. Report ONGC, 64 p.

Slichter, S. C., 1906. The under flow in Arkansas valley in Western Kansas. U.S.G.S. Water Supply Paper No. 153, pp. 61.

Soman, G. R., Kale, M. G., and Pawar, N. J., 1996. Diagenesis of Quaternary Karal (Beachrock) of coastal Maharashtra. Jour. Ind. Association of Sedimentologists, v. 15, no. 2, 143-154.

Thigale, S. S., 2004. An excursion through the watersheds from Konkan coast to upland western Maharashtra. Presidential Address, Earth System Science, 91st Session, Indian Science Congress, Chandigarh, 1-36.

Thigale, S. S., Pawar, N. J., Duraiswami, R. A., and Maggirwar, B., 1996. Hydrogeochemical characters of shallow aquifers in parts of coastal Maharashtra. Abst. National Seminar on tectonic Movements and their Geo-Environmental Impacts, (NETMOGEI), December 27-28, Mysore, pp 17.

Thigale, S. S., Pawar, N. J., Meshram, D. C., Duraiswami, R. A., Bhokre, M., 1998. Hydrologeological and hydrochemical studies on saline and freshwater interaction in Coastal aquifers, Maharashtra. Report Department of Science & Technology Project (No. ESS/23/198/94).

Thorstenson, D. C., Mackenzie, F. T., and Ristvet, B. L., 1972. Experimental vadose and phreatic cementation of skeletal carbonate sand. Jour. Sed. Petrology, v. 42, 162-167.

Tran, M. T., 2004. Multi-Objective Management of Saltwater Intrusion in Groundwater: Optimization under Uncertainty. Unpubl. Ph.D. Thesis, Delft University of Technology, the Netherlands.

Tucker, M., 1988. Techniques in Sedimentology. Blackwell Scientific Publication. Oxford London, pp. 394.

Umrikar, B. N., 2004. Hydrogeology of Guhagarh area, District Ratnagiri, Maharashtra with special reference to watershed management. Unpubl. Ph.D. thesis, University of Pune, India.

10

SALTWATER ENCROACHMENT IN REGIONAL DRAINAGE BASINS OF SOUTHERN LOUISIANA: AN UPDATE

Thomas P. Van Biersel and L. Riley Milner
Louisiana Geological Survey - Louisiana State University
3079 Energy, Coast and Environment Building, Baton Rouge, LA 70803, USA

ABSTRACT

The principal regional rivers of Southern Louisiana are exhibiting changes associated with the encroachment of saltwater within their estuaries and upstream. This migration upstream is likely to have future negative localized impacts on the recharge zones of Louisiana's principal coastal aquifer system, where surficial clays are thin or non-existent. The threatened areas of the state include the Chicot Aquifer near Lafayette in the southwest, the Mississippi River Alluvial Aquifer in the south-central, and the Norco, Gonzales, New Orleans, and Upper Ponchatoula aquifers of the Southern Hills Aquifer System along the north shores of Lakes Pontchartrain and Maurepas in the southeast.

Although time-biased corrected precipitation values for the southern third of Louisiana indicate that precipitation in the last 110 years has steadily increased, several rivers (including the Mermentau, Calcasieu and Tchefuncte Rivers) are showing a decreasing trend in water discharge overtime. This is in contradiction with the overall decrease of both ground and surface water usage in Southwest Louisiana. In Southeast Louisiana, the water usage trend is the opposite; showing increased usage of both ground and surface water. However, the primary source of surface water is the Mississippi River, which was not surveyed as part of this study.

Chloride concentrations in the coastal drainage basins indicate that saltwater is moving landward throughout Southern Louisiana. Risks to groundwater recharge zones are more evident along the north shore of Lake Pontchartrain, along the Vermilion River, and in the Mississippi River alluvial valley. In addition, saltwater appears to have migrated upstream of the saltwater barrier located north of Lake Charles along the Calcasieu River.

10.1. INTRODUCTION

The 1993 Nonpoint source pollution assessment report (DEQ, 1993) by the Louisiana Dept. of Environmental Quality (DEQ) identifies saltwater intrusion as occurring in the Atchafalaya River Basin, the Barataria Basin, the Lake Pontchartrain Basin, the Mermentau River Basin, the Vermilion-Teche River Basin, the Sabine River Basin and the Terrebonne Basin (Figure 1). Saltwater intrusion in coastal Louisiana is not a new issue. In the past 50 years, the U.S. Army Corps of Engineers (USACE) has installed and monitored saltwater barriers located on the Calcasieu and Mermentau Rivers. Within the Mississippi River channel, the USACE has been monitoring and mitigating the migration of a saltwater

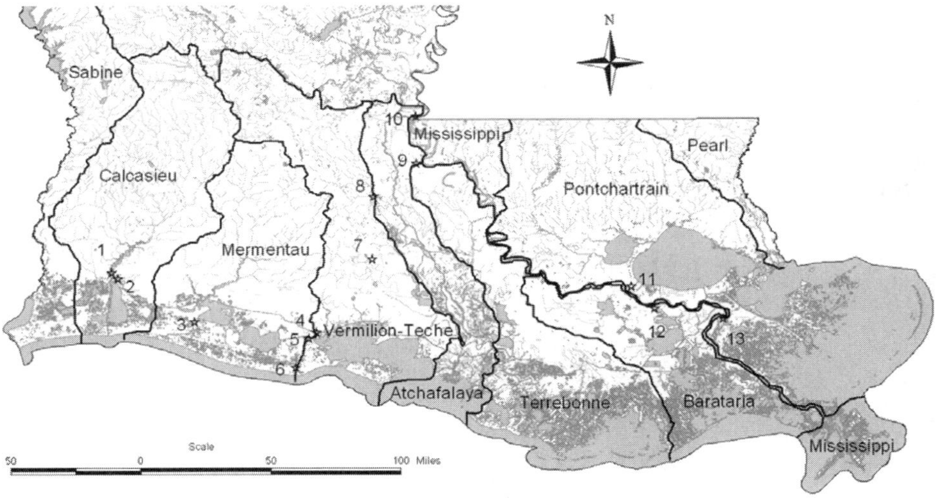

Fig. 1. Watershed of Southern Louisiana. The USACE control structure are (1) Calcasieu River and Pass salt water Barrier; (2) Calcasieu Lock; (3) Catfish Point Control Structure; (4) Leland Bowman Lock; (5) Schooner Bayou Control Structure; (6) Freshwater Bayou Lock; (7) Ruth Canal; (8) West Atchafalaya Basin Project; (9) Morganza Spillway; (10) Old River Control Structure; (11) Bonnet Carré Freshwater Diversion Structure; (12) Davis Pond Diversion Structure; and (13) Caernarvon Freshwater Diversion Structure. GIS layers are from the "Louisiana GIS CD: A Digital Map of the State" version 2.0 (Office of the Governor, 1999).

wedge up river for many years. These intrusions have caused problems for the City of Houma in the past, and may cause problems to the City of New Orleans in the future (DEQ, 1993). Similarly, a statistical analysis of salinity data collected in Louisiana's estuaries (Wiseman et al., 1990) shows that there are small, significant trends in salinity changes, although without spatial pattern.

The study focuses predominantly on the Pontchartrain, Calcasieu, Mermentau, and Vermilion-Teche basins (Figure 1). The Mississippi River, Atchafalaya, Terrebonne and Barataria Basins, because of the heavily engineered nature of these systems, will not be addressed in this paper. The flow in these basins is altered/controlled by the USACE through control structures (Figure 1) along the Mississippi (USACE, 1999, and USACE, undated). The Old River Control structures, and under some circumstances the Morganza Spillway, are used to divert the Mississippi River water into the Atchafalaya River. Similarly, Mississippi River water has been diverted into the Lake Pontchartrain Basin through the Bonnet Carré and Caernarvon Freshwater Diversion structures. The Barataria Basin receives Mississippi River water through the Davis Pond Freshwater Diversion structure. Other control structures built by the USACE divert freshwater throughout Southern Louisiana. Water from the Atchafalaya River is diverted into Bayou Teche, and, in turn, water from Bayou Teche (Ruth Canal) is diverted into the Vermilion River

(Saucier, 1998). On the Mermentau, five structures have been installed on the lower part of the drainage basin to prevent saltwater intrusion (Figure 1). We did not research the Sabine and Pearl River basins that border the State of Louisiana to the west and east, respectively.

The USACE monitors the salinity of the water in the Gulf Intracoastal Waterway (GIWW), Grand Lake and Lake Pontchartrain. Within the Mississippi River, the USACE has been monitoring the movement of a saltwater wedge which has migrated upriver from the mouth of the river (http://www.mvn.usace.army.mil/eng/saltwater/wedge_now.htm).

10.2. METHODOLOGY

The chloride data used in this study was obtained from the U.S. Geological Survey's National Water Information System (NWIS) web site (http://waterdata.usgs.gov/nwis/), the USACE office in New Orleans, and the DEQ's Water Quality Assessment Division. The database includes data collected between 1905 and present. The main difference between the USGS data and that of the DEQ and USACE comes from the sampling method. The USGS samples are composite samples of the channel, while the DEQ's and USACE's are grab samples for a specific depth, primarily ~1-foot. The stream discharge data was obtained from the USGS NWIS web site.

The data was plotted and linear trend lines generated to observe the long term change in chloride concentration. This information was then compared to surface water discharge data, water usage and precipitation data. As mentioned above, the surface water discharge data was obtained from the USGS. The gaging station on the subject river with the most continuous record was used to evaluate the long term changes in discharge. The monthly precipitation data used is the National Oceanic and Aeronautic Administration (NOAA) National Climatic Data Center's time-biased precipitation for the regions comprising Southern Louisiana, including the State Divisions # 6, 7, 8 and 9, the Florida Parishes, Southwest Louisiana, Lafayette and New Orleans, respectively (http://www.ncdc.noaa.gov/oa/climate/climatedata.html#monthly). An average quarterly precipitation for Southern Louisiana was calculated for the time period of interest. The 120 years of record was used to determine an arithmetic mean quarterly precipitation value, which was used to calculate a precipitation differential. This enabled the identification of extended periods of low precipitation. The precipitation data is presented on Figure 2. The water usage data was obtained from the USGS's Water Use in Louisiana 2000 (Sargent, 2002). The plot of the cumulative ground and surface water usage for southern Louisiana parishes is presented in Figure 3.

The southern most gaging and water sampling station were used for all the rivers of interest in this study. Chemical and flow hydrographs were prepared to identify trends in the data. Concurrently, aerial plots of the extent of saltwater intrusion pre- (1,096 DEQ and 10,040 USGS chloride values) and post-1980 (2,764 DEQ and 7,932 USGS chloride values) were generated using GIS software. The plots are presented on Figures 4 and 5. A summary of the observed changes in river discharge is included in Table 1. The percent change in discharge was calculated from a linear regression of the data set. A summary of the observed changes in chloride concentration is included in Table 2. The percent change in chloride concentration was calculated from a linear regression of the data set. Figure 6 locates the sampling and gaging stations identified on Tables 1 and 2.

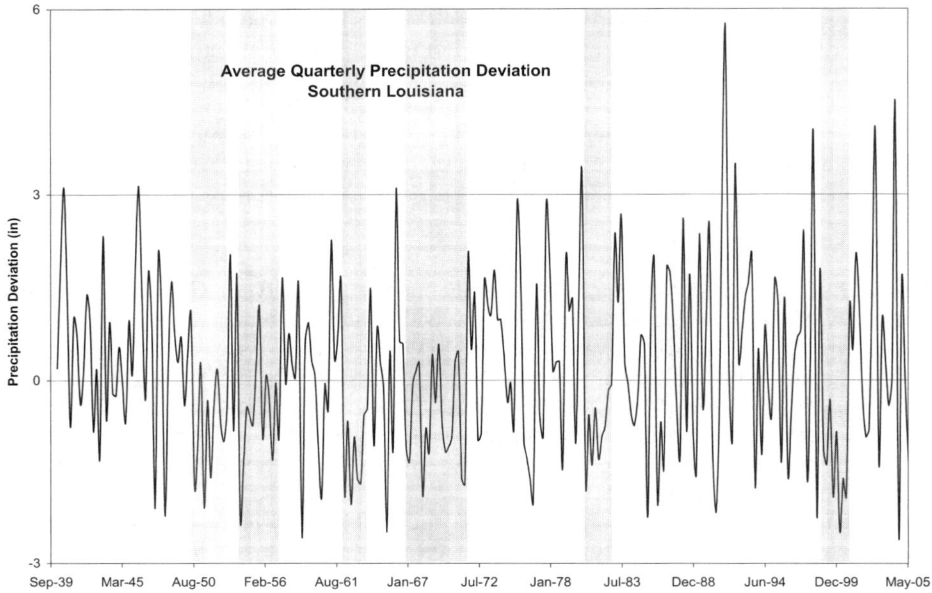

Fig. 2. Precipitation deviation for the mean quarterly precipitation for Southern Louisiana. The shaded areas represent prolonged periods (three successive quarters) at or below average precipitation.

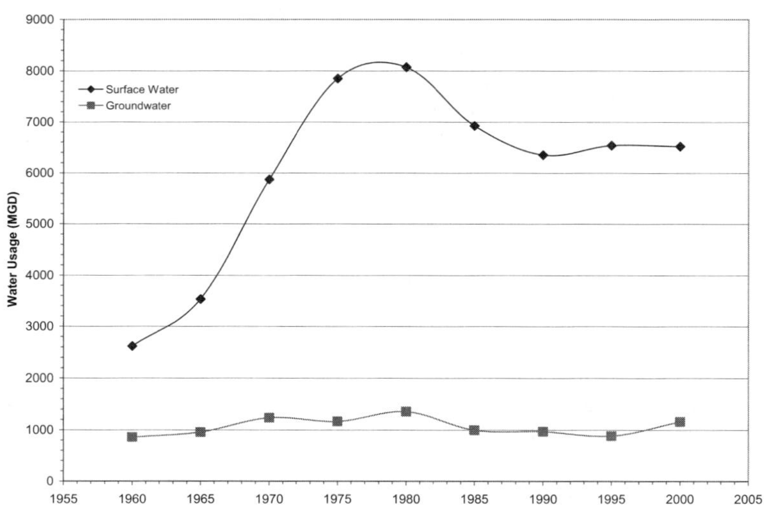

Fig. 3. Surface and ground water usage in southern Louisiana (data from Sargent, 2002).

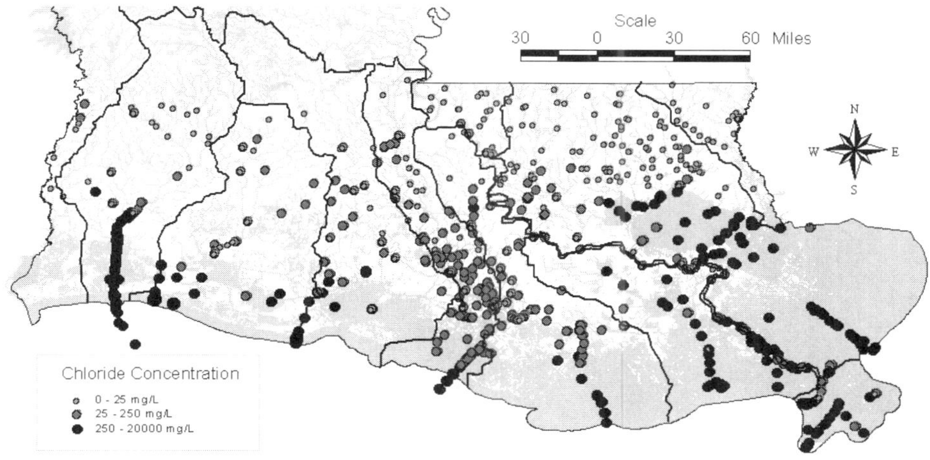

Fig. 4. Pre-1980 chloride concentration in surface water. GIS layers are from the "Louisiana GIS CD: A Digital Map of the State" version 2.0 (Office of the Governor, 1999).

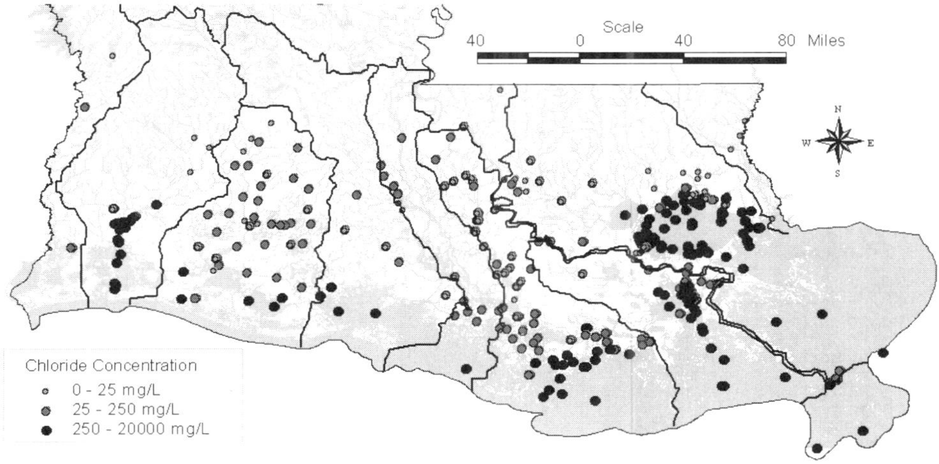

Fig. 5. Post-1980 chloride concentrations in surface water. GIS layers are from the "Louisiana GIS CD: A Digital Map of the State" version 2.0 (Office of the Governor, 1999).

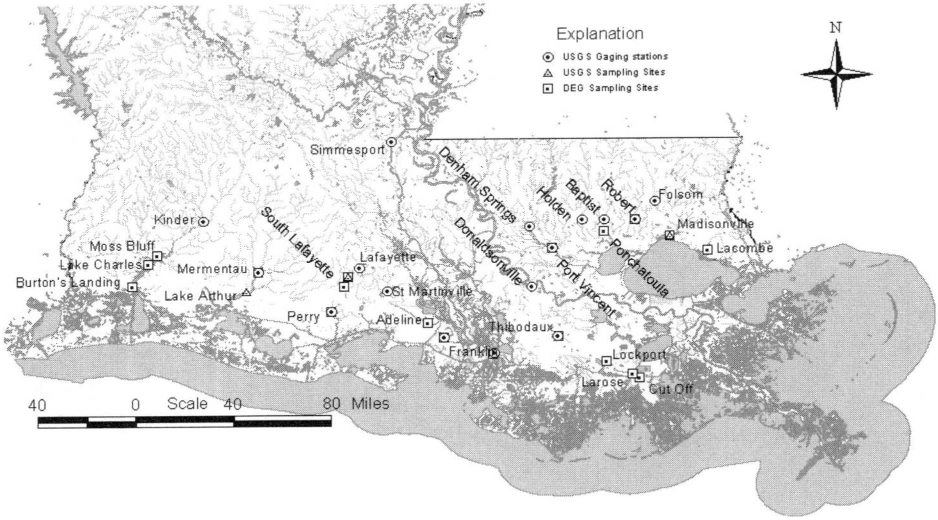

Fig. 6. Gaging stations and sampling sites referred in the text and Tables 1 and 2. GIS layers are from the "Louisiana GIS CD: A Digital Map of the State" version 2.0 (Office of the Governor, 1999).

Table 1. Summary of the observed changes in river discharge at selected USGS gaging stations

Basin	River	Gaging Station	Period of Record	Change in Flow
Calcasieu	Calcasieu	Kinder	80 yrs (1922-2002)	-6.4%
Mermentau	Mermentau	Mermentau	18 yrs (1984-2002)	-12%
Vermilion-Teche	Vermilion	Perry	29 yrs (1973-2002)	-160%
Vermilion-Teche	Vermilion	Lafayette	35 yrs (1967-2002)	+12%
Vermilion-Teche	Bayou Teche	St. Martinville	43 yrs (1959-2002)	-5%
Vermilion-Teche	Bayou Teche	Franklin	8 yrs (1984-1992)	+74%
Atchafalaya	Atchafalaya	Morgan City	7 yrs (1995-2002)	-9%
Terrebonne	Bayou Lafourche	Donaldsonville	25 yrs (1960-1985)	-41%
Terrebonne	Bayou Lafourche	Thibodaux	18 yrs (1984-2002)	+90%
Pontchartrain	Tchefuncte	Folsom	59 years (1943-2002)	-26%
Pontchartrain	Tangipahoa	Robert	64 yrs (1938-2002)	+15%
Pontchartrain	Amite	Denham Springs	64 yrs (1938-2002)	+33%
Pontchartrain	Amite	Port Vincent	18 yrs (1984-2002)	-32%
Pontchartrain	Tickfaw	Holden	62 yrs (1940-2002)	+41%
Pontchartrain	Natalbany	Baptist	59 yrs (1943-2002)	+8%

Table 2. Summary of the observed changes in chloride concentration at selected USGS and DEQ sampling sites

Basin	River	Gaging Station	Period of Record	Change in Conc.	Source
Calcasieu	Calcasieu	Kinder	51 yrs (1943-1994)	-6%	USGS
Calcasieu	Calcasieu	Moss Bluff	47 yrs (1958-2005)	-51%	DEQ
Calcasieu	Calcasieu	Lake Charles	27 yrs (1971-1998)	-13%	DEQ
Calcasieu	Calcasieu	Burton Landing	34 yrs (1971-2005)	+26%	DEQ
Mermentau	Mermentau	Mermentau	47 yrs (1958-2005)	-74%	DEQ
Mermentau	Mermentau	Mermentau	54 yrs (1949-2003)	+10%	USGS
Mermentau	Mermentau	Lake Arthur	51 yrs (1949-2000)	-9%	USGS
Vermilion-Teche	Vermilion	Perry	47 yrs (1958-2005)	-28%	DEQ
Vermilion-Teche	Vermilion	Perry	9 yrs (1965-1974)	+24%	USGS
Vermilion-Teche	Vermilion	Lafayette	17 yrs (1969-1986)	-45%	USGS
Vermilion-Teche	Vermilion	Lafayette	35 yrs (1963-1998)	-91%	DEQ
Vermilion-Teche	Vermilion	South Lafayette	7 yrs (1991-1998)	+221%	DEQ
Vermilion-Teche	Bayou Teche	Adeline	27 yrs (1978-2005)	-34%	DEQ
Vermilion-Teche	Bayou Teche	Franklin	45 yrs (1958-2003)	-64%	DEQ
Vermilion-Teche	Bayou Teche	Franklin	17 yrs (1975-1992)	-31%	USGS
Atchafalaya	Atchafalaya	Morgan City	27 yrs (1978-2005)	-32%	DEQ
Atchafalaya	Atchafalaya	Morgan City	43 yrs (1956-1999)	-38%	USGS
Terrebonne	Bayou Lafourche	Lockport	7 yrs (1991-1998)	+8548%	DEQ
Terrebonne	Bayou Lafourche	Larose	32 yrs (1972-2004)	+1873%	DEQ
Terrebonne	Bayou Lafourche	Cut Off	12 yrs (1978-1990)	+276%	DEQ
Pontchartrain	Tchefuncte	Madisonville	27 yrs (1978-2005)	+212%	DEQ
Pontchartrain	Tchefuncte	Madisonville	41 yrs (1958-1999)	+59%	USGS
Pontchartrain	Tangipahoa	Robert	47 yrs (1958-2005)	-18%	DEQ
Pontchartrain	Tangipahoa	Robert	56 yrs (1943-1999)	+35%	USGS
Pontchartrain	Amite	Port Vincent	23 yrs (1978-2001)	-49%	DEQ
Pontchartrain	Amite	Port Vincent	36 yrs (1958-1994)	-94%	USGS
Pontchartrain	Tickfaw	Holden	21 yrs (1955-1976)	+7%	USGS
Pontchartrain	Natalbany	Baptist	26 yrs (1943-1969)	+26%	USGS
Pontchartrain	Natalbany	Ponchatoula	10 yrs (1991-2001)	-36%	DEQ
Pontchartrain	Bayou Lacombe	Lacombe	10 yrs (1991-2001)	+6%	DEQ
Pontchartrain	Bayou Lacombe	Lacombe	30 yrs (1969-1999)	+532%	USGS

10.3. FINDINGS

The southwestern part of Louisiana is heavily farmed. Agriculture, including rice farming, taps both ground and surface water sources. Within this region of Louisiana, we focused on the Calcasieu, Mermentau and Vermilion Rivers, as well as Bayou Teche. The southeast portion of the state, also known as the Florida Parishes, encompasses the Lake Pontchartrain Basin. The Amite, Natalbany, Tangipahoa and Tchefuncte Rivers, and Bayou Lacombe, are all located in the Lake Pontchartrain Basin. The Lake has been becoming saltier (Figure 7) in the recent past, as the openings between Lakes Pontchartrain and Borgne have been altered. The Lake Pontchartrain Basin lies in southeastern Louisiana, which encompasses one of the fastest growing urbanized areas of Louisiana, and is a large user of ground water as its water supply.

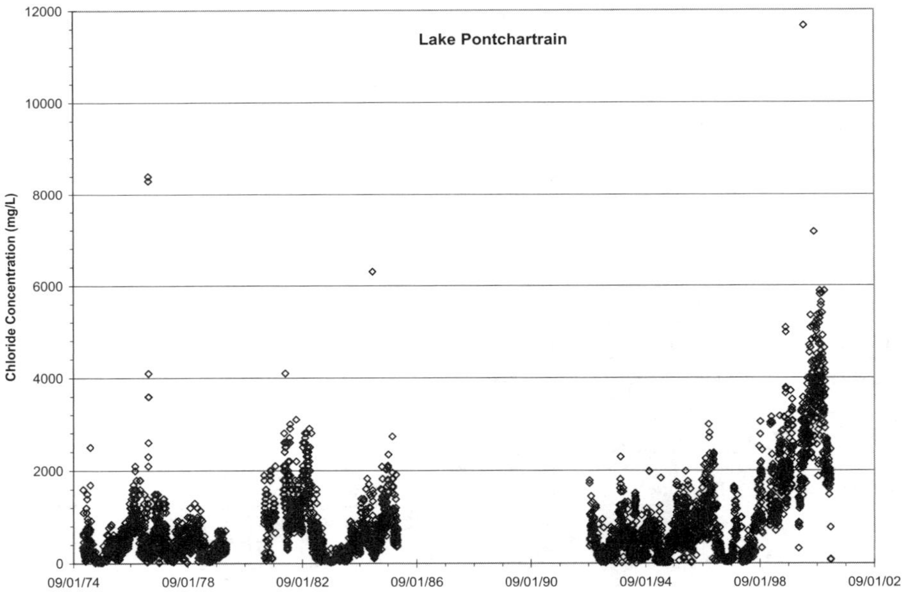

Fig. 7. USACE chloride data for Lake Pontchartrain.

10.3.1. Calcasieu River

The discharge of the Calcasieu River has slightly declined north of the saltwater barrier (Figure 1). Near Kinder, Louisiana a decrease of ~6.4 % has been recorded (Table 1). The chloride concentrations have remained relatively stable in the northern half of the basin. Directly north of the saltwater barrier, elevated concentrations have been measured near Moss Bluff (Figure 6), and farther upstream. The concentrations at Moss Bluff have been steadily decreasing (-51%), however, the highest concentration (5,160 mg/L) on record has occurred in the last 5 years. Downstream of the barrier, concentrations near Lake Charles have been decreasing, most likely the result of the addition of freshwater through the diversions. Closer to the mouth of the river, into Lake Calcasieu, at Burton's Landing, concentrations have been steadily increasing (Table 2).

10.3.2. Mermentau River

Similar to the Calcasieu River, the discharge of the Mermentau River exhibits a decrease (-12%) in the recent past. The chloride concentrations in Mermentau collected by the USGS indicate that concentrations are increasing (+10%), in comparison to those of the DEQ, which shows a decline (-74%). This may be indicative of a saltwater wedge moving up river that would not be detectable in the surface samples collected by the DEQ. Downstream, near Lake Arthur, a decrease in concentration can be observed, possibly the result of the freshwater structures installed by the USACE. Bayou Queue, which drains into the Mermentau between these two locations, is showing a decrease in discharge.

10.3.3. Vermilion River and Bayou Teche

The Vermilion River and Bayou Teche have been the subject of engineered alterations in the recent past. Although the discharge of the Vermilion in Lafayette has been helped (+12%) by freshwater diversions, by the time the river reaches Perry, the current flow trend shows a ~160% reduction in flow. The chloride concentrations have been helped by the addition of freshwater; however, no recent stream composite data is available for Lafayette and Perry. Surface samples show a general chloride decrease over time at Perry and Lafayette; however, a DEQ sampling site south of Lafayette recorded an increase in the last ten years. Bayou Teche exhibits a similar trend to that of the Vermilion River. Chloride concentration is generally lower (Table 2) and discharge shows a slight decrease (-5%) downstream from Ruth Canal (Figure 1), but an overall increase (+74%) near the outlet. This variability is the result of freshwater diversions in the basin.

10.3.4. Amite River

The Amite River is showing an increase in discharge, which is understandable as East Baton Rouge, Livingston and Ascension Parishes are rapidly growing. Near Port Vincent, 25 miles upstream from Lake Maurepas, chloride concentrations are decreasing. The Blind River, which is a tributary of the Amite, shows a similar decrease near Gramercy. Near the outlet of the Amite River in Lake Maurepas, the DEQ has recorded chloride concentrations as high as 1,327 mg/L. These concentrations are similar to the levels in Lake Maurepas, which have been increasing.

10.3.5. Natalbany River

The Natalbany River, like the Amite, discharges into Lake Maurepas. The discharge of the Natalbany near Baptist exhibits an 8% increase. Chloride concentrations measured prior to 1969 by the USGS exhibited a 26% increase; however there are no recent values. Near Ponchatoula, the DEQ data exhibits a 36% reduction in the past 15 years in surface samples.

10.3.6. Tangipahoa River

The Tangipahoa River discharges into Lake Pontchartrain. Flow measurements made near Robert show that discharge has increased by ~15%. However, composite samples' chloride concentration measured at that gaging station show a ~35% increase. The surface samples exhibit a slight decrease (~18%). This difference may again be due to a saltwater wedge migrating upstream of Lake Pontchartrain.

10.3.7. Tchefuncte River

The Tchefuncte River discharges into Lake Pontchartrain in Madisonville. The river's discharge, which is monitored by the USGS near the Folsom, shows a 26% decline over the past 59 years. At Madisonville, chloride concentrations have substantially increased, and anecdotal information indicates that saltwater fishes are being caught farther up river.

10.3.8. Bayou Lacombe

Bayou Lacombe discharges into Lake Pontchartrain between Mandeville and Slidell. Although the river is not gaged, DEQ water samples indicate that over the last 10 years chloride concentrations in the river have increased 6% at Lacombe, which is 4.5 miles upstream of the outlet. USGS samples collected earlier indicate a much larger increase (+531%); however, these samples were more temporally scattered.

10.3.9. Lakes Pontchartrain and Maurepas

Lake Pontchartrain is an estuarine lake connected to the Gulf of Mexico by narrow outlets to Lake Borgne. It is connected to Lake Maurepas by Pass Manchac. Ongoing monitoring of the lakes and surrounding marshes have shown that the salinity of the lakes is increasing (Figure 7). A similar trend can be seen in Lake Maurepas; however surficial sample data available for both Pass Manchac (between Lakes Maurepas and Pontchartrain) and the Rigolets (between Lakes Pontchartrain and Borgne) exhibit a decreasing trend. This contradiction is most likely an artifact of the sampling methods. Sampling during the 1980s and 1990s by the USGS indicate an increasing trend within the passes.

10.3.10. Mississippi River

As indicated earlier in this paper, a saltwater wedge is migrating up the Mississippi River channel. The latest information provided by the USACE indicates that the toe of the wedge has migrated approximately 42 miles upstream of the mouth of the river (http://www.mvn.usace.army.mil/eng/saltwater/wedge_now.htm). A survey of the chloride data available on the lower Mississippi indicates that there has been a general increase in concentrations between St Francisville and Belle Chasse; however, this increased salinity is likely of anthropogenic sources. In the delta, near Pointe a la Hache, sampling by the DEQ suggests that chloride concentrations are increasing, which is consistent with sampling by the USGS near Venice, which similarly exhibit an increase.

10.3.11. Atchafalaya River

Within the Atchafalaya Basin, chloride concentration decreased at the Melville, Krotz Springs, Simmesport and Morgan City sampling sites. These decreases are consistent with the significant increase in flow observed at Simmesport, which is located near the Old River Structures (Figure 1). Near Morgan City, the discharge exhibits a slight decrease in flow.

10.3.12. Bayou Lafourche

Bayou Lafourche is located in the Terrebonne Basin. It receives some Mississippi discharge through a pump/siphon station near Donaldsonville (http://www.lacoast.gov/reports/index.htm). Chloride concentrations in Bayou Lafourche show a slight increase at the Donaldsonville and Thibodaux sampling sites, and a significant increase at the Lockport, Larose and Cut Off sites. The Raceland site exhibits a slight decrease, while the Golden Meadow site, which is the closest to the Gulf of Mexico, exhibits a large decrease, as well as greatly scattered data.

10.4. CONCLUSIONS

Based on the results of this study, it can be determined that saltwater intrusion remains a problem in the drainage basins of southern Louisiana. In the southwest part of the state, the Calcasieu, Mermentau and Vermilion Rivers surveyed exhibit a decreasing trend in their discharge near their outlets. Only Bayou Teche exhibits a gain near its outlet. Increased salinity (elevated chloride levels) was observed in the outlet of the Calcasieu River. Although concentrations decreased north of Burton's Landings, the concentration remained high (>250 mg/L) north of the saltwater barrier. The Vermilion exhibits a similar trend, with increase near Perry and decrease upstream. The Mermentau, in comparison, exhibits a decrease in chloride concentration at its outlet near Lake Arthur, and an increase upstream near Mermentau. Within the Lake Pontchartrain Basin, both the Amite and Tchefuncte Rivers exhibit a decrease in discharge, while the Tickfaw, Tangipahoa and Natalbany exhibit a gain. The Tchefuncte, Bayou Lacombe and Tangipahoa exhibit increased chloride concentrations. The Natalbany and Tickfaw showed a decline, however, there was no sampling station near the mouth of these rivers.

There is no direct correlation between decreased surface water discharge and saltwater intrusion at all the stations (Figure 8). This observation is normal, since flow in most of the rivers, with the exception of those in the Lake Pontchartrain, have their flow manipulated to some extent. It can be observed that the highest increase in chloride concentration recorded was in Bayou Lafourche, which exhibits the largest flow increase (+90%). However, it should be noted that the increases are farther down stream than the last gaging station with substantial record. Some of the discrepancies in the correlations may also be the results of the use of both the surface grab (DEQ and USACE) and composite (USGS) samples. It is more likely that the composite samples detected saltwater intrusion than the surface grab samples. This could be due to the likelihood that the saltwater is moving up the river basins as a wedge along the channel bottom.

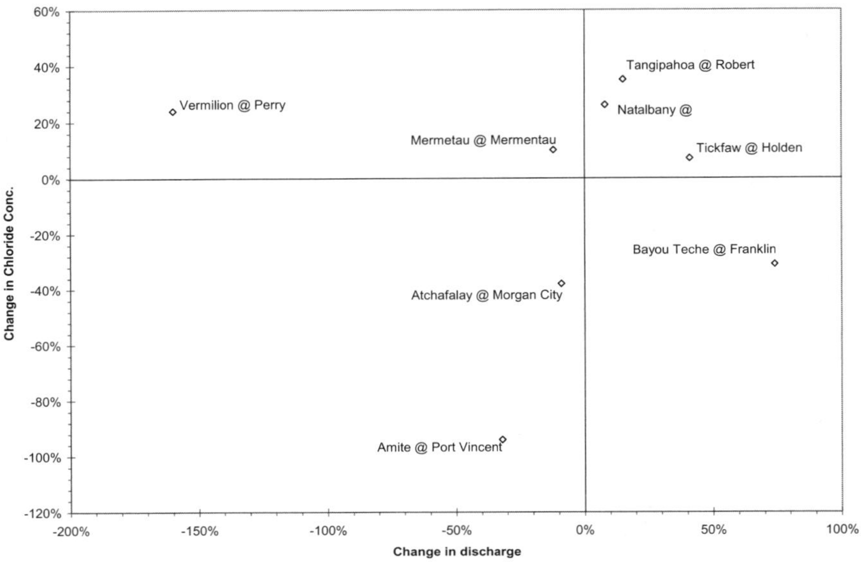

Fig. 8. Plot of the change in discharge versus the change in chloride concentration for station near the river outlet with both sets of data.

ACKNOWLEDGEMENTS

The authors would like to thank the USGS's NWIS, the USACE's office in New Orleans, NOAA's National Climatic Data Center and the DEQ's Water Quality Assessment Division for making their data available.

REFERENCES

DEQ (Louisiana Dept. of Environmental Quality), 1993. State of Louisiana Nonpoint source pollution assessment report November 1993, LDEQ Office of Water Resources' State of Louisiana Water Quality Management Plan Nonpoint Source Pollution Assessment Report, Volume 6, Part A.

Office of the Governor, 1999. Louisiana GIS CD: A Digital Map of the State version 2.0, State of Louisiana, Baton Rouge.

Sargent, P., 2002. Water Use in Louisiana, 2000, Louisiana Dept. of Transportation and Development Water Resources Special Report No. 15, pp. 133.

Saucier, M., 1998. Water Resources Development in Louisiana 1998, U.S. Army Corp of Engineer, New Orleans, pp. 177.

USACE, 1999. Old River Control, U.S. Army Corps of Engineers, New Orleans, pp. 14.

USACE, undated. Freshwater Diversion, U.S. Army Corps of Engineers, New Orleans, pp. 19.

Wiseman, W., Swenson, E., and Power, J., 1990. Salinity trends in Louisiana Estuaries, Estuaries 13(3): 265-271.

11

UPDATED GEOLOGY AND SALTWATER INTRUSION FOR THE CHICOT AQUIFER OF SOUTHWEST LOUISIANA

L. Riley Milner and Thomas P. Van Biersel
Louisiana Geological Survey, Louisiana State University
3079 Energy, Coast, and Environment Building, Baton Rouge, LA 70803, USA

ABSTRACT

The Chicot Aquifer has been designated by the USEPA as a "sole source aquifer." Groundwater quality issues identified in the last 10 years resulted in the need for a new geologic characterization of the aquifer, and of the ongoing saltwater intrusion. The Chicot Aquifer has historically been subdivided into several sub-aquifers. A recently developed geologic model indicates that in the northern parishes the freshwater sands of the Chicot are hydraulically connected. Moving to the southwest, the two upper freshwater sands are separated by a thick and more laterally-continuous, confining clay, while the deeper sands are separated by thinner and less continuous clay. In addition, these sands contain little freshwater. In the central portion of the aquifer, the confining clays are discontinuous, and the freshwater sands are lumped into the Chicot's undifferentiated area. In this portion of the Chicot, the freshwater zone includes discontinuous layers of clays; however the deeper portion of the aquifer contains no freshwater. To the east, the Chicot is subdivided to an upper and lower designation. However, this division is arbitrary, as there are no extensive confining clays in this portion of the aquifer. Overall, the base of the freshwater zone of the Chicot Aquifer becomes shallower to the east and south. Due to faulting and other structural features, there are anomalously deeper base locations of freshwater in the southeastern area of the Chicot Aquifer. Based on available chloride data, it can be observed that the fresh/saltwater interface has steadily migrated landward and upward into the aquifer.

11.1. INTRODUCTION

The Chicot Aquifer is a confined aquifer in southwestern Louisiana, and has been designated by the USEPA as a "sole source aquifer." Groundwater quality issues identified in the last 10 years, such as drought and increased demand, resulted in the need for a new geologic characterization of the aquifer. The drought in the late 1990's resulted in farmers finding saltwater flowing into their freshwater wells in St. Landry Parish, and some coastal areas of southwest Louisiana. This prompted the Louisiana Legislature to form a water commission and advisory group to address the water issues in the state. The Louisiana Geological Survey (LGS) and Louisiana State University's, Department of Civil and Environmental Engineering were asked by the Louisiana Department of Transportation and Development (DOTD) Water Resources Division to study the geologic characterization, and to model the Chicot Aquifer system in southwest Louisiana (Figure 1).

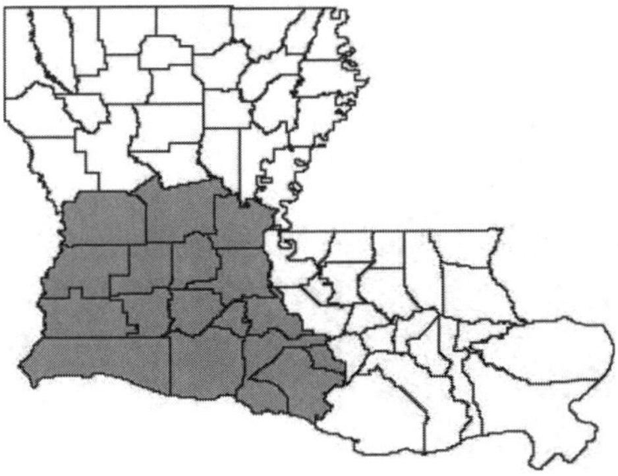

Fig. 1. Study area location map of Chicot Aquifer southwest Louisiana.

11.2. METHODOLOGY

Information and published data from LGS reports and bulletins, United States Geological Survey (USGS) technical reports and data files, DOTD technical reports and water-well registration database, Louisiana Department of Natural Resources (DNR) oil and gas well log library, and published reports in scientific journals and bulletins were used in this study. The geological characterization of the Chicot Aquifer was accomplished using a combination of 1,011 oil and gas exploration geophysical resistivity well logs, and 225 water supply geophysical and drillers' logs. The water-well logs were the primary source of information for mapping the upper 500 feet of the subsurface, because many of the oil and gas logs typically did not extend to the land surface. These logs were used to identify the hydrostratigraphic, lithologic, and hydrochemical facies (freshwater and saltwater) of the Chicot Aquifer. The freshwater/saltwater interface, or the beginning of the transition zone into more saline water, resulted in a significant reduction in resistivity displayed on the geophysical logs. The approximate position for the base of the freshwater was determined from the resistivity curve, using 40 ohms as the cut-off point for freshwater. This value, due primarily to the Formation Factor of the Chicot Aquifer, is a conservative TDS value of approximately 900. Each of the sand and clay units was identified with a minimum thickness of 20-feet. Sand or clay layers (20 feet thick or less) were grouped with the unit above and below the thinner layer.

The chloride and well head water level data used in this study were obtained from the USGS's National Water Information System (NWIS) web site (http://waterdata.usgs.gov/nwis/). The data from both data sets were then plotted, and a linear trend line generated to observe the long-term change in chloride concentration and head level. The monthly precipitation data used is the National Oceanic and Aeronautic Administration (NOAA) National Climatic Data Center's time-biased precipitation for Southwest Louisiana (http://www.ncdc.noaa.gov/oa/climate/climatedata.html#monthly).

11.3. GEOLOGIC CHARACTERIZATION

The USGS has classified the southwestern Louisiana Chicot Aquifer as part of the Coastal Aquifer System (CAS) (Renken, 1998). The CAS covers the Gulf Coast regions of Texas,

Louisiana, Mississippi, Alabama, and the western Florida panhandle. The CAS consists of a gulfward thickening, heterogeneous, unconsolidated- to poorly-consolidated wedge of discontinuous beds of gravel, sand, silt, and clay (Renken, 1998).

The Evangeline Aquifer of Pliocene and Miocene Epoch underlies the Quaternary Period alluvial deposits of the Chicot Aquifer. The Chicot Aquifer has a graveliferous texture that differentiates it from the underlying Evangeline Aquifer, which is composed of a deltaic sequence of fine-to-medium sand interbedded with silt, soft-to-moderately hard laminated clay, and local beds of coarse sand (Jones, et al., 1954). Thin clay beds separate the Chicot and Evangeline aquifers. However, in many areas, these clay beds are missing, and the uppermost sands of the Evangeline Aquifer are in direct contact with the lowermost sands and gravels of the Chicot Aquifer (Whitfield, 1975). The base of the Chicot Aquifer deposits are not readily distinguishable from the underlying Evangeline Aquifer in the southern part of the study area, because the early Pleistocene deposits are gradational from a graveliferous deposit into sand, silt, and clay gulfward (Jones, et al., 1954).

The Chicot Aquifer is interbedded with calcareous clay and other clays, silt, sand, and gravel. These dip southeast, ranging in thickness from a few feet to greater than 2,000 feet thick, forming a wedge to the Gulf Coast. The wedge then thins offshore into the deeper waters beneath the Gulf of Mexico. The dip of the Chicot Aquifer system is approximately 20 ft /mi. (Smoot, 1989). Fisk (1944) described the Chicot Aquifer sediments as cyclically deposited interglacial Pleistocene clastics. Each glacial cycle caused the sea level to drop as much as 400 feet below the present-day coastline. As the sea level dropped, coastal streams scoured their channels forming deep trenches. During interglacial cycles the sea level rose, which in turn caused a flattening of the stream channel gradients, and the filling of the scoured channels with clastic debris. This resulted in a regional expanse of braided-stream networks. The meandering stages of the Mississippi and Red Rivers covered the braided-stream deposits with levee, backswamp, and splay deposits. At the end of the Pliocene and well into the Pleistocene the Red River and Mississippi River had separate channels. However, upon reaching the Gulf Coastal Plain, both rivers formed a large amalgamated fluvial-dominated delta system. By Sangamon time, the Red River was permanently captured by the deeply-incised proglacial Mississippi River (Galloway et. al., 2000). The axes of the ancestral Mississippi River channels have shifted not only in geographic position, but also in direction, throughout the late Tertiary and Quaternary. By the beginning of the early Pleistocene, the channel system had moved westward into the central Acadia Parish region, leaving its sedimentary load to form the base of the Chicot Aquifer on top of the Evangeline Aquifer sediments (McFarlen and LeRoy, 1988).

The dominant geologic process in southwest Louisiana and the central Gulf Coast has been subsidence and deposition that have moved the coastline gulfward and landward by several hundreds of miles. This deposition of thousands of feet of sediment along the northern Gulf Coast has aided in the development of piercement plugs or salt domes. The greatest diameter of a salt dome structure in southwest Louisiana is seldom more than three miles, although radial faults associated with salt domes may extend several miles away from the dome, and upward to the surface. GeoMap Company's (2002) maps show a ridge of salt domes with an east-west trend through the center of the study area south of the Tepetate Fault System. The shallowest top of a salt dome along this trend is the Jennings Dome, located in west-central Acadia Parish, at 3,000 feet below mean sea level (MSL).

The GeoMap Company's (2002) maps show the presence of regional faulting with an east-to-west trend throughout the study area. Heinrich (2000) demonstrates the presence of a surface expression associated with a fault-line scarp in north-central Calcasieu Parish, south of DeQuincy, in northeastern Jefferson Davis Parish, south of Kinder and Elton, and across northern Acadia Parish, south of Eunice. This fault-line scarp is associated with a regional listric fault system called the Tepetate Fault System. The Tepetate is shown by Smoot, (1989), to be a groundwater flow barrier. Smoot (1989) shows this by a marked increase in depth of the base of freshwater on the high side of Tepetate Fault. The high side of the Tepetate Fault exhibits a depth to base of freshwater at greater than 2,000 feet below MSL. However, on the downthrown side the depth to the base of freshwater is as shallow as 500 feet below MSL. Gulfward, the base of freshwater increases to over 1,000 feet MSL on either side of the ridge of salt domes mentioned earlier, and becomes shallower toward the Gulf coast, eventually becoming all saltwater in the Chicot sediments. In the northeastern parishes of the study area (St. Landry, northeast Acadia, northern Lafayette, and northern St. Martin) a tongue of saltwater was found to be moving along the Tepetate Fault system. Also, in the eastern edge of the Chicot Aquifer, along the Atchafalya River basin, a shallow wedge of salt water has intruded the alluvial sediments into the Chicot Aquifer. Figure 2 is a contour map representing the base of freshwater in the Chicot Aquifer.

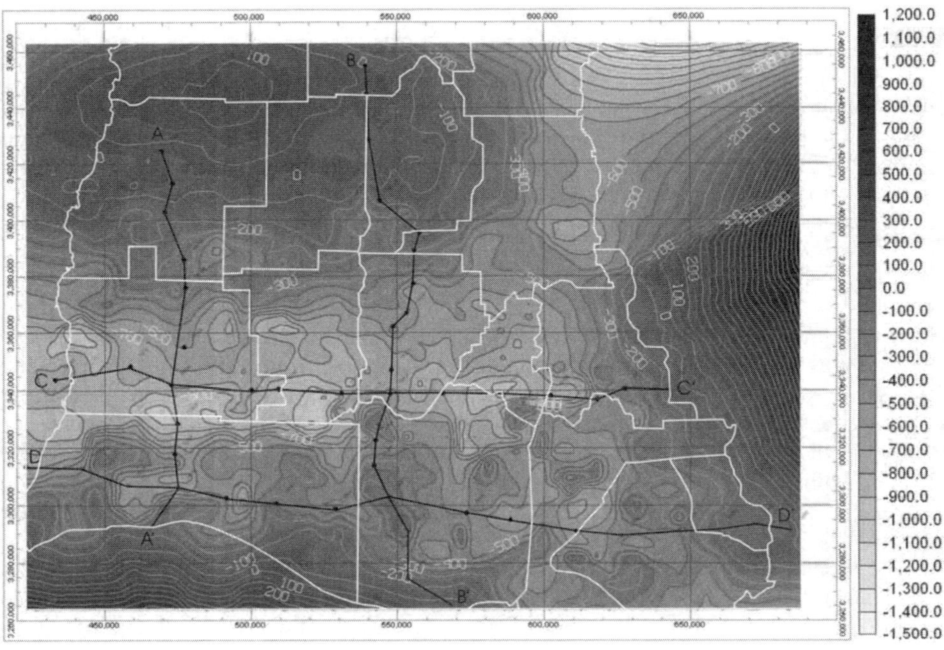

Fig. 2. Base of freshwater map in the Chicot Aquifer and cross section index. Plot was generated using RockWorks 2002.

11.4. CHICOT AQUIFER

The Chicot Aquifer has historically been subdivided into three sub-aquifers, and considered a confined aquifer. The surface confining clay ranges in thickness from the outcropping area in Vernon and Rapids parishes to the north, to over 500 feet thick in the Cameron and

Calcasieu parishes to the southwest. The majority of the surface confining clay is approximately 100 feet thick (Figures 3, 4, 5, and 6). In Lafayette and Vermilion parishes in

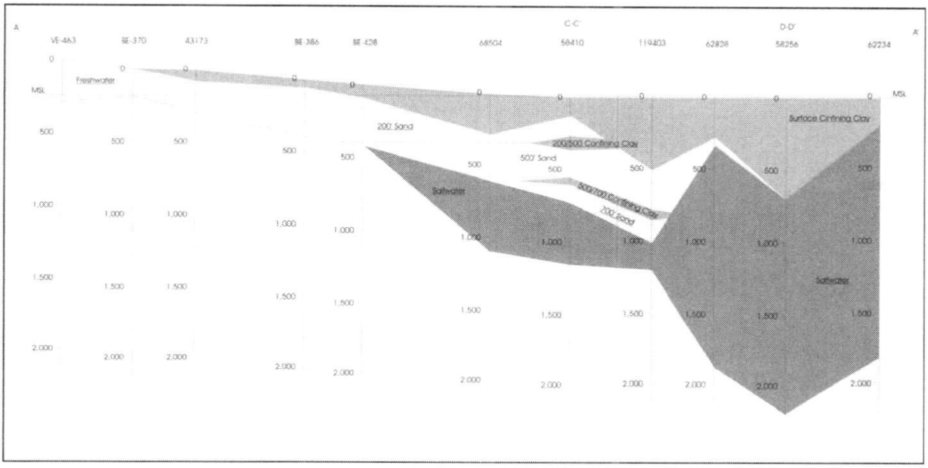

Fig. 3. Generalized stratigraphic N-S cross section A-A', generated by RockWorks 2002.

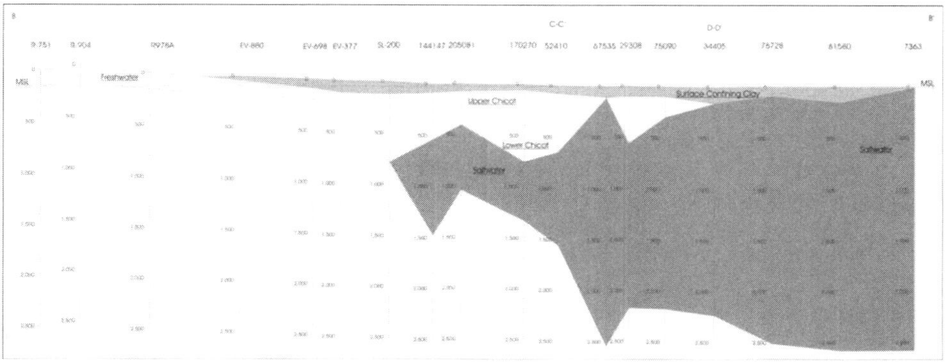

Fig. 4. Generalized stratigraphic N-S cross section B-B', generated by RockWorks 2002.

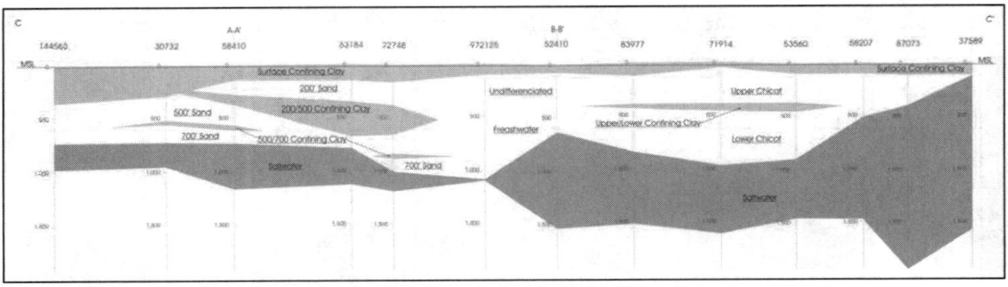

Fig. 5. Generalized stratigraphic W-E cross section C-C' B', generated by RockWorks 2002.

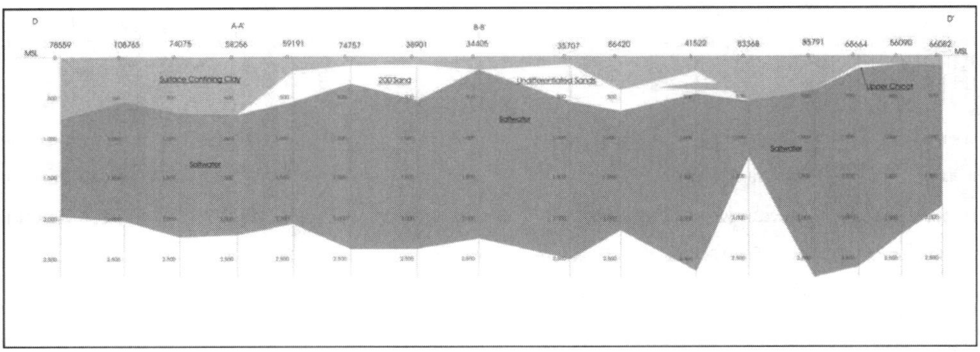

Fig. 6. Generalized stratigraphic W-E cross section D-D' B', generated by RockWorks 2002.

the Pleistocene Mississippi River Meander (PMRM) area the surface confining clay is less than 50 feet thick (Jones, et al, 1954) (Figure 5; at well no. 71914). This study has found the surface confining clay to be as thin as 10 feet in the PMRM area. In the northern parishes, the freshwater sands of the Chicot are hydraulically connected, with no confining clay, and designated as undifferentiated. In the west and southwest area of the Chicot, the two upper freshwater sands, "200-foot" sand and "500-foot" sand, are separated by a thick, laterally-continuous confining clay, referred to as the "200/500-foot"confining clay (Figures 3 and 5). The deeper "700-foot" sand is separated from the "500-foot" sand by thinner and less continuous clay, the "500/700-foot" confining clay. In the south-central area of the Chicot Aquifer (Jefferson Davis, eastern Cameron, into Vermilion parishes) the "500-foot" sand merges with the "700-foot" sand (Boniol, 1988) (Figure 5). In addition, these lower sands contain little or no freshwater (Figures 4 and 6). In the central portion of the Chicot, Jefferson Davis Parish, the aquifer is referred to as undifferentiated. The freshwater zone includes discontinuous layers of clay; however, the deeper portion of the aquifer contains no freshwater (Figures 4, 5, and 6). To the east, the Chicot is subdivided to an "Upper Chicot" and "Lower Chicot". This division is arbitrary, as there are no extensive confining clays in this portion of the aquifer (Figures 4, 5, and 6). Overall, the base of the freshwater zone of the Chicot Aquifer becomes shallower to the east and south. Due to faulting and other structural features, there are anomalously deeper base locations of freshwater in the southeastern area of the Chicot Aquifer (Figure 2).

The hydraulic conductivity of the Chicot Aquifer exhibits a broad range of values, both with depth and area. Vertically, hydraulic conductivity within the Chicot Aquifer is quite variable throughout the region. In Calcasieu and Cameron parishes, the average hydraulic conductivity in the "200-foot" sand is 119 ft/day and 70 ft/day in the "500-foot" sand while the "700-foot" sand is 50 ft/day. The "Upper Chicot" has an average hydraulic conductivity of 145 ft/day, and the "Lower Chicot" is 30 ft/day. The undifferentiated region is 79 ft/day. The overall hydraulic conductivity throughout the Chicot is an average of 91 ft/day (Carlson et al., 2003). Laterally, hydraulic conductivity is as variable as the vertical hydraulic conductivity. The average hydraulic conductivity in the northern region in Rapides Parish is 430 ft/day. To the south in St. Mary Parish it is only 20 ft/day. The central region in Acadia, Allen, Evangeline, Lafayette, southern Rapides, and Vermilion parishes has an average hydraulic conductivity of 177 ft/day. The average hydraulic conductivity from east to west is 58 ft/day and 67 ft/day, respectively (Carlson et al., 2003).

11.5. CHLORIDE CONDITIONS

For the purposes of this study the Chicot Aquifer has been divided into an *Upper Portion* and a *Lower Portion*. This division was set at approximately 400' to 500' below the ground surface. The USGS has conducted sampling for chemical analyses of the groundwater in the Chicot Aquifer system since the early 1940s. This has resulted in a data set of numerous well locations and chloride concentration values in mg/L. However, this data set is limited, due to the small number of samples collected at the same sampling location through the data set period (1940s to present). Of the thousands of analyses only 27 wells in the *Upper Portion*, and 68 wells in the *Lower Portion* of the Chicot Aquifer provide continuous sampling for 15 or more years (Tables 1 and 2). The majority of the wells in the *Upper Portion* are located in Acadia, Jefferson Davis, Cameron, and Vermilion parishes, and Calcasieu Parish in the *Lower Portion*.

The *Upper Portion* wells have a range in chloride change showing an increase of 230 mg/L in well CN-83, to a decrease in chloride concentration of 1,250 mg/L in well CN-81U. This decrease in well CN-81U is substantial; however, the decreased concentration is still well above the 250 mg/L EPA standard for drinking water at 350 mg/L (Table 1). The *Lower Portion* wells exhibit a maximum increase in chloride of 610 mg/L in well CU-767, well CU-173 exhibits a maximum decrease of 140 mg/L (Table 2). The chloride concentration was compared to the USGS head measurements collected during the same period as the chloride analysis. The two values are shown to track the other through time. As seen on Figure 7, a plot of head levels to chloride concentration at well SMN-109 in the *Upper Portion* indicates an increase of 45% in chloride concentration, while the head values dropped four feet in 38 years, an average of 0.1 foot per year. Wells SMN-57U and SMN-57L are located at the same location, and screened at different depths. SMN-57L was tested over a 29-year period, and the chloride concentration has increased 130 mg/L, and SMN-57U chloride concentration has decreased 70 mg/L over a 39 year period (Table 1).

Table 1. Upper portion of Chicot Aquifer chloride change.

Well #	Latitude	Longitude	Elev. (ft.msl)	Depth (ft)	Cl Change (mg/l)
CN- 83	29.7991	-93.3243	5	277	+230
VE- 630L	29.8422	-92.3424	5	528	+210
VE- 637L	29.8960	-92.1687	4	243	+160
CN- 80U	29.9797	-92.6365	5	453	+130
CN- 92	30.0180	-93.0324	6	443	+130
SM- 57L	29.7972	-91.6732	9	738	+130
VE- 637U	29.8958	-92.1686	4	198	+90
VE- 629U	29.8072	-92.3390	2	457	+40
SMN- 109	30.2178	-91.7111	11	375	+37
JD- 491	30.0858	-92.6824	10	377	+33
AC- 22	30.4238	-92.4057	45	300	+11
VE- 630U	29.8422	-92.3424	5	506	+10
AC- 452	30.1285	-92.4499	13	246	+4
SL- 48	30.5183	-92.4237	52	279	+4
AC- 75	30.3674	-92.3685	39	0	+2
VE- 639	29.6461	-92.4471	6	608	0
AC- 451	30.1280	-92.4474	14	293	-1
CU- 771	30.2269	-93.3085	18	241	-2
JD- 211	30.2285	-92.7521	30	320	-2
JD- 196	30.1349	-92.7210	22	198	-9
CN- 93	29.7861	-93.2954	4	360	-20
JD- 447	30.3477	-92.9435	0	262	-22
CN- 90	29.9411	-93.0800	3	396	-30
VE- 629L	29.8072	-92.3390	2	487	-40
SM- 57U	29.7972	-91.6732	9	638	-70
I- 93	30.0097	-91.7439	19	585	-173
CN- 81U	30.0238	-92.6404	5	448	-1250

Table 2. Lower portion of Chicot Aquifer chloride change.

Well #	Latitude	Longitude	Elev. (ftmsl)	Depth (ft)	Cl Change mg/l
CU- 767	30.17687	-93.2124	11	850	+610
CU- 970	30.09299	-92.9104	5	780	+540
CU- 694	30.1791	-93.331	15	556	+320
CN- 87	29.89022	-93.4018	8	804	+300
CU- 689	30.18326	-93.3177	15	531	+251
CU- 786U	30.1391	-93.2221	13	840	+161
CU- 661	30.19854	-93.2107	15	729	+142
SMN- 108	30.21798	-91.7112	11	505	+140
CN- 86L	30.02243	-93.5357	4	641	+120
CU-1023	30.22548	-93.2799	15	701	+110
CU- 754	30.22937	-93.2849	16	570	+105
CU- 788	30.14049	-93.4357	6	805	+100
CU- 615	30.22743	-93.2913	15	533	+97
CU- 960	30.17549	-93.3471	21	598	+92
CU- 560	30.1766	-93.3263	15	563	+75
CU- 840	30.22771	-93.2838	10	500	+60
CU- 869	30.23048	-93.2877	15	526	+56
CN- 88L	30.01549	-93.1585	9	804	+50
CU- 709	30.22437	-93.2852	15	524	+50
CU- 710	30.22882	-93.2813	15	684	+50
CU- 828	30.19715	-93.319	10	560	+50
CU- 583	30.22271	-93.2813	15	669	+49
CN- 88U	30.01549	-93.1585	9	666	+40
CU- 718	30.23909	-93.2188	15	500	+40
CU- 786L	30.1391	-93.2221	13	880	+40
CU- 708	30.17687	-93.2121	13	549	+37
CN- 86U	30.02243	-93.5357	4	535	+30
CU- 842	30.19687	-93.3257	12	575	+30
CU- 3	30.24048	-93.2182	10	686	+29
CU- 450	30.23993	-93.2643	10	523	+29
CU- 696	30.28548	-93.2877	20	494	+28
CU- 868	30.22798	-93.2949	15	520	+24
CU- 827	30.19298	-93.3199	10	560	+23

Well #	Latitude	Longitude	Elev. (ftmsl)	Depth (ft)	Cl Change (mg/l)
CU- 588	30.15632	-93.3415	12	586	+21
CU- 465	30.23548	-93.2715	10	520	+20
CU- 850	30.20021	-93.3221	10	548	+20
CU- 851	30.20361	-93.3214	10	555	+20
CU- 849	30.2016	-93.3071	10	564	+19
CU- 867	30.23104	-93.2918	15	500	+14
CU- 589	30.15493	-93.341	12	585	+8
CU- 683	30.19882	-93.2107	18	497	+6
CU- 79	30.23243	-93.2674	10	519	+5
CU- 630	30.15021	-93.2671	0	625	+5
CU- 463B	30.18521	-93.3424	17	516	+1
CU- 690	30.19576	-93.3277	10	601	+1
CU- 770	30.22687	-93.3085	18	490	+1
CU- 458	30.23965	-93.2685	10	509	0
CU- 635	30.2141	-93.5365	20	460	0
CU- 74	30.18548	-93.3338	15	545	-1
CU- 844	30.19687	-93.3257	12	430	-1
CU- 99	30.22465	-93.2807	15	667	-2
CU- 787	30.06493	-93.3507	4	734	-3
CU- 831	30.25298	-93.2835	6	497	-4
CU- 464	30.1916	-93.339	15	530	-7
CU- 692	30.18687	-93.3379	0	560	-7
CU- 959	30.17528	-93.3469	21	733	-10
CU- 712	30.22771	-93.2702	10	500	-11
CU- 664	30.18771	-93.3435	15	501	-12
CU- 90A	30.24409	-93.2693	10	584	-13
CU- 951	30.22465	-93.2993	18	500	-20
CU- 684	30.17687	-93.2127	14	548	-26
CU- 957	30.1891	-93.3196	17	500	-27
CU- 789	30.18354	-93.3335	18	640	-60
CU- 862	30.18021	-93.3263	15	550	-63
CU- 976	30.20826	-93.3085	10	550	-70
CU- 769	30.22667	-93.3083	18	642	-100
CU- 173	30.15854	-93.1543	24	726	-140

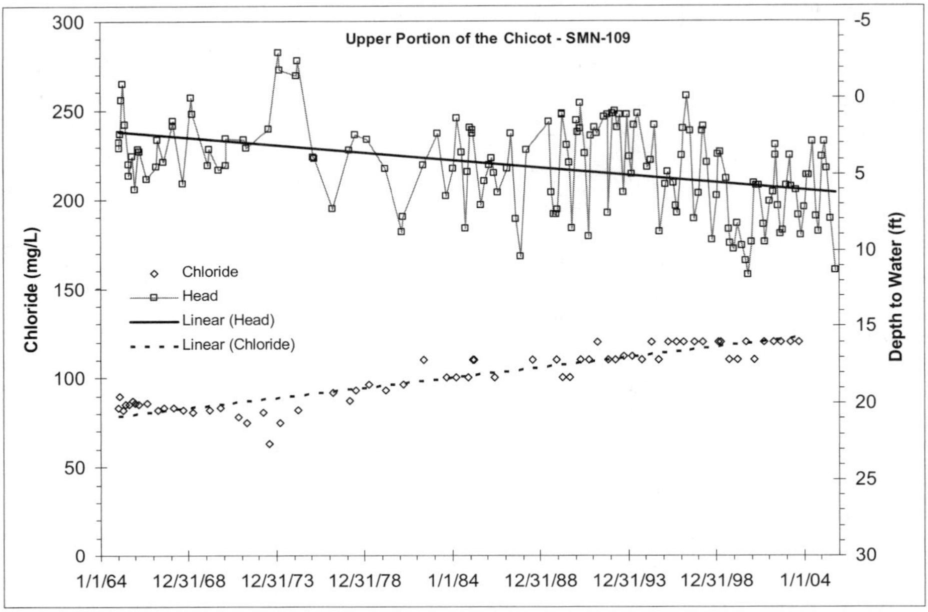

Fig. 7. Chloride concentration and head levels in well SMN-109.

The *Lower Portion* displays the same correlation as the *Upper Portion*. Well CU-786U which had an increase in chloride concentration of 233%, exhibits a six-foot drop in 31 years (Figure 8). Wells CU-786L is at the same location as CU-786U and screened at a deeper position also showed an increase in chloride concentration. However, well CU-786U, the shallowest has a much higher increase in chloride than CU-786L. Three other paired wells screened at different depths all showed an increase in chloride concentration in both screened positions. The decrease in head values could be related to precipitation variations and an increase in groundwater use, as shown on Figure 9.

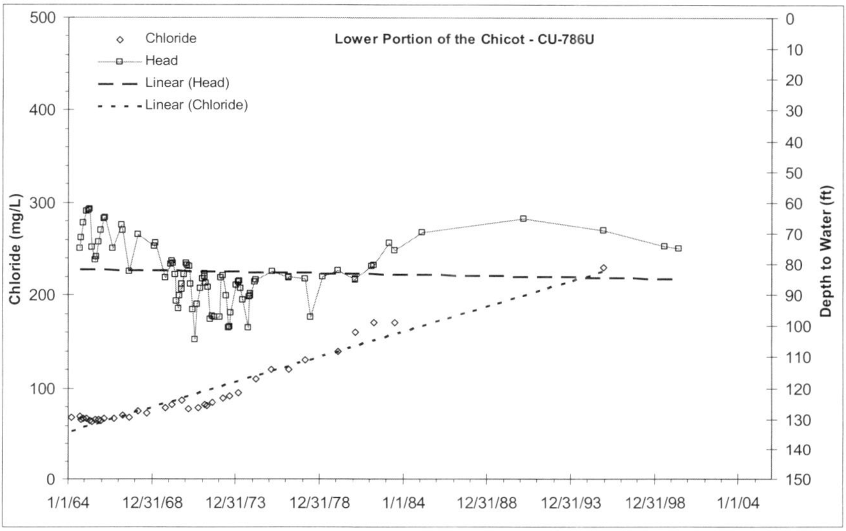

Fig. 8. Chloride concentration and head levels in well CU-786U.

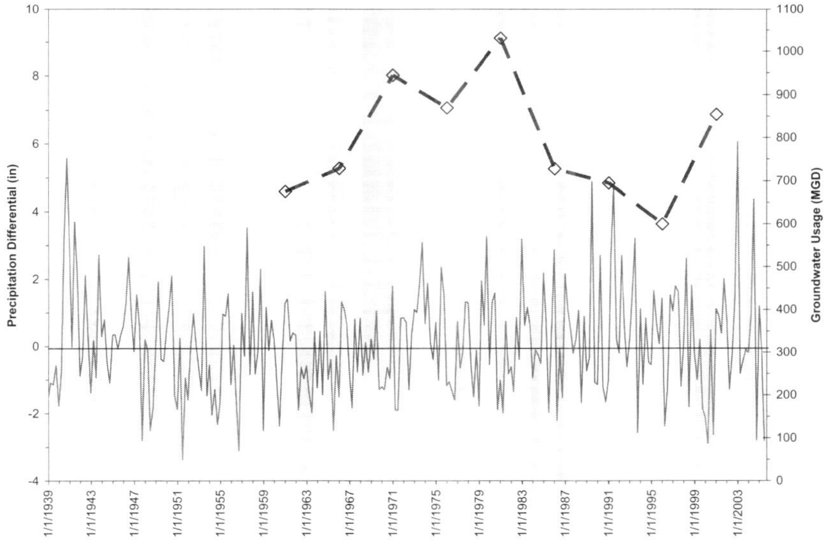

Fig. 9. Time-biased average quarterly precipitation variation and groundwater usage for southwest Louisiana. The time-biased data was obtained from the National Oceanic and Atmospheric Administration-National Climatic Data Center. The groundwater usage (dashed lines) is modified from Sargent (2002). The shaded areas represent prolonged periods (three successive quarters) at or below average precipitation

11.6. CONCLUSIONS

Saltwater is indeed moving in and out of the Chicot Aquifer System, due to drought and increased usage (Figure 9). In the *Upper Portion,* where data is available in the coastal parishes (Cameron, Vermilion, and St. Martin) saltwater is moving within the aquifer system, resulting in chloride concentrations continuing to gradually increase through time. The other areas of available data in the *Upper Portion* do show chloride concentration increasing and decreasing during drought and increased seasonal usage. However, there are periods of decreased usage, resulting in a decrease in the chloride concentration over time signifying a recovery of freshwater. The data available for the *Lower Portion* shows a dramatic increase in chloride concentration. This increase is not only a fluctuation due to drought and increased usage, but also indicates a continual increase in chloride concentration in the "500-foot" sand and the "700-foot" sand. The current available data does indicate such during periods of drought and increases in use. This finding provides evidence that saltwater is moving in and out of the Chicot Aquifer system during cycles of lower precipitation and increase in usage, and that in many cases, the chloride concentrations continue to increase over time. Two possible sources of saltwater in the Chicot Aquifer System are: (1) the movement of saltwater from shallow regional faulting, salt domes and associated radial faulting; and (2) saltwater intrusion from the Gulf of Mexico moving through the system in the surface rivers and in the subsurface movement of saltwater by tidal motion and decreased precipitation.

ACKNOWLEDGEMENTS

The authors would like to thank the USGS's NWIS, the Louisiana DOTD's Water Resources Division, the Louisiana DNR and NOAA's National Climatic Data Center for making their data available.

REFERENCES

Boniol, D., Autin, W. J., and Hanson, B. C., 1988. Recharge Potential of Louisiana Aquifers: A Supplement to the State Aquifer Recharge Map and Atlas Plates, Open-File Series No. 88-07, Louisiana Geological Survey, Baton Rouge, LA, pp. 57.

Carlson, D., Milner, L. R., Hanson, B., Wilson, C., Rahman, A., and Hartano, S., 2003. Evaluation of Aquifer Capacity to Sustain Short-Long-Term Groundwater Withdrawal from Point Sources in the Chicot Aquifer for Southwest Louisiana: Part1, Louisiana Geological Survey, p. 117.

Galloway, W. E., Ganey-Curry, P.E., Li, X., and Buffler, R. T., 2000. Cenezoic depositional history of the Gulf of Mexico basin. AAPG Bull. 84(11): 1743-1774.

GeoMap Company, 2002. Southern Louisiana Executive Reference Map 330 2002 edition: GeoMap Company, 1 sheet.

Fisk, H.N., 1944. Geological investigation of the alluvial valley of the lower Mississippi River. War Department Corps of Engineers, Mississippi River Comm., Vicksburg, Mississippi, 78p.

Heinrich, P. V., 2000. DeQuincy Fault Scarp, Beauregard and Calcasieu Parishes, Louisiana. Basin Research Institute Bull. 9: 38-50.

Jones, P. H., Turcan, A. N., Jr., and Skibitzke, H. E., 1954. Geology and Groundwater Resources of Southwestern Louisiana. Department of Conservation and Louisiana Geological Survey, Geological Bulletin, no. 30, 285 p.

Louisiana Geological Survey Stratigraphic Charts of Louisiana, Folio Series No. 8, 2000, web site http://www.lgs.lsu.edu/lacharts.pdf.

McFarlan, E., and LeRoy, D.O., 1988. Subsurface Geology of the Late Tertiary and Quaternary Deposits, Coastal Louisiana. Gulf Coast Association of Geological Societies Transactions 38: 421-433.

Renken, R. A., 1998. Ground Water Atlas of the United States, Segment 5, Arkansas, Louisiana, Mississippi. US Geological Survey Hydrologic Atlas 730-F, p. 28, web site http://capp.water.usgs.gov/gwa/index.html.

Sargent, P., 2002. Water Use in Louisiana, 2000. Louisiana Department of Transportation and Development Water Resources Section, Special Report No. 15, pp. 133.

Smoot, C. W., 1989. Louisiana Hydrologic Atlas Map No. 4, Geohydrologic Sections of Louisiana, U.S. Geological Survey, 1 sheet.

Whitfield, Jr., M. S., 1975. Geohydrology of the Evangeline and Jasper Aquifers of Southwest Louisiana. Department of Conservation and Louisiana Geological Survey, Water Resources Bulletin No. 20, 72 p.

12

SALINISATION OF THE NORTHERN COASTEL AREA OF THE NETHERLANDS DUE TO LAND SUBSIDENCE AND SEA LEVEL RISE

Perry de Louw and Gualbert Oude Essink
TNO Geological Survey of The Netherlands
P.O. Box 80015, 3508TA Utrecht, The Netherlands

ABSTRACT

Exfiltration of brackish to saline groundwater takes place at the northern coastal area of the Netherlands (Northwest Friesland). This situation patterns for the coastal part of the Netherlands that is situated under the sea level. Land subsidence in the study area due to salt exploitation and future sea level rise is increasing the exfiltration of brackish groundwater causing serious salinisation problems for agriculture. Surface water will not be suitable for irrigation anymore and salt damage to crops can occur when salt groundwater reaches the root zone.

Salinisation problems can be controlled by means of fine tuning of surface water management. The most sustainable water management plan of this coastal area was composed with the help of numerical models and field measurements. That implies that surface water levels have to be increased at most parts of the area to limit the salinisation of surface water and root zone. Other measures such as increasing the distance between ditches and drain tubes also appear to be effective to increase fresh rainwater lenses in order to avoid salt damage to crops.

12.1. INTRODUCTION

At the northern coastal area of the Netherlands (Northwest Friesland) salt is exploitated at a depth of 2000 meters causing a land subsidence up to 40 centimeters. To compensate the relative rise of the groundwater table, the local Water Board Wetterskip Fryslan lowered the surface water level equal to this land subsidence. As a consequence, deep brackish to saline groundwater cones up towards the surface and salinisation of the shallow groundwater and surface water takes place. The water board decided the make alternative plans in order to reduce the salinisation of the shallow water system. Salinisation of the root zone and surface water in this area is a major problem for agriculture. Surface water will not be suitable for irrigation anymore and salt damage to crops can occur when salt groundwater reaches the root zone.

TNO was asked to calculate the effects of different scenarios on the salinisation of the groundwater and surface water system. The model code MOCDENS3D was used and field measurements of the brackish-fresh groundwater interface were executed. In this paper the method, results and conclusions of this investigation are described.

12.2. DESCRIPTION OF INVESTIGATION AREA

The study area is located in the province of Friesland (The Netherlands) at the northern coast of the Waddenzee (Figure 1). During the Holocene, before the embankment and isolation from the sea, this area was an estuary where marine clay and fine sand were deposited resulting in a cover layer of at least 8 meter to 25 meter. Below this cover layer a sequence of aquifers and semi-permeable layers of clastic sediments is found. Groundwater on a regional scale is flowing through these layers. The hydraulic head in the first aquifer at most of the area is higher than the surface water level and phreatic groundwater level in the cover layer. Therefore groundwater from the first and second aquifer is exfiltrating in the polders where surface water levels of –1.0 to –2.0 meters below sea level are maintained.

This exfiltrating groundwater is brackish to salt. It has its origin from transgressions during the Holocene when seawater was infiltrated to great depth. Nowadays intrusion of seawater is taking place but at slower rates. However, the intrusion will be accelerated due to future sea level rise.

All available measurements of the chloride concentration of groundwater were inter- and extrapolated to produce a 3-dimensional chloride map. This was needed as initial condition for the density dependent groundwater flow simulations (TNO, 2005). Figure 2 shows the chloride concentration of a depth of 25 meter below sea level. At this depth the concentration varies from 2500 mg/l inland to 6000 mg/l at the coastal zone.

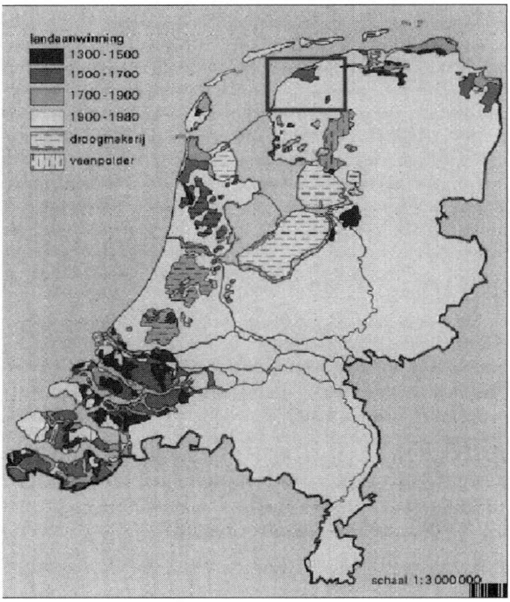

Fig. 1. Location of the study area and the reclamation period of polders in the Netherlands.

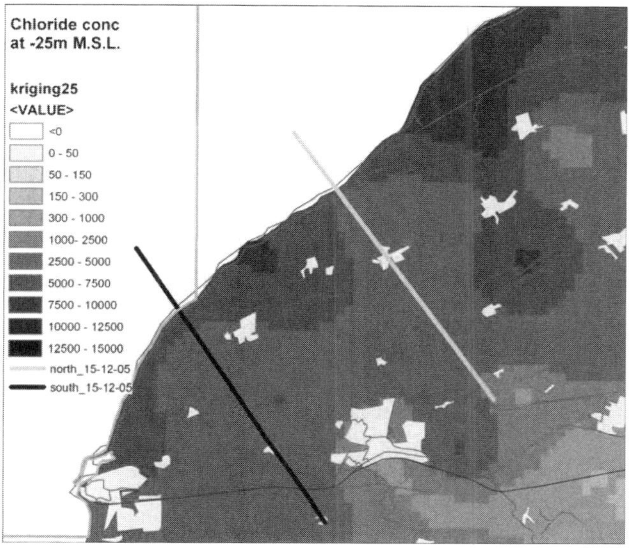

Fig. 2. Chloride concentration (mg/l) of groundwater at a depth of 25 meters below sea level.

12.3. METHOD

The model code MOCDENS3D (Oude Essink, 2001, 2004) was used to simulate density dependent groundwater flow and coupled salt transport at two different scales. For two transects perpendicular to the coastline the regional groundwater system is modeled. Various local scale models are set up to simulate the dynamics of the fresh-brackish groundwater interface.

The regional model is used to calculate the effect of surface water level changes, sea level rise and land subsidence on groundwater level, exfiltration rates, salinisation of groundwater and surface water system at a regional scale.

The local model is used to calculate the effects of exfiltration rate, surface water level, drainage characteristics on the dynamics of the fresh-brackish groundwater interface.

12.3.1. Description of MOCDENS3D

To quantify changes in this coastal groundwater flow regime, the processes transient variable-density groundwater flow and coupled solute transport in three dimensions are modeled. The modular variable-density groundwater flow and solute transport code MOCDENS3D has been used. It is based on MODFLOW (MCDonald & Harbaugh, 1988) and MOC3D (Konikow et al., 1996), which has been adapted for density differences. MODFLOW (U.S. Geological Survey public domain) is the most widely used computer code for groundwater flow in porous media. Advective and hydrodynamic dispersive solute transport processes through porous media are modeled by a particle tracking technique in combination with the finite difference method, respectively. As chloride is the major conservative negative ion in Dutch coastal groundwater, the discussion about salinisation is focused on that predominant solute.

12.3.2. Regional Transect Modeling

For two transects perpendicular to the coastline the following geohydrological characteristics were put into the model.

- Length of transect = 10km, thickness of groundwater system = 300m
- Geohydrological schematization from surface to depth (k=hor. permeability, d=thickness of layer)
    - Cover layer of peat and clay; k=0.3m/d; d=8m (transect 1) en d=21m (transect 2)
    - First aquifer; k=2m/d; d=6m (transect 1) en d=4m (transect 2)
    - Semi permeable layer; k=0.06m/d; d=10m (transect 1) en d=20m (transect 2)
    - Second aquifer =25m/d; d=6 (transect 1) en d=8m (transect 2)
    - Semi permeable layer =0.1m/d; d=0m (transect 1) en d=2m (transect 2)
    - Third aquifer =26m/d; d=175m (transect 1) en d=150m (transect 2)
    - Fourth aquifer =20m/d; d=95m (transect 1) en d=95m (transect 2)
    - Anisotropy is 1/3

- Salt transport
    - Effective porosity = 30%
    - Longitudinal dispersivity: α_L=1m; α_{TH}=α_{TV}=transversal disp.=1/10*α_L
    - Molecular diffusion: D_{mol}=8.64*10^{-5} m^2/d.

- Model discretisation:
    - 200 columns, cell size Δx=50m
    - 100 layer, upper 50 layers d= Δz=1m, lower 50 layers d= Δz=5m
    - The flow time step to recalculate flow equation is set to 1 month
    - simulated period is 30 years
    - 9 particles each model cell to calculate advective transport
    - hydrostatic pressure at vertical boundaries
    - impermeable base at -300m below sea level

12.3.3. Local Model

For two different soil types the following geohydrological characteristics were put into the model:

- Length of transect = 170m, thickness of groundwater system = 10m
- Geohydrological schematization from surface to depth (k=hor. permeability, d=thickness of layer)
    - Soil type silty clay:
        - layer 1=0.05m/d, thickness=1.4m
        - layer 2=0.2m/d, thickness =1.6m
        - layer 3=0.5m/d, thickness =7.0m
    - Soil type clay:
        - layer 1=0.05m/d, thickness =1.4m
        - layer 2=0.5m/d, thickness =8.6m

    - Anisotropy is 1/3

- Salt transport
    - Effective porosity = 30%
    - Longitudinal dispersivity: α_L=1m; α_{TH}=α_{TV}=transversal disp.=1/10*α_L
    - Molecular diffusion: $D_{mol}=8.64*10^{-5}$ m^2/d.

- Model discretisation:
    - 170 columns, cell size Δx=1m
    - 50 layers, thickness Δz=0.2m
    - The flow time step to recalculate flow equation is set to 10 days
    - 9 particles each model cell to calculate advective transport
    - Hydrostatic pressure at vertical boundaries
    - Fixed head at lower model boundary, 10m below sea level (varies with case)

12.4. RESULTS

12.4.1. Regional Model

The results of transect 1 will be described. At the **present situation** (2000 AD) the groundwater becomes more saline with depth and the chloride concentration of the groundwater decreases with distance to the coast of the Waddenzee (Figure 3). Saline groundwater reaches the surface at ditches where exfiltration rates are high due to low surface water levels. Even higher exfiltration rates are found at locations where the hydraulic resistance for vertical flow is reduced because of the present of sandy sediments in the cover layer (creek deposits). Due to land subsidence caused by salt exploitation, the surface water level is lowered to guarantee sufficient dry conditions for agricultural activities. In the **autonomous situation** it is estimated that the sea level will rise 18 centimeters for the next 30 years. Exfiltration rates of deeper and more saline groundwater increase as a result of higher hydraulic heads. The model results show that the saline front is moving inland and to the surface (Figure 3). At polders where water managers maintain low surface water levels this process is even faster. In the autonomous situation after a period of 30 years salinisation of groundwater and surface water will take place at higher magnitude and in a larger area. The effect of sea level rise reaches to a distance of 8 kilometers to the coast.

To compensate wetting effects of land subsidence due to salt exploitation and to reduce salinisation by land subsidence and sea level rise, the water board is redesigning the surface water system into a more sustainable water system. Drainage patterns and surface water levels have to be changed to achieve this. The regional model is used to simulate different water management scenarios. The following two scenarios will be discussed:

- **Scenario 1**: increase of the surface water level in most of the area, except for a polder called Roptavaart where surface water levels are lowered dramatically in order to attract saline groundwater so the surrounded polders receive less saline groundwater (surface water level Polder Roptavaart is –250 centimeter below sea level).

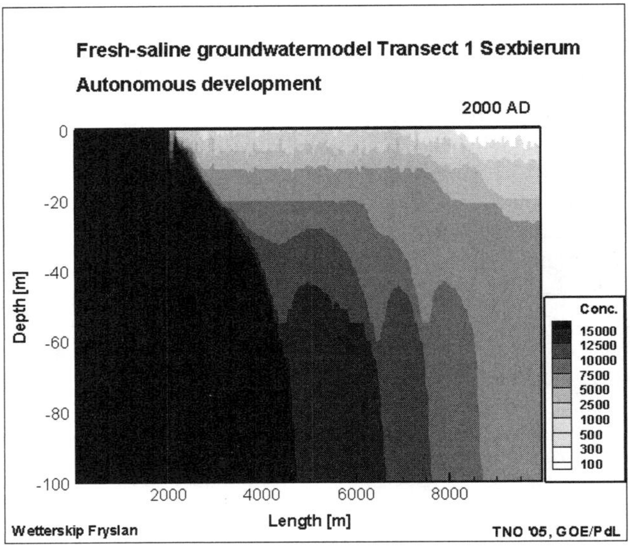

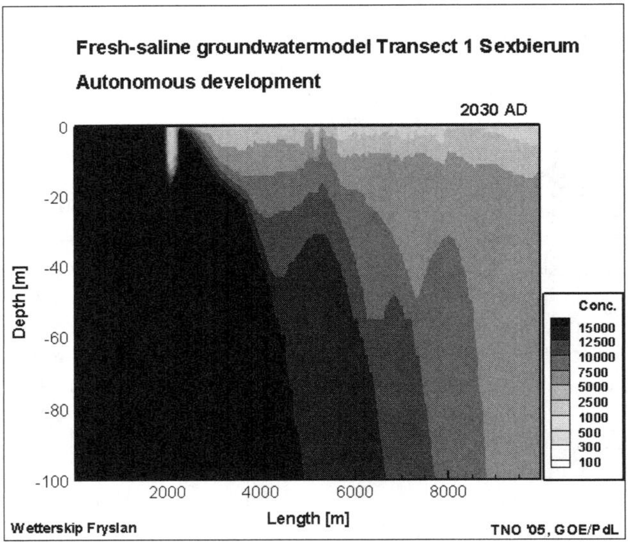

Fig. 3. Calculated chloride concentration of groundwater at present situation and after 30 years autonomous situation.

- **Scenario 2**: same as scenario 1, except for Polder Roptavaart where surface water levels are at higher elevation in order to reduce salinisation of ground water and surface water in this area (surface water level Polder Roptavaart is –110 centimeter below sea level). In this situation hydrological conditions are no longer suitable for agricultural activities in the Polder Roptavaart.

Figure 4 shows the results of these two scenarios after a period of 30 years. Extreme upconing of saline groundwater in Polder Roptavaart is clearly visible for scenario 1.

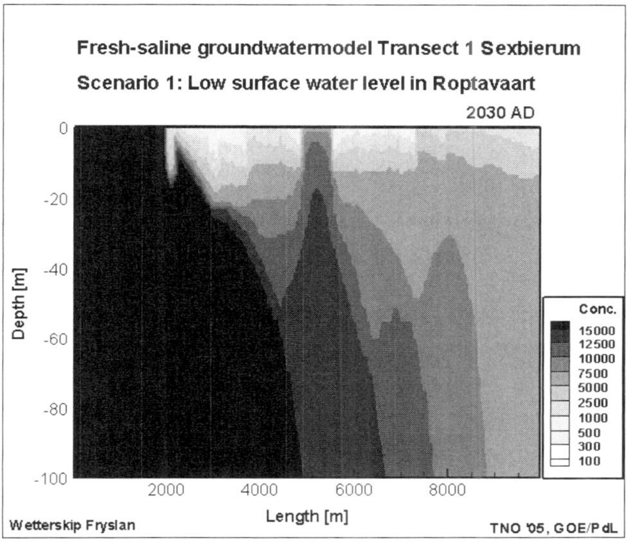

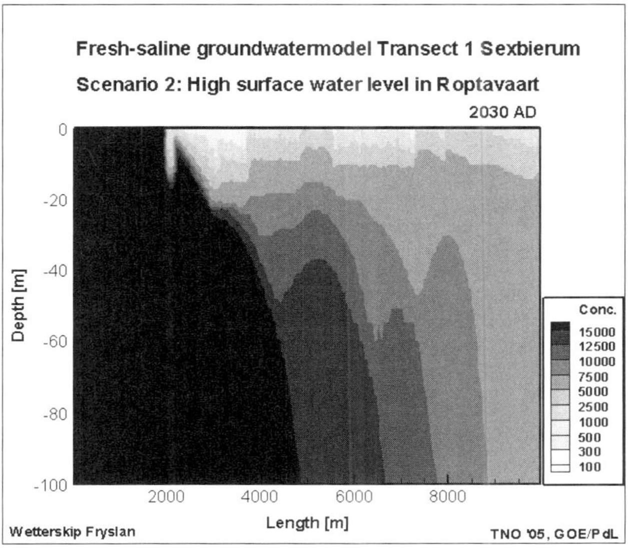

Fig. 4. Calculated chloride concentration of groundwater scenario 1 and 2 after 30 years.

However, the reduction of salinisation in the surrounding of Polder Roptavaart is small as can also be seen in Figure 5. The reduction of exfiltration due to extreme low surface water levels in Polder Roptavaart reach only 200-300 meter.

Table 1. Model results of different scenarios. For the present situation the current values are given, for the different scenarios the difference relative to the present scenario is given. The values are averaged for the whole transect after 30 years.

	Present situation	**Sea level rise 18cm next 30 years**	**Sea level rise and land subsidence, autonomous situation**	**Scenario 1**	**Scenario 2**
	Current value	Effect	Effect	Effect	Effect
Surface water level	-1.1m below sea level	+0m	-0.30m	+0.10m	+0.26m
Exfiltration rate	0.12mm/d	+0.03mm/d	+0.08mm/d	-0.01mm/d	-0.05mm/d
Cl-conc of shallow groundwater	1700mg/l	+320mg/l	+970mg/l	-232mg/l	-492mg/l
Chloride loads to surface water	600kg/ha/yr	+300kg/ha/yr	+1070kg/ha/yr	+1614kg/ha/yr	-316kg/ha/yr

As a consequence of sea level rise the salinisation of the surface will increase from 600 to 900 kg/ha/yr (Table 1). Together with the land subsidence and accompanying surface water level reduction the salinisation process will increase with 1070 kg/ha/yr to an average chloride load of 1670 kg/ha/yr. Scenario 2 will reduce the chloride loads to the surface water while scenario 1 leads to the highest average salinisation. The extreme low surface water levels at Polder Roptavaart cause these high average chloride loads. However, the advantage is that the highest chloride loads are concentrated at one polder.

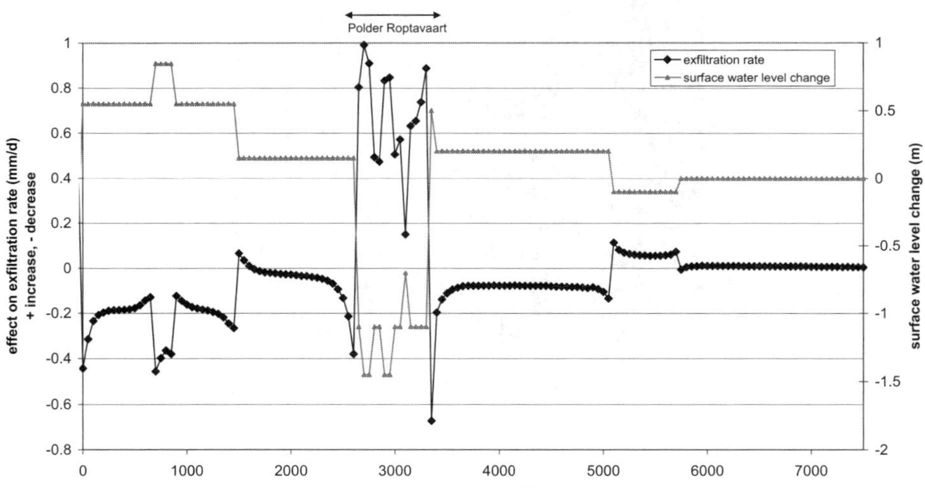

Fig. 5. Calculated effect of surface water level changes on the exfiltration rate presented in transect 1 perpendicular at the coast (scenario 1).

12.4.2. Local Model

In these brackish groundwater areas near the coast agricultural is possible because of the existence of fresh rainwater lenses. The yearly rainfall surplus in the Netherlands is on average 300 mm/y. In exfiltration areas this amount will be finally discharged to the surface water system. However, the rainwater is temporally stored in the soil, forming fresh rainwater lenses on top of the brackish groundwater (see Figure 6). The lens tends to be thicker in winter when more precipitation and less evapotranspiration is taking place than in summer. The exfiltration rate, drainage density, drainage level, hydraulic conductivity of cover layer, meteorological conditions control the dynamics of the fresh rainwater lens.

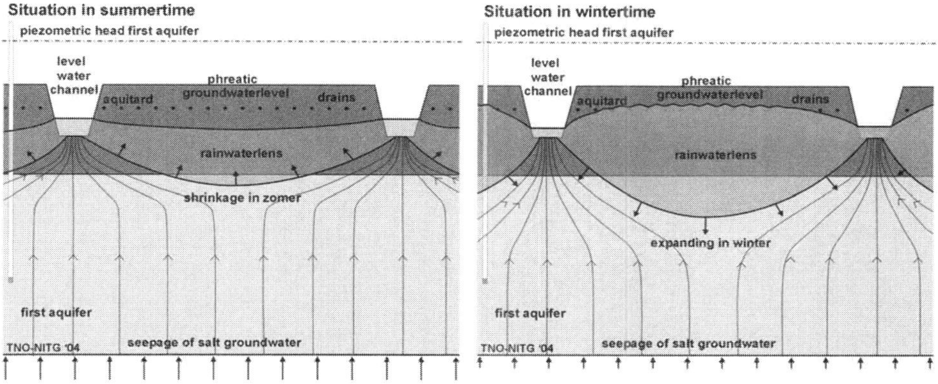

Fig. 6. The dynamics of the fresh rainwater lens during summer and winter.

Measurements in the study area show a large spatial variation of the thickness of the fresh rainwater lens. Two examples of measurements of the interface between brackish groundwater and rainwater are shown in Figure 7. The electrical conductivity of the groundwater was measured at depth increments of 10cm. The two profiles are situated only 200 meters from each other while the thickness of the fresh rainwater lens differs much. At parcel 1 the lens varies from 100 cm to 0 cm. At the lowest point in the parcel only brackish groundwater is found. At parcel 2 no brackish groundwater was found while the measurements reach a depth of 3 meters. Both the land surface and surface water level at parcel 2 is 0.5 meter higher than parcel 1. At parcel 2 probably no exfiltration but infiltration of rainwater takes place while the measurements show that brackish groundwater is exfiltrating at the center of the ditch bottom. At the sides of the ditch fresh rain water is exfiltrating, drained from the parcel.

For different cases the dynamics of the fresh rainwater lens was modeled with MOCDENS3D. The calculated temporal variation of the thickness of the lens is shown in Figure 8. The effect of the meteorological conditions is clear. The thickness varies from 1.5-3 meter in wet years to 0.5-1.5 meter in dry years.

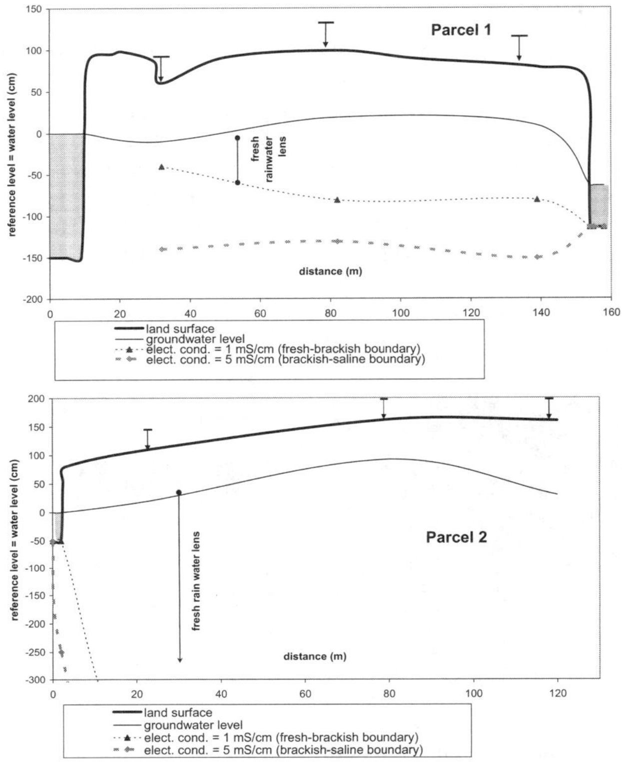

Fig. 7. Measurements of the interface between fresh rainwater and brackish groundwater (August 20th, 2005).

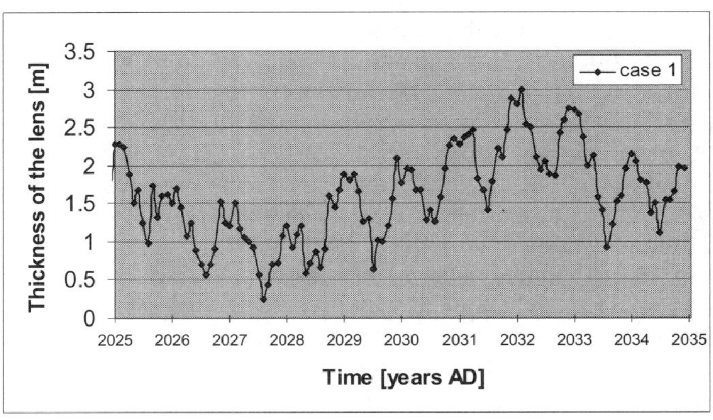

Fig. 8. Calculated temporal variation of thickness of fresh rainwater lens.

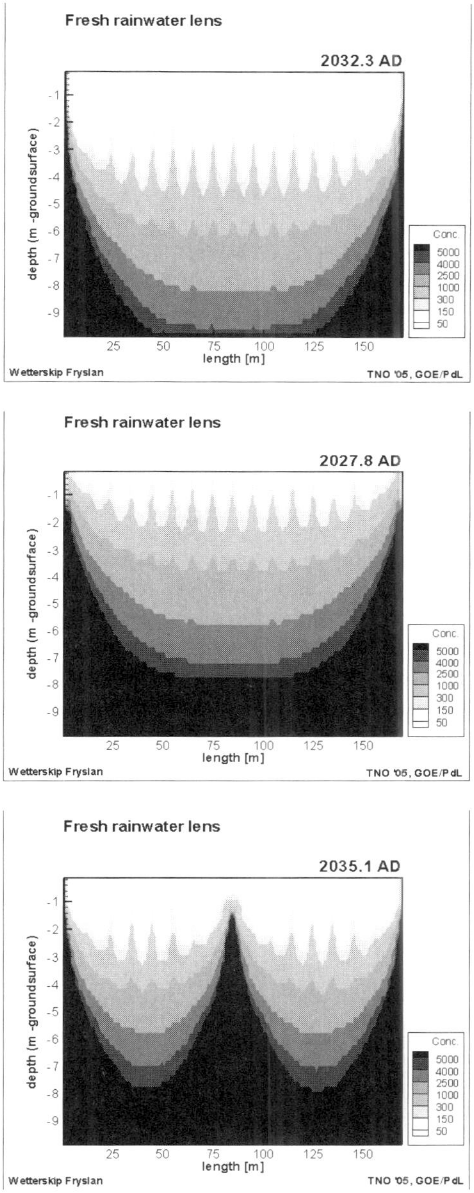

Fig. 9. Calculated chloride concentration of groundwater between two ditches at two different periods (bottom: winter, middle: summer). The top picture shows the winter situation in case of reduced distance between the ditches.

Saline groundwater exfiltrates at the ditches, while in the parcel the brackish groundwater is pushed down by the infiltrating rainwater. The effect of drain tubes can clearly be seen in Figure 9 where brackish groundwater is attracted. Higher drainage density (reducing distance between ditches) leads to thinner rainwater lenses as can be seen in Figure 9c.

Table 2. Calculated thickness of fresh rainwater lens and chloride concentration for different cases.

Case	Exfiltration rate	Surface water level change	Drainage level change and distance between drain tubes	Distance between ditches	Drainage resistance	Initial chloride conc. of groundwater	Thickness of fresh rainwater lens (m)					Chloride concentration of exfiltrated groundwater mg/l	
	mm/d	m	m	m	d	mg/l	average	winter	summer	max	min	-1.1m	-2.7m
1	0.3	0	0 + 10m	170	5	7500	1.60	1.86	1.35	3.00	0.24	3684	7480
4	0.3	-0.5	0 + 10m	170	5	7500	1.21	1.45	0.98	2.57	0.08	5290	7500
7	0.3	+0.5	+0.5m + 10m	170	5	7500	6.53	6.81	6.24	8.21	5.09	500	500
10	0.3	0	+0.5m + 6m	170	5	7500	5.60	5.89	5.32	7.44	3.89	2256	7305
13	0.3	0	0 + 10m	85	5	7500	1.21	1.44	0.97	2.52	0.07	3852	7485
19	0.3	0	0 + 10m	170	5	5000	1.50	1.75	1.25	2.84	0.24	3151	4999
20	0.3	0	0 + 10m	170	5	4000	1.46	1.71	1.21	2.83	0.20	2704	4000
16	0.3	0	0 + 10m	170	10	7500	2.61	2.87	2.34	4.11	1.01	3287	7450
17	0.3	0	0 + 10m	170	1	7500	0.63	0.83	0.42	1.85	0.00	4138	7496
18	0.3	0	0 + 10m	170	5	7500	6.64	6.91	6.36	8.42	4.90	467	456
2	0.1	0	0 + 10m	170	5	7500	5.35	5.62	5.08	7.26	3.57	680	1507
5	0.1	-0.5	0 + 10m	170	5	7500	4.07	4.35	3.80	5.86	2.48	3822	7500
8	0.1	+0.5	+0.5m + 10m	170	5	7500	6.25	6.53	5.98	7.87	4.80	500	500
11	0.1	0	+0.5m + 6m	170	5	7500	6.12	6.40	5.85	7.95	4.42	628	1403
14	0.1	0	0 + 10m	85	5	7500	4.52	4.77	4.26	6.17	2.88	695	1508
21	0.1	0	0 + 10m	170	5	5000	5.35	5.62	5.08	7.24	3.58	693	1601
23	0.1	0	0 + 10m	170	5	4000	5.37	5.64	5.10	7.28	3.61	705	1640
3	0.5	0	0 + 10m	170	5	7500	0.13	0.22	0.05	1.10	0.00	5625	7500
6	0.5	-0.5	0 + 10m	170	5	7500	0.12	0.19	0.04	1.02	0.00	6615	7500
9	0.5	+0.5	+0.5m + 10m	170	5	7500	6.92	7.20	6.64	8.40	5.39	495	498
12	0.5	0	+0.5m + 6m	170	5	7500	5.25	5.53	4.96	7.02	3.55	3497	7499
15	0.5	0	0 + 10m	85	5	7500	0.10	0.17	0.03	0.94	0.00	5539	7500
22	0.5	0	0 + 10m	170	5	5000	0.12	0.21	0.04	1.08	0.00	4135	5000
24	0.5	0	0 + 10m	170	5	4000	0.13	0.21	0.04	1.10	0.00	3437	4000

For 24 cases per soil type the dynamics of the fresh rainwater lens were simulated and the results are summarized in Table 2. The (initial) exfiltration rate has major impact on the thickness of the lens. At average values of 0.3 mm/d the lens has an average thickness of 1.60 m. Reducing this values to 0.1 mm/d leads to an enormous increase of the thickness to average values of 5.35 m while increasing the exfiltration rate to 0.5 mm/d the lens is disappearing in the summer and has average values of 0.13m.

Water management can control the thickness of the lens as can be seen in Table 2. Increasing the surface water level with 0.5 m (cases 7-9) will increase the lens significantly because the hydraulic head difference is increased and therefore the exfiltration rate. Exfiltration of brackish groundwater in the ditches (case 13) will also be reduced by this measure. Reducing the distance between ditches will reduce the fresh rainwater lens from 1.60 m to 1.21 m in the 0.3 mm/d exfiltration rate case. More rainwater will be drained and the average groundwater level is lowered causing higher exfiltration rates.

12.5. CONCLUSIONS

Salinisation of the surface water and the root zone is taking place due to exfiltration of brackish groundwater. Sea level rise and land subsidence will accelerate this process significantly. Model results showed that within a period of 30 years the chloride loads to the surface water will be almost three times the present loads and fresh rainwater lenses will reduce in size increasing salt damage to crops.

Salinisation of the shallow water system can be controlled by means of surface water management and so the water board decided the make alternative plans. The most sustainable water management plan of this coastal area was composed with the help of numerical models and field measurements. That implies that surface water levels have to be increased at most parts of the area to limit the salinisation of surface water and root zone. Other measures such as increasing the distance between ditches and drain tubes also appear to be effective to increase fresh rainwater lenses in order to avoid salt damage to crops.

REFERENCES

Konikow, L. F., D. J. Goode, et al., 1996. A three-dimensional method-of-characteristics solute-transport model (MOC3D). U.S.G.S. Water-Resources Investigations Report 96-4267: 87.

McDonald, M. G., and Harbaugh, A. W., 1988. A modular three-dimensional finite-difference ground-water flow model, U.S.G.S. Techniques of Water-Resources Investigations, Book 6, Chapter A1, 586 pp.

Oude Essink, G. H. P., 2001. Salt Water Intrusion in a Three-dimensional Groundwater System in The Netherlands: a Numerical Study. Transport in Porous Media 43(1): 137-158.

Oude Essink, G. H. P., 2004. Modelling 3D density dependent groundwater flow at the Island of Texel, The Netherlands (chapter 4). Coastal Aquifer Management-Monitoring, Modeling and Recent Practices. A. H.-D. Cheng and D. Ouazar, CRC Press: 75-92.

TNO-NITG, 2005. Chloride-concentration at the bottom of the Holocene aquitard in The Netherlands (in Dutch), Oude Essink, G.H.P., Goes, B. & Houtman, H., NITG rapport 05-056A.

13

TOWARDS A NATIONAL MONITORING NETWORK 'SALT WATER INTRUSION' IN THE NETHERLANDS

Roelof Stuurman and Gualbert Oude Essink
TNO Geological Survey of The Netherlands
P.O. Box 80015, 3508TA Utrecht, The Netherlands

ABSTRACT

Due to the European Water Framework Directive (EWFD) the Netherlands need to monitor salt water intrusion. Till now only the public water supply companies were used to monitor salt water intrusion around their pumping stations. Nowadays, because of the EWFD monitoring will also focus on regional groundwater bodies in general and on wetlands. Land reclamation, drainage and intensive exploitation of groundwater have triggered or enhanced the flow of various stagnant, saline groundwater bodies in the Netherlands, whereas changes in land use contribute to raised salinities of originally fresh waters. A monitoring procedure will be proposed based on: (1) the sources of salt water, (2) the salt water intrusion processes (flow systems), (3) possible monitoring methods (wells, geophysical) and (4) economy (costs). The many possible sources of salt to groundwater urge for careful mapping and determination of the vulnerability of the water system. The groundwater flow systems are classified according to possible salinization processes (space, time). The Dutch experience, the method of design, the proposed monitoring network and the first results will be presented.

13.1. OBJECTIVES

The Dutch ministry of Housing, Spatial planning and Environment (VROM) is active with the implementation of the European Water Framework Directive (EWFD). Part of this EWFD pays attention to the topic 'salt water intrusion'. The ministry wants to know the starting points of a future, EWDF proof, salt-water intrusion monitoring network. As such, the following topics are examined: (1) The starting points of the actual monitoring programs, (2) The requirements of the EWFD and (3) The practice of circumscription of groundwater bodies in relation to differences in salt content.

13.2. RESEARCH ACTIVITIES

To fulfill the objectives the following activities are executed (Stuurman, 2006):

- The EWFD is studied in relation to its consequences for salt water intrusion monitoring,

- Based on meetings and interviews with water managers, their perception towards the need and implementation of salt water intrusion monitoring is studied,
- The hydrological processes concerning salt water intrusion is studied,
- The fresh-water salt-water interface is mapped,
- An inventory has been made of the Dutch state of the art in relation to monitoring salt water intrusion and methods of monitoring,
- An inventory has been made of all existing 'salt watchers' and monitoring wells eventually useful for salt water intrusion monitoring,
- A uniform method for network design is proposed.

13.3. SALT WATER INTRUSION IN THE EUROPEAN WATER FRAMEWORK DIRECTIVE

On 23 October 2000, the "Directive 2000/60/EC of the European Parliament and of the Council establishing a framework for the Community action in the field of water policy" or short the EU Water Framework Directive (or even shorter the WFD) was finally adopted. The increasing demand by citizens and environmental organizations for cleaner rivers and lakes, groundwater and coastal beaches has been evident for a considerable time. The new European Water Policy will make polluted waters to be clean again, and will ensure that clean waters are kept clean. In achieving these objectives, the roles of citizens and citizens' groups is crucial. This is why a new European Water Policy urges citizens to be more involved. In response to this, the Commission presented a Proposal for a Water Framework Directive with the following key aims:

- expanding the scope of water protection to all waters, surface waters and groundwater
- achieving "good status" for all waters by a set deadline
- water management based on river basins
- "combined approach" of emission limit values and quality standards
- getting the prices right
- getting the citizen involved more closely
- streamlining legislation

There is a number of objectives in respect of which the quality of water is protected. The key objectives at European level are general protection of the aquatic ecology, specific protection of unique and valuable habitats, protection of drinking water resources, and protection of bathing water. All these objectives must be integrated for each river basin. It is clear that the last three - special habitats, drinking water areas and bathing water - apply only to specific bodies of water (those supporting special wetlands; those identified for drinking water abstraction; those generally used as bathing areas). In contrast, ecological protection should apply to all waters: the central requirement of the Treaty is that the environment is protected to a high level in its entirety.

The presumption in relation to the quality of groundwater is that it should not be polluted at all. For this reason, setting chemical quality standards may not be the best approach, as it gives the impression of an allowed level of pollution to which Member States can fill up. A few of such standards have been established at European level for particular issues (nitrates, pesticides and biocides), and these standards must always be adhered to. But for general protection, we have taken another approach. It is essentially a precautionary one. It comprises a prohibition on direct discharges to groundwater, and (to cover indirect discharges) a requirement to monitor groundwater bodies so as to detect changes in chemical composition, and to reverse any antropogenically induced upward pollution trend. Taken together, these approaches should ensure the protection of groundwater from all contamination, according to the principle of minimum anthropogenic impact.

Quantity is also a major issue for groundwater. Briefly, the issue can be put as follows. There is only a certain amount of recharge into a groundwater body each year, and of this recharge, some is needed to support connected ecosystems (whether they are surface water bodies or terrestrial systems such as wetlands). For good management, only that portion of the overall recharge that is not needed by ecology can be abstracted. This is called the sustainable resource, and the Directive limits abstraction to that quantity. One of the innovations of the Directive is that it provides a framework for integrated management of groundwater and surface water for the first time at European level (EU, 2006).

Article 3 of the EWFD (about a minimum list of pollutants and threshold values) also deals with 'parameters indicative of saline or other intrusions'. This is the only passage concerning saline intrusions: With regard to saline concentrations resulting from human activities, Member States may decide to establish threshold values either for sulphate and chloride or for conductivity. In an additional directive "on the protection of groundwater against pollution" (EU, January 2006) which will be incorporated into the EWFD, more specific attention has been devoted on salt water intrusion and monitoring. For investigating purposes whether the conditions for good groundwater chemical status referred to in Article 4(2)(b)(ii) and (iii) are met, Member States will, where relevant and necessary, and on the basis of relevant monitoring results and of a suitable conceptual model of the body of groundwater, assess:

(a) the amounts and the concentrations of the pollutants being, or likely to be, transferred from the body of groundwater to the associated surface waters or directly dependent terrestrial ecosystems;

(b) the likely impact of the amounts and concentrations of the pollutants transferred to the associated surface waters and directly dependent terrestrial ecosystems;

(c) the extent of any saline or other intrusions into the body of groundwater; and

(d) the risk from pollutants in the body of groundwater to the quality of water abstracted, or intended to be abstracted, from the body of groundwater for human consumption.

For each European country due to this EWFD it became necessary to describe the main surface water basins and groundwater bodies. Regarding to saline intrusions it became important to map and monitor the salt-fresh interface. Salinization due to pollution will be part of a surface water and groundwater monitoring system.

13.4. SALT WATER INTRUSION PROCESSES IN THE NETHERLANDS

Water management of the low-lying coastal areas of The Netherlands faces serious difficulties due to the salinization of the subsoil. Due to several causes, Dutch coastal groundwater systems are threatened by an intrusion of saline water as well as fresh water, see Figure 1. Figure 1 shows a polder which is an area that is protected from water outside the area and which has a controlled water level. In the dune area, surface water flows into the aquifer by natural recharge and infiltration through high lying pools. Groundwater from the freshwater lens is extracted for drinking water supply. Low-lying polders in the hinterland, that attract brackish seepage, have to be flushed to guarantee good water quality. Locally, also freshening of the groundwater resources is taking place between high-lying lakes and low-lying polders. The salt originates from marine deposits of Pliocene to Early-Pleistocene origin as well as from transgressions of the sea during the Holocene (Post, 2003). Though locally also freshening of the fresh groundwater resources is taking place, the salinization affects water management sectors such as agriculture, domestic and industrial water supply. Water courses in low-lying areas are flushed frequently to dispose the excess of salt. In addition, there is the danger of salt damage to crops due to a high salt load from the deeper subsoil.

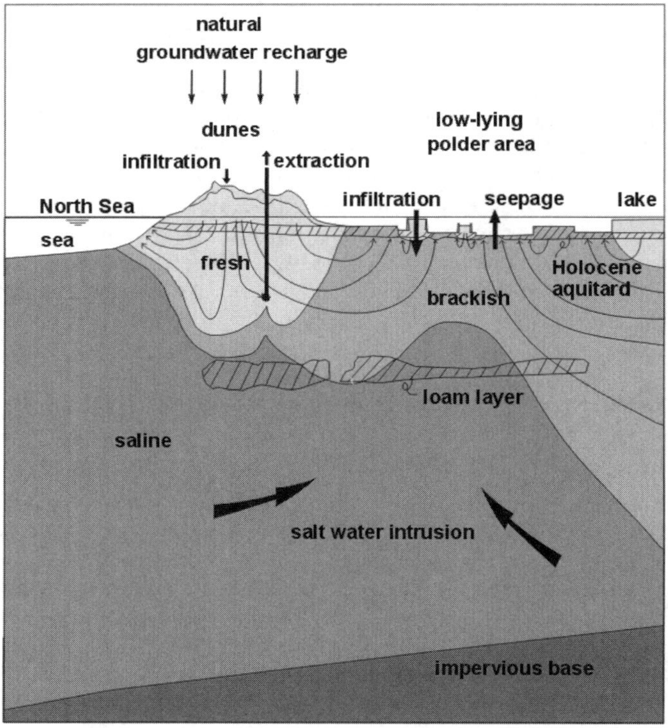

Fig. 1. Schematization of the coastal hydrogeological situation in The Netherlands. Vertical axes: some 200m; horizontal axes: some 50 kilometer.

The physical processes under consideration are advection, dispersion, molecular diffusion and possibly chemical osmosis. The complicated distribution of fresh, brackish and saline groundwater is related to the diversity of local flow systems associated with man-made topography during the past millennium (Figure 2). The salinization problem is in fact caused by two processes of anthropogenic origin that has been going on for nearly a millennium. The first –drainage of peaty and clayey soil by digging channels and building dikes– is a slow and continuous process, leading to land subsidence by peat oxidation as well as compaction and shrinkage of clay. The greatest land subsidence of this kind occurred in the western and northern parts of The Netherlands. The second process – land reclamation – causes a relatively abrupt change in the land surface level, creating the well-known Dutch polders. The last centuries, these processes have lead to relative large groundwater velocities, up to several tens of meters per year.

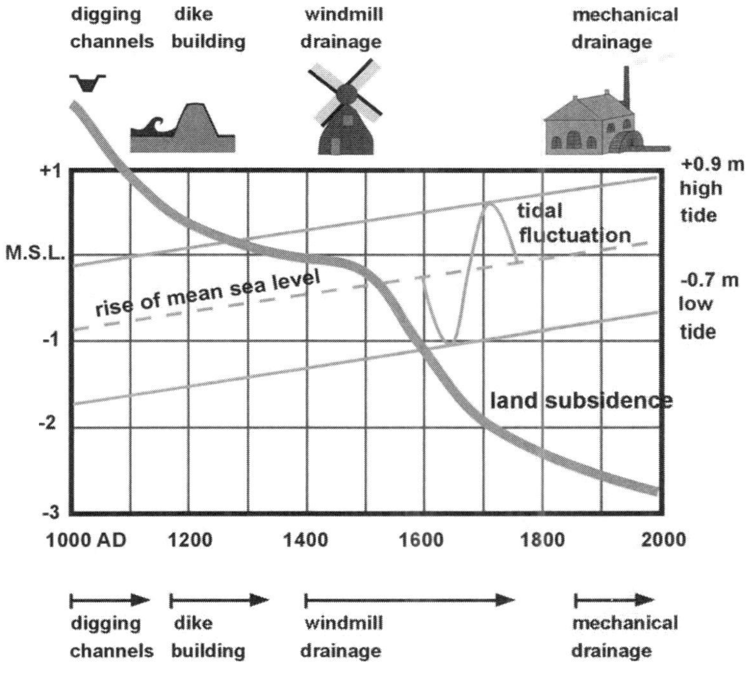

Fig. 2. Lowering of the ground surface in The Netherlands (schematic). Modified from: Atlas van Nederland: deel 15 Water, 1986. A significant head difference exists between the high mean sea level and the controlled surface water level in the polders, causing salt water intrusion.

Analyses of historical chloride concentration as well as numerical density dependent groundwater flow models demonstrate these findings (Oude Essink, 2005). A future sea level rise will lead to a larger inflow of saline groundwater to the coastal aquifers. However, the change in volume of freshwater lenses is probably negligible, as the supposed increase in natural groundwater recharge will probably compensate the effect of sea level rise.

Nowadays, many low-lying polders are currently experiencing upward groundwater flow (seepage) with a high salinity. For instance, up coning of saline groundwater occurs under the low-lying polder Groot-Mijdrecht (Figure 3). This small low-lying polder in the Province of Utrecht is one of the deepest polders of The Netherlands, with phreatic water levels of -5.60m to -6.25m M.S.L. The Holocene resistance layer is thin. Fresh surface water infiltrates from the adjacent high-lying lake Vinkeveense Plassen at the right and flows under the dike towards the deep polder, with velocities up to several mm/day. In the central part of the polder seepage is brackish to saline. The chloride concentrations of this seepage is several thousands of mg Cl^{-}/l, causing a severe salt load from this polder to the surface water management system of the entire region (Oude Essink en Schaars, 2003).

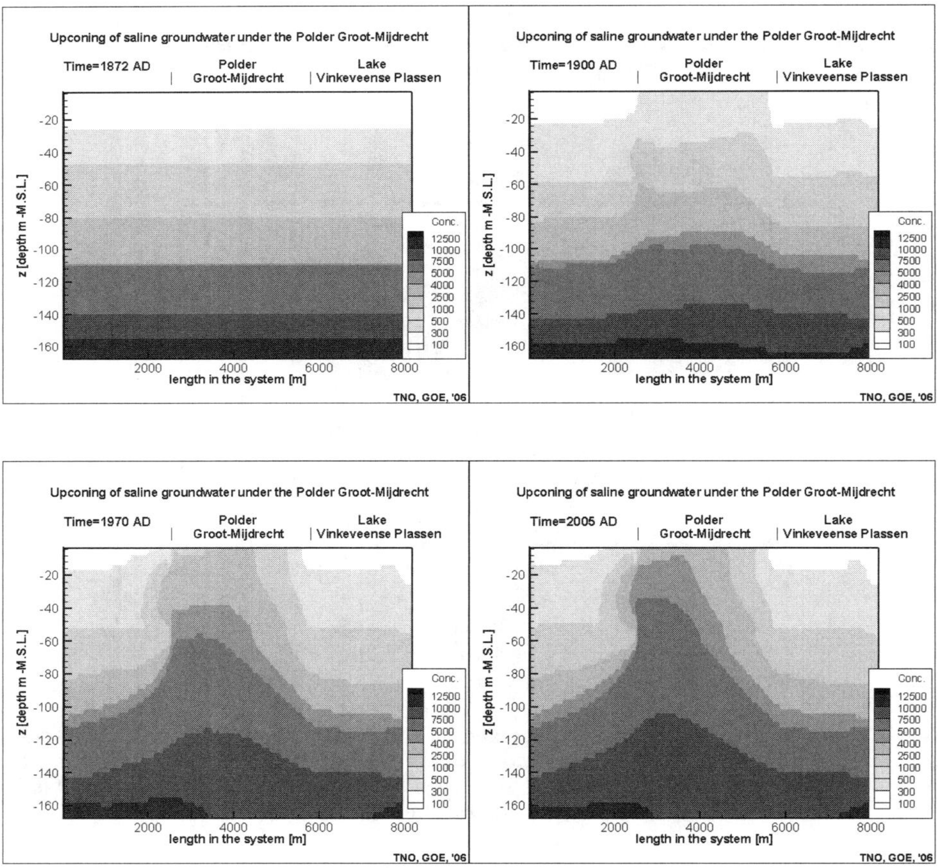

Fig. 3. Example of a numerical model simulating the upconing of brackish groundwater under the low-lying Polder Groot-Mijdrecht (reclaimed in the years 1872-1877).

In relation to the EWFD it is important to understand the status of a groundwater system. Irreversible salinizations due to historical activities are accepted but new activities may not create saline intrusions. Continuing salinization in existing deep polders will be accepted. In Tables 1 and 2 the main situations are summarized.

Table 1. Based on numerous Dutch case-studies on saline-fresh processes, the following situations have been classified:

Process	Cause
1. Irreversible human activity (in hydraulic head distribution) from the past	*Reclamation of polder areas*
	Lowering in phreatic groundwater level during excavation of sand, peat or clay
2. Irreversible human activity (in chemical composition) from the past	*An historical event has displaced saline groundwater in such a way that recovery after stopping the event will not occur*
3. (Still) active irreversible human activity	*Land subsidence due to gas- and salt mining*
	Land subsidence due to water level management
4. (Still) active reversible human activity	*Upconing: Groundwater abstraction (recovery takes 10 x as long as salinization process of upconing)*
	Reversed groundwater flow in an area with a vertical salt-fresh interface
5. Historical natural cause	*Salinization from closed sea inlets, estuaries or rivers*
6. Active natural cause	*Landward advancing sea due to sea level rise (causing the origin of tidal inlets)*
	Landward advancing salt water tongue in (infiltrating) open river months
7. Combinations	

Table 2. Saline-fresh processes can be subdivided into:

a. In equilibrium	Will account for small systems, i.e. along dikes and small dune areas
b. Not in equilibrium, expanding salt water intrusion	Often occurring, reclaimed polder areas
c. Not in equilibrium, expanding fresh (surface) water intrusion	Along the boundaries of (deep) polders and high-lying lakes
d. Not in equilibrium, expanding, recovering	Closed or reduced groundwater extractions (i.e. large dune areas)

13.5. SALINIZATION: CHLORIDE POLLUTION OR SALT WATER INTRUSION?

The EWFD is not really clear about the definition of 'saline intrusion'. The main process is salinization. In this article salinization is subdivided into 'chloride pollution' and 'saline intrusion'. Chloride pollution is an man-made process and can have many sources: artificial (KCl) or animal manure used in agriculture, road salt, local point sources, raised marine construction sand, brine due to man-made osmosis of brackish water (Stuyfzand and Stuurman, 1994). Figure 4 shows the chloride content of shallow groundwater beneath agricultural areas due to the use of the fertilizer KCl. Natural chloride concentrations due to evaporating rainwater would be approximately 10-18 mg/l . In general the chloride content can be 3-10 times higher. There are many locations were the chloride content is even higher than 100 mg/l. For example in areas were groundwater was recharged by river Rhine water. During the years '70 and '80 industrial and mining activities in France and Germany increased the chloride content into 300 mg/l. These high chloride concentrations can have a negative influence on ecology.

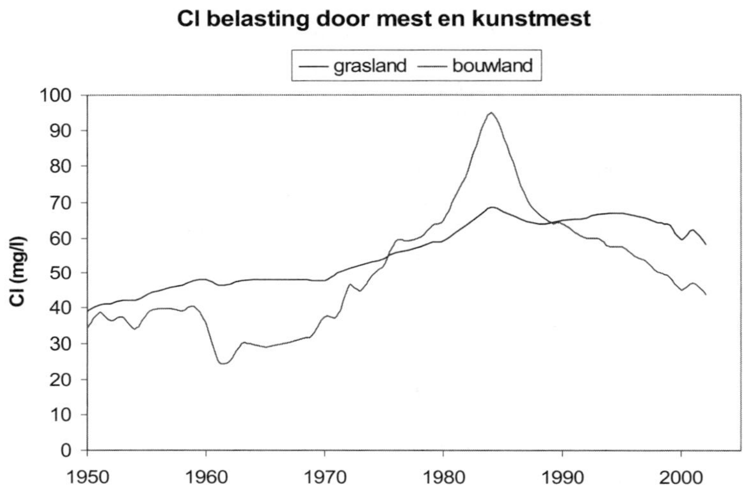

Fig. 4. The chloride content of shallow groundwater in the province of Noord-Brabant (grasland= meadow; bouwland=agricultural area).

Saline intrusion is caused by human activities or by natural processes (Table 1). As a result the volume of the fresh groundwater body decreases. The source of salt water (chloride) is always seawater. This can be original formation water from the time of deposition. It can intrude from the sea or estuaries and it can originate from recent of historical flooding.

13.6. DUTCH EXPERIENCES WITH SALT WATER INTRUSION MONITORING

In the Netherlands exists a groundwater quality monitoring using hundreds of monitoring wells (depth < 25 m). In general these wells are sampled every year. So monitoring chloride pollution is covered well. On the contrary the experiences with salt water intrusion monitoring are very poor, based on a literature survey and interviews with water managers. Nevertheless, Table 3 summarizes monitoring instruments, used in The Netherlands. Only around groundwater abstractions for drinking water purposes, frequent monitoring of raw water quality is normal as part of the production process.

Table 3. A summary of monitoring instruments in The Netherlands used for salt water intrusion monitoring.

1. Raw water	Groundwater quality for drinking water is monitored daily
2. Hydraulic head observations	-Decreasing trends can be an indication for salt water intrusion, -In the region Zeeland monitoring of the hydraulic head of groundwater below the fresh water lenses is common
3. Salt guards	Common around drinking water abstractions
4. Groundwater quality	The national and regional groundwater quality monitoring networks (observation < 25 m) are useful in the polder areas
5. Cone penetration Test (CPT)	In the region Zeeland the extension of fresh water lenses will be monitored yearly using CPT
6. Electro-magnetic borehole survey	In the region Brabant for the first time EM bore hole survey is used as a monitoring tool and compared with 30-40 year old bore hole surveys.
7. Geo-electrical (VES) measurement	Repeating VES-measurement are sometimes in use
8. Geo-electrical (ERT) measurement	Possible this method will be used in the northern region Fryslân. This method is using two wells. The result is a two-dimensional conductivity profile
9. Periodic conductivity measurements	At some places automatic periodic groundwater conductivity measurements are used
10. Variable-density modeling	Numerical modeling of density dependent groundwater flow and coupled solute transport in combination with monitoring can be an effective tool

Near some abstraction sites, vulnerable to salt water intrusion, permanent electrode systems for groundwater quality monitoring ("salt guards") are used. Such a electrode system consists of a 26-strand cable, onto which 26 helical electrodes are attached in pairs (Figure 5). Both ends of the electrodes of the electrodes are fixed at the cable. The distance of two electrodes forming an observation point is usually 20 cm. The intervals between electrode pairs may vary, depending on the local conditions and the requirements, but

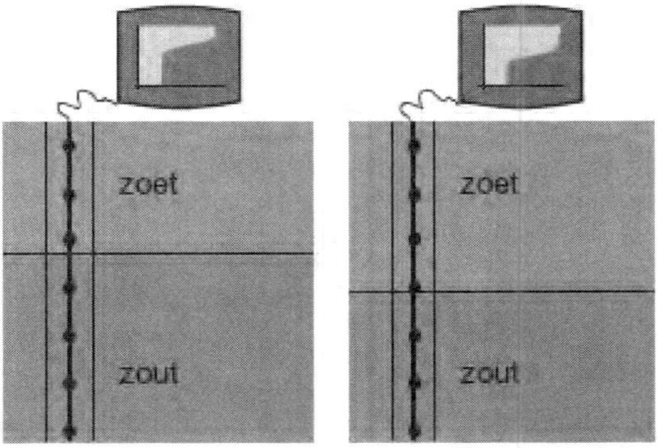

Fig. 5. Monitoring fresh/saline interface using permanent electrode systems (salt guards) for groundwater quality monitoring (zoet=fresh; zout=salt).

should not be less than 1 m. For salt water intrusion monitoring in the Netherlands 538 of these salt guards are installed (146 near a drinking water abstraction). However, the monitoring results are difficult to recover. A promising relatively cheap method is electro-magnetic borehole logging in existing PVC monitoring wells.

Ten methods of salt water intrusion monitoring can be distinguished (Table 3) but only 3 methods are general used; (1) raw water analysis (only interesting for operational drinking water management), (2) deep multi-screen observations wells and (3) salt guards. As part of the project an inventory was made of these potential monitoring locations for a national monitoring network (Table 4). Brackish water is the zone between 150-1000 mg/ chloride. The salt water boundary is defined at 1000 mg/l chloride. In Figure 6 these locations are combined with the map of the Netherlands.

Table 4. Number of potential monitoring locations in the Netherlands.

method	*Outside drinking water Protection zone*	*Inside drinking water protection zone*	*total*
Multi-screen wells (brackish)	976	375	1351
Multi-screen wells (salt)	991	268	1259
Salt guards	386	146	532

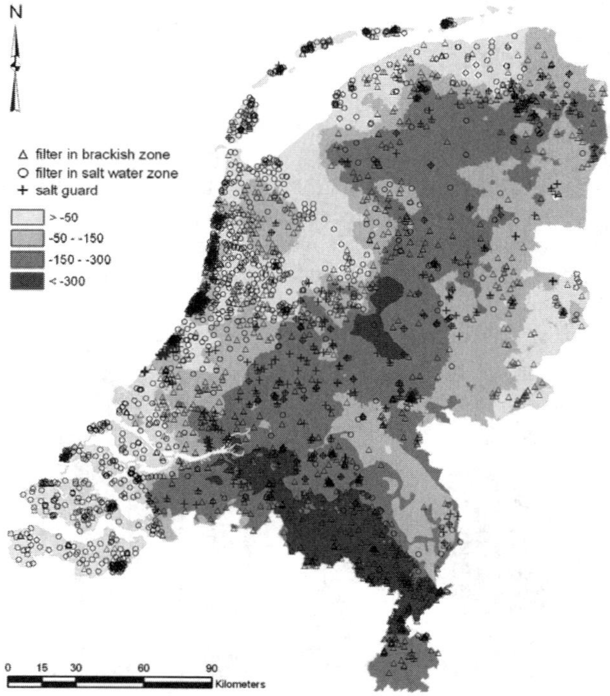

Fig. 6. Depth to brackish/salt groundwater interface and the suitable monitoring locations.

13.7. STARTING POINTS FOR A NATIONAL MONITORING NETWORK

Based on this study the following starting points for monitoring salt water intrusion can be identified. The EWFD is not very clear in its statutory obligation in relation to the requirements of a monitoring network. Of course the data collection around pumping stations for drinking water needs to be more detailed than for regional groundwater bodies. Observation locations should be at places were there is significant movement of the fresh-salt interface. The density and depth of observation wells and the monitoring frequency depends on the vulnerability of protected objects (wetlands, drinking water body). Also the costs and benefits should be taken into account.

At first site there are plenty existing monitoring wells and salt guards which could be explored into a national monitoring system. The design of a monitoring network depends strongly on conceptual models of the groundwater bodies (Tables 2 and 3). The monitoring data should be added into a central database. It should be clear how and when the monitoring results should be reported. The EU demands reports at an interval of six years. This interval is, in general, more than enough in relation to the velocity of potential salt water intrusion in regional groundwater bodies. However, around pumping stations for drinking water supply, the monitoring frequency should be higher.

REFERENCES

Atlas van Nederland, Deel 15 Water, (in Dutch), Ven, G.P., van de et al. (eds), Staatsuitgeverij, 's-Gravenhage, 1986.

EU, 2006. WFD. Introduction and overview. The key issue of a new directive (http://europa.eu.int/comm/environment/water/water-framework/overview.html)

Oude Essink, G. H. P., 2001. Salt Water Intrusion in a Three-dimensional Groundwater System in The Netherlands: a Numerical Study. Transport in Porous Media 43(1): 137-158.

Oude Essink, G.H.P., F. Schaars, 2003. Impact of climatic change on the groundwater system of the water board of Rijnland, The Netherlands. Proc. 17th Salt Water Intrusion Meeting, Delft. The Netherlands.

Oude Essink, G. H. P., 2004. Modelling 3D density dependent groundwater flow at the Island of Texel, The Netherlands (chapter 4). Coastal Aquifer Management-Monitoring, Modeling and Recent Practices. A. H.-D. Cheng and D. Ouazar, CRC Press: 75-92.

Post, V. E. A., 2003. Groundwater salinization processes in the coastel area of the Netherlands due to transgressions during the Holocene. Amsterdam, the Ntehrlands, Vrije Universiteit.

Stuyfzand, P.J., R.J. Stuurman, 1994. Recognition and genesis of various brackish to hypersaline groundwaters in the Netherlands. Paper 13th Salt Water Intrusion Meeting, Villasimius, Sardinia, Italy.

Stuurman, R.J., G.H.P. Oude Essink, B. van der Grift, 2006. Monitoring slat water intrusion in relation to EWFD. TNO-report (in prep.).

TNO-NITG, 2005. Chloride-concentration at the bottom of the Holocene aquitard in The Netherlands (in Dutch), Oude Essink, G.H.P., Goes, B. & Houtman, H., NITG rapport 05-056-A.

Section 3:

COASTAL WATER EUTROPHICATION

14

RIVERINE NUTRIENT INPUTS TO LAKE PONCHARTRAIN, LOUISIANA, USA

Philip Saksa and Y. Jun Xu
School of Renewable Natural Resources
Louisiana State University Agricultural Center, Baton Rouge, LA 70803, USA

ABSTRACT

The Amite, Tickfaw, and Tangipahoa Rivers represent 72% of the freshwater basin drainage into Lake Pontchartrain, a large oligohaline estuary located north of the city of New Orleans in southeastern Louisiana, USA. The amount of nutrient loads carried by the river discharge may have profound effects on aquatic ecosystems in the lake and surrounding wetlands. This study assessed more than two decades (1978-2001) of total nitrogen (TN), total phosphorus (TP), and total organic carbon (TOC) loadings to investigate seasonal and interannual variations of nutrient fluxes in the three river basins. Long term discharge and water quality data were used in the study to quantify mass inputs of the nutrients to Lake Pontchartrain. The results show that on average, the three rivers annually discharged 4,133 Mg TN, 688 Mg TP, and 22,942 Mg TOC. Large variation in annual loading was found for TN ranging from 659 to 6,469 Mg, TP ranging from 124 to 1,146 Mg, and TOC ranging from 3,680 to 38,342 Mg. The more urbanized Amite River basin showed the highest fluxes of nitrogen (0.78 g m^{-2} yr^{-1}), phosphorus (0.14 g m^{-2} yr^{-1}) and organic carbon (4.8 g m^{-2} yr^{-1}), when compared with those in the rural Tickfaw River basin (in the same order: 0.56, 0.07, and 3.76 g m^{-2} yr^{-1}) and the more agricultural Tangipahoa River basin (0.71, 0.11, and 2.8 g m^{-2} yr^{-1}). Monthly loading varied extensively over the year, ranging from 128 Mg to 650 Mg for TN, 22 Mg to 120 Mg for TP and 634 Mg to 4,088 Mg for TOC. Highest loadings of the nutrients occurred during the winter and spring months from January through April and the lowest loadings occurred in the months from August to November. This large variation in loading closely followed the seasonal hydrologic conditions in the basins, with a positive correlation of the nutrient loads to the river discharge.

14.1. INTRODUCTION

Lake Pontchartrain is a 1,619 km^2, oligohaline estuary in southeastern Louisiana. The Amite River, Tangipahoa River, and Tickfaw River are located north of the lake, covering southeastern Louisiana and part of southern Mississippi. With 72% of the freshwater drainage area into Lake Pontchartrain, they have the largest potential impact on the levels of nutrient concentration in the lake waters. Both the Amite and Tickfaw Rivers initially drain into Lake Maurepas which connects to the Ponchartrain through Pass Manchac. Managing these watersheds for water quality maintenance and improvement is essential for the health of the lake and its aquatic systems.

Draining an area of 8,728 km^2, water quality in these three rivers is largely affected by land use activities. The Amite River basin is the largest, covering 4,822 km^2, with the

Tangipahoa and Tickfaw watersheds following at 2,010 km^2 and 1,896 km^2 respectively. Although the area is largely forested, increased urbanization, particularly in the Amite River basin, and agriculture in the Tangipahoa and Tickfaw River basins, has led to an increase in runoff and other changes in the flow of water through the basin (LDEQ, 2000). Jordan and others (2003) showed that more developed watersheds had increased annual discharge and variability, with higher peak rates, over less developed watersheds. Sand and gravel mining adjacent to the Amite and Tangipahoa Rivers, as well as throughout all three river basins, has led to additional river impairment (LPBF, 2005).

Increases in nitrogen and phosphorus loadings are the most common sources of eutrophication in an estuary (Nixon, 1995). Both nitrogen and phosphorus can be limited in estuaries at different times of the year, so increased loading of both nutrients can cause decreases in dissolved oxygen, increases in chlorophyll concentration, and are detrimental to submerged aquatic vegetation (Conley, 2000). Turner and others (2002) found that during an average year, most of the nitrogen loading into Lake Ponchartrain originates from rivers on the north shore. Excess TOC also causes reduced levels of dissolved oxygen and eutrophication in these coastal estuary systems.

Anthropogenic activities, such as increases in population and intensive agriculture, have additional significant effects on eutrophication levels in many coastal waterbodies. This study looks at the seasonal and interannual loadings of TN, TP and TOC over a 24-year period (1978-2001). Long term trends of these nutrient loadings are also analyzed. This information is necessary for resource managers and policy makers to develop plans to effectively manage land use, reducing the amount of nutrient and sediment runoff to Lake Pontchartrain.

14.2. METHODS

14.2.1. Study Area

The Amite, Tickfaw and Tangipahoa Rivers drain a total land area of 8,728 km^2, covering ten Louisiana parishes and three Mississippi counties (Figure 1). The elevation within the area varies from sea level in the south to 150 m above sea level in the north. Annual mean air temperature for the area is 19 °C, varying from 12 °C in January to 28 °C in July, and annual average precipitation in the study area is about 1600 mm, ranging from 1108 mm to 2178 mm. The highest monthly average precipitation in the area occurs in July (159 mm), while the lowest is in October (86 mm). The amount of urban land cover in the region is relatively low, 5.3% for the Amite, 1.1% for the Tickfaw, and 4.2% for the Tangipahoa River Basins (Wu and Xu, in press). These estimates are based on land cover data circa 1980 however, and may not be accurate throughout the entire time period of our study.

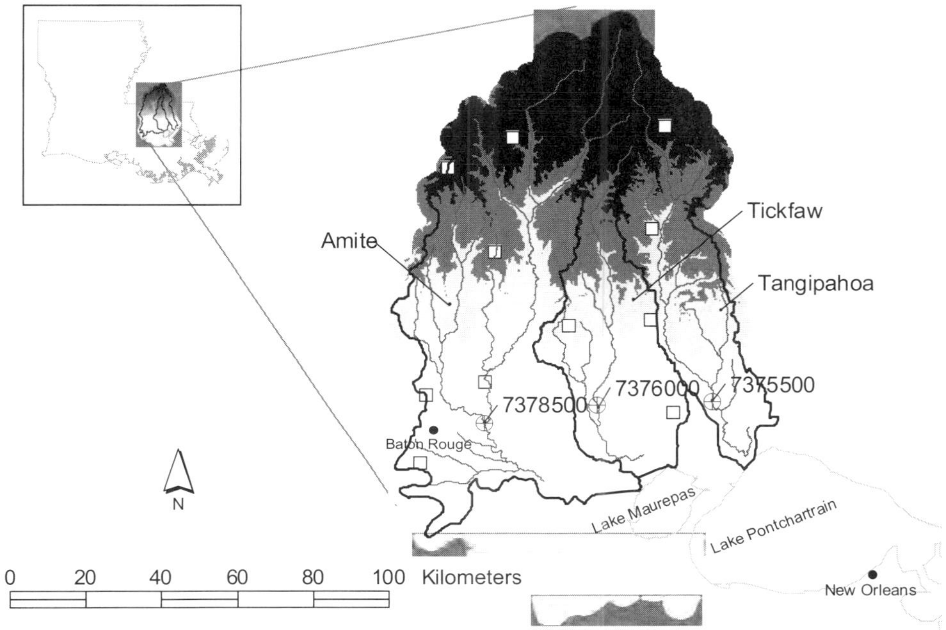

Fig. 1. The Amite, Tickfaw and Tangipahoa River basins and water quality monitoring stations (USGS station number; adopted from Wu and Xu, in press).

14.2.2. Data Collection & Analysis

Water quality data for each of the three rivers was obtained from three monitoring locations by the Louisiana Department of Environmental Quality: the Amite River at Port Vincent, the Tickfaw River at Holden, and the Tangipahoa River at Robert. The discharge data for the Amite River was obtained from a different station near Denham Springs, about nine miles north of Port Vincent. Monitoring locations on the Amite River did not have both long term discharge and water quality data available at the same site. The effective drainage area above each monitoring station, retrieved from the USGS National Water Information System (NWISWeb) and represented by the water quality data, is 3,315 km^2 for the Amite, 1,673 km^2 for the Tangipahoa, and 639 km^2 for the Tickfaw River basins. Monthly and annual discharge were obtained from a 62-year surface hydrology study on the basins by Wu and Xu (in press).

Nutrient loadings were calculated based on monthly recorded concentration levels and the river discharge. The data was then analyzed to determine monthly and annual characteristics from each individual drainage basin, as well as the collective input into Lake Ponchartrain. Complete discharge data was not available for 1978, so this year was not used in annual calculations. Neither was complete discharge data available for the years of 1988 and 1989 on the Tickfaw only, and therefore, these years were also excluded from this watershed's annual calculations. There were a number of months where water quality data was also unavailable. Therefore, a log-linear regression model was developed to estimate nutrient loadings for the missing months of data. The general model for each river is defined as:

$$\ln(S_i(t)) = b_0 + b_1 \ln(Q_{mon}(t)) + \varepsilon(t)$$

where S(t) represents total monthly loading (Mg), Q_{mon} is total monthly discharge (m^3), i is the type of nutrient, and $\varepsilon(t)$ is an error term assumed to be normally distributed.

The regression was calculated using SAS Statistical Software (SAS Institute Inc., 1996). The fitted parameters used to estimate the missing loadings are summarized in Table 1. Total nitrogen was derived from the summation of total Kjeldahl nitrogen and nitrate/nitrite nitrogen, but regression analysis for these nutrients was performed separately to create more accurate estimates of the missing data.

Table 1. Parameters and R-squares from log-linear regression.

		$NO_3 + NO_2$	TKN	TP	TOC
	b_0	3.23	3.94	2.17	3.67
Amite	b_1	1.10	1.15	1.16	1.27
	R-square	0.62	0.83	0.81	0.83
	b_0	-0.98	0.51	0.04	2.05
Tangipahoa	b_1	1.36	1.31	1.26	1.33
	R-square	0.68	0.55	0.64	0.68
	b_0	2.14	1.82	2.64	1.93
Tickfaw	b_1	1.18	1.27	1.11	1.39
	R-square	0.70	0.76	0.77	0.79

14.3. RESULTS

14.3.1. Hydrology

Long-term characteristics of the discharge (1940-2002) from the Amite, Tickfaw and Tangipahoa have been discussed in detail by Wu and Xu (in press). Their major findings are summarized briefly below. On average, the three rivers discharged ($\pm$ standard error) 0.27 $\pm$0.04 km^3 freshwater monthly and 3.29 $\pm$0.15 km^3 freshwater annually into the lake estuarine system, with the highest inflow from the Amite River (0.16 $\pm$0.03 km^3 mon^{-1}, and 1.91 $\pm$0.09 km^3 yr^{-1}) and the lowest inflow from the Tickfaw River (0.03 $\pm$0.00 km^3 mon^{-1}, and 0.34 $\pm$0.02 km^3 yr^{-1}).

During this study's period, from 1978 to 2001, the Amite River showed an average annual discharge of 2.18 $\pm$0.16 km^3, ranging from 1.0 km^3 to 6.1 km^3. The Tangipahoa and Tickfaw Rivers had respective average annual discharges of 1.12 $\pm$0.07 km^3, ranging from 0.4 km^3 to 1.9 km^3, and 0.38 $\pm$0.03 km^3, ranging from 0.09 km^3 to 0.6 km^3. The average annual discharge from all three rivers into Lake Ponchartrain was 3.67 $\pm$0.26 km^3. Over that same time period, the Amite River had a mean monthly discharge of 0.18 $\pm$0.01 km^3, ranging from 0.07 km^3 to 0.35 km^3. The Tangipahoa and Tickfaw Rivers had respective mean monthly discharges of 0.09 ± 0.005 km^3, ranging from 0.05 km^3 to 0.16 km^3, and 0.03 ± 0.002 km^3, ranging from 0.01 km^3 to 0.06 km^3. The total mean monthly discharge into the lake from all three rivers was 0.30 ± 0.02 km^3, ranging from 0.13 km^3 to 0.55 km^3 (Figure 2). The monthly discharge was highest in the months January through April, corresponding with the wet season for the area. Lowest discharges were found in the months of August to November, during the dry season. There was a large amount of seasonal variability, with high flow periods discharging about five times the amount in low flow periods for the Amite and Tickfaw Rivers, and more than three times the amount for the Tangipahoa.

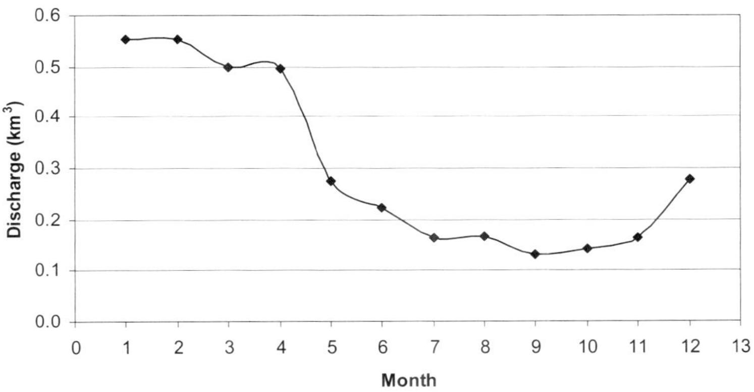

Fig. 2. Seasonal trend of the long-term discharge (1978-2001) from the Amite, Tickfaw and Tangipahoa Rivers.

14.3.2. Organic Carbon Loadings

Organic carbon showed the highest mass loading in the rivers, when compared with nitrogen and phosphorus. Annually, the Amite, Tangipahoa and Tickfaw Rivers discharged on average 15,780 Mg, 4,758 Mg and 2,404 Mg of organic carbon into Lake Ponchartrain (Figure 3). Interannual variation in the TOC loadings was high, ranging from 2,361 Mg to 30,449 Mg for the Amite River, 988 Mg to 7,742 Mg for the Tangipahoa River, and 331 Mg to 4,410 Mg for the Tickfaw River. The average TOC loading from the three rivers combined to the lake was 22,942 Mg annually, with a range of 3,680 to 38,342, a more than ten-fold variation during the twenty-four years.

The Amite River had monthly mean TOC loadings (Figure 4) of 1,294 Mg, with a range of 419 Mg to 2,845 Mg. Average monthly TOC loadings for the Tangipahoa and Tickfaw Rivers were 390 Mg, ranging 129 Mg to 788 Mg, and 198 Mg, ranging from 41 Mg to 539 Mg, respectively. The mean monthly TOC loadings from all three rivers to Lake Ponchartrain totaled 1,882 Mg, ranging from 634 Mg to 4,088, more than a six-fold variation. Seasonal loading was consistent with seasonal discharge trends.

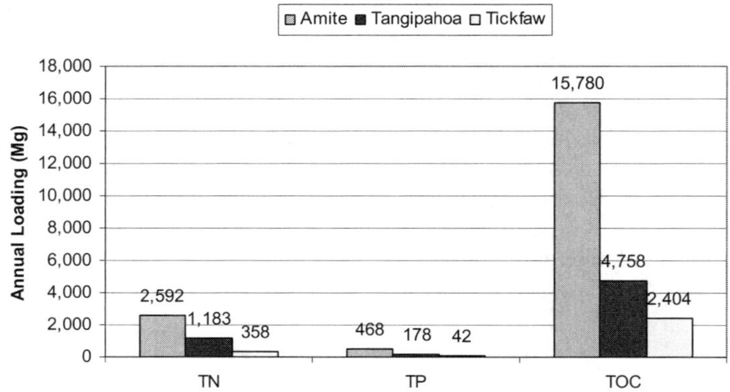

Fig. 3. Annual averages of total nitrogen (TN), total phosphorous (TP) and total organic carbon (TOC) loadings from three upper Lake Pontchartrain basins.

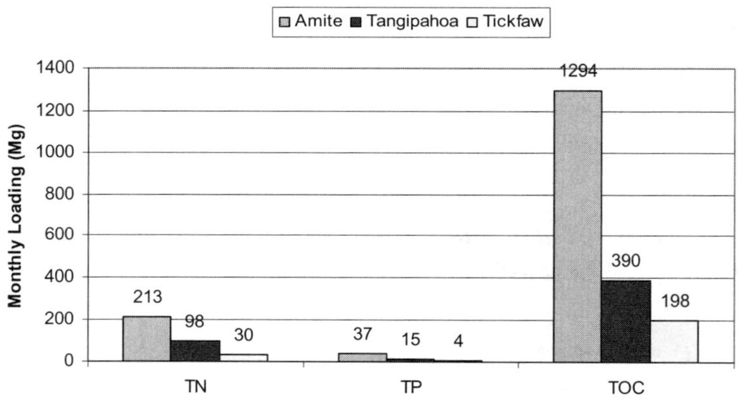

Fig. 4. Monthly averages of total nitrogen (TN), total phosphorous (TP) and total organic carbon (TOC) loadings from three upper Lake Pontchartrain basins.

14.3.3. Nitrogen Loadings

Long-term average annual total nitrogen loading for the Amite River was 2,592 Mg, with a range of 466 Mg to 4,183 Mg (Figure 3). The Tangipahoa and Tickfaw Rivers each had average annual TN loadings of 1,183 Mg, ranging from 160 Mg to 2,236 Mg, and 358 Mg, ranging from 33 Mg to 574 Mg, respectively. Average TN loading to the lake from all three rivers totaled 4,133 Mg annually, with a range of 659 Mg to 6,469 Mg, a variation of almost ten-fold.

Monthly average TN loading for the Amite River was 213 Mg, with a range of 78 Mg to 402 Mg (Figure 4). Mean monthly TN loadings for the Tangipahoa and Tickfaw Rivers were 98 Mg, ranging 29 Mg to 185 Mg, and 30 Mg, ranging from 9 Mg to 75 Mg, respectively. The three rivers combined had a mean monthly loading of 341 Mg to Lake Ponchartrain, ranging from 128 Mg to 650 Mg, a five-fold variation. The seasonality of TN loading remained consistent with the hydrologic regime, with the high period starting one month earlier in December and lasting through April, and the low period occurring from August to November.

14.3.4. Phosphorus Loadings

Average annual total phosphorus loading (Figure 3) in the years 1978-2001 for the Amite River was 468 Mg, varying from 81 Mg to 719 Mg. Annual TP loadings from the Tangipahoa and Tickfaw Rivers averaged 178 Mg and 42 Mg, ranging from 34 Mg to 364 Mg and from 9 to 78 Mg, respectively. The average annual TP loading to the lake from all three rivers totaled 688 Mg, ranging from 124 Mg to 1,146 Mg, a variation of more than four-fold during the two decades.

Monthly mean TP loading (Figure 4) for the Amite River was 39 Mg, with a range of 13 Mg to 78 Mg. The Tangipahoa and Tickfaw Rivers had average monthly loadings of 15 Mg, ranging 4 Mg to 33 Mg, and 4 Mg, ranging from 1 Mg to 9 Mg respectively. The three rivers combined had a mean monthly loading of 57 Mg to Lake Ponchartrain, ranging from 22 Mg to 120 Mg, a variation of more than five-fold. The seasonality of total phosphorus loading was the same as discharge, with a high period of January through February, and a low period of August through November.

14.4. DISCUSSION

Nutrient loading was positively correlated with discharge levels using r-square values (Table 1). This positive correlation, along with direct comparison of monthly averages, confirms that seasonal loading generally follows, and even depends upon, seasonal discharge with high rates during the months of January – April and low rates during the months of August – September. Coefficients of variation in monthly loading averages (TN, TP, TOC) were determined for the Amite (7.4%, 8.4%, 8.1%), Tangipahoa (7.7%, 9.0%, 7.6%), and Tickfaw (7.8%, 8.3%, 10.1%) Rivers. Monthly load variation was generally even among the rivers, however, variation was largest in the Tangipahoa River for TP while the Tickfaw River showed the highest monthly TN and TOC variation.

Long term total loadings of nitrogen, phosphorus and carbon into Lake Ponchartrain over the 24-year period (Figure 5) did not produce any major discernable trends. From their study on the rivers' long-term discharge (1940-2002), Xu and Wu (in press) reported a 24-year high flow period from 1975-1998 that largely overlaps this study. This may indicate that the mass loadings of nitrogen, phosphorus and carbon may have increased due to the increased river discharge. The coefficients of variation in annual loading averages (TN, TP, TOC) were also determined for the Amite (7.7%, 8.2%, 8.9%), Tangipahoa (8.3%, 9.5%, 8.8%), and Tickfaw (9.4%, 8.8%, 10.7%) Rivers. Annual variability was generally higher than monthly variability, and once again was evenly distributed among the three rivers. However, variability was still highest in the Tickfaw River for TN and TOC, and remained the highest in the Tangipahoa River for TP as well.

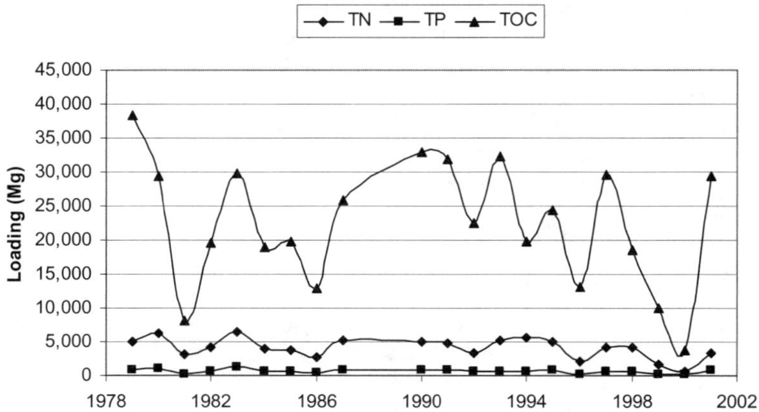

Fig. 5. Interannual variation of total nitrogen (TN), total phosphorous (TP) and total organic carbon (TOC) loadings from three upper Lake Pontchartrain basins (no 1988-89 data available for the Tickfaw).

The annual runoff rates for the Amite and Tangipahoa Rivers were 573 mm and 629 mm respectively, with the Tickfaw at 534 mm (Wu, 2005). The average annual TN fluxes from the Amite and Tangipahoa rivers totaled 0.78 g m^{-2} and 0.71 g m^{-2} respectively, with the flux from the Tickfaw at 0.56 g m^{-2} yr^{-1}. Average annual TP fluxes from the Amite and Tangipahoa rivers measured 0.14 g m^{-2} and 0.11 g m^{-2} respectively, with the Tickfaw at 0.07 g m^{-2} yr^{-1}. Lastly, the average annual TOC fluxes in the Amite and Tangipahoa Rivers were 4.8 g m^{-2} and 2.8 g m^{-2} respectively, with the annual flux from the Tickfaw at 3.8 g m^{-2} (Table 2).

After being adjusted to account for the difference in watershed area, it appears clear that the Amite and Tangipahoa Rivers contributed higher nitrogen and phosphorus yields per unit land area that the Tickfaw River. Research in the Mississippi River basin by Turner and Rabalais (2003) found a direct relationship of erosion and nutrient concentrations to land use changes - particularly urbanization, cropland, and pasture. The Amite showed the highest fluxes for all parameters studied. Gravel mining along the river, and the increase in

Table 2. Annual averages of total nitrogen (TN), phosphorus (TP), and organic carbon (TOC) fluxes in three Gulf coastal river basins.

	Amite	Tangipahoa	Tickfaw
Area (km^2)[1]	3,315	1,673	639
TOC (g m^{-2})	4.8	2.8	3.8
TP (g m^{-2})	0.14	0.11	0.07
TN (g m^{-2})	0.78	0.71	0.56

[1]Drainage area above water quality monitoring location

urbanization that Wu (2005) found, likely resulted in higher erosion rates. These higher fluxes in the basin with the most urbanization are also concurrent with the findings by Jordan and others (2003) in a study on the Patuxent River estuary, which showed increases in annual mean concentrations of TN, TP, and TOC with increases in the proportion of developed land. The Tangipahoa showed slightly lower TN and TP fluxes than the Amite, corresponding with the highest percentage of agriculture among all the basins, also documented by Wu (2005).

Goolsby and others (2001) conducted a comprehensive study on nitrogen fluxes from 42 interior basins within the Mississippi-Atchafalaya River basin from 1980 to 1996, and they found annual TN fluxes typically ranging from 1.5 g m^{-2} to 3.1 g m^{-2} in the upper basin region, with watersheds exceeding 6 g m^{-2} during years of high streamflow. In another study on nutrient loading from the Upper Mississippi River basin during 1973 to 1994, Turner and Rabalais (2004) reported a TN flux of 0.8 g m^{-2} yr^{-1} - 1.1 g m^{-2} yr^{-1}, TP flux of 0.068 g m^{-2} yr^{-1} - 0.076 g m^{-2} yr^{-1}, and carbon flux of 2.1 g m^{-2} yr^{-1} – 2.5 g m^{-2} yr^{-1}. Compared to these studies, in our watersheds the TN fluxes are lower, TP ranges from an equal flux in the Tickfaw to being doubled in the Amite, and the carbon fluxes are all higher. In this comparison, the doubled TP flux and nearly double TOC flux in the Amite River basin are noteworthy. A study by Howarth and others (1991) showed a disproportionately large amount of sediment and organic carbon inputs to the Hudson River estuary system originating from urban and suburban areas. In the Amite River basin, Wu (2005) found that population density nearly doubled from 1960 to 1980, rising from 4.8 persons km^{-2} to 9.2 persons km^{-2}. During the same period, the Tangipahoa and Tickfaw River basins experienced little population change, rising from 5.7 to 6.2 persons km^{-2} and 1.1 to 1.3 persons km^{-2}, respectively. After 1980 population density in all three basins remained relatively steady, however, land use during 1978-2001 may have continued to be affected by the previous population increase, resulting in the large TP and TOC fluxes in the Amite River basin.

14.5. CONCLUSIONS

All three watersheds investigated showed higher carbon fluxes when compared with upper Mississippi River basins, emphasizing the elevated role of carbon in subtropical systems. Along with a generally higher flux in carbon due to regional and climatic differences, this study also suggests that population increase in Gulf Coast river basins can greatly affect the level of nutrient loadings. We believe that further investigation of land use changes associated with changes in population is needed to fully understand impacts of urbanization on stream chemistry in coastal waters.

ACKNOWLEDGMENT

The Louisiana Department of Environmental Quality provided long-term water quality data.

REFERENCES

Conley, D. J.. 2000. Biogeochemical nutrient cycles and nutrient management strategies. Hydrobiologia 410: 87-96.

Goolsby, D. A.. Battaglin, W. A., Aulenbach, B. T., and Hooper, R. P.. 2001. Nitrogen Input to the Gulf of Mexico. Journal of Environmental Quality 30:329-336.

Howarth R. W., Fruci, J. R., and Sherman, D.. 1991. Inputs of Sediment and Carbon to an Estuarine Ecosystem: Influence of Land Use. Ecological Applications 1 (1): 27-39.

Jordan, T. E., Weller, D. E., and Correll, D. L., 2003. Sources of Nutrient Inputs to the Patuxent River Estuary. *Estuaries* 26: 226-243.

LDEQ, 2000. Louisiana's Nonpoint Source Management Plan. Office of Environmental Assessment, Louisiana Department of Environmental Quality, Baton Rouge. 334 p.

LPBF, 2005. Comprehensive Habitat Management Plan for the Lake Ponchartrain Basin. Lake Ponchartrain Basin Foundation Nov 18, 2005:25-34.

Nixon, S. W., 1995. Coastal Marine Eutrophication: A Definition, Social Causes, and Future Concerns. Ophelia 41:199-219.

SAS Institute Inc.. 1996. SAS/STAT user's guide, Version 6.12, 4th ed. SAS Institute Inc., Cary, NC, USA.

Turner, R. E., Dortch, Q., Justic, D., and Swenson, E. M., 2002. Nitrogen loading into an urban estuary: Lake Pontchartrain (Louisiana, USA). Hydrobiologia 487 (1):137-152.

Turner, R. E., and Rabalais, N. N.. 2003. Linking landscape and water quality in the Mississippi River Basin for 200 Years. Bioscience 53:563-572.

Turner, R.E., and Rabalais, N.N., 2004. Suspended sediment, C, N, P, and Si yields from the Mississippi River Basin. Hydrobiologia 511 (1): 79-89.

Wu, K., 2005. Long-term freshwater input and sediment load from three tributaries to Lake Pontchartrain, Louisiana. Ph.D. Dissertation, Louisiana State University, Baton Rouge, Louisiana, 138p.

Wu, K., and Xu, Y.J., in press. Evaluation of the applicability of the swat model for coastal watersheds in southeastern Louisiana. Journal of American Water Resources Association.

Xu, Y. J., and Wu, K., in press. Seasonality and interannual variability of freshwater inflow to a large oligohaline estuary in the Northern Gulf of Mexico. Estuarine, Coastal, and Shelf Science.

15

ASSESSING FLOODPLAIN-SWAMP WATER QUALITY REHABILITATION IN THE ATCHAFALAYA RIVER BASIN, LOUISIANA

Aaron M. Podey, William E. Kelso, and D. Allen Rutherford
Louisiana State University Agricultural Center
123 RNR Bldg., Baton Rouge, LA 70803, USA

ABSTRACT

The hydrology of the Atchafalaya River Basin (ARB) has a long history of human induced alterations, including mainstem channelization, canal dredging, and levee construction. Effects of these alterations on water quality and movement during the flood pulse have resulted in efforts to "enhance, protect and conserve" the ARB. Multiple rehabilitation projects are being planned and executed to alleviate hypoxic conditions and unnatural hydrology in this system. Large scale floodplain-swamp rehabilitation projects are rare and methods for determining their success have not been developed. We have monitored areas of proposed water management projects to compare pre- and post-project conditions. In August 2005 the first project was completed by the state of Louisiana, which included dredging a natural bayou, piling the spill along the edge, and leaving 60-m gaps at floodplain elevation to allow water to flow into the adjacent swamp. We used several data analyses based on sampling at 12 sites in the Bayou Postillion area to determine the success of this project, including multiple analysis of variance and discriminate function analysis. We were able to distinguish between pre- and post-construction conditions ($p < 0.0001$), with habitat and water quality improvements at 5 of the 12 sites, and water quality improvements at an additional 5 sites. Continued monitoring of these rehabilitation projects is necessary to determine their long-term effectiveness given the increasing prevalence of exotic macrophytes in the ARB, and the amount of sediment being deposited each year during the flood pulse.

15.1. INTRODUCTION

The Atchafalaya River Basin (ARB) is a large floodplain swamp that surrounds the Atchafalaya River, a 225 km distributary of the Mississippi River that flows to Atchafalaya Bay in the Gulf of Mexico (Reuss 1988). Overflow habitats in the ARB are subject to frequent and prolonged flooding from the Atchafalaya River, with the annual flood pulse typically beginning in December and lasting until June. Although floodwaters reconnect the diverse aquatic habitats that are prevalent throughout the lower floodplain, anthropogenic alterations to the Basin's historic hydrology have significantly altered ARB water circulation and physicochemistry.

Similar to the lower Mississippi River (Harmar et al., 2005), the ARB has a long and unique history of hydrologic alteration and management. Historically, the ARB encompassed 8,345 km^2 of bottomland forests, bayous, lakes, canals, and swamps that stretched eastward from Bayou Teche to the Mississippi River and Bayou La Fourche (Figure 1). In the 1830's, the U. S. Army Corps of Engineers (USACE) connected the Mississippi, Red, and Atchafalaya rivers (Reuss, 1988) and cleared a 48-km logjam that had effectively blocked flow down the Atchafalaya for decades (Comeaux, 1970). This began a period of increasing discharge down the Atchafalaya River, as well as mainstem hydrologic modifications by the USACE to ensure efficient movement of water to the Gulf of Mexico. On the Basin floodplain, oil and gas exploration began in the early part of the twentieth century, which led to extensive dredging of canals for access and for laying pipelines (Reuss, 2004). After the Mississippi River Flood of 1927 the USACE was charged with flood protection in the lower Mississippi River valley, and the ARB became an emergency floodway, designed to convey one half of a three million cfs "Project Flood" to the Gulf (McPhee, 1995). In 1928 the USACE began a program to enhance the floodway characteristics of the ARB by straightening the main channel, closing distributaries to the floodplain, dredging extensively and creating levees and gates to control and direct water flow. These activities compartmentalized the ARB, which now consists of about one quarter of the original basin area (Sabo et al., 1999a). Water flowing into the Atchafalaya River is controlled by the Old River Control complex (low sill structure, auxiliary structure, and Sidney A. Murray hydroelectric plant), with the Atchafalaya River receiving an average of 30 percent of the combined flows from the Mississippi and Red rivers.

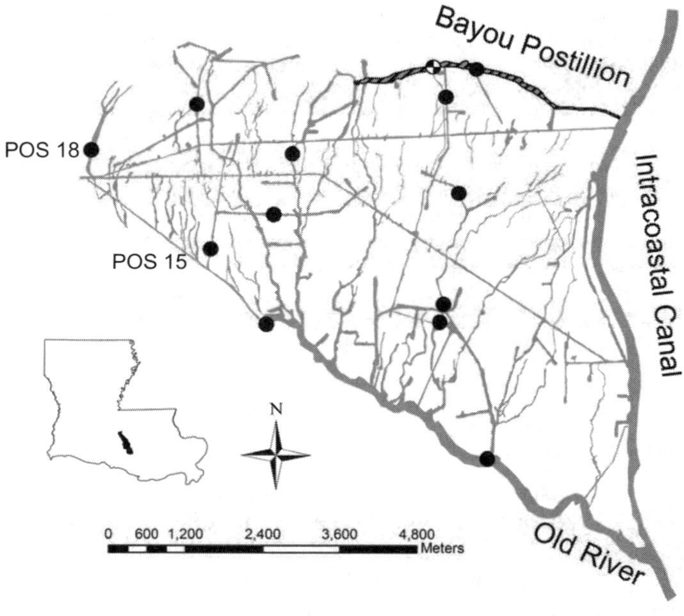

Fig. 1. Location of the study area and the 12 sites sampled from 2001-05 in the Bayou Postillion area of the Atchafalaya River Basin, Louisiana (⊕ staff gauge, ● sample sites). Hatched area indicates water management construction area.

The external ARB levees, internal flood control levees, main channel alterations, oil and gas exploration canals, and sediment inputs from the Mississippi and Red rivers have resulted in highly altered hydrology on the ARB floodplain. The natural, northwest to southeast flow patterns within the ARB are generally evident only at extremely high river stages. The majority of access and pipeline canals run east/west across the floodplain, and most are bordered by uninterrupted spoil banks on the north and south sides. These spoil banks impede floodplain water circulation and help to further compartmentalize the floodway into numerous hydrolocially distinct units, at least at moderate and low river stages. Many of these areas have only one or two source water inlets, which leads to stagnant conditions characterized by higher temperatures and lower dissolved oxygen concentrations during the latter stages of the flood pulse. Levee construction and channel training associated with management of the mainstem river has led to a deeply incised main channel and increasing riparian elevations, which, together with several distributary closures, have impeded movement of water onto the floodplain, except at high river stages.

Effects of these alterations on water quality and movement during the flood pulse have resulted in efforts to "conserve, restore, and enhance" the ARB (Atchafalaya Basin Program, 2002). Conserving the ARB is important because it functions as a nitrogen sink (Xu, 2006), supports extremely high diversities of terrestrial and aquatic organisms, provides a diversity of economically important recreational opportunities, is one of only six accreting deltas in the world (Ouder, 2003), is America's largest river basin swamp (Lambou, 1990) and is of considerable historic significance to Louisiana's Native Americans and Cajun cultures (Atchafalaya Basin Promotion and Research Board, 1999). In 1982, the USACE produced a feasibility study and Environmental Impact Statement for the Atchafalaya Basin Floodway System, LA Project. Three years later the plan for the Atchafalaya River Basin Floodway was completed (Public Law 99-88), and in 1996, the Atchafalaya Basin Advisory Committee was created, with the Louisiana Department of Natural Resources (LADNR) as the lead agency, which led to the adoption of the State Master Plan for the ARB two years later. The plan was designed to improve public access, purchase and lease environmental easements, create recreation opportunities, and conserve, restore and enhance the natural habitat of the ARB (Atchafalaya Basin Promotion and Research Board, 1999).

Multiple projects are being planned to alleviate hypoxic conditions and restore natural hydrology in this complex ecosystem. Since 2001 we have monitored water quality and habitat at locations throughout the lower ARB in an effort to capture baseline conditions in areas where rehabilitation projects were most likely to occur. In August 2005, the State of Louisiana completed its first large scale water management project in the Bayou Postillion area of the ARB (Figure 1). This project consisted of dredging a natural bayou, depositing the spoil along the edges, and leaving gaps at floodplain elevation to allow water inflow to the adjacent swamp. Large-scale floodplain-swamp rehabilitation projects are rare, and methods for determining their success have not been developed. The goal of this study was to determine a method to evaluate the success of the Bayou Postillion dredging project that can be used for future projects in the ARB.

15.2. METHODS

15.2.1. Site description

Bayou Postillion (666544.01 m Easting and 3321076.86 m Northing; UTM Zone 15 North) is located on the east side of the ARB, and this small sub-basin encompasses an area of about 4,000 ha. The area is bounded by Old River (south and west), the Intracoastal Canal (east), and Bayou Postillion (north; Figure 1). Bayou Postillion receives sediment-rich water from the Intracoastal Canal, which resulted in the long-term aggradation throughout the bayou and partial closing of the bayou mouth during the 1990s. This resulted in less freshwater inundation, lower flows, higher water temperatures and lower dissolved oxygen levels in the surrounding area, particularly at intermediate and low water stages during the latter part of the flood pulse. It also caused a reversal of flow direction at certain water stages, where water flowed north into the area from Old River.

Most of the water management project in Bayou Postillion was completed during the period from 22 January through 11 July 2005, and consisted of dredging 4,400 m of the bayou west from the Intracoastal Canal with a bucket dredge. Dredged material was placed along the edges of the bayou, with gaps left in the spoil banks to maintain a connection with the floodplain under normal water levels (Figure 2). The purpose of the project was to restore the north-south flow of water into the area south and west of Bayou Postillion, to increase flow velocities, reduce surface vegetation densities, lower water temperatures and increase dissolved oxygen concentrations.

Fig. 2. Aerial photo of the Bayou Postillion rehabilitation project, Atchafalaya River Basin, Louisiana, July 2005. Photo courtesy of E. Gammons, Hydro Consultants, Inc.

15.2.2. Data collection

We began collecting water quality and habitat data in July 2001, and have continuously monitored 12 sites within the 4,000 ha area at biweekly intervals. At each site we record Secchi disk depth (0.5 cm), flow (cm/s), flow direction, and percent cover, by species, of aquatic vegetation. We also record temperature (°C), pH, specific conductance (mS/cm), dissolved oxygen (DO, mg/l and percent saturation), depth (m), and turbidity (NTU) at the surface and bottom (and at mid-column if the water depth is greater than one m) with a Hydrolab® Quanta water quality monitoring system. Local river stage data is recorded on each monitoring trip from a non-rectified staff gauge located in the bayou. From the dissolved oxygen data, we also calculated the percent of the water column that was hypoxic (DO < 2.0 mg/l; Constant et al., 2003), and DO differential (DOD; surface minus bottom; Davidson et al., 1998) as an index of water column stratification.

15.2.3. Statistical analysis

River stage in the ARB is seasonally dynamic, with most of the poor water quality in the Basin occurring in the latter portion of the flood pulse at intermediate and lower river stages (Fontenot et al., 2001). Therefore, we only used data from the median water level and below (1.2 m at the staff gauge in Bayou Postillion) for our analyses. All physicochemical variables were analyzed for univariate normality and homogeneity of variance with the Shapiro-Wilk test, and all data were subsequently $\log_e(X+1)$ transformed to approach or achieve normality. We reduced the number of variables used in the analyses by eliminating one member of variable pairs that exhibited a Pearson correlation in excess of 0.65 (Proc CORR, SAS Institute Inc., 2002). Bottom, middle and surface DO and percent saturation, and column hypoxia were highly correlated ($r > 0.7$), as were bottom, middle and surface temperature ($r > 0.95$), bottom, middle and surface pH ($r > 0.65$), bottom, middle and surface specific conductance ($r > 0.95$), and bottom, middle and surface turbidity and Secchi disk depth ($r > 0.7$). Of variables that were highly correlated, we chose to include the surface variables in our analyses because we felt that surface conditions were the most biologically important to aquatic organisms, i.e., if surface conditions were poor, it was likely that the entire water column was marginally habitable. Subsetting the original variables left us with 10 variables for the subsequent analyses (Table 1).

We used principle components analysis (PCA, Proc FACTOR, SAS Institute Inc. 2002) with a varimax rotation to further reduce the number of physicochemical variables. To interpret the new principle components (PC) we examined the Pearson product-moment correlations (loadings > 0.6) between the original physicochemical variables and each component. Factor scores were calculated for each site/date and a multiple analysis of variance (MANOVA, Proc GLM, SAS Institute Inc. 2002) was used to determine if mean pre- and post-construction scores were significantly different ($\alpha = 0.05$). Post-MANOVA least-squares means were used to determine if the overall (sites combined), and individual site scores changed between pre- and post-construction time periods. As an alternative analysis technique, we ran a canonical discriminant function (CDFA, Proc CANDISC, SAS Institute Inc., 2002) on the site scores to determine if we could separate pre- and post-construction conditions ($\alpha = 0.05$), and to determine which components were more important in discriminating pre- and post-construction conditions.

Table 1. Principle component loadings of the physicochemical variables (units) from the Bayou Postillion area of the Atchafalaya River Basin, LA, on the first three components.

Physicochemical variable	Principle component		
	1	2	3
Water velocity (cm/s)	0.80		
Flow direction (N, NE, E, SE, S, SW, W, NW)	0.84		
Surface vegetation coverage	-0.67		
Surface temperature (°C)		-0.60	
Surface dissolved oxygen (mg/l)		0.82	
Surface pH		0.69	
Surface specific conductance (μS/cm)		0.64	
Surface turbidity (NTUs)		0.66	
Depth (m)	0.67		
Dissolved oxygen differential (surface–bottom DO mg/l)			0.89

15.3. RESULTS

We visited the Bayou Postillion area 125 times between 15 July 2001 and 15 December 2005, and the biweekly, year- round sampling at the 12 sites resulted in 1,092 samples, 529 of which were below median staff gauge level and were used in the analyses. A wide range of physicochemical conditions were observed, e.g., DO (0.12 to 14.58 mg/l), temperature (9.93 to 32.96°C), flow velocity (0 to 29.9 cm/sec), pH (2.36 to 8.49), specific conductance (0.001 to 0.852 mS/cm), turbidity (0 to 1580 NTU), and percent saturation of oxygen (1.52 to 166.3%).

Principle components analysis yielded three components with eigenvalues over 1.0 (3.78, 1.67 and 1.29) that together explained 68% of the variance in the original data set. We interpreted PC1 as reflecting more of the physical habitat conditions at the 12 sites based on positive loadings for velocity, southerly flow, deeper water, and a negative loading for vegetative cover (Table 1). We interpreted PC2 as a water quality component, based on positive loadings for DO, specific conductance, pH and turbidity, and a negative loading for temperature. Only DOD loaded highly on PC3. MANOVA and comparisons of least-square mean scores for each site revealed significant differences between pre- and post-construction conditions for both PC1 and PC2 ($P < 0.0001$), however, PC3 was not significant between the two time periods ($P = 0.5728$). Five of the 12 sites showed significant post-construction improvement in physical habitat conditions on PC1, and these sites and five additional sites also exhibited improved post-construction water quality conditions on PC2 (Figure 3). The CDFA based on the PCA site scores was able to successfully classify pre- and post-construction conditions in Bayou Postillion ($P < 0.001$). Analyses of the relative importance of the three components in classifying pre- and post-construction conditions, based on pooled within-class standardized canonical coefficients, revealed that PC2 scores exhibited the greatest post-construction changes (0.948), followed by PC1 (0.799), with negligible changes in PC3 (-0.154).

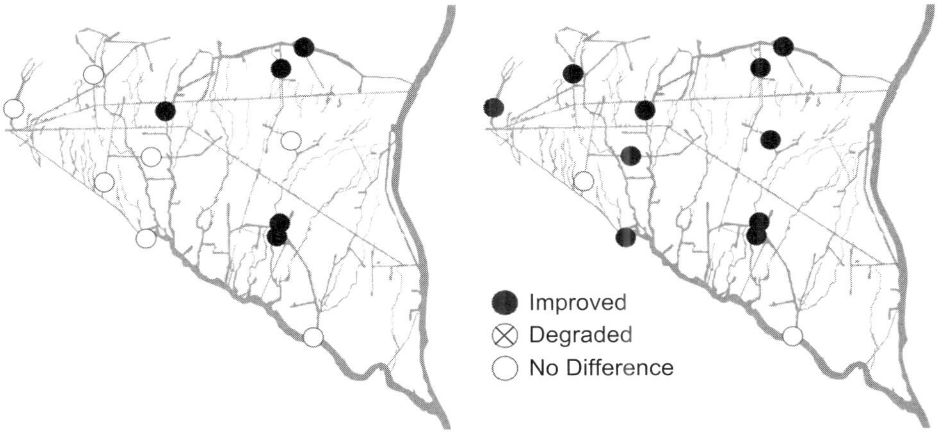

(a) (b)

Fig. 3. Sites of improved, degraded, or no difference on principle component 1 (map a, flow direction and velocity, surface vegetation and depth) and 2 (map b, dissolved oxygen, temperature, pH, specific conductance and turbidity).

15.4. DISCUSSION

Alteration of the hydrology, morphology and ecological function of large river systems is evident throughout the world (Stanford et al., 1996; Jungwirth et al., 2002), and the inherent difficulty of large-scale rehabilitation projects (Shields and Gore, 1995), particularly in floodplain ecosystems characterized by seasonal changes in river stage, temperature, water quality, and biotic productivity (Schiemer et al., 1999; Amoros and Bornette, 2002) make the implementation and monitoring of these projects problematic. In the ARB, this natural variability combined with the effects of anthropogenic changes in river-floodplain connectivity (e.g., Dister et al., 1990), hydrology, and sedimentation rates over the last century have resulted in significant and pervasive changes in aquatic habitat conditions (Sabo et al., 1999a, b), and rehabilitation of this system will necessarily be a spatially extensive and temporally prolonged effort. However, based on our analyses, we are encouraged by the post-construction improvements in flow velocity and water quality that were evident in the Bayou Postillion area during the latter stages of the flood pulse. These results exemplify not only the potential success of water management activities, but also the ability of our monitoring efforts to assess the effectiveness of these projects in improving water flow and quality in adjacent aquatic habitats.

Our knowledge of the ecology and management of large river systems has increased significantly over the last two decades, and understanding the effects of the lateral and temporal aspects of hydrologic connectivity (Amoros and Bornette, 2002) on biotic productivity is particularly important in the ARB. The annual ARB flood pulse typically begins in December, with peak discharge in the Atchafalaya River in April, declining to low water conditions in June and July. However, within this general pattern of rising and falling water levels, circulation and physicochemistry (particularly DO) within the floodplain swamp have been complicated by channel training activities, distributary closures, and levees associated with floodplain canal excavations (Sabo et al., 1999a and b), which have resulted in extended periods of stagnant floodplain inundation each year. Widespread hypoxic conditions develop throughout the flooded swamp when water temperatures rise above 18°C (Constant et al., 2003), and this hypoxic water subsequently spreads throughout the lower ARB as river stage declines. Most of the water management projects planned for the ARB will be similar to the Bayou Postillion project, and will focus on reconnecting the interior swamp to sources of river water input, improving circulation of water across the floodplain during spring stage increases in the Atchafalaya River, and facilitating drainage of the swamp as river stages fall. It is anticipated that these hydrologic modifications will reduce the spatial and temporal magnitude of late spring hypoxia, which has been found to have significant impacts on the abundance and distribution of ARB zooplankton (Davidson et al., 1998), as well as larval (Fontenot et al., 2001) and adult fishes (Rutherford et al., 2001).

Although the geomorphology of large rivers adjusts to engineered modifications in discharge regimes and longitudinal and cross-sectional profiles over various time scales (Harmar, 2005), many alterations such as channelization and levee construction significantly alter river functions (e.g., limit inundation of floodplains; Steiger et al., 1998). These activities reduce the abundance and connectivity of terrestrial, semi-aquatic, and aquatic habitats in the main channel and its associated floodplain (Hohensinner et al., 2004), and ultimately result in a loss of plant and animal diversity that may not recover even under satisfactory water quality conditions (Aarts et al., 2004). In the ARB, similar floodplain-source water disconnections have occurred on a number of spatial scales ranging from the mainstem Atchafalaya River to excavated canals and small natural bayous, and most of the planned water management projects are designed to re-establish some of these connections and improve water quality, flow velocity, the diversity of lateral floodplain channel types (Tockner et al., 1999), and biotic richness and productivity. Increased water circulation at five of the 12 sites, evidenced by higher post-construction site scores along PC1, and improved water quality at 10 of the 12 sites, indicated by higher post-construction site scores along PC2, indicate that hypoxic conditions were less prevalent during 2005, and flow-related habitat complexity increased in these floodplain channels. We believe that the positive physicochemical changes resulting from this dredging project affected an area of approximately 2,700 ha south and west of Bayou Postillion, about 68% of the area of this sub-basin (Figure 4).

Although small in areal extent, this project demonstrates that quantifiable changes in system function can be realized in areas adjacent to active water management projects in the ARB. Re-watering of a small forested wetland in Florida resulted in significant increases in groundwater levels within one year, and forest regeneration and increased avian and piscine diversities within three years (Weller, 1995). Changes in flow and water quality in the Bayou Postillion area were evident throughout 2005, and we anticipate that these effects will enhance aquatic habitat conditions and biotic diversity in succeeding years, similar to patterns of enhanced fish diversity evident in Missouri River floodplain habitats after successive high flood years from 1982-1996 (Galat et al., 1998). However, long-term improvements in system function will depend on rates of sediment accretion in the Bayou Postillion area during future spring flooding, and the continuing negative effects of exotic macrophytes on canal and bayou water quality. The ARB macrophyte community is currently dominated by hydrilla *Hydrilla verticillata*, water hyacinth *Eichhornia crassipes*, and common salvinia *Salvinia minima*, and these species often cover the majority

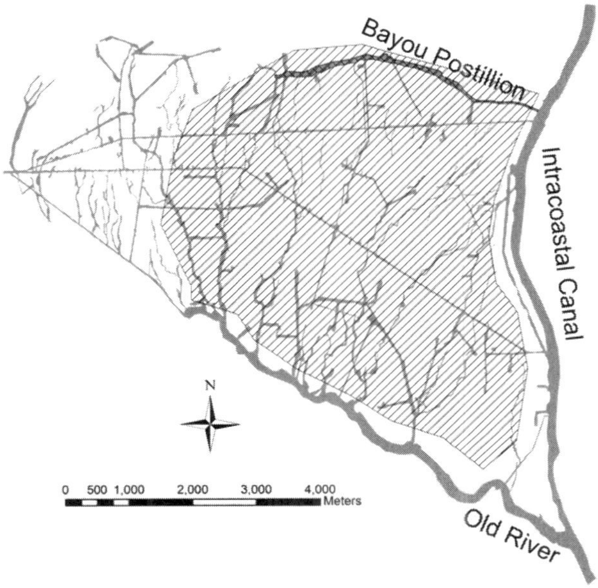

Fig. 4. Potential area of impact of the rehabilitation work in the Bayou Postillion drainage of the Atchafalaya River Basin, Louisiana.

of the water surface (and water column in the case of hydrilla) in many areas of the ARB as river stages fall, resulting in water column hypoxia and minimal habitat conditions for fishes and invertebrates. Although improvements in flow velocity and water quality may be evident during the flood pulse and may promote increased reproduction of ARB zooplankton and fishes, potential improvements at lower river stages later in the year may be obviated by excessive macrophyte growth. This relationship may have been evident at the POS15 and POS18 sites (Figure 1), both of which supported large stands of all three macrophytes during 2005 and exhibited no significant improvement in post-construction flow velocities.

Although many rehabilitation projects have been initiated in smaller riverine systems throughout the world, large river projects are much less common, and constitute a much more daunting task. Projects in the Kissimmee River, Florida, the Rhine River in central Europe, and the Drau River in Austria, however, show that with proper planning and clear objectives, restoration of historical hydrology, geomorphology, and floodplain connectivity is feasible (Jungwirth et al., 2002). In the ARB, rehabilitation efforts are being directed mostly at re-establishing flow patterns and wetlands function of the floodplain swamp (e.g., Buijse et al., 2002), with floodplain hydrologic modifications designed to take advantage of spring floodwater inputs from the Atchafalaya River and Intracoastal Waterway. Water management activities are scheduled to begin in 2006 in the Buffalo Cove area, a 14,000-ha sub-basin in the western ARB. Results of the Bayou Postillion project and a previous levee-breeching project in Buffalo Cove indicate that the ecological benefits of these activities may be immediately evident in water velocity and water quality profiles, and we are confident that our monitoring methods and data analyses can be used in future projects to assess short-term and long-term benefits to the ARB floodplain.

ACKNOWLEDGEMENTS

We gratefully acknowledge the financial and logistical support of the Louisiana Department of Natural Resources, the U.S. Army Corps of Engineers, and the LSU Agricultural Center. We also acknowledge A. Piehler, M. Krumrine, B. Heimann, J. Lang, A. Williamson, A. Commagere, M. Kaller, R. Walley, J. Fisher, T. Mason, M. Engel, J. Colon-Gaud, and G. Constant for their help in data collection. This manuscript was approved for publication by the Director of the Louisiana Agricultural Experiment Station as manuscript 06-40-0116.

REFERENCES

Aarts, B. G. W., Van Der Brink, F. W. B., and Nienhuis, P. H., 2004. Habitat loss as the main cause of the slow recovery of fish faunas of regulated large rivers in Europe: the transversal floodplain gradient. River Research and Applications 20:3-23.

Amoros, C., and Bornette, G., 2002. Connectivity and biocomplexity in waterbodies of riverine floodplains. Freshwater Biology 47:761-776.

Atchafalaya Basin Program, 2002. 2002 Report to the Governor and Legislature. Louisiana Department of Natural Resources, Baton Rouge, Louisiana.

Atchafalaya Basin Promotion and Research Board, 1999. Atchafalaya Basin floodway system Louisiana project, executive summary of the State Master Plan. Louisiana Department of Natural Resources, Baton Rouge, Louisiana.

Buijse, A. D., Coops, H., Staras, M., Jans, L. H., Van Geest, G. J., Grifts, R. E., Ibelings, B. W., Oosterberg, W., and Roozen, F., 2002. Restoration strategies for river floodplains along large lowland rivers in Europe. Freshwater Biology 47:889-907.

Comeaux, M. L., 1970. The Atchafalaya River raft. Louisiana Studies 9: 217-227.

Constant, G. C., Kelso, W. E., and Rutherford, D. A., 2003. Method for evaluating the physicochemical changes associated with water management projects in the Atchafalaya Basin. Report to the Louisiana Department of Natural Resources, Atchafalaya Basin Program. SRNR, LSU AgCenter, Baton Rouge, Louisiana.

Davidson, N. L., Kelso, W. E., and Rutherford, D. A., 1998. Relationships between environmental variables and the abundance of cladocerans and copepods in the Atchafalaya River Basin. Hydrobiologia 379: 175-181.

Dister, E., Gomer, E., Obrdlik, P., Petermann, P., and Schneider, E., 1990. Water management and ecological perspectives of the Upper Rhine's floodplains. Regulated Rivers: Research and Management 5:1-15.

Fontenot, Q. C., Rutherford, D. A., and Kelso, W. E., 2001. Effects of environmental hypoxia associated with the annual flood pulse on the distribution of larval sunfish and shad in the Atchafalaya River Basin, Louisiana. Transactions of the American Fisheries Society 130: 107-116.

Galat, D. L., Fredrickson, L. H., Humburg, D. D., Bataille, K. J., Bodie, J. R., Dohrenwend, J., Gelwicks, G. T., Havel, J. E., Helmers, D. L., Hooker, J. B., Jones, J. R., Knowlton, M. F., Kubisiak, J., Mazourek, J., McColpin, A. C., Renken, R. B., and Semlitsch, R. D., 1998. Flooding to restore connectivity of regulated, large-river wetlands. Bioscience 48:721-733.

Harmar, O. P., Clifford, N. J., Thorne, C. R., and Biedenharn, D. S., 2005. Morphological changes of the lower Mississippi River: geomorphological response to engineering intervention. River Research and Applications 21:1107-1131.

Hohensinner, S., Habersack, H., Jungwirth, M., and Zauner, G.., 2004. Reconstruction of the characteristics of a natural alluvial river-floodplain system and hydromorphological changes following human modifications: the Danube River (1812-1991). River research and applications 20:25-41.

Jungwirth, M., Muhar, S., and Schmutz, S., 2002. Re-establishing and assessing ecological integrity in riverine landscapes. Freshwater Biology 47:867-887.

Lambou, V. W., 1990. Importance of Bottomland Hardwood Forest Zones to Fishes and Fisheries: The Atchafalaya Basin, A Case History. Pages 125-193 in J. G. Gosselink, L. C. Lee, and T.A. Muir, editors. Ecological processes and cumulative impacts: illustrated by bottomland hardwood wetland ecosystems. Lewis Publishers, Inc., Chelsea, Michigan.

Ouder, P., 2003. Hands-on: human activities continue to shape the Atchafalaya Basin. Coast and Sea 11:7-12.

Reuss, M. R., 1988. The challenge of a vital resource: along the Atchafalaya. Environment 30: 6-44.

Reuss, M. R., 2004. Designing the bayous: the control of water in the Atchafalaya Basin 1800-1995. Texas A&M University Press, College Station, Texas.

Rutherford, D. A., Gelwicks, K. R., and Kelso, W. E., 2001. Physicochemical effects of the flood pulse on fishes in the Atchafalaya River Basin, Louisiana. Transactions of the American Fisheries Society 130: 276-288.

Sabo, M. J., Bryan, C. F., Kelso, W.E., and Rutherford, D. A., 1999a. Hydrology and aquatic habitat characteristics of a riverine swamp: I. Influence of flow on water temperature and chemistry. Regulated Rivers: Research and Management 15: 505-523.

Sabo, M. J., Bryan, C. F., Kelso, W. E., and Rutherford, D. A., 1999b. Hydrology and aquatic habitat characteristics of a riverine swamp: II. Hydrology and the occurrence of chronic hypoxia. Regulated Rivers: Research and Management 15: 525-542.

SAS Institute Inc., 2002. SAS OnlineDoc® 9. SAS Institute Inc., Cary, North Carolina.

Schiemer, F., Baumgartner, C., and Tockner, K., 1999. Restoration of floodplain rivers: The 'Danube Restoration Project'. Regulated Rivers: Research and Management 15:231-244.

Shields, J. A., and Gore, F. D., Jr., 1995. Can large rivers be restored? BioScience 45: 142-152.

Stanford, J. A., Ward, J. V., Liss, W. J., Frissell, C. A., Williams, R. N., Lichatowich, J. A., and Coutant, C. C., 1996. A general protocol for restoration of regulated rivers. Regulated Rivers: Research and Management 12:391-413.

Steiger, J., James, M., and Gazelle, F., 1998. Channelization and consequences on floodplain system functioning on the Garonne River, SW France. Regulated Rivers: Research and Management 14:13-23.

Tockner, K., Schiemer, F., Baumgartner, C., Kum, G., Weigand, E., Zweimuller, I., and Ward, J. V., 1999. The Danube Restoration Project: species diversity patterns across connectivity gradients in the floodplain system. Regulated Rivers: Research and Management 15:245-258.

Weller, J. D., 1995. Restoration of a south Florida forested wetland. Ecological Engineering 5:143-151.

Xu, Y. J., 2006. Organic nitrogen retention in the Atchafalaya River Swamp. Hydrobiologia 560: 133-143.

16

ORGANIC CARBON FLUXES FROM THE ATCHAFALAYA RIVER INTO THE GULF OF MEXICO

Y. Jun Xu and Abhijit Patil
School of Renewable Natural Resources
Louisiana State University Agricultural Center, Baton Rouge, LA 70803, USA

ABSTRACT

The transport of riverine organic carbon to the oceans is a significant component of the global carbon cycle, and exerts crucial impacts on marine ecosystems and habitats. The total amount and timing of the carbon fluxes are subject to a number of factors including land surface process, climate variation, and anthropogenic activities. This study investigates long-term seasonal and inter-annual variability of the total organic carbon (TOC) discharge and its removal in the Atchafalaya River Basin. By utilizing 25 years (1978-2002) of the river's discharge and water quality data, monthly and annual TOC fluxes were estimated and their temporal relationship with the river's hydrological conditions were analyzed. The results showed that, annually, TOC fluxes from the Atchafalaya ranged from 1,093,625 Mg, varying from 642,555 Mg in the dry year of 2000 to 1,466,981 Mg in the wet year of 1979. Monthly TOC fluxes from the river averaged 91,135 Mg, with the highest values from March to June (132,582 Mg mon^{-1}) and lowest values from August to November (49,117 Mg mon^{-1}). The large variation in both monthly and annual TOC fluxes was closely correlated to the level of river discharge. Furthermore, the results showed that the annual TOC discharge from the Atchafalaya was about 16% lower than the annual average TOC inflow (1,303,943 Mg) into the basin, displaying the critical role that coastal river swamps can have as a net sink of organic carbon.

16.1. INTRODUCTION

The total riverine input of organic carbon to the world's oceans has been estimated to be approximately 4 x 10^{14} g yr^{-1}, which is about four times larger than the Aeolian flux of particulate organic carbon or burial flux in the sediment (Hedges 1992). In another study by Meyback (1982), an even higher annual discharge of dissolved organic carbon (DOC) to the oceans (about 2.5 x 10^{15} g) was estimated. Malcolm and Durum (1976) reported a 2 x 10^{12} g DOC discharge from the Mississippi - Atchafalaya River system, while Lambou and Hern (1983) calculated a 11.2 x 10^{8} kg (1.12 x 10^{12} g) total organic carbon (TOC) flowing into the Atchafalaya River annually, based on their organic carbon concentration data from river water samples collected between July 1976 and June 1977. These estimates have never been confirmed by other studies and they contradict each other in some ways. As anthropogenic induced land use change has dramatically increased over the past few decades, organic carbon fluxes from world's rivers into the oceans are expected to rise. However, little is known about the recent decadal transport of organic carbon and its variability as it travels from the land to the oceans.

Existing ambient water quality discharge standards have effectively curbed point source pollution, but the nonpoint source discharges from agricultural land, small-scale industries, and urban areas exert another tremendous pressure on the natural systems. These nonpoint source discharges from the land contain high amounts of nutrients, sediments, and carbon, which have been attributed to the widespread eutrophication of the waters around the world's coastal regions during the second half of the 20th century (Justic et al., 2005). The rapid increase in the size of the hypoxic zone observed in the Northern Gulf of Mexico during the recent decade (Rabalais and Turner, 2001; Rabalais et al., 2002) is a prime example of how human's activities inland can alter the chemical, biological and ecological conditions of the oceans.

Although the need is growing, developing mitigation measures for nonpoint source pollution is a challenging task. Treatment wetlands are considered to be among the most viable options for reducing pollutants and nutrients from household discharges. The principle idea is to utilize processes, such as sedimentation and denitrification, to lower suspended solid and nitrogen concentrations in the overflowing waters. In artificially-controlled treatment systems, these processes may be maximized through mechanical, chemical and biological techniques. Under natural systems, however, the processes are entirely dependent on environmental conditions. Organic carbon is either consumed by the aquatic community, deposited in the benthic zone, or transferred into the atmosphere. Carpenter and Pace (1997) found that peatlands play a significant role in DOC export in regional redistribution of terrestrial carbon. The amount of DOC exported from peatlands may depend upon interaction between the flow of water and the production and consumption of DOC within the peatlands (Paster et al., 2003). It has been also reported that some natural surface flow wetlands release nutrients instead of removing them (e.g., Devito et al., 1989; Soranno et al., 1996). Mulhollands and Kuenzler (1979) found that the export of carbon per unit of runoff from streams with considerable swamp drainage was higher than export rates from upland watersheds. Researchers have attributed nutrient removal efficiency to a variety of factors including watershed size, mass loading rate, water depth, or a combination of climate, soil characteristics and plant communities (Kadlec and Knight, 1996).

In general, data on organic carbon fluxes in large rivers with extensive floodplains and backwater swamps is scarce. Such information can be very useful for developing management plans to retain riverine nutrients. With this in mind, this study was conducted to quantify organic carbon fluxes and investigate their seasonal and annual variations in a large river-swamp system - the Atchafalaya River Swamp Basin. The major goal of the study was to establish a mass carbon input-output budget. By doing so, the carbon retention rate of the Atchafalaya and the relationship between the carbon input-output budget and the river's hydrologic conditions were investigated.

16.2. METHODS

16.2.1. The Atchafalaya River Basin

The Atchafalaya River is the largest distributary of the Red, Black, and Mississippi Rivers, formed by the diversion near Simmesport, Louisiana. The Atchafalaya carries one-third of the Mississippi River's water and is confined by levees on both east and west sides to serve as an alternate floodway. The confined floodway area, about 18 miles wide and 150 miles long, is widely known as the Atchafalaya River Swamp Basin (Figure 1) and is the largest freshwater swamp in the United States.

Climatic conditions in the Atchafalaya River Basin are humid, subtropical with typically mild winters, hot summers and abundant rainfall (about 1500 mm per year). Soils in the high and intermediate areas on the natural levees range from very fine sandy loams to silty loams, to silty clay loams (Murphy et al., 1977). Poorly drained loamy and clayey soils are present in the lower positions on the natural levees and in back-swamps located on the floodplain of the Atchafalaya River. Surface hydrology of the basin plays a vital role in the diversity, health and dynamics of the swamp ecosystems.

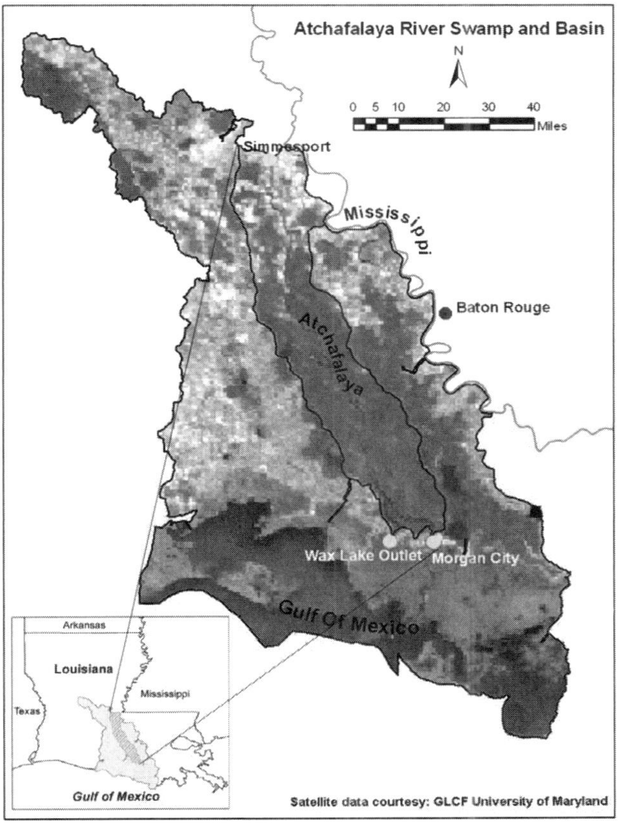

Fig. 1. The Atchafalaya River Swamp Basin.

The Atchafalaya River Basin is predominantly comprised of wooded lowland and cypress-tupelo swamps, with freshwater marshes in the lower distributary area (LDEQ 1993). According to the Louisiana Non-Point Source Assessment Report (LDEQ 2000) the basin is 19% forested, 15% agricultural lands, with wetlands covering 45% of the area and the remaining 21% containing surface water. This large swamp basin is home to a variety of wildlife that includes more than 170 bird species, more than 40 mammalian species and more than 100 finfish and shellfish species (Demas et al., 2002). Major uses of the basin have been recreation and crawfish farming.

16.2.2. Data Collection and Analysis

In this study, the Atchafalaya River Basin is considered to be a black-box system, with a well-defined watershed boundary. Long-term (1978-2002) river discharge data at the upper-basin location, Simmesport, and two down-basin outlets, Morgan City and Wax Lake Outlet, were obtained from the New Orleans District Office of the U.S. Army Corps of Engineers and the Louisiana District Office of the U.S. Geologic Survey in Baton Rouge. Monthly water quality data for the same 25-year period was obtained from the Louisiana Department of Environmental Quality. The LDEQ ambient monitoring program measures various water quality parameters, of which TOC was selected for the study.

The TOC fluxes at the basin's inlet (Simmesport) and outlets (Morgan City and Wax Lake Outlet) were calculated using discharge and water quality data. Using all existing TOC records (inlet: n=142, outlet: n=242), two log-linear regression functions were developed to estimate daily TOC input (Equation 1) and output (Equation 2) as given below:

$$ln(S) = 1.181\, ln(Q) - 3.077 \tag{1}$$

$$ln(S) = 1.164\, ln(Q) - 2.751 \tag{2}$$

where S is the daily TOC loading in grams and Q is the daily discharge in liters. The regression coefficients (r^2) obtained for Equation 1 and Equation 2 were 0.81 and 0.77, respectively.

The estimated daily TOC loadings were totaled to produce monthly and annual TOC loadings. A total TOC mass inflow-outflow budget between the upper-basin and down-basin locations was established for the 25-year period and annual TOC removal rates (TOC_R) were calculated by:

$$TOC_R = \frac{TOC_{in} - TOC_{out}}{TOC_{in}} \times 100 \tag{3}$$

where TOC_{in} is the TOC mass input in the overflowing waters at Simmesport and TOC_{out} is the TOC mass output in the overflowing waters at Morgan City and Wax Lake Outlet.

The statistical analysis was performed using the SAS software package (SAS Institute, 1996). Linear regression was used to determine relationships between TOC removal and basin's hydrological conditions.

16.3. RESULTS AND DISCUSSION

16.3.1. River discharge and TOC concentrations

From 1978 to 2002, the total inflow at Simmesport averaged 1.722 x 10^{10} m^3 mon^{-1}, varying from to 0.804 x 10^{10} m^3 mon^{-1} to 2.592 x 10^{10} m^3 mon^{-1}, whereas the total outflow from Morgan City and Wax lake Outlet averaged 1.638 x 1010 m^3 mon^{-1}, varying from 0.840 x 10^{10} m^3 mon^{-1} to 2.410 x 10^{10} m^3 mon^{-1}. The seasonal flow trend showed a higher flow from March to June and a lower flow from August to November (Figure 2). During the wet season (December to June) discharge at the upstream location was significantly higher than that at the downstream location, suggesting that a percentage of water was being retained in the basin from winter to late spring. However in the dry season, especially in the month of September, the discharge at the downstream location was slightly higher than that of the upstream location, meaning that water retained during the wet season flowed out during the dry periods.

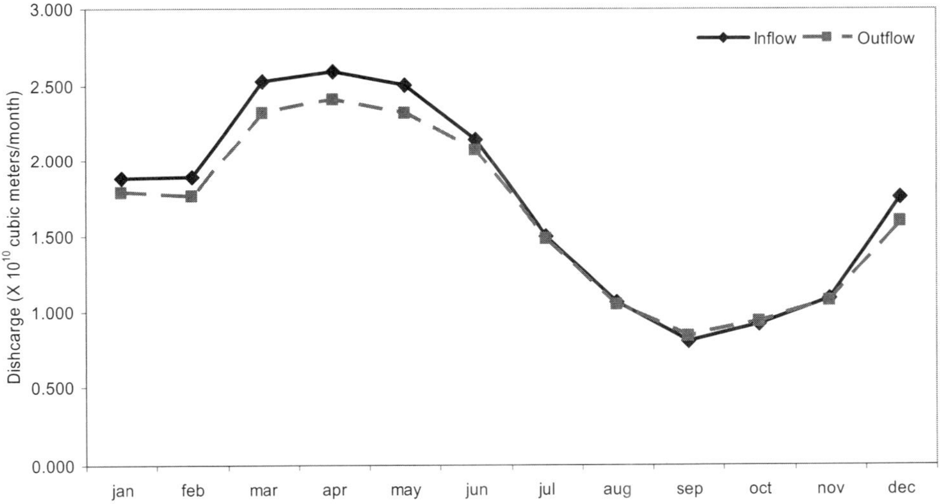

Fig. 2. Monthly average inflow at Simmesport and outflow at Morgan City and Wax Lake Outlet of the Atchafalaya River Swamp Basin.

TOC concentrations at Simmesport varied from 1.50 to 18.00 mg L^{-1}, with an average concentration of 6.47 mg L^{-1} over the past 25 years. This was significantly higher than the average TOC concentration in the waters discharged from the basin at Morgan City - 5.67 mg L^{-1}, varying from 1.5 to 19.00 mg L^{-1}. Interestingly, the long-term data did not show a clear trend of change in TOC fluxes in the Atchafalaya River Basin from 1978 to 2002, as was expected (data not presented).

Monthly averages of TOC concentrations varied from 4.52 mg L^{-1} in September to 7.51 mg L^{-1} in March at Simmesport, and from 4.28 in September to 7.14 mg L^{-1} in April at Morgan City. Higher TOC concentrations were observed from February through June and lower TOC concentrations were found between late summer and winter. This seasonal change in TOC concentration coincided with the seasonal trend of discharge in the Atchafalaya River Basin, demonstrating the impact of agricultural practices during the spring in the Upper Mississippi River Basin on carbon fluxes to the Gulf of Mexico.

16.3.2. TOC mass input-output balance and removal rates

The annual average of TOC input at Simmesport was 1,303,943 Mg, varying from 742,024 Mg in the dry year 2000 to 1,947,767 Mg in the wet year 1993 (Table 1). The annual average of TOC output at Morgan City and Wax Lake Outlet was 1,093,625 Mg, varying from 642,555 Mg in 2000 to 1,466,981 Mg in 1979, another wet year (Table 1).

Table 1. Annual discharge and TOC input and output of the Atchafalaya River Basin.

Year	Discharge [x10^{11} m^3]		TOC [Mg]		TOC Input-Output	Removal Rate
	Inflow	Outflow	Input	Output	[Mg]	[%]
1978	1.804	1.738	1111745.831	947291.893	164453.938	14.79
1979	2.718	2.539	1800912.939	1466981.395	333931.544	18.54
1980	1.667	1.628	999037.281	867189.064	131848.217	13.20
1981	1.394	1.385	810033.132	719237.956	90795.176	11.21
1982	2.095	1.977	1306851.721	1087551.278	219300.442	16.78
1983	2.615	2.452	1732435.396	1415318.040	317117.356	18.30
1984	2.360	2.225	1518448.198	1257000.966	261447.232	17.22
1985	2.267	2.142	1448497.682	1205345.488	243152.193	16.79
1986	1.987	1.949	1228122.621	1073040.073	155082.548	12.63
1987	1.559	1.537	932842.517	819269.871	113572.646	12.17
1988	1.436	1.378	869618.464	744474.904	125143.560	14.39
1989	2.194	1.963	1395154.653	1093175.813	301978.840	21.64
1990	2.353	2.009	1519985.100	1120023.126	399961.975	26.31
1991	2.478	2.217	1619675.849	1257969.805	361706.043	22.33
1992	1.803	1.608	1089250.753	857736.709	231514.044	21.25
1993	2.959	2.520	1947767.196	1440198.408	507568.787	26.06
1994	2.129	2.027	1349561.338	1123228.868	226332.470	16.77
1995	1.961	2.083	1220083.276	1165626.021	54457.254	4.46
1996	2.157	2.040	1352575.749	1135227.577	217348.172	16.07
1997	2.295	2.258	1484860.791	1292721.623	192139.168	12.94
1998	2.310	2.330	1473299.491	1317595.427	155704.064	10.57
1999	1.897	1.882	1185819.261	1051744.484	134074.778	11.31
2000	1.292	1.234	742024.442	642555.196	99469.246	13.41
2001	1.907	1.919	1178477.537	1065103.984	113373.553	9.62
2002	2.035	2.096	1281495.080	1175017.060	106478.020	8.31
Mean	**2.067**	**1.965**	**1303943.052**	**1093625.001**	**210318.051**	**15.48**

Monthly TOC fluxes from the river averaged 91,135 Mg. Seasonally, the TOC fluxes were highest from March to June (average: 132,582 Mg mon^{-1}) and lowest from August to November (average: 49,117 Mg mon^{-1}) (Figure 3). The difference between the monthly TOC input at Simmesport and the monthly TOC output at Morgan City and Wax Lake Outlet showed a similar trend: the largest difference (32,659 Mg) was observed in March and the smallest (2,301 Mg) was observed in September (Figure 3).

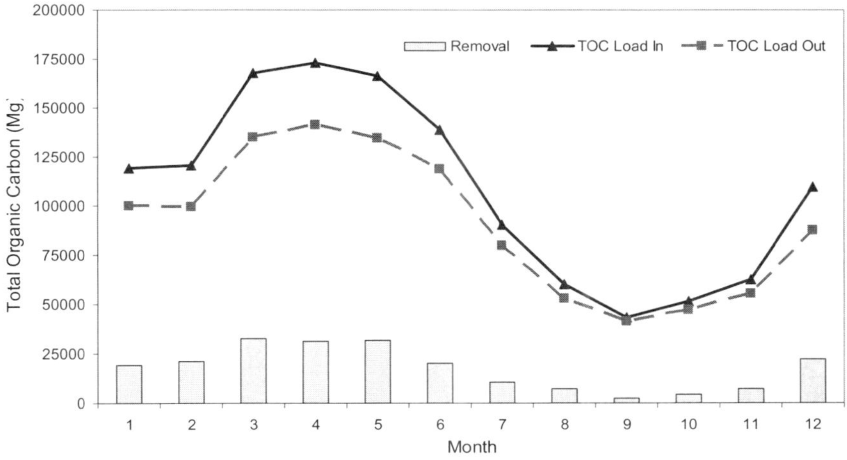

Fig. 3. Monthly averages of TOC input, output, and input-output balance in the Atchafalaya River Basin.

Over the 25 years from 1978 to 2002, about 32.6 million Mg of TOC entered the Atchafalaya river basin, whereas approximately 27.3 million Mg of TOC discharged into the Gulf of Mexico from the basin. The 5.3 million Mg difference between the input and output represented a 16% retention reduction by the basin. The annual retention rates, however, fluctuated largely, varying from 4.5% in 1995 to 26% in 1990.

The amount of TOC retained in the basin (input – output) was positively correlated with the river's discharge (Figure 4). In both the monthly and annual relationships, an inflow threshold was found where the retention appeared to increase. The threshold was about 2 x 10^{-10} m^3 in the monthly retention (Figure 4, top) and was about 2 x 10^{-11} m^3 in the annual retention (Figure 4, bottom).

The riverine organic matter was observed to be soil-derived and refractory (Dittmar and Kattner, 2003). It was suspected that terrestrial organic carbon must undergo an estuarine mixing process (Hedges et al., 1997). Many studies have demonstrated that microbial degradation and photochemical oxidation have played an important role in the removal of terrestrial DOC in coastal waters (Benner et al., 1995; Miller and Moran, 1997; Opsahl and Benner, 1998; Coffine and Cifuentes, 1999; Raymon and Bauer, 2001). The sedimentation and chemical absorption mechanism may also remove nutrients in the river and wetland systems (Watson et al., 1989). Using terrestrially derived lignin phenols as bio-markers, Benner and Opsahl (2001) found that photo-oxidation was the major removal process of the terrestrial DOC, and that the photochemical reactions mainly occurred in higher salinity waters, while flocculation and microbial degradation were more dominant in terrestrial DOC removal from low saline waters in the Mississippi River estuary. The organic carbon retention found in this study may have also been influenced by these factors.

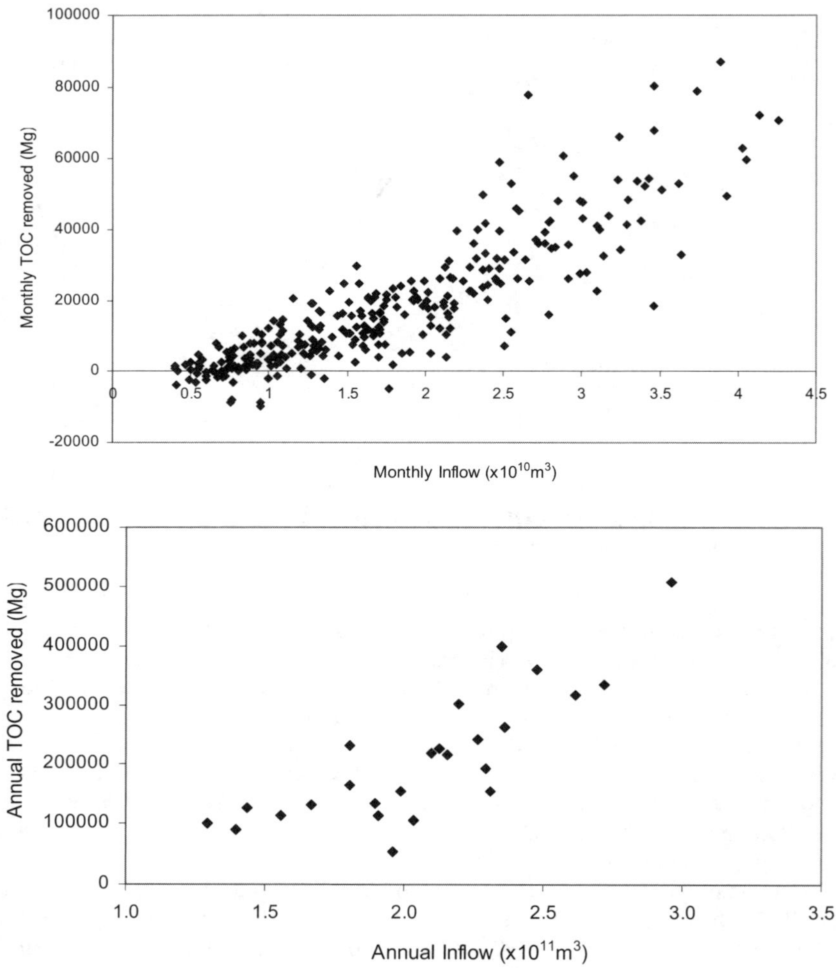

Fig. 4. Relationships of monthly (top) and annual (bottom) TOC retention in the Atchafalaya with the river's inflow.

16.4. CONCLUSIONS

This study yielded the first insight into the nature of organic carbon fluxes from the Atchafalaya River Swamp Basin into the Gulf of Mexico. It is also the first comprehensive assessment on carbon retention capacity for this large river-swamp system in North America. The major findings can be summarized as follows:

1. The Atchafalaya River introduces a large amount of organic carbon (1,093,625 Mg or 1.09 x 10^{12}) each year into the Gulf of Mexico;

2. On average, the Atchafalaya River Swamp Basin retains about 16% of the organic carbon from the overflowing waters; and
3. Both the loading and retention of total organic carbon are highest from March to June, lowest from August to November, and are positively correlated to the river's discharge.

REFERENCES

Benner, R., Opsahl , S., Chin-Leo, G., Richey, J.E., and Forsberg, B.R., 1995. Bacterial metabolism in the Amazon river system. Limnology and Oceanography 40: 1262-1270.

Benner, R., and Opsahl, S., 2001. Molecular indicator of sources and transformations of dissolved organic matter in Mississippi River plume. Organic Geochemistry 32:597-611.

Carpenter, S.R. and Pace, M.L., 1997. Dystrophy and eutrophy in lake ecosystem: Implications of fluctuating inputs. Okios 78: 3-14.

Coffin, R.B., and Cifuentes , L.A., 1999. Stable isotope analysis of carbon recycling in the Perdido estuary, Florida. Estuaries 22: 917-926.

Demas, C.R., Brazelton, S.R., and Powell, N.J., 2002. The Atchafalaya Basin - river of trees. U.S. Geological Survey, Fact Sheet FS-021-02.

Devito, K.J., Dillon, P.J., and Lazerte, B.D., 1989. Phosphorus and nitrogen retention in 5 precambrian shield wetlands. Biogeochemistry 8 (3):185-204.

Dittmar, T., and Kattner G., 2003. The biogeochemistry of the river and shelf ecosystem of the Arctic Ocean: a review. Marine Chemistry 83: 103-120.

Hedges, J.I., 1992. Global biogeochemical cycles: progress and problems. Marine Chemistry 39: 67-93.

Hedges, J.I., Keil, R.G., and Benner, R., 1992. What happens to terrestrial organic matter in the ocean? Organic Geochemistry 27 (5/6): 195-212.

Malcolm, R.L. and Durum, W. H., 1976. Organic carbon and nitrogen concentration and annual carbon load of six selected rivers in United States. USGS Water Supply paper 1817F.

Justic, D., Rabalais, N.N., and Turner, R.E., 2005. Coupling between climate variability and coastal eutrophication: Evidence and outlook for the northern Gulf of Mexico. Journal of Sea Research 54 (1): 25-35.

Kadlec, R.H., and Knight, R.L., 1996. Treatment wetlands. Lewis Publishers, New York.

LDEQ, 1993. State of Louisiana Nonpoint Source Pollution Assessment Report November 1993 (http://nonpoint.deq.state.la.us/assess32.html).

LDEQ, 2000. Louisiana's Non-point Source Management Plan. Volume 6, State of Louisiana Water Quality Management Plan. Baton Rouge, Louisiana.

Meybeck, M., 1982. Carbon, nitrogen and phosphorous transport by world rivers. American Journal of Science 282: 401-450.

Miller, W. L., and Moran, M.A., 1997. Interaction of photochemical and microbial processes in the degradation of refractory dissolved organic matter from a coastal marine environment. Limnology and Oceanography 42: 1317-1324.

Mulholland, P.J., and Kuenzler, E.J., 1979. Organic-carbon export from upland and forested wetland watersheds. Limnology and Oceanography 24 (5): 960-966

Murphy, K. E., Touchet, B. A., White, A. G., Daigle, J. J., and Clark, H. L., 1977. Soil survey of St. Martin Parish, Louisiana: USDA Soil Conservation Service, Washington, D.C.

Pastor, J., Solin, I., Bridgham, S.D., Updegraff, K., Harth, C., Weishampel, P., and Dewey, B., 2003. Global warming and the export of dissolved organic carbon from boreal peatlands. Oikos 100:380-386.

Raymond, P.A., and Bauer, J.E., 2001. Use of 14C and 13 C natural abundance for evaluating riverine estuarine and coastal DOC and POC sources and cycling: a review and synthesis. Organic Geochemistry 59:469-485.

Rabalais, N.N., and Turner, R.E. (Eds.), 2001. Coastal Hypoxia: Consequences for Living Resources and Ecosystems. Coastal and Estuarine Studies 58, American Geophysical Union, Washington DC.

Rabalais, N.N., Turner, R.E., and Scavia, D., 2002. Beyond science into policy: Gulf of Mexico hypoxia and the Mississippi river. BioScience 52 (2): 129-142.

Soranno, P.A., Hubler, S.L., Carpenter, S.R., and Lathrop, R.C., 1996. Phosphorus loads to surface waters: A simple model to account for spatial pattern of land use. Ecological Applications 6 (3): 865-878.

Watson, J.T., Reed, S.C., Kadlec, R.H., Knight, R.L., and Whitehouse, A.E., 1989. Performance expectations and loading rates for constructed wetlands. In Constructed wetlands for wastewater treatment, 319-351. D. A. Hammer, eds. Chelsea, MI: Lewis Publishers.

17

SPATIOTEMPORAL RELATIONS OF CARBON TO NITROGEN RATIO IN ATCHAFALAYA RIVER BASIN

April Mason and Y. Jun Xu
School of Renewable Natural Resources, 227 RNR Building
Louisiana State University Agricultural Center, Baton Rouge, LA 70803, USA

ABSTRACT

Eutrophication caused by excess nutrients such as nitrogen has resulted in detrimental effects on biota and has created a dead zone with depleted oxygen in the Gulf of Mexico. There has been concern about water quality in the lower Mississippi River Basin because studies have shown the river delivers high levels of nitrogen into the Gulf. Most nitrogen transported by rivers to the oceans is associated with organic matter, making the organic carbon to nitrogen ratio for particulate or dissolved pools in those waters a critical parameter in understanding carbon and nutrient cycling. This study analyzed twenty-five years of historical water quality data from the Atchafalaya River Basin to determine space-time characteristics of the organic carbon to nitrogen ratio. Total organic carbon (TOC), total Kjeldahl nitrogen (TKN), and nitrate plus nitrite nitrogen were used to develop TOC:nitrate/nitrite, TOC:TKN, and TOC:total nitrogen (TKN+NO_3+NO_2) ratios. Monthly and yearly average ratios at an upstream location (Simmesport) and a downstream location (Morgan City) were examined and compared to assess spatial and temporal trends. From the Mississippi River diversion near Simmesport, water travels about 190 km downstream to Morgan City, so this data was also used to detect any changes in TOC and nitrogen concentrations as the river flowed through the basin.. The study found that the average TOC:NO_3+NO_2 and TOC:TKN ratios at Simmesport were 9.23 and 7.01, and that the TOC:NO_3+NO_2 and TOC:TKN ratios at Morgan City averaged 7.45 and 8.23, respectively. While the concentrations of TOC and TKN showed a similar temporal trend with the river's discharge, no seasonal difference in the ratios between both locations was found. TOC to total nitrogen monthly ratios were significantly higher at Morgan City than Simmesport. When examining TOC and nitrogen concentrations individually, it was found that the average TOC and TKN concentrations at Simmesport (6.47 and 0.97 mg L^{-1}) decreased downstream at Morgan City (5.67 and 0.74 mg L^{-1}). However, the average nitrate/nitrite concentration at Morgan City (0.94 mg L^{-1}) was slightly higher than that of Simmesport (0.89 mg L^{-1}). The higher nitrate/nitrite concentrations at the downstream location mainly occurred during the summer months of June through August, indicating increased biological effects on riverine nitrogen transformation in the basin.

17.1. INTRODUCTION

Despite the reduction in point source pollution attained by discharge standards, nonpoint source (NPS) pollution continues to be a major problem for many waterbodies across the nation. NPS pollution with excessive sediment and nutrient runoff from agricultural lands, small scale industries, and urban areas has resulted in the eutrophication of rivers and lakes, which is also believed to be the cause of the dead zone in the Gulf of Mexico. More specifically, as increasing amounts of nitrogen are released into the soil, the terrestrial environment becomes saturated, resulting in an escalation of nitrogen delivery to streams (Starry et al., 2005).

The dead zone in the Gulf of Mexico is an area of hypoxia from the Mississippi River birdfoot delta extending up to 20,000 km^2 west, ending at the upper Texas coast (Rabalais et al., 2001). Hypoxia for the Gulf of Mexico is classified as water with dissolved oxygen (DO) concentrations of less than 2 mg L^{-1}. At these low DO concentrations, trawlers tend not to catch shrimp or demersal fish in nets. Continental shelf fishery in the Gulf of Mexico is approximately 25% of the US total, making hypoxia both an ecological and an economic problem (Mitsch and Gosselink, 2000). The size of the dead zone has increased two-fold since 1985. During 1985-1992, the average midsummer hypoxic zone was 8,000-9,000 km^2, which increased to an average of 16,000-20,700 km^2 during 1993-2001 (Rabalais and Turner 2001; Rabalais et al., 2002). The hypoxic zone appears to be a seasonal feature, potentially corresponding with seasonal increases in discharge and temperatures (Eadie et al., 1994). This increase in hypoxic zone is attributed to an increase in riverbourne nutrients that favor algal blooms, aggravate oxygen depletion, and alter marine food webs (Goolsby et al., 2001; Rabalais et al., 1996, 1998, and 2002).

The primary sources of freshwater input into the Gulf are from the Mississippi and Atchafalaya Rivers, contributing 95% (Goolsby et al., 2001; Rabalais et al., 2002). This river system drains about 41% of the continental United States (Eadie et al., 1994; Goolsby et al., 2001). Considering this vast drainage area, and that a large percentage contributes to the Gulf of Mexico, concern about excess nutrients in this river system has been increasing. Over the past few decades, the increased nitrate loads are believed to have increased primary productivity in the Gulf's shelf waters (Trefry et al., 1994). Much of the nitrogen reaching the Atchafalaya comes from northern agricultural states (Starry et al., 2005; Strauss et al., 2004) including Minnesota, Iowa, Illinois, Indiana, and Ohio (Goolsby et al., 2001), placing further importance on alternative means for reducing nitrogen loading, without impacting agriculture practices. The best mitigation measure for such large nitrogen discharges is the implementation of treatment wetlands. It has been found that N processing by headwater streams may alter N load to downstream systems (Starry et al., 2005). The diversion of 30% of the Mississippi River's discharge to the Atchafalaya River basin is one example of an existing method.

Most nitrogen transported by rivers to oceans is associated with organic matter, making the organic carbon to nitrogen ratio for particulate or dissolved pools in those waters a critical parameter in understanding carbon and nutrient cycling. Organic carbon is indirectly related to the oxygen availability in water (Thunell et al., 2000). Organic carbon fluxes can also be affected by land surface processes, climate variation, and anthropogenic activities - land use changes, for example, have increased organic carbon fluxes from the Atchafalaya (Xu and Patil, in this book). The majority of organic matter in the Gulf originates from nutrient inputs, creating increased primary production, and the Mississippi River, which deposits 2 x 10^{12} g of particulate organic carbon annually (Eadie et al., 1994). Organic carbon from increased primary production further enhances oxygen consumption and the hypoxic dead zone in the Gulf of Mexico (Trefry et al., 1994).

Sedimentation and denitrification, which are important processes for lowering the sediment and nitrogen concentration, can occur in natural wetlands such as the Atchafalaya

River Basin. The organic carbon is either consumed by the aquatic community, deposited in the benthic zone, or transformed into the atmosphere. There is a tendency for increased C to inhibit nitrification, which is the process in which ammonium is converted to nitrate (Starry et al., 2005). It affects nitrification by changing microbial dynamics, so when C is abundant (i.e. a high C:N ratio), heterotrophic microbes outcompete autotrophic nitrifying bacteria for ammonium (Starry et al., 2005). Adjustment of C to N ratio has been an effective measure in controlling inorganic nitrogen in aquaculture facilities and this adjustment is one of the most cost effective methods available (Avinmelech, 1999). It is important to see if the same is true for lotic systems.

The objectives of this study were to determine if changes exist in carbon to nitrogen ratios between the upstream location at Simmesport and the downstream location at Morgan City, and investigate seasonal and interannual variations in carbon to nitrogen ratios with the river's hydrologic conditions. Our overall goal in the study was to discern nitrogen and carbon dynamics in this large river swamp.

17.2. METHODS

17.2.1. Study Site

The Atchafalaya River Basin, located in south central Louisiana, flows south-southeast into the Gulf of Mexico (Figure 1). It is a distributary of the Black, Red, and Mississippi Rivers. From the Old River control structure at Simmesport, Louisiana, 30% of the Mississippi River's flow is directed to the Atchafalaya. Because of this diversion, the Atchafalaya also receives much of the pollutants and nutrients from the Mississippi River. The river is well confined by levees on the north, south, and west sides. The 4,678 km^2 drainage basin is the largest contiguous freshwater river swamp in the United States. It is a surface flow swamp with some freshwater marshes in the lower distributary area. Bottomland hardwood and cypress-tupelo forests dominate the basin.

The Atchafalaya River Basin's climate is humid, subtropical with mild winters, hot summers and abundant rainfall (1500 mm per year). Very fine sandy loam to silty loam soils are present in the higher and intermediate areas on the natural levees. In the lower positions on the natural levees and in backswamps on the floodplain of the river, poorly drained loamy and clayey soils are present. Surface hydrology is important for the diversity, health and dynamics of swamp ecosystems. The basin is home to a variety of wildlife including 170 bird species, more than 40 mammalian species, and more than 100 finfish and shellfish species (Demas et al., 2002). According to the Louisiana Non-Point Source Assessment Report (LDEQ, 2000), the basin consists of 19% forests, 15% agriculture, 43% wetlands and 21% water. Sherburne Wildlife Management Area, located in the upper third of the basin, is a 4,767 ha area managed by Louisiana Department of Wildlife and Fisheries. The U.S. Fish and Wildlife Service also established the Atchafalaya National Wildlife Refuge on 6,159 ha adjacent to the Sherburne Wildlife Management Area.

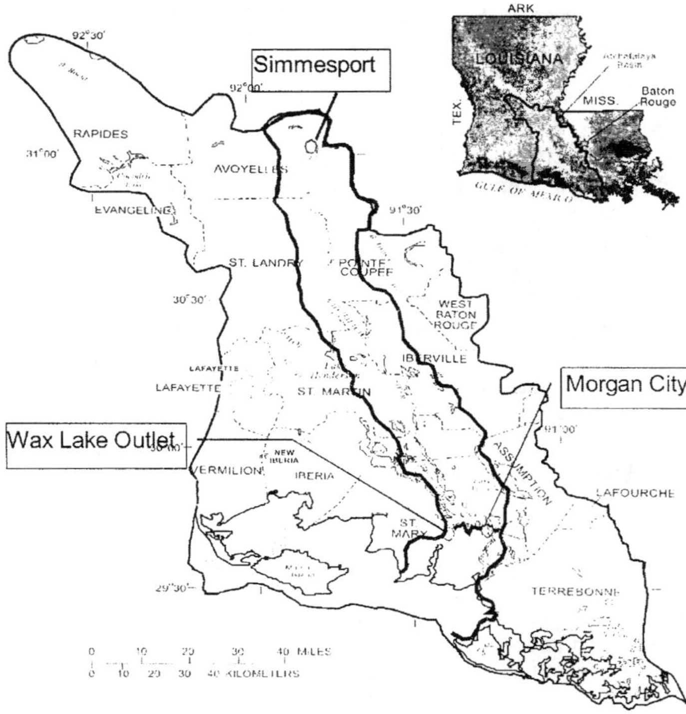

Fig. 1. Map of the Atchafalaya River Basin (adopted from Xu, 2006).

17.2.2. Data Collection

Long-term historical data on water quality, including monthly concentrations of nitrate-nitrite, TKN, and TOC, were collected to develop TOC:nitrate/nitrite, TOC:TKN, and TOC:total nitrogen (TKN+NO_3+NO_2) ratios. Data was collected from 1978-1990 at Simmesport, Louisiana and 1978-2004 at Morgan City, Louisiana. These data sets were obtained from the Louisiana Department of Environmental Quality (LDEQ). The basin was treated as a black box system, with Simmesport as the input, and Morgan City, located 190-km downstream, as the output (Figure 1). With the extensive levee system acting as impermeable barriers, this seems to be an accurate assumption.

17.2.3. Data Analysis

Ratios were calculated by dividing monthly TOC concentrations by total nitrate/nitrite, TKN, and total nitrogen. Total nitrogen is a sum of TKN concentrations and nitrate/nitrite concentrations. Ratios were averaged for monthly and yearly concentrations so that seasonal and temporal changes could be seen. A t-test was performed comparing the ratios of the two sites, with a p value of 0.05 to verify the significance of the results.

17.3. RESULTS AND DISCUSSION

17.3.1. Carbon and Nitrogen Concentrations

Average total organic carbon concentration at Simmesport was 6.32 mg L^{-1} (Table 1) ranging from 1.50 mg L^{-1} to 18.00 mg L^{-1}. Average TOC significantly decreased downstream at Morgan City to 5.67 mg L^{-1} (1.50-19.00 mg L^{-1}). There is an estimated 16% retention by the basin (Xu and Patil, in this book). Neither location showed any clear changes in TOC concentrations from 1978 to 2002. This was surprising, considering the findings of an overall increase in TOC fluxes due to land use changes.

Table 1. Average concentrations of carbon and nitrogen and the carbon to nitrogen ratios at the upstream and downstream locations in the Atchafalaya River Basin (* denotes the statistical significance at a 5% level).

	Simmesport Upstream	Morgan City Downstream	*Δ%*
TOC (mg L^{-1})	6.32	5.67	*10.2**
Nitrate/nitrite (mg L^{-1})	0.89	0.82	*7.9*
TKN (mg L^{-1})	0.97	0.94	*3.1**
Total N (mg L^{-1})	1.86	1.75	*5.9*
TOC:nitrate/nitrite	9.23	7.45	*19.3**
TOC:TKN	7.01	8.24	*-17.5**
TOC: total N	3.68	3.70	*-0.5*

Average nitrate/nitrite concentration was 0.89 mg L^{-1} at Simmesport (Table 1), ranging from 0.08 to 2.56 mg L^{-1}. Downstream at Morgan City the average nitrate/nitrite concentration was higher at 0.94 mg L^{-1} (0.17 mg L^{-1}-2.38 mg L^{-1}). This increase was present during the summer months indicating increased biological effects on riverine nitrogen transformation in the basin. Temperatures are highest during the summer months, which increases nitrification. Nitrification is the process by which ammonium is converted to nitrate, which is easily mobilized in water (Bernot and Dodds, 2005). This creates a problem for lotic systems, allowing nitrate to travel from the initial source.

Average TKN concentration at Simmesport was 0.97 mg L^{-1} (0.31 mg L^{-1}-2.80 mg L^{-1}). TKN was significantly lower at Morgan City averaging 0.82 mg L^{-1} (0.29 mg L^{-1} - 8.36 mg L^{-1}). There was a temporal trend relating TKN concentrations with discharge, however there were no seasonal changes in TKN concentrations. From 1978 to 2002 the total inflow at Simmesport averaged 6,559 m^3 s^{-1}. The total outflow from Morgan City averaged 6,228 m^3 s^{-1} (Xu, 2006). There was a seasonal trend of higher flow from March to June and a lower flow from August to November. During December to June, the typical wet season, Simmesport had significantly higher discharge than Morgan City indicating water was retained in the basin during this time period. The opposite was true for the dry season. The discharge at Simmesport was significantly lower than Morgan City discharge indicating retained water flowed out during dry weather (Xu, 2006).

17.3.2. Carbon : Nitrogen Ratios

C:N yearly average ratios did not show a clear trend over the past twenty plus years (Figures 2-4). There was fluctuation, which is expected from year to year, however no apparent trend. TOC to total nitrogen monthly ratios were significantly higher downstream at Morgan City (3.70 ranging from 0.49 to 12.80) than at Simmesport (3.68 ranging from 1.27 to 9.20) (Figure 4). Despite the decrease in TOC downstream, there was also a decrease in nitrogen, namely TKN, causing an increased ratio. The study found that the average TOC:NO_3+NO_2 and TOC:TKN ratios at Simmesport were 9.23 (1.63-36.19) and 7.01 (2.20-19.63), and that the TOC:NO_3+NO_2 and TOC:TKN ratios at Morgan City averaged 7.45 and 8.23, respectively (Figures 2 and 3). An increase in TOC:TKN at Morgan City is due to the significant decrease in TKN at Morgan City. There is a decrease downstream in the TOC:NO_3+NO_2 ratio attributing to an increase in nitrate/nitrite concentrations. As shown in Figure 3, TOC:NO_3+NO_2 decreases during the summer months, attributing to the increase in nitrate/nitrite during this time from the biological effects.

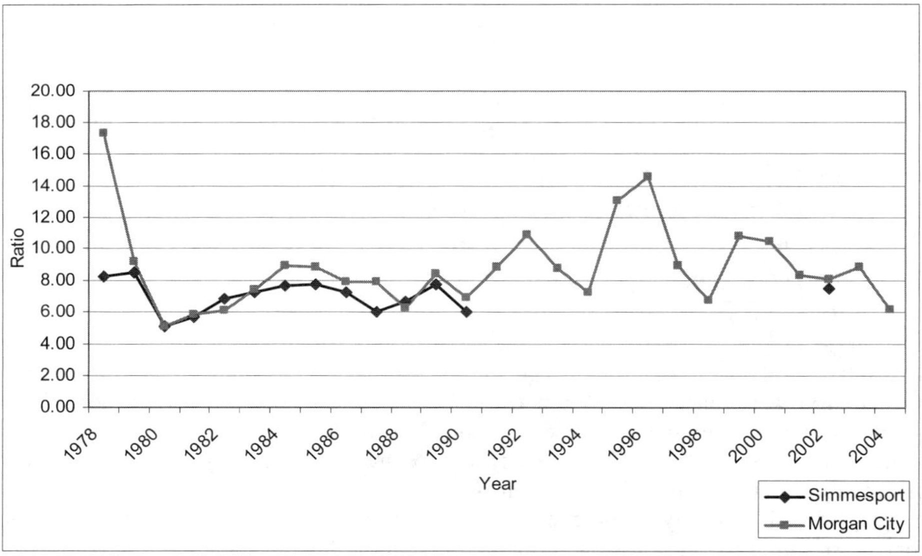

Fig. 2. Yearly averages of TOC:TKN concentrations from 1978-2004 at Simmesport, La and downstream at Morgan City, La in the Atchafalaya River.

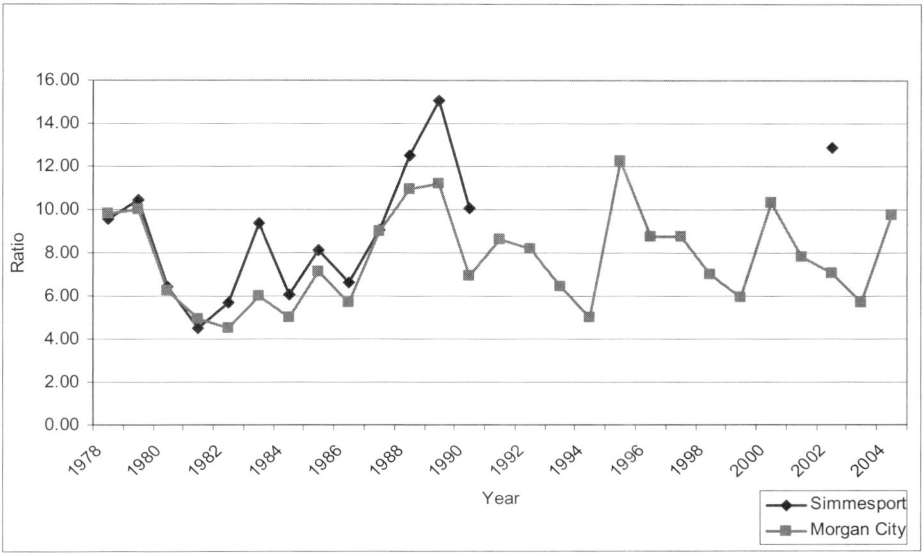

Fig. 3. **Yearly averages of TOC: nitrate/nitrite concentrations from 1978-2004 at Simmesport, La and downstream at Morgan City, La in the Atchafalaya River.**

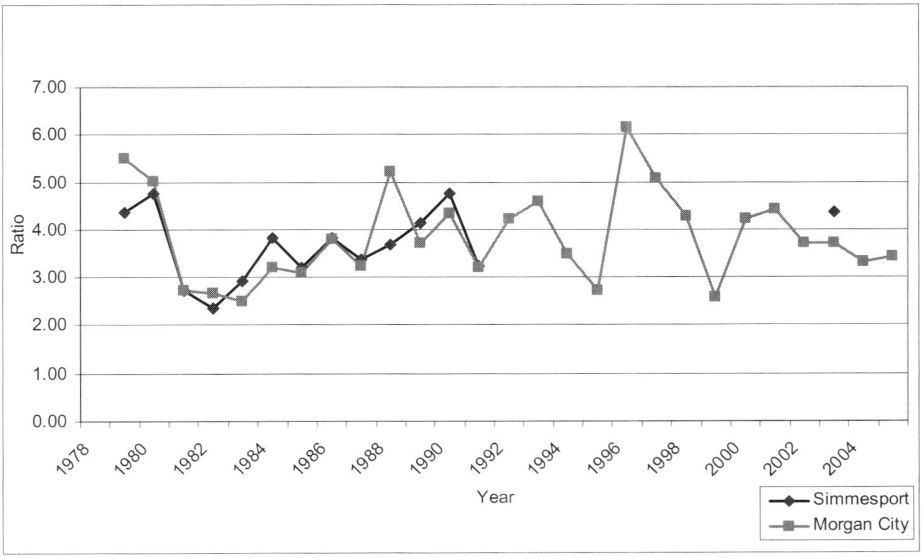

Fig. 4. **Yearly averages of TOC:total N concentrations from 1978-2004 at Simmesport, La and downstream at Morgan City, La in the Atchafalaya River.**

Rates of nitrification are affected by temperature and dissolved oxygen availability (Starry et al., 2005). It is expected that there is a seasonal trend in C:N ratios, in which C:N is lowest during the summer and spring and highest during fall and winter due to the seasonal temperature change. Other research has seen an increased in C in the fall attributed to leaf litter resulting in increased C:N ratios (Starry et al., 2005). Studies have found that high DOC could significantly reduce nitrification rates, resulting in lower nitrate concentrations (Strauss et al., 2002). It appears in this system that as C is lowest in the summer, nitrate/nitrite increases since nitrification is not inhibited. Strauss et al. (2002) adjusted C:N ratios in an in-lab experiment finding that low C:N ratios increased nitrification (resulting in higher nitrate concentrations), whereas high C:N ratios inhibited nitrification. There was a cyclic seasonal pattern in nitrogen concentrations found in the Upper Mississippi River (UMR) watershed (Strauss et al., 2004). This increased nitrification found during spring and summer was driven by higher temperatures, and the subsequent growth of nitrifying bacteria. Since much of the N reaching the Atchafalaya is from the UMR, it is expected that the Atchafalaya would experience these same seasonal variations as well. The hypoxia in the Gulf of Mexico is also a seasonal event, where the widest spread of the hypoxic zone occurs during the summer.

17.4. CONCLUSIONS

TOC and TKN concentrations decreased as the Atchafalaya flowed towards the Gulf of Mexico, indicating the river basin's ability to retain riverborne organic nutrients. Downstream at Morgan City, nitrate/nitrite increased during the summer months corresponding with temperature, suggesting increased nitrogen transformation in the basin. It appears that the Atchafalya River Basin is effectively holding TOC since although fluxes are changing as a result of land use changes, the concentrations in the Atchafalaya remain relatively constant. Organic carbon to nitrogen ratios help to better explain carbon and nutrient cycling in a complex system. This valuable information could supply further insight into dealing with the hypoxic dead zone in the Gulf of Mexico.

ACKNOWLEDGEMENTS

The research was supported in part by the National Science Foundation/Louisiana Education Quality Support Fund through the contract: NSF/LEQSF(2005)-PFUND-12. The authors would like to thank those who helped edit this manuscript. We would also like to thank Louisiana Department of Environmental Quality for use of long term historical data.

REFERENCES

Avnimelech, Y., 1999. Carbon/nitrogen ratio as a control element in aquaculture systems. Aquaculture 176:227-235.

Bernot, M.J., and Dodds, W.K., 2005. Nitrogen retention, removal, and saturation in lotic ecosystems. Ecosystems 8:442-453.

Demas, C.R., Brazelton, S.R., and Powell, N.J., 2002. The Atchafalaya Basin-river of trees. U.S. Geological Survey, Fact Sheet FS-021-02

Eadie, B.J., McKee, B.A., Lansing, M.B., Robbins, J.A., Mertz, S., and Trefry, J.H., 1994. Records of nutrient-enhanced coastal ocean productivity in sediments from the Louisiana Continental Shelf Estuaries 17:754-765.

Goolsby, D.A., Battaglin, W.A., Aulenbach, B.T., and Hooper, R.P., 2001. Nitrogen input to the Gulf of Mexico. Journal of Environmental Quality 20:329-336.

LDEQ, 1993. State of Louisiana Nonpoint Source Pollution Assessment Report. November 1993 (http://nonpoint.deq.state.la.us/assess32.html).

Mitsch, W.J., and Gosselink, J.G., 2000. Wetlands. Third Edition. John Wiley &Sons.

Rabalais, N. N., Turner, R. E., Justic, D., Dortch, Q., Wiseman, Jr., W. J., and Sen Gupta, B. K., 1996. Nutrient changes in the Mississippi River and system responses on the adjacent continental shelf. Estuaries 19:386-407.

Rabalais, N.N., Tirmer, R.E., Wiseman, Jr., W.J., and Dortch, Q., 1998. Consequences of the 1993 Mississippi River Flood in the Gulf of Mexico. Regulated Rivers: Research & Management 14:161-177.

Rabalais, N.N., and Turner, R.E. (Eds), 2001. Coastal Hypoxia: Consequences for living resources and ecosystems. Coastal and Estuarine Studies 58, American Geophysical Union, Washington DC.

Rabalais, N.N., Turner, R.E., Wiseman, Jr., W.J., 2001. Hypoxia in the Gulf of Mexico. Journal of Environmental Quality 30:320-329.

Rabalais, N.N., Turner, R.E., Dortch, Q., Justic, D., Bierman, Jr., V.J., and Wiseman, Jr., W.J., 2002. Nutrient-enhanced productivity in the northern Gulf of Mexico: past, present and future. Hydrobiologia 475/476:39-63.

Starry, O.S., Valett, H.M., and Schreiber, M.E., 2005. Nitrification rates in a headwater stream: influences of seasonal variation in C and N supply. Journal of the North American Benthological Society 24: 753-768.

Strauss, E.A., Mitchell, N.L., and Lamberti, G.A., 2002. Factors regulating nitrification in aquatic sediments: effects of organic carbon, nitrogen availability, and pH. Canadian Journal of Fisheries and Aquatic Sciences 59 (3):554-563.

Strauss, E.A., Richardson, W.B., Bartsch, L.A., Cavanaugh, J.C., Bruesewitz, D.A., Imker, H., Heinz, J.A., and Soballe, D.M., 2004. Nitrification in the Upper Mississippi River: patterns, controls, and contribution to the NO_3- budget. Journal of the North American Benthological Society 23 (1):1-14.

Thunell, R.C., Varela, R., Llano, M., Collister, J., Muller-Karger, F., and Bohrer, R., 2000. Organic carbon fluxes, degradation, and accumulation in an anoxic basin: Sediment trap results from the Cariaco Basin. American Society of Limnology and Oceanography. 45:300-308

Trefry, J.H., Metz, S., Nelsen, T.A., Trocine, R.P., Eadie, B. J., 1994. Transport of particulate organic carbon by the Mississippi River and its fate in the Gulf of Mexico. Estuaries 17:839-849.

Xu, Y.J., and Patil, A., 2006. Organic Carbon Fluxes from the Atchafalaya River into the Gulf of Mexico. In Y.J. Xu and V.P. Singh (Eds.): Coastal Environment and Water Quality, Section 3. Water Resources Publications, Highland Ranch, CO.

Xu, Y.J.,2006. Organic nitrogen retention in the Atchafalaya River Swamp. Hydrobiologia 560: 133-143.

Section 4:

BEST MANAGEMENT PRACTICES

18

REDUCING CRAWFISH AQUACULTURE EFFLUENT IMPACT THROUGH BMPS

Landon D. Parr, Robert P. Romaire, and W. Ray McClain
Aquaculture Research Station, Louisiana State University Agricultural Center
Baton Rouge, Louisiana 70820, USA

ABSTRACT

Nearly 53,000 ha of farmland are used to cultivate freshwater crawfish in Louisiana. Many of these ponds discharge effluent into watersheds in south-central and southwest Louisiana. Crawfish pond effluent can potentially contribute to the degradation of a stressed ecosystem (i.e., watershed) by exceeding the ecosystem's ability to assimilate the effluent. A study was conducted in six commercial crawfish ponds to evaluate Best Management Practices (BMPs) that would reduce water use and effluent release. The study site was located in the Cole Gully sub-watershed within the Plaquemine Brule watershed, Acadia Parish (County), southwest Louisiana. Effluent quality was characterized during the culture phase and final drawdown phase of the 2003-2004 crawfish production cycle. The highest concentrations and mass loading of solids, nutrients, and organic matter were found during final drawdown with the last 20% of the pond's volume having the highest concentration of solids, nutrients, and oxygen demand. BMPs that reduced water use and effluent were: maintaining water storage capacity during the culture phase, reducing water exchanges during the culture phase, and allowing the last 20% of the pond's volume to evaporate during final drawdown. Discharging through low gradient vegetated ditches during final drawdown reduced mass loading into receiving streams.

18.1. INTRODUCTION

Freshwater crawfishes are exploited for food worldwide and most of the United States crawfish production is derived from ponds. Nearly all of the USA's 53,000 ha of farm-raised crawfish (*Procambarus clarkii and P. zonangulus*) production are located in southwestern and south-central Louisiana (McClain and Romaire, 2004) in water basins identified by environmental regulatory agencies as being impaired. As with many types of pond aquaculture, crawfish producers occasionally discharge water from ponds to maintain good water quality and increase production. Water is also discharged from ponds when precipitation exceeds pond storage capacity. Ponds are drained annually in late spring or early summer to conform to the natural reproductive cycle of crawfish within subsurface burrows and to establish a forage crop. Methods for producing crawfish in Louisiana can best be described by two major approaches, crawfish monocropping (single-crop crawfish) or crawfish-agricultural crop rotational practices (Eversole and McClain, 2000). Crawfish is the only crop harvested in monoculture systems, and production typically occurs in ponds constructed solely for crawfish. Rotational systems that include crawfish almost always include rice production and sometimes soybeans. In these rotational systems, crawfish are either rotated with rice in the same physical location year after year (rice-crawfish-rice), or crawfish are rotated to fields in a different location each year to conform to typical field rotations of crops (rice-crawfish-soybeans or rice-crawfish-fallow). Presently, it is estimated that nearly two-thirds of Louisiana's crawfish aquaculture area uses a rice-crawfish rotational system.

Aquaculture operations nationwide have come under increased scrutiny because of potential environmental degradation caused by the discharge of water from production facilities. Crawfish pond effluent can potentially contribute to the degradation of a stressed ecosystem (i.e., watershed) by exceeding the ecosystem's ability to assimilate the effluent. The impact and contribution of crawfish aquaculture effluents on stream water quality in impaired waterbasins in Louisiana is not known. Nutrients and solids in effluents from 19 commercial crawfish ponds in Louisiana, representing different production systems in south-central and southwest Louisiana, were determined in the early 1990's but the volume of effluents and their contribution to total daily maximum load (TMDL) in receiving waters were not determined (Orellana, 1992). Parr (2002) developed water budgets and seasonal effluent mass loading models from research conducted in experimental crawfish ponds, and from this research and other empirical observations a set of best management practices (BMPs) for minimizing effluent discharge and mass loading from crawfish ponds were developed (Lutz et al., 2004) for adoption and evaluation by commercial crawfish farmers. BMPs have been determined to be an effective and practical means for reducing point and non-point water pollutants to levels compatible with environmental quality goals. In addition to providing a highly valued and desirable seafood product, crawfish ponds serve as favorable wetland habitat to many species of waterfowl, wading birds, and furbearers. Often land that is marginal for traditional row crops is used in crawfish production. Integration of crawfish aquaculture with traditional agricultural land uses serves as a practical means of land and water conservation.

The purpose of this project was to assess the contribution and impact of effluents from three commercial crawfish farms on stream water quality in southern Louisiana, with special focus on the Cole Gully sub-watershed, Bayou Plaquemine Brule, Acadia Parish (County), and to evaluate effluent management and culture practices (in the form of BMPs) that will improve stream water quality and conserve natural resources (especially water).

18.2. METHODS

18.2.1. Site location and Best Management Practices

Three pairs of adjacent, leveed ponds (fields), six ponds total, were selected as study and monitoring sites during the 2003-2004 crawfish production cycle. All sites were located in the Cole Gully sub-segment watershed basin, Bayou Plaquemine Brule watershed, in Acadia Parish (County), southwest Louisiana. The six ponds (fields) were hydrologically isolated from each other. The cooperators agreed to manage one of each pair of ponds according to conventional crawfish farming (non-BMP) practices for rice-crawfish rotations systems (Table 1) and abide by a set of BMPs on the companion pond, thereby providing side-by-side comparison of BMP and conventional practices. The Bieber farm (B) had two paired ponds: 70B (BMP, 28.60 ha) and 37B (non-BMP, 14.96 ha). The Reiner farm (R) had two paired ponds: E8R (BMP, 3.36 ha) and W8R (non-BMP, 3.06 ha). The Hundley farm (H) had two paired ponds: Z35H (BMP, 14.03 ha) and Z22H (non-BMP, 8.68 ha).

Non-BMP ponds employed standard crawfish management procedures (Avery and Lorio, 1996). One pond at each site location adopted BMPs associated with two phases

Table 1. Summary of standard operating practices for rice-crawfish-rice and rice-crawfish-soybean (fallow) systems in Louisiana (modified from McClain and Romaire, 2004).

Months	Rice-Crawfish-Soybeans or Rice-Crawfish-Fallow Rotation
July – August	Harvest rice crop and manage for rice re-growth
September – October	Flood pond (field) for crawfish in October and monitor/manage water quality
November – December	Monitor/manage water quality
January – February	Harvest crawfish 2 - 4 days per week according to CPUE* and markets
March – April	Harvest crawfish 3 - 4 days per week according to CPUE* and markets
May – June	Drain pond/field and plant soybeans (or harvest crawfish until CPUE* is not feasible, drain and leave fallow)
July - . . .	Harvest soybeans in October, plant rice in March/April, then repeat cycle

* CPUE = Crawfish catch per unit trap effort

of the crawfish production cycle: (1) in-pond phase and (2) final drawdown phase. The in-pond phase is from the time crawfish farmers fill their ponds with water in the fall until they remove their crawfish traps in preparation for draining the field in spring or early summer. The final drawdown phase is from the time the crawfish farmer begins to drain the crawfish pond in spring or summer by lowering or removing the drainage structure until no water remains in the pond. Two BMPs were identified for the in-pond phase and four BMPs for the final drawdown phase and were to be implemented by the three cooperators.

The two in-pond crawfish BMPs adopted by each cooperator were as follows:

1. Partially filling crawfish ponds in the fall to target about 10 cm of additional water storage capacity so as to reduce unscheduled precipitation overflow. The non-BMP or current recommended water management practice is to completely fill the pond to the top of the drainpipe providing zero water storage capacity.

2. Reduce the frequency of water exchanges (flushing) used during low oxygen episodes because many, and perhaps most, commercial crawfish ponds cannot add sufficient oxygenated water rapidly enough to have a corrective effect on low dissolved oxygen within the pond. The non-BMP and current recommended water management practice is to flush ponds with aerated water when dissolved oxygen levels decline below 2 or 3 mg/L.

The four final drawdown crawfish pond BMPs were as follows:

3. Cease crawfish harvest activities 1 to 2 weeks prior to final drawdown. Operation of crawfish harvesting boats results in suspension of solids and we hypothesized that cessation of harvesting activities 1 to 2 weeks prior to final drawdown could

potentially reduce the level of solids, nutrients, and organic matter released in final drawdown effluent. The non-BMP would allow for crawfish harvesting immediately prior to final drawdown.

4. Drain the pond from the top of the water column down as slowly as practical to potentially reduce scouring and release of unconsolidated sediments from the pond bottom. The non-BMP or standard operating procedure allows for rapid draining from the surface on down or removal of the drop drain pipe which drains the pond rapidly from the bottom.
5. Allow the remaining 20% of the pond water volume to evaporate when a flooded crawfish pond will remain fallow (no crop grown, either terrestrial or aquatic) following the crawfish crop. The standard operation practice (non-BMP) recommends complete discharge of water from the pond through the drainage structure during final drawdown.
6. Discharge pond effluent during final drawdown through wide, vegetated or semi-vegetated ditches, with moderate to slight gradient where possible, to enhance the settling of suspended solids before the effluent reaches major drainage systems. Sediments that accumulate in the drainage ditch can be used to rebuild adjacent pond levees and roads.

Only one of the three cooperators had a drainage system that was heavily vegetated with slight gradient. The other two cooperators shared a common drainage ditch that was lightly vegetated with moderate gradient. BMP recommendations 1, 2, and 6 were applicable to fall and winter 2003/2004 and BMP recommendation 3, 4, 5, and 6 were applicable immediately prior to and during the final drawdown phase in late spring and summer 2004.

18.2.2. Sampling, Water Analysis, Water Budgets, and Mass Loading

Effluent inflow into the discharge pipe of each of the six ponds was instrumented with a Thel-Mar volumetric weir coupled with an area-velocity flow meter (ISCO Model 750) which activated and controlled an automated water sampler (ISCO Model 6712) to periodically withdraw and composite water samples from the pipe. All pond effluent passed through these drainpipes. The sampler was programmed so that the composite sample was a flow-weighted average of the total discharge. The cooperating farmers flushed non-BMP ponds for oxygen management at their own discretion. Although we were unable to estimate the frequency and the volume of water associated with flushing, the quantity was reflected in the volume of water released through the drains during the in-pond phase of the production season. Precipitation was measured with rain gauges (ISCO model 674) placed at each paired pond site.

Ponds were monitored weekly throughout the study for collection of data and maintenance of instrumentation. Not every discharge event could be measured because of temporary equipment malfunction. Precipitation and flow data from the instrumentation were downloaded to an RTD datalogger (ISCO Model 581). During final drawdown, effluent samples were collected at 100, 80, 60, 40, 20, and 5 percent of estimated pond volume. At 60% estimate pond volume, five grab water samples were collected over a distance of 0 (discharge pipe) to 1,340 meters in a heavily vegetated receiving ditch with slight gradient and from a lightly vegetated receiving ditch with moderate gradient. Water samples were placed on ice and transported to Louisiana State University for analysis of the following parameters (Clesceri et al., 1998): total suspended solids (TSS), volatile suspended solids (VSS), total phosphorus (TP), total nitrogen (TN), and 5-day biochemical oxygen demand ($CBOD_5$). The number of sampling events in which effluent samples were collected for analysis ranged from 9 to 15 for the "in-pond" phase and 4 to 6 for the "drawdown" phase. Water discharged as overflow and final drawdown was determined directly from data collected from the flow meters at each of the six pond sites. Mass loading for each discharge event was determined as mass loading = [(concentration x volume / area)].

18.3. RESULTS

The effluent discharged from the six commercial crawfish ponds during the 2003-2004 crawfish season from overflow and final drawdown is presented in Table 2. Overflow included precipitation overflow through the drainpipes, water exchanges for water quality improvement, and water released from levee breaches/breaks caused by a combination of wave action and burrowing activities of rodents, primarily muskrat and nutria. An average

Table 2. Water discharged as effluent (ha-m/surface ha) from three commercial crawfish farming operations subject to effluent BMP and standard operating (non-BMP) practices, 2003-2004 crawfish production season.

	Non-BMP			BMP		
Farm	**Overflow**	**Drawdown**	**Total**	**Overflow**	**Drawdown**	**Total**
B	0.111	0.037	0.148	0.195	0.012	0.207
H	0.131	0.121	0.252	0.206	0.105	0.311
R	0.195	0.106	0.300	0.203	0.084	0.287
Average	0.146	0.088	0.234	0.201	0.067	0.269

of 0.251 ha-m of effluent per surface ha was discharged into Cole Gully stream from the six ponds, with 69% of the water release associated with overflow, including levee breaks, and 31% with final drawdown. The average water released as overflow from non-BMP ponds was 0.146 ha-m/surface ha and for BMP ponds, 0.201 ha-m/surface ha. The higher overflow discharge from BMP ponds was attributed to several events which occurred in the BMP ponds from April 2004 through final drawdown and were not associated with BMP implementation. These events included a levee breach which allowed water to flow from a non-BMP pond into an adjacent BMP pond, a flushing activity in a BMP pond to remove a nuisance floating aquatic fern (*Azolla* spp.), and malfunction of an inflow water valve that resulted in significant water addition to a BMP pond. Effluent discharge during final drawdown for non-BMP ponds averaged 0.088 ha-m/surface ha, and for BMP ponds, 0.067 ha-m/surface ha, resulting in a 23.7% reduction in effluent released in final drawdown from BMP ponds.

The relationship between additional water storage capacity and effluent discharge from January through March (in-pond phase) of the crawfish production cycle is illustrated in Figure 1. Added storage capacity in the ponds and non-flushing of ponds for water quality

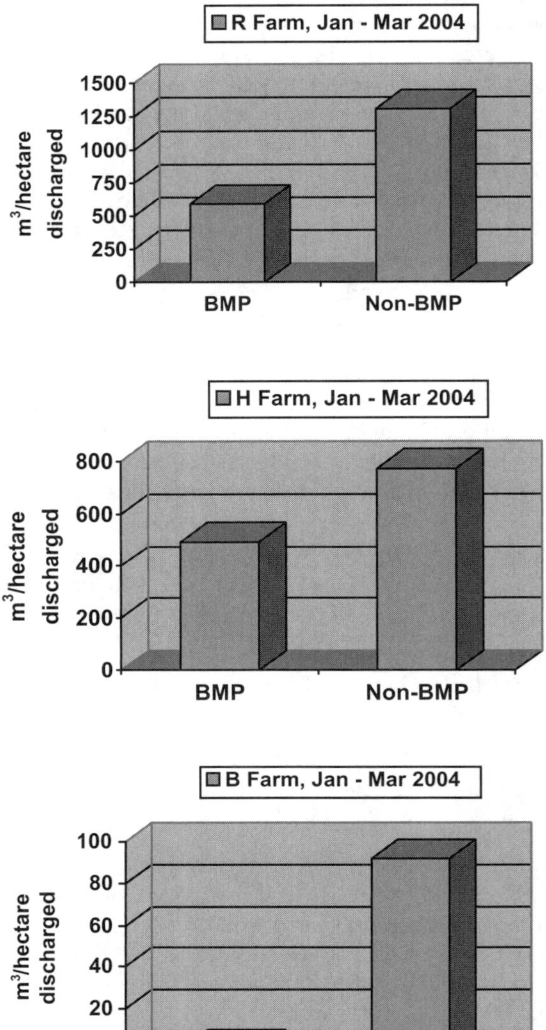

Fig. 1. Effluent overflow discharged from January through March from three commercial crawfish farms with 10 cm pond water storage capacity (BMP) compared to 0 cm water storage capacity (non-BMP).

improvement for crawfish significantly decreased the amount of water released from overflow from January through March. Mean concentrations of TSS, VSS, TN, and BOD were significantly higher in drawdown effluent than in overflow effluent (Table 3).

Interestingly, the concentrations of TP did not follow this trend with concentrations slightly higher in overflow effluent. Mass loading of TSS, VSS, TN, TP, and BOD was about five to 10 times higher per discharge event in drawdown effluent than overflow effluent (Table 4), and this observation conforms with findings reported by Parr (2002) from experimental crawfish ponds. Because of levee breaks that occurred in all ponds (BMP and non-BMP) during the 2003-2004 crawfish production season, it was not possible to provide a valid comparison of mass loading from conventionally managed ponds and BMP ponds.

Table 3. Mean concentration of water quality variables (mg/L/discharge event) for commercial crawfish ponds.

	Overflow (Mean)	Overflow (Range)	Drawdown (Mean)	Drawdown (Range)
Total Suspended Solids (TSS)	190	70-427	592	319-1186
Volatile Suspended Solids (VSS)	32	15-56	90	31-175
Total Phosphorus (TP)	0.59	0.09-3.17	0.47	0.21-0.83
Total Nitrogen (TN)	4.08	2.48-6.00	8.47	5.23-11.10
Biochemical Oxygen Demand ($CBOD_5$)	6.73	2.35-10.7	16.73	9.40-30.45

Table 4. Mean mass loading (kg/ha/discharge event) for commercial crawfish ponds.

	Overflow (Mean)	Overflow (Range)	Drawdown (Mean)	Drawdown (Range)
Total Suspended Solids (TSS)	79	7.9-240	1133	120-2473
Volatile Suspended Solids (VSS)	14	1.2-43	156	28-364
Total Phosphorus (TP)	0.22	0.01-0.91	1.0	0.06-2.1
Total Nitrogen (TN)	1.40	0.191-4.62	11.2	2.48-18.7
Biochemical Oxygen Demand ($CBOD_5$)	3.40	0.084-8.20	24.2	8.78-34.8

Mean concentrations of TSS, VSS, TN, TP, and BOD generally increased as the ponds were dewatered in late spring/summer (Table 5). Generally, the highest increase in solid, nutrient, and organic matter concentrations occurred with 20% or less volume remaining in the ponds. The spike in solid, nutrient, and organic matter concentrations are largely attributed to an increase in scouring of the pond bottom as the remaining water concentrated around the discharge pipes. The concentration of TSS in drawdown effluent decreased substantially at 268 meters from the discharge pipe in the heavily vegetated receiving ditch (Table 6). An apparent decrease in water velocity caused from the vegetation and slight gradient of the ditch allowed heavier solids to settle out of the water. No further decrease in TSS concentrations in the heavily vegetated ditch was observed after 536 meters. Concentrations of TSS also decreased in the lightly vegetated receiving ditch over a distance of 536 meters, but the decrease was not as high as that observed in the heavily vegetated ditch.

Table 5. Mean concentration (mg/L) of total suspended solids (TSS), volatile suspended solids (VSS), total nitrogen (TN), total phosphorus (TP), and 5-day biochemical oxygen demand ($CBOD_5$), weighted for volume discharged, during final drawdown from six commercial crawfish ponds, Cole Gully, Acadia Parish, Louisiana, 2004.

Water volume remaining in pond (%)	TSS	VSS	TN	TP	$CBOD_5$
100	260	50	7.926	0.45	9.83
80	324	59	7.238	0.42	15.30
60	404	72	7.531	0.43	14.23
40	457	70	9.224	0.41	19.81
20	1082	146	11.868	0.59	27.08
5	1372	176	11.639	0.71	19.35

Table 6. Total suspended solids (TSS) concentrations in heavy vegetated receiving ditch and a non-vegetated receiving ditch receiving effluent during drawdown phase in commercial crawfish ponds.

Distance from Discharge Pipe (m)	TSS concentration (mg/L) in Heavily Vegetated Drainage Ditch	TSS concentration (mg/L) in Lightly Vegetated Drainage Ditch
0	2,027	1,490
268	464	*
536	276	987
804	226	704
1,072	244	*
1,340	*	587

*Not sampled.

18.4. DISCUSSION

The relatively small and narrow levees of the commercial ponds utilized for rice-crawfish rotational production systems were subject to frequent breaks from wave erosion and burrowing activities by rodents thereby releasing effluents. All ponds, whether a BMP or non-BMP pond, were subject to levee breaks, and breaks compromised the ability to effectively implement and monitor some BMPs in this study. Crawfish aquaculture in Louisiana is highly integrated into the rice production industry, with rice being the primary crop and crawfish a secondary crop (McClain and Romaire, 2004). Consequently, smaller levees (height and width) that are constructed to meet the requirements for rice production rather than for crawfish production are subject to a higher frequency of failure. Ponds constructed solely for the purpose of crawfish production require larger perimeter levees (width and height) than is recommended for rice production. A BMP to minimize water release from rice fields used for crawfish production would be to construct perimeter levees according to specification outlined in Avery and Lorio (1996) and implement effective control measures for nuisance wildlife that damage levees from burrowing activity.

Although levee failures did compromise to some extent the objectives of this study, several of the BMPs when used in concert showed promise to reduce effluent. Providing crawfish pond storage capacity by not filling ponds to the top of the drainpipe (structures) to reduce unscheduled effluent discharge during precipitation reduced unintentional effluent discharge until such time that precipitation exceeded water loss from evaporation, transpiration, and seepage. Although the period of time at which storage capacity is eliminated depends on a number of factors, modeling of crawfish ponds (Parr 2002) and empirical observations in this study indicate water storage capacity in an average southern Louisiana year is not likely to be exhausted until mid to late winter or early spring. The addition of water storage capacity to reduce incidental discharge in most cases would require no, or minor, modifications to pond levee heights or drainage structures because the biological need to have water at the recommended depth of 35 to 45 cm for crawfish production is not necessary until late February/March when intensive harvesting of crawfish begins.

Elimination or significant reduction of water exchanges, or pond flushing to improve water quality, for crawfish production has significant potential to reduce effluent. Although early recommendations in crawfish aquaculture recommended frequent water exchange to maintain satisfactory water quality to optimize crawfish production (Avery and Lorio, 1996) more recent studies indicate that frequent water exchanges do not significantly increase overall crawfish production (Huner, 2004). In addition, crawfish have a relatively high tolerance to hypoxia (McClain, 1999). Many crawfish farmers in southern Louisiana do not have sufficient pumping capacity or properly designed aeration systems to effectively correct or improve oxygen deficits present in crawfish ponds, nor do many measure oxygen concentrations in ponds on a routine basis. Thus, many farmers likely pump and flush water through ponds for the general well being of the crawfish when it is not required. Not only does the practice of water exchange increase effluent, it also increases energy consumption and production cost with apparently minimal benefit to crawfish production (Huner, 2004). Exchanging water to improve oxygen levels in crawfish ponds should be implemented when low oxygen conditions are verified by direct measurement in ponds and when influent water is of high quality.

Parr (2002) reported that mass loading of nutrients and solids in experimental crawfish ponds were highest during the final drawdown, particularly with the first few percent of water discharged, and the remaining 20% of water discharged. These findings were further substantiated from the commercial ponds used in this study. Where possible, allowing the last 20% of water in the pond to evaporate at the end of the crawfish production cycle can reduce solid, nutrient, and organic matter mass loading. Although this practice is not always possible depending on the production scheme used by a crawfish farmer, if the ponds (fields) are allowed to remain fallow following the crawfish crop, as is often the case in rice-crawfish rotational production schemes, this may be a practical BMP that can be implemented by crawfish producers with no detriment or cost to the producer.

When crawfish pond effluent released during final drawdown was allowed to flow through a heavily vegetated ditch the concentration of suspended solids was substantially reduced. Vegetated filter strips are commonly used to reduce suspended solids and nutrient loads in wastewater and effluent in agricultural, industrial, and municipal applications. In aquacultural applications, cost of constructing filter strips for suspended solids removal may be cost prohibitive (Kousk and Engle, 1994). However, many crawfish producers, by virtue of site location of their farming operations, discharge pond effluent into relatively shallow ditches with moderate to slight gradient. These ditches in time establish thick stands of emergent and submergent vegetation through which effluent must flow before it reaches the terminal destination in watersheds. Although sediments will accumulate in drainage ditches, and over time must be removed to allow for continued drainage, sediments can be removed by dredging and used to rebuild adjacent pond levees and roads which deteriorate over time. Often local governmental entities will provide this service at little or no cost to the producer to enhance drainage of local watersheds.

Empirical observations on cessation of crawfish harvesting activities 1 to 2 weeks prior to final drawdown indicated that this BMP is less certain to be effective than was originally hypothesized. As a general rule, late in the season prior to final drawdown crawfish ponds have little or no standing vegetation remaining. Wave action on windy days is not suppressed by standing vegetation and suspension of bottom sediments in the water column appears to be higher during this time than at any other period of the production season. Additionally, bottom foraging activities by crawfish can apparently contribute significantly to the presence of suspended solids late in the season prior to final drawdown even when crawfish harvesting activities cease. However, the recommendation of cessation of harvesting activities is an easily adoptable BMP with little or no cost incurred by the crawfish producer and should have no apparent negative biological consequence to subsequent years' crawfish production. A more plausible and acceptable BMP for crawfish producers would be to suspend harvesting activities 1 to 2 days prior to initiation of final drawdown which should provide sufficient time for heavier solids to settle depending on weather conditions.

The suggested BMP of draining the pond from the top of the water column down as slowly as practical did not appear to be an effective BMP based on our observations, but does require additional investigation. We hypothesized that slowly draining the pond from the top down could potentially reduce scouring and the release of unconsolidated sediments from the pond bottom. However, general observations in the relatively shallow (<0.5 m) crawfish ponds suggest that unconsolidated sediments will be released in the effluent regardless of whether the pond is drained from the top or drainage drop pipes are removed to effect rapid draining of the crawfish pond.

Concurrent with this study, a survey of water use and discharge practices by crawfish farmers in Acadia parish (county) was made. Results of the survey showed that there is a relatively high adoption rate of recommended crawfish effluent BMPs evaluated in this study by crawfish farmers in that parish. A majority of crawfish farmers attempt to maintain water storage capacity in crawfish ponds to capture precipitation during the in-pond phase of the crawfish production cycle at least until such time that precipitation reduces storage capacity to a zero level. In addition, most do not flush or exchange water during the in-pond phase of the crawfish production cycle in an attempt to correct oxygen deficiencies during hypoxic conditions thereby decreasing effluent. Many crawfish farmers discharge effluent during final drawdown through vegetated ditches. Retention of the remaining 20% of water in a pond during final drawdown, if the field remains fallow following the crawfish crop, does not currently have a relatively high adoption rate among farmers. The agronomic implications on other crops when retaining the remaining 20% of final drawdown and allowing it to evaporate must be investigated before farmers will consider adopting this practice as a crawfish effluent BMP.

REFERENCES

Avery, J. and W. Lorio, 1996. Crawfish production manual. Publication number 2637, Louisiana State University Agricultural Center, Baton Rouge, LA.

Clesceri, L., Greenberg, A., and Eaton, A. (Editors), 1998. Standard methods for the examination of water and wastewater. 20th Edition. APHA, AWWA, WEF.

Eversole, A., and McClain, R.W., 2000. Crawfish culture. Pages 185-198. In: Robert R. Stickney, editor. Encyclopedia of Aquaculture. John Wiley and Sons, Inc., New York, New York, USA.

Huner, J. V., 2004. Effects of selected forage species on water quality. Unpublished draft of final report "Final Model Sustainable Farm for Practical Water Quality Management", Louisiana Department of Environmental Quality Contract No. 72-6000820.

Lutz, C.G., Romaire, R.P., and Sanders, F., 2004. Aquaculture production best management practices (BMPs). Publication number 2894, Louisiana State University Agricultural Center, Baton Rouge, LA.

Kousk, P., and Engle, C., 1994. Cost of alternative effluent treatments for catfish production. Publication 467, Southern Regional Aquaculture Center, Stoneville, MS.

McClain, W.R., 1999. Effects of hypoxia on growth and survival of the crayfish *Procambarus clarkii*. Freshwater Crayfish 12: 121-133.

McClain, W. R., and Romaire, R., 2004. Crawfish culture: A Louisiana aquaculture success story. World Aquaculture 35(4): 31-35, 60-61.

Orellana, F. X., 1992. Characterization of effluents from commercial crawfish ponds in south Louisiana. Master's thesis. Louisiana State University, Baton Rouge, LA, 98 p.

Parr, L. D., 2002. Water discharge models, seasonal effluent mass loading, and best management practices for crawfish ponds. Master's thesis. Louisiana State University, Baton Rouge, LA, 107 p.

19

CHALLENGES FOR IMPROVING SURFACE WATER QUALITY IN AN AGRICULTURAL WATERSHED IN LOUISIANA

D. D. Poudel
Department of Renewable Resources, University of Louisiana at Lafayette
PO BOX 44650, Lafayette, LA 70504, USA

ABSTRACT

With Louisiana's unique landscape, extensive hydromodification, and insufficient erosion control measures in croplands, one of the state's great environmental challenges is to improve surface water quality in its agricultural watersheds. Surface water quality not only affects indigenous vegetative and aquatic life but also affects the human population that is dependent upon its natural resources. In order to isolate the challenges to improving surface water quality, Dissolved Oxygen (DO), temperature, conductivity, pH, and turbidity were monitored in Bayou Plaquemine Brule Watershed in southwestern Louisiana from April 2002 to March 2005. Grab samples were collected for laboratory determination of Total Nitrogen, Total Phosphorus (TP), Nitrate/Nitrite-N, 5-day Biological Oxygen Demand (BOD_5), Total Suspended Solids (TSS), and Soluble Reactive Phosphate from October 2002 to September 2004. Results indicated that strong spatial and seasonal variation of surface water quality exists across the watershed. For example, tributaries showed 51% higher TSS values as compared to the main channel, and average summer (March – November) DO values were 38% lower as compared to the average winter (December – February) DO values. Also, elevated TSS and TP concentrations were found following rainstorm events. Identification of 'hot spots' across the watershed of nonpoint source pollution, more accurate quantitative relationships between landuse types and nutrient loading, and the implementation of appropriate Best Management Practices (BMPs) for nonpoint source pollution control are suggested.

19.1. INTRODUCTION

Very flat landscapes, intensive agriculture, extensive hydromodification, significant drainage water, and residential areas characterize Louisiana's agricultural watersheds. Major agricultural land use products in the state include rice, soybeans, sugarcane, crawfish, catfish, cotton, cattle, and corn. Rice and crawfish production are common in lowland areas, while sugarcane, soybeans, cotton, and cattle are major agricultural products in upland areas. Catfish farming is also practiced in the upland areas in northern Louisiana. In general, agricultural practices are characterized by high inputs, a lack of sufficient conservation tillage practices, intensified agriculture, and year-round production. These practices lead to a large amount of fertilizer and other chemical-use. Coupled with high annual averages of precipitation, the elevated levels of TSS and nutrient loadings in agricultural runoff are characteristics of Louisiana agricultural watersheds. These accelerated amounts of TSS and nutrient loadings in the bayous and streams of Louisiana's agricultural watersheds have resulted in the impairment of these water bodies, hence the listing of the majority of them in LDEQ's 303 (d) lists.

In order to address the issues of nonpoint source pollution and TMDLs, a water quality monitoring study was launched in the Bayou Plaquemine Brule watershed in 2001. The specific objective of this study was to identify the challenges to surface water quality improvement in an agricultural watershed.

19.2. MATERIALS AND METHODS

Seven sites were identified for water quality monitoring in the 234,400 acres Bayou Plaquemine Brule watershed in southwestern Louisiana (Figure 1). Since April 2002, Dissolved Oxygen (DO), temperature, turbidity, conductivity, and pH were measured twice per month in the stream with multi-probe YSI Sonde probes that were attached to hand-held data loggers (YSI Sonde models 6820 with 650MDS). Measurement was always performed from the marked point on the bridge at three different depths (20%, 60%, and 80% from the surface of the water). All sampling was done from 10:00 AM to 2:00 PM. Water samples have been collected by the grab method from all sampling locations on a biweekly basis since October 2002. One sampling event consisted of a regular/background sampling, while another consisted of an after rain event sampling. Each composite sample consisted of 2 L and was stored in glass or plastic bottles. Sample bottles were then placed in an ice chest with ice packs and transported to the laboratory. In the laboratory, water samples were prepared for BOD_5, TSS, PO_4-P, TP, NO_3/NO_2-N, and TN, and were analyzed according to Standard Methods (Clesceri et al., 1998). Excel spreadsheets and SAS program were used for simple regression analyses, Student's t-test, and the calculation of simple statistics.

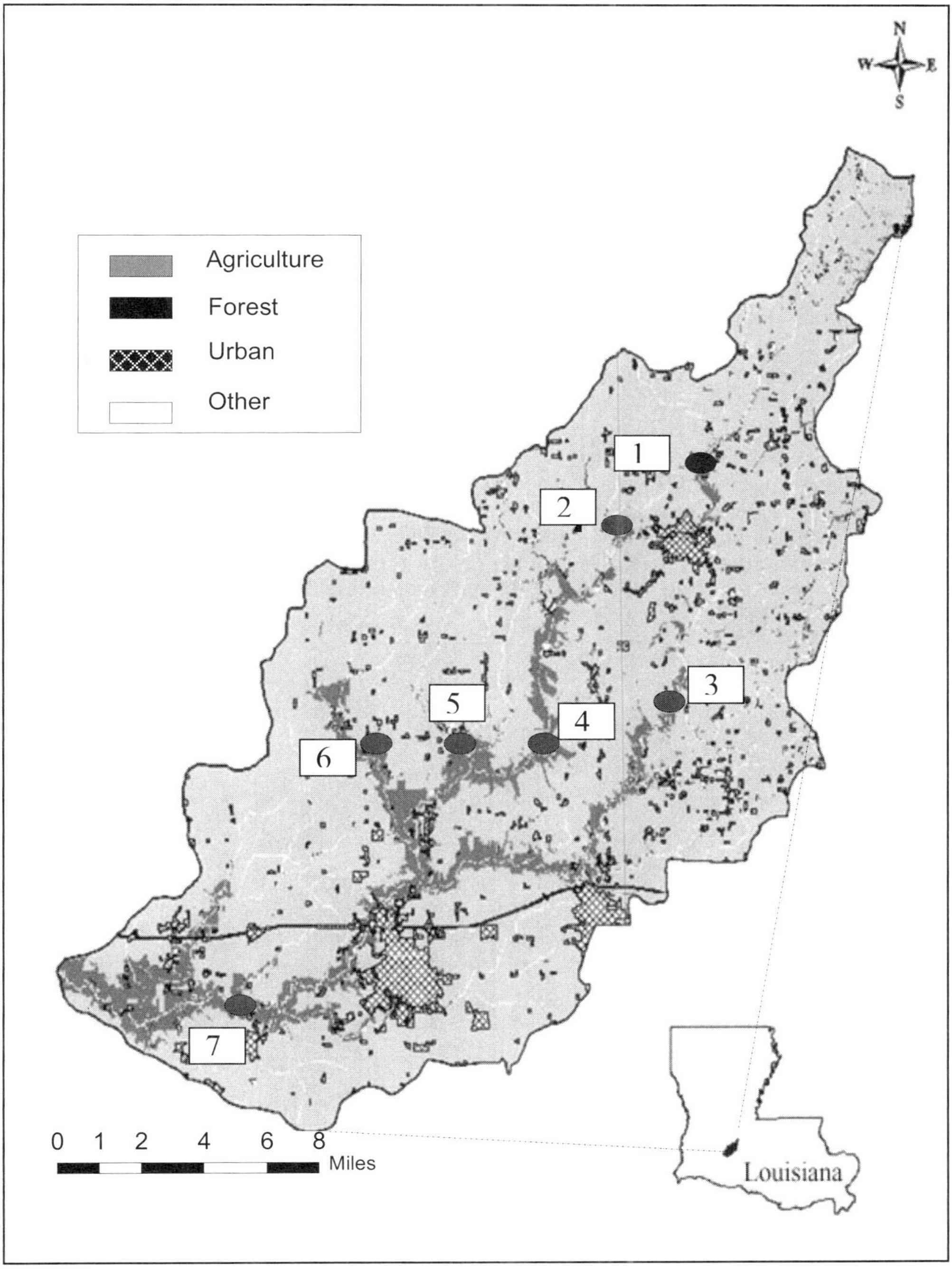

Fig. 1. Locations of Bayou Plaquemine Brule water quality monitoring sites.

19.3. RESULTS AND DISCUSSION

19.3.1. Seasonal variability

The State had set the in-stream standard level of DO for Plaquemine Brule at 5 mg L^{-1} between December and February and 3 mg L^{-1} between March and November (LDEQ, 2000). Large seasonal variations existed in surface water DO levels in Bayou Plaquemine Brule. On average, DO levels ranged between 4 and 5 mg L^{-1} during the months of April through September, and between 6 and 9 mg L^{-1} for the remainder of the months. Also, surface water temperatures were above 20°C from April through October, with peak temperatures occurring in July. While analyzing water quality data between the winter months (December – February) and the summer months (March – November), all parameters but DO showed an increase in concentration during the summer months compared to the winter months, while DO values for the summer months were significantly lower compared to those for the winter months (Table 1). In the summer, TN and TP concentrations increased by 58% and 44%, respectively, compared to the winter. The summer turbidity and TSS values were also higher by 29% and 36%, respectively, compared to the winter values. Higher nutrient concentration, TSS, and turbidity values in the summer suggest that surface water quality in Bayou Plaquemine Brule is clearly associated with agricultural activities that are carried out during the summer months.

Table 1. Summer (March – November) and winter (December – February) months Surface Water Quality in Bayou Plaquemine Brule Watershed (April 2002 – May 2005).

	Winter		Summer		% increase/decrease
	N	mean	N	mean	from winter to summer
TSS (mg L^{-1})	133	81 (±14)[b]*	389	111 (±7)[a]	36
TN (mg L^{-1})	133	2.15 (±0.07)[b]	293	3.4 (±0.1)[a]	58
TP (mg L^{-1})	133	0.28 (±0.01)[b]	293	0.40 (±0.01)[a]	44
Temp (°C)	131	12.5 (±0.27)[b]	401	23 (±0.24)[a]	84
Conductivity (mS cm^{-1})	131	0.45 (±0.02)[b]	401	0.55 (±0.02)[a]	21
Turbidity (NTU)	131	236 (±20)[b]	401	305 (±23)[a]	29
DO (mg L^{-1})	131	8.4 (±0.15)[a]	395	5.3 (±0.11)[b]	-38
pH	117	7.3 (±0.06)[b]	387	7.5 (±0.02)[a]	2

*Numbers in parentheses are ± standard error of mean. Values across the rows with different letters are significantly different at 0.05 probability by Student's t-test.

Water temperatures were negatively related to Dissolved Oxygen levels (Figure 2). As the temperature dropped, the solubility of oxygen dropped also; hence, higher Dissolved Oxygen values were found in the winter. In a forest system, greater diurnal variation of stream temperatures were reported in those streams where clear-cut harvesting, rather than where patch-cutting, was practiced (Lynch et al., 1984). Clearance of streamside vegetation affected the stream temperature, which in turn altered the balance of net radiation, turbulent heat exchange across the water surface boundary, and stream/streambed heat exchange

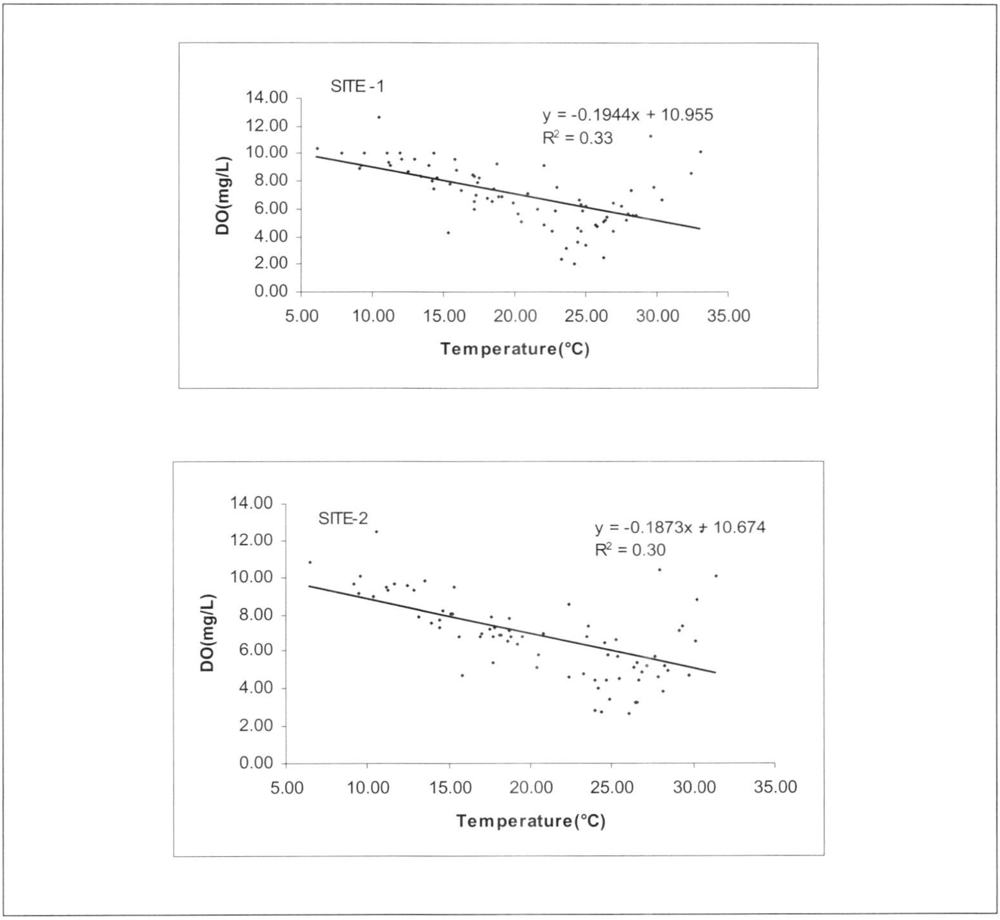

Fig. 2. DO vs. temperature regression lines for site 1, 2, 4 and 7 in Bayou Plaquemine Brule Watershed Water Quality Monitoring Project.

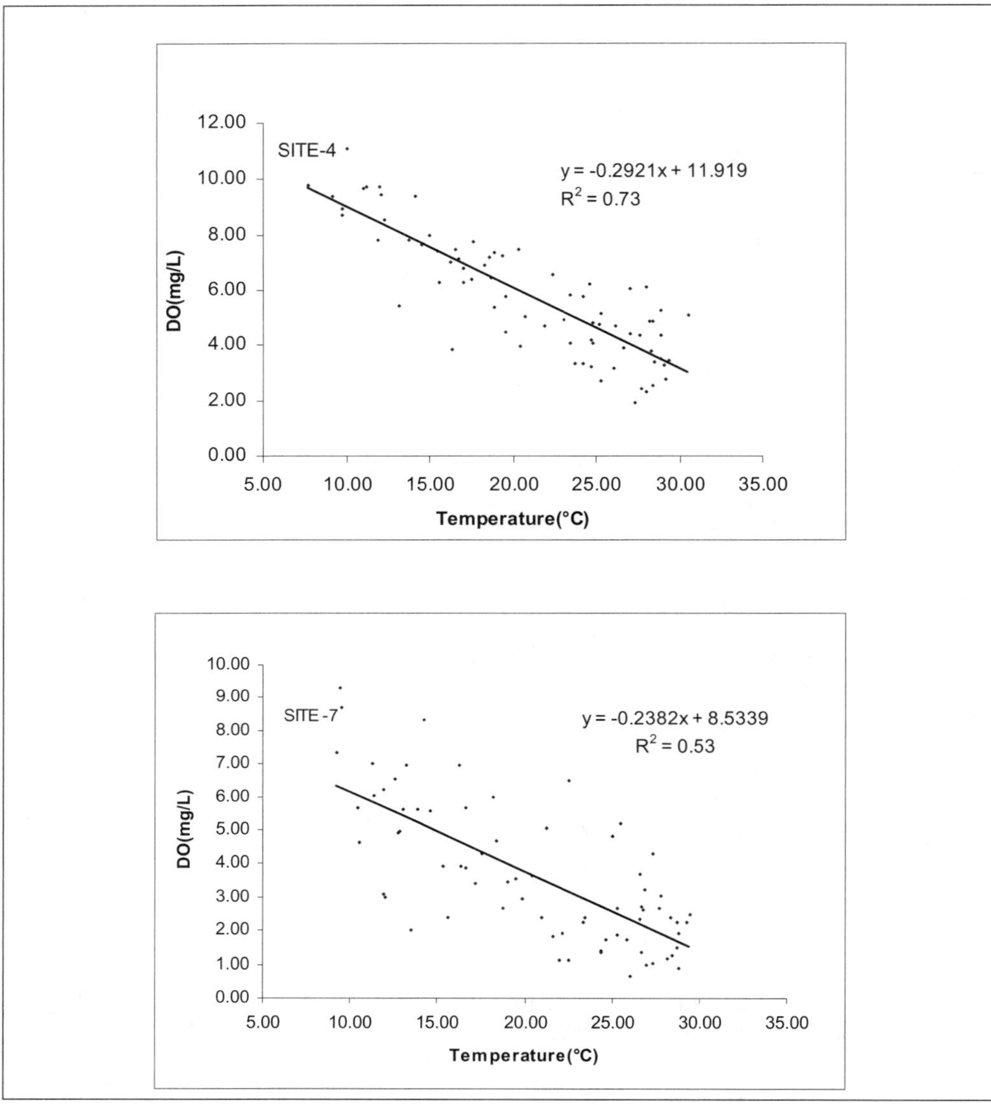

Fig. 2. Continued.

(Sridhar et al., 2004). As changes in stream temperature affected the biochemical processes in the water, the DO and aquatic life in a water body are also affected. Tate et al. (2005) have reported a diurnal pattern of DO levels in a stream in California. They found highest levels in mid- to late-afternoon due to oxygen production by aquatic plants, and lowest levels before dawn due to the plant's nighttime respiration. Lower discharge rates, return of warm drainage water to the stream, and reduction of riparian buffer canopy, are some of the factors related to low DO in a stream (Tate et al., 2005). Several researchers (Randall et al., 2003; Tate et al., 2005) have reported seasonal variation in nutrient losses. Randall et al, 2003, reported 69% of annual nitrate losses in April, May, and June, from their corn-soybean rotation research in Minnesota. Tillage practices have significant effects on surface runoff. Truman et al, 2005, reported that no-till plots have at least two times less runoff and four times less sediment delivery compared to conventional till plots in their research in central Alabama.

Conservation tillage practices in agricultural lands help to improve surface water quality by lowering TSS loading rate in an agricultural watershed. Relatively higher DO levels may be possible in those stream areas which have sufficient shade and cooler water in the summer as compared to those stream areas where no vegetation cover exists. Therefore, one of the challenges for the improvement of surface water quality in this watershed appears to be the implementation of BMPs in agricultural lands and other areas, and the establishment of stream vegetation across the watershed.

19.3.2. Tributaries and watershed reaches

Dissolved Oxygen (DO) levels were higher for upstream locations along the main channel at Sites 1, 2, and 4 as compared to the downstream location at Site 7 (Table 2). This may be due to both the cumulative effects of oxygen-demanding substances on DO levels at the lower reaches of the watershed and the relatively faster moving water upstream. Different land use types impact surface water quality in radically different ways. As a result, as land

Table 2. Average TSS, TN, TP, Conductivity, Turbidity, DO, BOD_5, and pH for upper reaches, tributaries, and lower reaches of the watershed (April 2002 – May 2005).

	Upper Reaches (Sites 1, 2, and 4)	Tributaries (Sites 3, 5, and 6)	Lower Reaches (Site 7)
TSS (mg L^{-1})	93 (±11)*	127(±23)	58
TN (mg L^{-1})	2.92(±0.12)	3.23(±0.20)	2.63
TP (mg L^{-1})	0.40(±0.03)	0.33(±0.02)	0.32
Conductivity (mS cm^{-1})	0.48(±0.01)	0.60(±0.05)	0.42
Turbidity (NTU)	244(±37)	333(±65)	285
DO (mg L^{-1})	6.6(±0.37)	6.3(±0.34)	3.6
BOD5 (mg L^{-1})	4.8(±0.18)	5.4(±0.23)	4.8
pH	7.4(±0.02)	7.5(±0.05)	7.4

*Numbers in parentheses are ± standard error of mean.

use types vary across watersheds, variance occurs in surface water quality across them as well. Tufford et al. (2003) monitored nutrient concentrations in forested, wetland creek, urban creek, and urban ponds, and reported that DO, N, and NH_4 concentrations were greater for forested streams than for urban streams, while NO_3/NO_2-N and TP concentrations were greatest for urban creek. They found higher TN, NH_4, DO, N, and TP in the summer, while NO_3/NO_2-N was found higher in the winter. Stream temperature generally increased from the upper reaches to the lower reaches in a watershed (Tate et al., 2005), but in this watershed Site 7 always had lower water temperatures compared to the rest of the sites. This variation is believed to be due to a large and deep-water body in Site 7 compared to the monitoring sites in upper reaches. These findings suggest that one of the major challenges for surface water quality improvement in this watershed is understanding quantitative relationships between landuse types and nonpoint source pollution, and the identification of "hot spots" for nonpoint source pollution control across the watershed.

19.3.3. Rainstorms and flooding

Rain events also have a significant effect on surface water quality. In particular, TSS and TP concentrations were elevated following rainstorm events (Table 3). Besides elevated TSS and TP levels, large rain events that produce flooding, particularly those following a drought, have been found to be very disastrous to the quality of surface water in many other ways. The large rain events of Hurricane Lilly in 2002 and Hurricane Rita in 2005 resulted in flooding of the Plaquemine Brule watershed, and a distinct difference in surface water

Table 3. Regular and rain event based surface water quality (April 2002 – May 2005).

	Regular (Background)		Rain Event		% increase or decrease from Regular Event
	n	mean	n	mean	
TSS (mg L^{-1})	341	86 (±8)[b*]	181	134 (±10)[a]	55
TN (mg L^{-1})	287	3.05 (±0.10)	139	2.93 (±0.11)	-4
TP (mg L^{-1})	287	0.33 (±0.01)[b]	139	0.41 (±0.02)[a]	22
Temperature (°C)	356	20.41 (±0.36)	176	20.40 (±0.4)	-0.04
Conductivity (mS cm^{-1})	356	0.64 (±0.02)[a]	176	0.28 (±0.02)[b]	-56
Turbidity (NTU)	356	282 (±26)	176	298 (±19.5)	5
DO (mg L^{-1})	350	6.04 (±0.15)	176	6.05 (±0.15)	0.2
BOD_5 (mg L^{-1})	336	5.16 (±0.2)	181	4.87 (±0.24)	-6
pH	335	7.5 (±0.02)[a]	169	7.35 (±0.04)[b]	-2

* Number in parentheses represents ± standard error of mean. Values across the rows with different letters are significantly different at 0.05 probability by Student's t-test.

quality and fish kills resulted from these storms (Table 4). Hurricane Lilly brought more rain than did Hurricane Rita and heavily flooded the watershed. However, no fish kill occurred following Hurricane Lilly. Alternatively, Hurricane Rita occurred after a drought, and many fish were killed. A likely explanation for this drought condition phenomenon is that water discharge and upstream continuous flow decreases during drought. This decrease causes much grass and other vegetation to die in the channel. As a result, organic litter accumulates in the stream, and due to elevated temperatures litter decomposition is enhanced. In tandem, increased temperature, decreased water flow, and increased organic debris in the stream results in a very low DO during such a drought. Consequently, fish kills occur. If flooding occurs after such a long drought spell (as did after Hurricane Rita), washouts of the bank of the streams or of stockpiles of organic matter from the bottom of the stream, contribute to extremely low DO, which ultimately also results in fish kills. Furthermore, during Hurricane Rita, crawfish ponds were virtually dry and the bottom of the ponds had accumulated dead vegetation, which amounted to more oxygen-demanding substances in the floodwater. This further resulted in low DO and fish kills. Any subsequent floodwater that flowed after that initial flooding did not have a similar effect. Following Hurricane Rita, water quality in Bayou Plaquemine Brule was restored in a relatively short duration.

Table 4. Comparison of surface water quality in Bayou Plaquemine Brule watershed following two big rainstorms (Hurricane Lilly and Hurricane Rita). (n=7)

	Following Hurricane Lilly (10-29-02)	Following Hurricane Rita (09-27-05)
Temperature (°C)	22.5 (±0.18)[b*]	28.5 (±0.30)[a]
Conductivity (mS cm^{-1})	0.1 (±0.02)[b]	0.30 (±0.05)[a]
pH	7.0 (±0.27)[b]	7.8 (±0.07)[a]
Turbidity (NTU)	144.1 (±24.78)[a]	36.5 (±5.8)[b]
DO (mg L^{-1})	8.02 (±0.56)[a]	0.67 (±0.09)[b]
TSS (mg L^{-1})	39 (±12)	59 (±7)
TP (mg L^{-1})	NA	0.62 (±0.05)
TN (mg L^{-1})	NA	3.5 (±0.29)
NO_3/NO_2-N (mg L^{-1})	0.37 (±0.06)[a]	1.9 (±0.15)[b]
PO_4 (mg L^{-1})	0.24 (±0.04)[b]	0.55 (±0.06)[a]
BOD_5 (mg L^{-1})	2.67 (±0.56)[b]	11 (±1.4)[a]

* Number in parentheses represents ± standard error of mean. Values across the rows with different letters are significantly different at 0.05 probability by Student's t-test.

NA - Data not available.

Lack of sufficient riparian buffer in Bayou Plaquemine Brule and its tributaries is another major reason for elevated pollution, especially during rainstorms. Riparian buffers help to remove TSS and nutrients from surface runoff (Lowrance et al., 1988). Lee et al. (2003) reported a removal of 95% of the sediment, 80% of the total nitrogen, and 78% of the total phosphorus in surface runoff by switchgrass buffer, while the switchgrass/woody buffer removed 97% of the sediment, 94% of the total N, and 91% of the total P in surface runoff in Iowa. In Bayou Plaquemine Brule watershed, maintaining sufficient flow during the drought condition, installing BMPs in crawfish ponds and other areas, and establishing riparian buffer across the watershed, appears to be another major challenge for surface water quality improvement.

19.3.4. Litter/trash

Excessive litter/trash in the water body has been found to be a major challenge in surface water quality improvement in Bayou Plaquemine Brule watershed. Refrigerators, televisions, pesticide bottles, Styrofoam, beer bottles/cans, glass, tires, batteries, plastic bags, clothes, dead animals, and much other debris and organic matter were identified in the water body. This litter not only affects water quality, but it also degrades the aesthetic beauty of the water body. Litter control is a major challenge, and the implementation of a strong public awareness program on littering and its negative impacts on surface water quality in the watershed is needed.

19.4. CONCLUSIONS

Surface water quality for Bayou Plaquemine Brule decreases from upstream to downstream, and its tributaries have relatively higher TSS values than the main channel. Higher TSS values in the summer months indicated that much nonpoint source pollution occurs during these months, which is apparently associated with agricultural practices. A large seasonal variation in nutrient concentration exists. For the majority of the year, Dissolved Oxygen levels were found to be greater than the threshold level set by LDEQ. As quite turbid waters and elevated discharge rates were often observed during days without rain events, it is obvious that agricultural drainage played a major role in surface water quality in this watershed. Identification of "hot spots" for nonpoint source pollution, implementation of BMPs (primarily in rice, soybean, and crawfish production), establishing riparian buffer and stream vegetation, and litter control are identified as major challenges for surface water quality improvement in the Bayou Plaquemine Brule watershed.

ACKNOWLEDGEMENTS

Sincere appreciation is given to the Louisiana Department of Environmental Quality for funding the Bayou Plaquemine Brule Watershed Water Quality Monitoring Project. Assistance from Dr. Chang-Yoon Jeong, and Ms. Colette Anzalone of the University of Louisiana at Lafayette, and Mr. Jack Haller of NRCS-Crowley for site selection and subsequent data collection, is highly appreciated.

REFERENCES

Clesceri, L.S., Greenberg, A.E., and Eaton, A.D., 1998. Standard Methods for the Examination of Water and Wastewater, 20th Edition. APHA, AWWA, WEF.

LDEQ, 2000. Louisiana's nonpoint source management plan, Vol. 6. Office of Environmental Assessment, LDEQ.

Lowrance, R.R., McIntyre, S., and Lance, C.L., 1988. Erosion and deposition in a field/forest system estimated using cesium-137 activity. Journal of Soil and Water Conservation 43(2):195-199.

Lee, K.H., Isenhart, T.M., and Schultz, R.C., 2003. Sediment and nutrient removal in an established multi-species riparian buffer. Journal of Soil and Water Conservation 58(1): 1-7.

Lynch, J.A., Rishel, G.B., and Corbett, E.S., 1984. Thermal alteration of stream draining clear-cut watersheds: Quantification and biological implications. Hydrobiologia 111:161-169.

Randall, G.W., Vetsch, J.A., and Huffman, J.R., 2003. Nitrate losses in subsurface drainage from a corn-soybean rotation as affected by time of nitrogen application and use of nitrapyrin. Journal of Environmental Quality 32:1764-1772.

Sridhar, V., Sansone, A.L., LaMarche, J., Dublin, T., and Lettenmair, D., 2004. Prediction of stream temperature in forested watersheds. Journal of the American Water Resources Association 40(1):197-213.

Tate, K.W., Lile, D.F., Lancaster, D.L., Porath, M.L., Morrison, J.A., and Sado, Y., 2005. Graphical analysis facilities evaluation of stream-temperature monitoring data. California Agriculture 59(3): 153-160.

Truman, C.C., Shaw, J.N., and Reeves, D.W., 2005. Tillage effects on rainfall portioning and sediment yield from an ultisol in central Alabama. Journal of Soil and Water Conservation 60(2):89-98.

Tufford, D.L., Samarghitan, C.L., McKellar, H.N., Jr., Porter, D.E., and Hussey, J.R., 2003. Impacts of urbanization on nutrient concentrations in small southeastern coastal streams. Journal of the American Water Resources Association 39(2): 301-312.

20

WATER QUALITY CHANGES ACROSS AN URBAN-RURAL GRADIENT IN WESTERN GEORGIA

Jackie F. Crim and B. Graeme Lockaby
School of Forestry and Wildlife Sciences
Auburn University
602 Duncan Drive, Auburn, AL 36849, USA

ABSTRACT

The Middle Chattahoochee Watershed in western Georgia is undergoing rapid urbanization. As the conversion to developed land spreads, so does the threat of water quality degradation. Consequently, anthropogenic influences jeopardize the biotic integrity of Georgia's streams. Our study area near Columbus, GA, lies near the Piedmont/Coastal Plain transition. The urban development impacting Piedmont streams may cause impairments downstream into the Coastal Plain physiographic region. Eighteen subwatersheds representing an array of land uses were selected for study. Land use categories included pine forests, mixed species forests, rural, developing, and urban. Grab samples are being examined for nutrients and sediment. Preliminary data indicate differences in water quality among varying land uses and reveal elevated levels within urbanized watersheds. This study will help to clarify the effects of land development on the physicochemical properties of streamwater.

20.1. INTRODUCTION

Conversion of forested land to suburban and urbanized landscapes is occurring at extreme rates, especially in the Southeastern United States. The expansion of urbanization poses a threat to stream ecosystems. The status of 23% of the nation's total rivers and streams was assessed in 1998 (USEPA, 2000). Over 290,000 miles of the assessed rivers and streams were impaired and did not meet water quality standards. Of those miles, approximately 120,000 were impaired by urbanization (USEPA, 2000). This makes urbanization second only to agriculture as the leading cause of stream impairment. Water quality management will become increasingly important as the human population continues to expand and the conversion of natural lands to urban areas increases.

One of the major contributors of water quality impairment caused by urbanization stems from increased impervious surfaces. Impervious surfaces prevent natural pollutant processing by decreasing infiltration and increasing surface runoff. The elevated velocity and surface runoff increases erosion (Finkenbine et al., 2000; Bledsoe and Watson, 2001; Rosi-Marshall, 2004). Past research consistently shows a strong correlation between the imperviousness of a watershed and the health of its receiving stream (Arnold and Gibbons,

1996). The average threshold value of imperviousness at which degradation first occurs is approximately 10% (Bledsoe and Watson, 2001).

Managing land use in a watershed is crucial to protecting drinking-water supplies, recreational resources, and stream ecosystem health. However, the effect of land use on streams is difficult to determine (Landers et al., 2002). It is important to study the effects of land use change at the local scale, as conversion of landscapes can have serious implications on aquatic systems. Low-order streams can make up to 85% of the total stream distance within a watershed (Peterson et al., 2001). The effects of land use practices are most easily seen in small streams because of their high abundance and susceptibility to human disturbance. For instance, as nitrogen inputs to streams increase, streams lose the ability to retain and transform nitrogen, transporting it downstream with consequent increases in eutrophication (Peterson et al., 2001). Small stream research is essential for evaluation of cumulative effects of land-use practices from upland areas to downstream river systems.

From a land use perspective, agricultural activities have been identified as major sources of nonpoint source pollutants (sediments, animal wastes, nutrients, and pesticides) and are known to impact water quality (USEPA, 2000). Urban areas are also key in generating large amounts of nonpoint source pollution from urban area runoff and storm sewer discharge (Basnyat et al., 2000; USEPA, 2000). Increased urbanization brings with it several environmental stressors including but not limited to increased flow (Bledsoe and Watson, 2001), nutrients (Emmerth and Bayne, 1996; Rose, 2002; Zampella, 1994), heavy metals (Callender and Rice, 2000), sediment (Finkenbine et al., 2000), and bacteria (Frick et al., 2001).

Excessive nutrients and sediment can cause diverse problems such as toxic algal blooms, loss of oxygen, fish kills, and loss of biodiversity. Nutrient inputs can include fertilizers, wastewater, and animal wastes. A study in New Jersey found concentrations of calcium, magnesium, nitrate, ammonium, and phosphorus were positively correlated with a watershed disturbance gradient of increasing land use intensity and wastewater flow (Zampella, 1994). Sediment is a major pollutant both for its effects on aquatic ecology and because of the fact that many other pollutants, such as heavy metals, can attach to eroded soil particles (Arnold and Gibbons, 1996). Channel erosion due to urbanization can become a predominant source of excess sediment to downstream reaches and result in degradation of biotic quality (Paul and Meyer, 2001). Construction sites are a critical area of concern for urban nonpoint source pollution. Erosion rates for construction sites can be extremely high. This eroded material contributes to sedimentation of waterbodies as well as to eutrophication (Carpenter et al., 1998).

As the human population continues to increase, the need for research concerning the ecological implications of urbanization will only grow. The effects of urbanization are far-reaching and need to be evaluated on the local watershed level to determine management practices that will keep the structure and function of waterbodies intact. This research focuses on eighteen small watersheds in west-central Georgia. Physicochemical water quality data are being collected from September 2004 until January 2006. Results shown here are as of May 2005. Additional information from an earlier study of these watersheds is described in Schoonover et al. (2005) and Helms et al. (2005). Eventually an overall assessment of the watersheds combining the studies will be conducted.

20.2. STUDY AREA

Eighteen subwatersheds of the Middle Chattahoochee Watershed ranging in size from ~300 to 2500 ha were selected for study. All watersheds resided within the Piedmont ecoregion. The Piedmont contains a mosaic of land uses including cropland, pasture, forest, and urban areas. Western Georgia, primarily Muscogee, Harris, and Merriweather counties, was chosen for study due to its rapidly expanding population in recent years. Muscogee County is highly urbanized as a result of the city of Columbus. Harris County is predominately forested. It does, however, reflect developing land use as the city of Columbus continues to expand outward. Harris County experienced a 33% increase in population between 1990 and 2000 and a 9% increase from 2000 to 2003 (U.S. Census Bureau, 2004). Merriweather County, on the other hand, is primarily rural and growth has been relatively stable over the last 10 years. Growth from Columbus occurs primarily in the northeast direction as a result of the Chattahoochee River to the west and Fort Benning military reservation to the southeast.

Impervious surface area was used as the primary measure of urbanization. In order to examine urban influences on water quality, the watersheds were chosen to represent a gradient of impervious surface from 1-40%. However, a gap existed between 4 and 13% and 13 and 23% impervious surface. Due to the difficulty in obtaining a full range of impervious surface, watersheds were also chosen to reflect a range of primary land uses including pine forests, mixed forests, rural, developing, and urban. Land cover classification was generated using GIS and remote sensing techniques based on a Landsat TM aerial view from March, 2003. Greater detail of methods concerning land cover classification can be found in Lockaby et al. (2005).

20.3. METHODS

20.3.1. Field Techniques

Physicochemical water quality parameters were sampled between September 2004 and May 2005. Samples were collected bimonthly during the winter and spring months from November to March. These months are optimal for water chemistry sampling because of increased stream flow due to high precipitation and low evapotranspiration (Lockaby et al., 1993). Sampling occurred monthly during the remainder of the year.

Grab samples were collected prior to other data collection to ensure no contamination would occur from persons wading in the stream. Before each collection, polypropylene bottles were conditioned by rinsing three times with stream water. Tissue culture flasks were used to detect low-level concentrations of cations and anions. These flasks were rinsed and filled with deionized water and then stored at 4°C for at least 24 hours. During sampling, these flasks were emptied and rinsed three times before taking a sample. Samples were kept on ice and then stored at 4°C until analyzed.

Stream discharge was recorded in order to determine nutrient and sediment loads. This involved measuring depth and velocity along transects across the stream channel. Stream depth was measured every 10, 20, or 50 cm, depending upon stream width. Increments were chosen in order to provide a minimum of 10 readings. Velocity was then measured at the mid-point of each depth using a Marsh-McBirney flowmeter. Total stream discharge (Q) was calculated using the following equation: $Q=\Sigma$(width of each increment*mean depth of each increment*velocity of each increment) (Gore, 1996).

20.3.2. Lab Analysis

Water samples were analyzed within five days after collection. Anions and cations (NO_3^-, Cl^-, SO_4^-, Na^+, NH_4^+, K^+) were analyzed using the Dionex DX-120 ion chromatograph (Sunnyvale, CA). Phosphorus was measured using the molybdate-blue method (Murphy and Riley, 1962; Wantanabe and Olsen, 1965). Total dissolved solids (TDS) were determined using a Fisher Accumet AB30 conductivity meter (Fisher Scientific, Pittsburgh, PA). Total suspended solids (TSS) were measured using filtration methods outlined by the Environmental Protection Agency (USEPA, 1999).

20.3.3. Statistical Analysis

Statistical software SAS Version 9.1 was used for all statistical analyses. Watersheds were separated into treatments containing < and > 10% impervious surfaces. Significant differences between treatments were determined using PROC GLM and the Student-Newman-Keuls Test. Land use impacts on water quality were determined by examining Pearson's correlation coefficients among land use percentages within the watersheds. Significance was determined using a 95% confidence level.

20.4. RESULTS AND DISCUSSION

The average ion concentrations in both treatments were generally lower in stormflow than baseflow, suggesting a dilution effect, with the exception of ammonium in the urban watersheds (Table 1). The higher ammonium in the stormflow of the urban watersheds may be attributed to leaky sewer systems. Much of Columbus, GA is still using a combined sewer overflow (CSO) system. Near-stream manhole covers were commonly displaced during storms. In general, the urbanized watersheds showed a greater dilution effect, most likely resulting from a larger proportion of impervious surfaces increasing overland flow. There was a slight increase in sulfate in the stormflow of the less urbanized watersheds. This may be attributed to the desorption of sulfate from the soil (Huntington et al., 1994).

Table 1. Nutrient and sediment concentration comparisons above and below 10% impervious surfaces.

Variable	Baseflow; Mean ± SE		Stormflow; Mean ± SE	
(ppm)	<10% IS	>10% IS	<10% IS	>10% IS
TDS	36.96 ± 1.44	**62.61 ± 2.53**	29.37 ± 1.33	**44.47 ± 2.73**
TSS	3.85 ± 0.3	5.11 ± 1.15	20.93 ± 2.32	15.77 ± 2.16
Cl	4.35 ± 0.26	**10.15 ± 1.12**	2.97 ± 0.15	**5.08 ± 0.39**
NO_3	1.38 ± 0.2	**2.33 ± 0.37**	1.12 ± 0.15	1.44 ± 0.1
SO_4	2.17 ± 0.18	**8.08 ± 1.05**	2.36 ± 0.12	**5.66 ± 0.24**
Na	5.16 ± 0.31	**6.67 ± 0.48**	4.42 ± 0.41	5.25 ± 0.69
NH_4	0.06 ± 0.01	**0.13 ± 0.04**	0.05 ± 0.01	**0.24 ± 0.05**
K	2.26 ± 0.09	**3.39 ± 0.21**	2.4 ± 0.2	2.94 ± 0.34
P	0.27 ± 0.02	0.25 ± 0.03	0.22 ± 0.03	0.15 ± 0.03

Values in bold are significant at alpha = 0.05.

In contrast to the concentration data, the load data revealed few significant differences between less-urbanized and urbanized watersheds (Table 2). Load data generally provides a better comparison of the watersheds. The watersheds are standardized since load data takes into account the size of the watershed. Ammonium loads were higher in the urbanized watersheds during both baseflow and stormflow. This is most likely explained by leaky sewer systems.

Table 2. Nutrient and sediment load comparisons above and below 10% impervious surfaces.

Variable (g/d/ha)	Baseflow; Mean ± SE		Stormflow; Mean ± SE	
	<10% IS	>10% IS	<10% IS	>10% IS
TDS	380.75 ± 67.78	300.66 ± 27.51	1953.55 ± 272.79	2160.72 ± 409.67
TSS	40.72 ± 7.33	26.8 ± 5.08	2046.21 ± 675.2	976.04 ± 214.08
Cl	52.41 ± 11.22	56.28 ± 10.64	192.52 ± 25.59	240.82 ± 50.81
NO_3	15.34 ± 3.08	14.17 ± 3.44	77.55 ± 15.21	85.54 ± 19.88
SO_4	32.67 ± 8.61	47.4 ± 9.74	184.5 ± 27.46	291.56 ± 51.83
Na	52.71 ± 10.32	30.59 ± 2.94	254.34 ± 34.99	212.09 ± 38.88
NH_4	0.44 ± 0.07	**1.35 ± 0.57**	5.68 ± 1.78	**13.47 ± 2.83**
K	20.97 ± 3.12	15.49 ± 1.43	131.77 ± 14.32	128.67 ± 24.03
P	3.5 ± 0.8	1.32 ± 0.26	**9.39 ± 1.24**	4.56 ± 1.17

Values in bold are significant at alpha = 0.05.

A highly urbanized watershed (54.7% impervious surface), Peachtree Creek in Atlanta, GA, was studied by Rose (2002). In comparison to our urban watersheds, Rose found similar concentrations of sodium, potassium, sulfate, and nitrate during both flows. Lenat and Crawford (1994) also found similar baseflow TDS and ammonium concentrations (67 and .11 mg/L) in an urbanized watershed in the North Carolina Piedmont.

Correlations between land use and sediment and nutrient variables are illustrated in Table 3. In both of the forested land uses, all the variables significantly correlated with land use were negatively correlated. In general, the more forested a watershed was, the less sediment and nutrients were contributed to the stream. Conversely, much of the parameters were positively correlated with the amount of urban land area within the watershed. High nutrient concentrations in urbanized areas may be attributed to faulty sewer systems, non-point source pollution discharges, and lawn care fertilizers. TDS, chloride, and sulfate revealed the strongest positive correlations with urban land use. Rose (2002) suggests high concentrations of sulfate and chloride may originate from rock weathering, evaporation within the soil horizon, or outside pollutants. Because our watersheds have similar mineralogy, it is not unlikely that pollution is the cause of increased nutrients in the urban areas. Not surprisingly, nitrate had the highest correlation with agricultural land use. This may be attributed to fertilizer and cattle waste. Agriculture was the only land use positively correlated with TSS. Beyers et al. (2005) found that unfenced pastures contributed significant TSS concentrations to Georgia streams during both baseflow and stormflow. Two streams in this study were frequented by cattle, causing severe erosion to stream banks.

Table 3. Pearson's correlation coefficients; values in bold are significantly correlated with land use

	Urban		Agriculture		Pine		Mixed	
Variable	**Base**	**Storm**	**Base**	**Storm**	**Base**	**Storm**	**Base**	**Storm**
TDS	**0.63**	**0.55**	-0.12	0.04	**-0.42**	**-0.43**	**-0.5**	**-0.49**
TSS	0.08	-0.14	0.15	**0.27**	**-0.2**	0.04	-0.09	-0.09
Cl	**0.57**	**0.61**	0.19	**0.31**	**-0.51**	**-0.56**	**-0.55**	**-0.67**
NO_3	0.18	0.16	**0.54**	**0.64**	**-0.29**	-0.26	**-0.43**	**-0.49**
SO_4	**0.61**	**0.8**	-0.07	-0.02	**-0.45**	**-0.6**	**-0.4**	**-0.58**
Na	**0.24**	0.14	-0.18	-0.1	-0.07	-0.03	-0.2	-0.14
NH_4	0.12	**0.4**	0.14	**0.25**	**-0.29**	**-0.5**	-0.11	**-0.39**
K	**0.46**	0.19	**0.23**	0.18	**-0.46**	-0.14	**-0.46**	**-0.25**
P	-0.03	-0.13	0.12	-0.08	-0.01	0.14	-0.07	0.13

Values in bold are significant at alpha = 0.05.

20.5. CONCLUSIONS

In the Georgia Piedmont, eighteen watersheds representing an array of land uses were sampled for nutrient and sediment concentrations and loads. Nutrient concentrations were generally higher in the urbanized watersheds during both baseflow and stormflow. With the exception of ammonium, average ion concentrations were lower in stormflow than baseflow, suggesting a dilution effect. Most parameters were negatively correlated with the percentage of forest in the watersheds. In contrast, most water quality parameters were positively correlated with urbanized land use.

Data from this study will be used to generate water quality prediction models for the Southern Piedmont. Data will also provide comparisons to a previous study in which we will evaluate the effects climatic variables had on water quality parameters.

ACKNOWLEDGMENTS

Funding for this research was provided by the Center for Forest Sustainability through the Peaks of Excellence Program at Auburn University.

REFERENCES

Arnold Jr., C. L. and Gibbons, C. J., 1996. Impervious surface coverage. Journal of the American Planning Association 62:243-258.

Basnyat, P., Teeter, L., Lockaby, B. G., and Flynn, K., 2000. The use of remote sensing and GIS in watershed level analyses on non-point source pollution problems. Forest Ecology and Management 128:65-73.

Bledsoe, B. P., and Watson, C. C., 2001. Effects of urbanization on channel instability. Journal of the American Water Resources Association 37:255-270.

Byers, H.L., Cabrera, M.L., Matthews, M.K., Franklin, D.H., Andrae, J.G., Radcliffe, D.E., McCann, M.A., Kuykendall,

H.A., Hoveland, C.S., and Calvert II, V.H., 2005. Phosphorous, sediment, and *Escherichia coli* loads in unfenced stream of the Georgia Piedmont, USA. Journal of Environmental Quality 34:2293-2300.

Callender, E., and Rice, K. C., 2000. The urban environmental gradient: anthropogenic influences on the spatial and temporal distributions of lead and zinc in sediments. Environmental Science & Technology 34:232-238.

Carpenter, S. R., Caraco, N. F., Correll, D. L., Howarth, R. W., Sharpley, A. N., and Smith, V. H. 1998. Nonpoint pollution of surface waters with phosphorus and nitrogen. Ecological Applications 8:559-568.

Finkenbine, J. K., Atwater, J. W., and Mavinic, D. S., 2000. Stream health after urbanization. Journal of the American Water Resources Association 36:1149-1160.

Gore, J. A., 1996. Discharge measurements and streamflow analysis. Lamberti, F.R. and Hauer, G.A., editors. Methods in stream ecology. Academic Press, San Diego, CA. Pages 53-74.

Helms, B.S., Feminella, J.W., and Pan, S., 2005. Detecton of biotic responses to urbanization using fish assemblages from small streams of western Georgia, USA. Urban Ecosystems 8:39-57.

Huntington, T. G., Hooper, R. P., and Auenbach, B. T., 1994. Hydrologic processes controlling sulfate mobility in a small forested watershed. Water Resources Research 30:283-295.

Landers, M. N., Ankcorn, P. D., McFadden, K. W., and Gregory, M. B., 2002. Does land use affect our streams? A watershed example from Gwinnett County, Georgia, 1998-2001. *in*. U.S. Geological Survey. Water-Resources Investigations Report 02-4281.

Lockaby, B. G., Zhang, D., McDaniel, J., Tian, H., and Pan, S., 2005. Interdisciplinary research at the Urban-Rural interface: The Westga project. Urban Ecosystems 8:7-21.

Murphy, J., and Riley, J.P., 1962. A modified single solution method for the determination of phosphate in natural waters. Analytical Chemistry 27:31-36.

Paul, M. J., and Meyer, J. L., 2001. Streams in the urban landscape. Annual Review of Ecology and Systematics 32:333-365.

Rose, S., 2002. Comparative major ion geochemistry of Piedmont streams in the Atlanta, Georgia region: possible effects of urbanization. Environmental Geology 42:102-113.

Rosi-Marshall, E. J., 2004. Decline in the quality of suspended fine particulate matter as a food resource for chironomids downstream of an urban area. Freshwater Biology 49:515-525.

Schoonover, J. E., Lockaby, B. G., and Pan, S., 2005. Changes in chemical and physical properties of stream water across an urban-rural gradient in western Georgia. Urban Ecosystems 8:107-124.

USEPA, 1999. Standard operating procedure for the analysis of residue, non-filterable (suspended solids) water. Method 160.2NS. United States Environmental Protection Agency Region 5 Central Regional Laboratory. Chicago, IL.

USEPA, 2000. The quality of our nation's waters. EPA 841-S-00-001.

Wantanabe, F.S., and Olsen, S.R., 1965. Test of an ascorbic acid method for determining phosphorus in water and $NaHCO_3$ extracts from soil. Soil Science Society of America Journal 29:677-678.

Zampella, R. A., 1994. Characterization of surface water quality along a watershed disturbance gradient. Water Resources Bulletin 30:605-611.

21

REDUCTION OF RIVERINE NITRATE LOADS THROUGH FIELD TARGETING OF BEST MANAGEMENT PRACTICES, PLACEMENT OF WETLANDS, AND IN-STREAM DENITRIFICATION: A MODELING / SIMULATION PROJECT

Timothy W. Appelboom
USDA-ARS Soil and Water Unit
4115 Gourrier Ave. Baton Rouge, LA 70808, USA

James L. Fouss
USDA-ARS Soil and Water Unit
4115 Gourrier Ave. Baton Rouge, LA 70808, USA

ABSTRACT

Increased agricultural production has lead to a reduction in water quality. In the case of the Mississippi River Basin increased nutrient loads linked to agricultural production has contributed to the hypoxic zone along the Louisiana Gulf coast. The objectives of this study are to: 1) gain a better understanding of in-stream denitrification, controlled drainage, and wetland functions on reducing agriculture derived nitrate in northeastern Louisiana surface waters, and 2) incorporate that information into the Annualized Agricultural Non-Point Source Pollutant Loading Model (AnnAGNPS) to simulate different placements of best management practices (BMPs) and wetlands within a watershed to both maximize nitrate load reductions at the outlet while minimizing costs. These objectives will be accomplished using the computer simulation model AnnAGNPS along with economic evaluations of each land use scenario applied to the Cabin-Teele sub-watershed in northeastern Louisiana. During each simulation, BMPs and wetland diversions will be placed at selected locations within the sub-watershed to determine how their location effects total nitrate load at the outlet. The expected results of this study are to: 1) demonstrate that in-stream denitrification and wetland diversions can reduce agricultural nitrogen loads within the farms drainage network; and 2) demonstrate that modeling is an important tool for prioritizing field Best Management Practices and wetland placements within a watershed to maximize the reduction of agricultural drainage water nitrates while minimizing costs.

21.1. INTRODUCTION

This paper is a project plan for a study to be conducted in the Cabin-Teele sub-watershed in northeastern Louisiana whose results should lead to guidelines to economically reduce agricultural nitrate losses to surface waters on the watershed scale which eventually contribute to hypoxia in the Gulf of Mexico.

As time progressed and populations increased, human activity relating to agricultural production has focused on enhancing biological systems to increase yields through increased land use conversion, fertilization, irrigation, drainage, and fossil fuel inputs. An example of this increased production is the Mississippi River Basin. In the past 100 years, the Mississippi River Basin (Figure 1) has become one of the world's most important agricultural regions. The increase in production is due to a significant change of land use (loss of forests, wetlands, and riparian zones to agriculture) (Rabalais et al., 2002), rapid expansion of drainage (increasing from less than 1 million hectares in 1900 to currently over 28 million hectares) (Mitsch et al., 2001), and dramatic increase in the use of nitrogen fertilizers (increasing from less than 0.5 million metric tons per year in the 1940s to over 7 million metric tons per year in 1999) (Goolsby et al., 1999).

Fig. 1. Mississippi River Basin (Source: National Park Service, U.S. Department of the Interior).

Many of these activities have had adverse effects on the environment as they have influenced the equilibrium of the natural cycles (i.e. carbon, sulfur, nitrogen and phosphorous) on every continent (EEA, 1995; Erisman et al., 1998). In the above example, long term data at St. Francisville, Louisiana show that nitrate concentrations in the Mississippi River have more than doubled since 1954 to 1960 (Goolsby et al., 1999) while the annual flux of nitrogen has tripled since 1955 to 1.6 million metric tons (61% in the form of nitrate which 74 % has been attributed to agricultural losses) (Rabalais et al., 2002).

Nitrogen has been recognized as a limiting nutrient in flooded soils, lakes, streams, estuaries, and oceans (Mitsch and Gosselink, 1993; Erisman et al., 1998; Binkley et al., 1999). This increase in surface water nitrogen can result in algal blooms (Stanley, 1983;

Beman, et al., 2005), hypoxia (Rabalais et al., 1999), eutrophication (Mitsch and Gosselink, 1993; Erisman et al., 1998; Binkley et al., 1999; Boesch, et al., 2001, Beman, et al., 2005), fish kills (Rabalais et al., 1999; Whitall and Pearl, 2001), and nitrate concentrations exceeding Environmental Protection Agency guidelines for drinking water (10 ppm) (EPA, 2005).

It has been shown that agricultural nitrate losses can be reduced at the field edge prior to entering surface waters through implementation of Best Management Practices (BMPs) such as; nutrient management (Power et al., 2001; EFOTG, 2004), riparian buffers (Cooper and Gilliam, 1987; Mitsch et al., 2001), tillage management (EFOTG, 2004), cover crops (EFOTG, 2004), and controlled drainage (Gilliam et al., 1979; Evans et al 1989; Evans et al., 1991; Skaggs et al., 1994; Fausey et al., 2004). Once nitrates enter surface waters, nitrates can be removed through the in-stream process of denitrification (Chatarpaul et al., 1979; Hill and Sanmugadas, 1985; Seitzinger, 1988). It has also been shown that drainage water nitrate loads can be reduced by routing them through natural or constructed wetlands (Mitsch and Gosselink, 1993; Mitsch et al., 2001; Dinnes, et al., 2002; Day et al., 2004).

Though BMPs have been shown to be effective in reducing nitrate losses from agricultural fields, BMPs are not universally applicable to all fields. For example, controlled drainage requires relatively flat topography to be economically practical. If the land contains too much relief, the number of control structures becomes cost prohibitive. Also in cases where the land is heavily drained for increased production, riparian buffers are bypassed by the drainage network rendering them ineffective. This implies that only BMPs which are applicable to a specific field should be considered for implementation on that field.

As we transition from looking at edge of field effects to watershed effects it is becoming clearer that the location of BMP implementation within the watershed is as important as the BMP itself. A BMP implemented on one field may have a much greater effect on the health of the overall watershed than the same BMP implemented on a different field. An example of this would be the implementation of a BMP on a field near the outlet of a watershed verses one near the upper reaches of the watershed. The effect on the nitrate load at the watershed outlet would be impacted greater by the BMP being implemented near the outlet. This is due to a large portion of the nitrate being removed through the in-stream process of denitrification within the drainage network as it moves from the upper reaches of the watershed to the outlet. Almost all the nitrate from the field near the outlet makes it to the outlet. This gives a higher benefit to cost ratio resulting in a larger reduction in the nitrate load coming from the watershed at a lower cost.

21.2. OBJECTIVES

The objectives of this research are to: 1) gain a better understanding of in-stream denitrification, controlled drainage, and wetland functions on reducing agriculture derived nitrate in northeastern Louisiana surface waters, and 2) incorporate that information into the Annualized Agricultural Non-Point Source Pollutant Loading Model (AnnAGNPS) to simulate different placements of applicable BMPs and wetlands within a watershed to both maximize nitrate load reductions at the outlet while minimizing costs.

21.3. BEST MANAGEMENT PRACTICE DESCRIPTIONS

Tillage management, or conservation tillage, manages the amount of crop residue which remains in the field after harvest to reduce surface erosion and increase infiltration. This practice also increases soil organic matter, soil organism populations, improves soil tilth, and productivity (EFOTG, 2004). Tillage management is currently implemented in most of the fields of the Cabin-Teele sub-watershed.

Cover crops are usually grass, small grains, or legumes grown in a field between crop harvest and planting the following season. They are primarily grown to control erosion, add fertility and organic material to the soil, improve soil tilth, and increase infiltration and aeration of the soil (EFOTG, 2004). Cover crops also help improve water quality by serving as a nitrogen sink during the non-cropping production (winter) months, thus reducing leachable nitrogen in the soil. In the Cabin-Teele sub-watershed, cover crops consist usually of volunteer vegetation.

Controlled-drainage is the use of open ditch (Figure 2), or subsurface field drainage systems, or series of field drainage systems including collector canals, to hold water in the field by the use of a control mechanism (Evans et al, 1995; Skaggs, 1999). Controlled-drainage is used to: maintain a predetermined water table in the field to enhance crop production (Skaggs, 1999); reduce the volume of water leaving the field to reduce downstream flooding; and reduce nitrate loss from the drained cropland (particularly subsurface flow to drainage ditches) (Gilliam et al., 1979; Evans et al 1989; Evans et al., 1991; Skaggs et al., 1994; Fausey et al., 2004). The Cabin-Teele sub-watershed is composed entirely of an open ditch drainage network so controlled-drainage will be implemented using flash-board risers in the ditches (Figure 3).

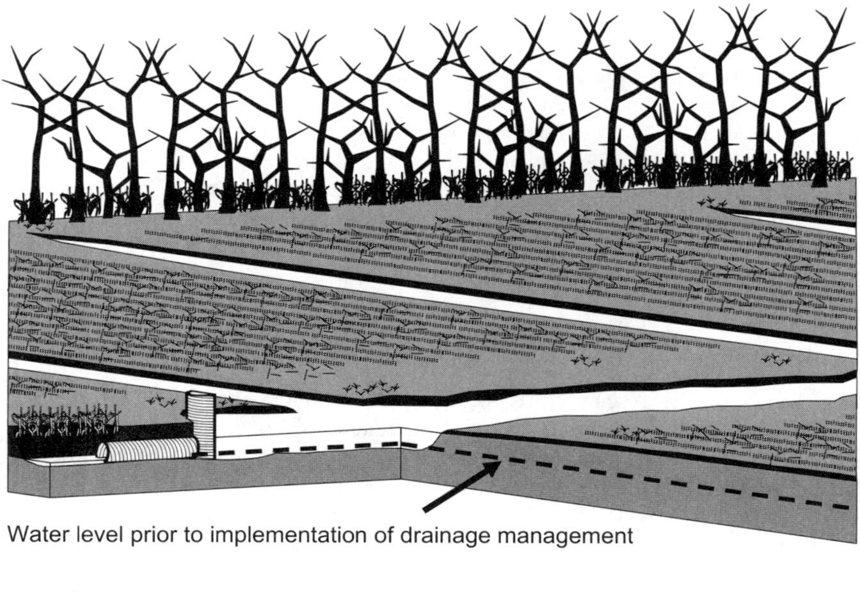

Fig. 2. Control drainage implemented on an open ditch drainage network.

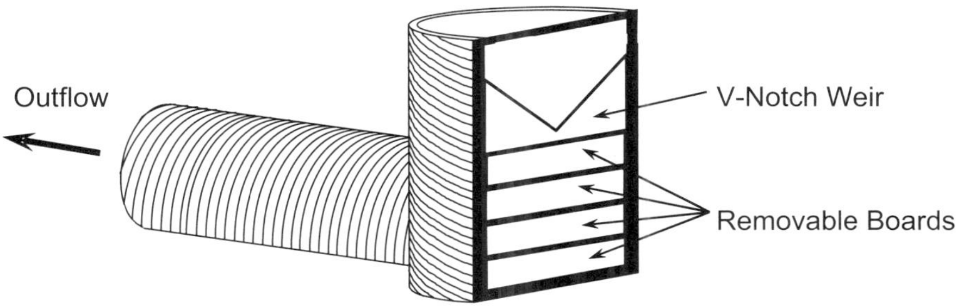

Fig. 3. Flash-board riser structure showing removable boards and V-notch weir for controlling water level and discharge.

A flash-board riser is a spillway structure, usually a V-notched weir, with removable sections allowing the spillway to be raised or lowered (Figure 3). By raising the spillway, the water table in the field is raised and water is held in the soil profile as well as within the field ditches. By lowering the spillway, the water table in the field lowers, draining the soil profile allowing equipment access to the field for planting or harvest operations. Using an assumption for the drainable porosity in the Cabin-Teele sub-watershed of 6% (based on personal discussion with Dr. J. Fouss, Ag. Eng. and Dr. B. Grigg, Soil Scientist of the ARS Soil and Water Research Laboratory, Baton Rouge, LA), for each foot the flashboard riser is raised, about 0.72 inches of rainfall can be stored onsite within the soil profile. This translates to a reduction of runoff and drainage discharge from the field as well as its associated nutrient load (primarily nitrates) (Gilliam et al., 1979; Evans et al 1989; Skaggs et al., 1994; Thomas et al., 1995). Nitrate losses to the drainage network are also reduced as the water is held within the soil profile for longer periods of time under anaerobic conditions resulting in increased denitrification (Gilliam et al., 1979; Evans et al., 1989; Thomas et al., 1995; Drury et al., 1997).

21.4. WETLAND DIVERSION

A wetland diversion is the diversion of a portion of flow from an original channel to flow through a wetland area, then reintroducing the flow back into the original channel down stream. Nutrient loads in the diverted flow are reduced via the wetland processes of sediment denitrification (Mitsch and Gosselink, 1993; Xue et al., 1999; Bachand and Horne, 2000), burial (Xue et al., 1999; Day et al., 2004), and plant uptake (Mitsch and Gosselink, 1993; Xue et al., 1999; Matheson et al., 2002). This results in a lower nutrient load for the diverted flow as it is reintroduced into the original channel. Wetland diversions in the Cabin-Teele sub-watershed will utilize Wetland Reserve Program (WRP) areas already established along the drainage network.

21.5. IN-STREAM DENITRIFICATION

In-stream denitrification occurs in the anaerobic portions of the sediments of streams, rivers, lakes, estuaries, and oceans. This process also occurs in the drainage networks of agricultural lands. Nitrate is transported from the water column into the sediments through advection in the upper layer and diffusion in the lower layers where they are denitrified to di-nitrogen gas (Hynes, 1983; Triska et al., 1989; Jensen et al., 1994). The longer the flow path, or residence time in the drainage network, the more nitrate will be denitrified (Johnson, 1991; Jansson et al, 1994; Dahm et al., 1998). This implies that the further from the outlet nitrate enters the drainage network, the smaller portion of that nitrate will reach the outlet. Substantial in-stream denitrification is expected in the Cabin-Teele sub-watershed drainage network due to the long residence times expected (up to 3 days travel time).

21.6. SITE DESCRIPTION

The Cabin-Teel sub-watershed is located along the eastern edge of Madison Parish, Louisiana (Figure 4). It is bound to the west by Walnut Bayou and Roundaway Bayou, and to the north, east, and south by the Mississippi River levee system (Figure 5). The sub-watershed consists of 17,752 total hectares with 13,413 hectares (76%) in agriculture, 2,293 hectares (13%) in the WRP (Wetland Reserve Program), 1,420 hectares (8%) in agro-forestry, 449 hectares (2%) in the CRP (Conservation Reserve Program), and 177 hectares (1%) in other land use. The main drainage network within the sub-watershed consists of Cypress Bayou (also known as Willow Slough in the southern portion of the sub-watershed) draining the northern and central portion of the sub-watershed, with Harper Bayou and Gar Bayou draining the eastern and western portions of the sub-watershed, respectively (Figure 5). All three branches join just prior to the discharge point in the southwest corner of the sub-watershed known as the Basin, a part of Roundaway Bayou. After discharge from the Cabin-Teele sub-watershed, water drains through Roundaway Bayou-Vidal Cutoff, Vidal Bayou, and Mill Bayou to the Tensas River.

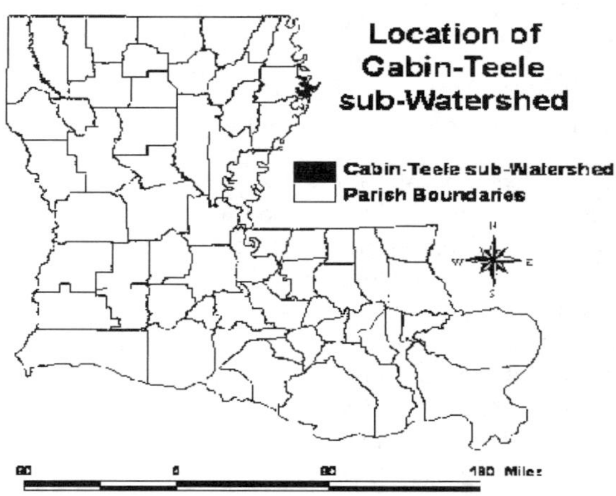

Fig. 4. Location of the Cabin-Teele Sub-Watershed (Source: USDA –NRCS).

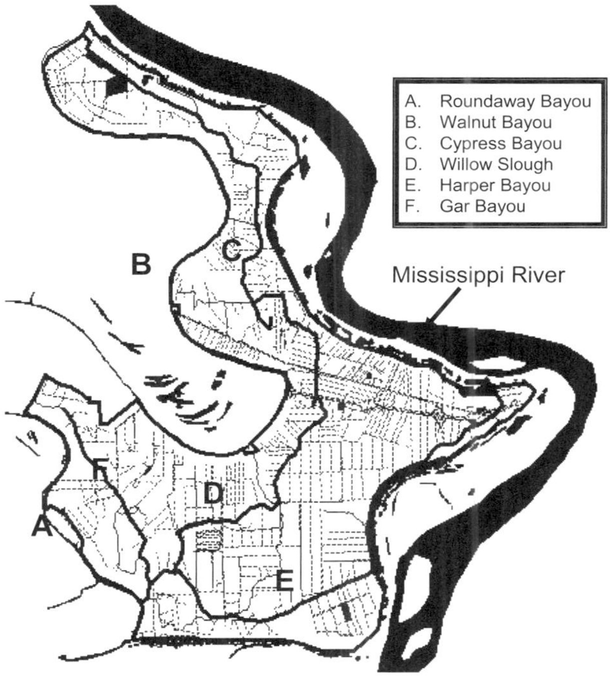

Fig. 5. Cabin-Teele Site Map and Drainage Network.

The Cabin-Teele sub-watershed offers many opportunities for BMP, controlled-drainage, and wetland application implementation. There are over 700 agricultural fields in which BMPs could be implemented ranging in size from one to 310 acres, with slopes ranging from 0.01 to 12.0 percent. The large number of similar fields allows for replication of the each BMP within similar soil, crop, acreage, and slope.

The design of the drainage network lends itself very well to the implementation of controlled-drainage using flash-board risers. It is constructed entirely of excavated canals and field ditches on relatively flat topography. Even though each field is flanked on either side by a drainage ditch, series of fields are connected along a single "header" canal. This allows for one structure to effectively control the drainage from a single field to a number of fields (Figure 2). Again the large number of fields and drainage canals allows for replication of this practice for our future field research studies.

There are a number of sites within the sub-watershed that allow for the implementation of the wetland practices of grassed waterways, constructed/natural wetlands, and flow diversions. In the case of the grassed waterways, these areas already exist. This eliminates the need for their construction as well as the time for construction and establishment to mature vegetation. The grasses waterways are shallow swales which are normally wet enough that their cultivation has become undesirable and have been allowed to grow over with primarily grasses and sparse trees or have been placed in the Wetland Reserve Program (WRP) program. These areas exist alongside fields which also have excavated ditches for easy comparison of nutrient and sediment loads. The constructed/natural wetlands will use either large areas that are already in wetlands or have been created as wetlands through the WRP Program or small wetlands that will be created along the drainage network. A constructed wetland will consist of a widening of the drainage canal or ditch where wetland plant species will be planted. There are a number of sites within the sub-watershed that make it possible to replicate these practices for our future field research activities. The flow diversion practice will be conducted in an area where a side-by-side comparison is possible. This location was a broad shallow stream bed which, when channelized to improve drainage, was divided into two zones (one, a deeper dredged channel, and the other a broader shallow flood prone area) by the placement of the dredge material. The flood prone area is ideal for wetland restoration as it becomes inundated naturally during most rainfall events due to its low elevation. A diversion cut made in the dredge material at the upper end of the section, and a reintroduction cut made in the dredge material at the lower end of the section allow for drainage water to be diverted from the main channel, through the constructed wetland allowing for the side-by-side comparison. By using this site, all environmental factors and flow distances will be the same for both the diverted flow as well as the channelized flow. There are also several potential sites where flow can be diverted through WRP lands, but the flow length will be different distances in these locations.

An automated NOAA weather station exists at the Tallulah-Vicksburg Regional Airport located in the center of the sub-watershed at Mound, Louisiana. All the required weather input data for the AnnAGNPS model will be obtained from this weather station. Prior year's weather data inputs for the model are available back to 1996 and have been provided by the Southeastern Regional Climate Center located on the Louisiana State University campus (http://www.srcc.lsu.edu/).

The soils of the Cabin-Teele sub-watershed are Mississippi River alluvial deposits. Eleven soil map units representing different phases of four different soil series are present in the sub-watershed. The detailed soil survey map for the watershed can be found in Figure 6.

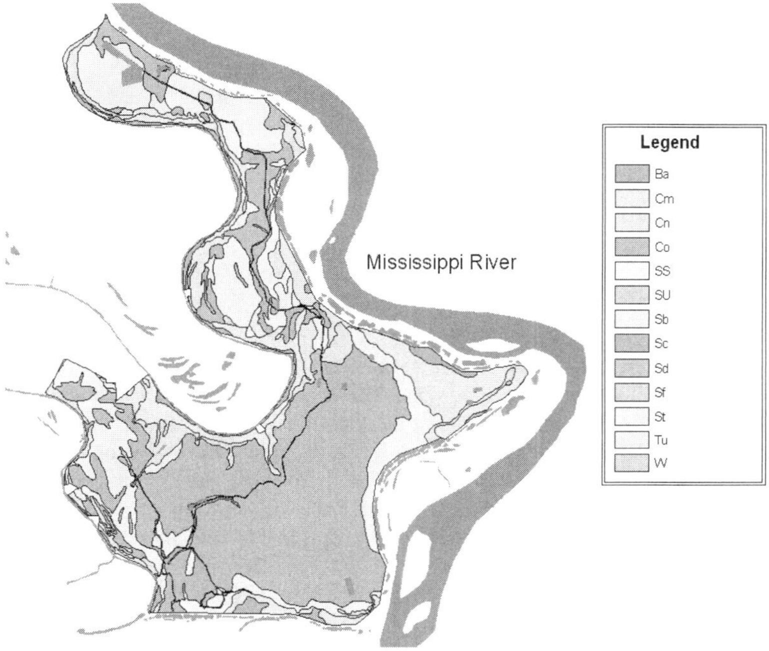

Fig. 6. Soils map of the Cabin-Teele Sub-Watershed.

21.7. SELECTION OF SUB-WATERSHED

The Cabin-Teele sub-watershed was selected because it is typical of the lower Mississippi river valley (alluvial soils, naturally leveed streams, and a large percentage (76%) of land converted from bottomland hardwood to agricultural row crops) while still retaining the unique feature of having a single drainage outlet. It has a broad representation of agricultural crops (corn, cotton, rice, sorghum, soybeans, millet, wheat, sunflower, and timber) allowing for the implementation and evaluation of Best Management Practices (BMPs) on a wide array of crops. It has been classified as having impaired waters (nutrient, pesticide, and sediment) so that any watershed benefits of BMPs will be reflected in an improvement in the overall water quality of the sub-watershed.

The Cabin-Teele sub-watershed has the additional benefits of having a broad land owner and multi-agency interest and participation in conservation programs (Wetland and Conservation Reserve Programs – WRP and CRP), BMP implementation, and previous research conducted within the sub-watershed. Organizations having shown interest or worked in the Cabin-Teele sub-watershed include; USDA Agricultural Research Service (ARS), USDA Natural Resources and Conservation Service (NRCS), Ducks Unlimited, Fifth Louisiana Levee District, Louisiana Department of Agriculture and Forestry (LDAF), Louisiana Department of Environmental Quality (LDEQ), Louisiana Department of Wildlife and Fisheries (LDWF), Louisiana Cotton Producers Association, Louisiana

Governors Office of Coastal Activities, Louisiana Hypoxia Working Group, Louisiana State University Agricultural Center, Louisiana State University Coastal Ecology Institute, Louisiana State University Cooperative Extension Service, Louisiana State University School of the Coast and Environment, Mississippi River Basin Alliance (MRBA), Madison Parish Soil and Water Conservation District (MSWCD), Nature Conservancy, Northeast Delta Resource Conservation & Development (RS&D), United States Army Corps of Engineers (USACE), United States Environmental Protection Agency (US-EPA), and United States Geological Survey (USGS).

21.8. MODEL DESCRIPTION

AnnAGNPS is a continuous simulation, watershed scale, pollutant loading model (Cronshey and Theurer, 1998). The watershed is subdivided into homogenous fields of similar soil type, land use, and management. The model simulates the movement of surface water, sediment, nutrients, and pesticides traveling from the fields and subsequent transport through the watershed. Runoff from the fields is calculated using runoff curve numbers. Sediment loss is determined using the Revised Universal Soil Loss Equation (RUSLE) (Renard et al.,1997). AnnAGNPS uses daily climate information and a daily time step for calculations. Output parameters (water, sediment, nutrients, and pesticides) can be selected for any single or multiple watershed sources (specific reaches and cells). A source accounting feature indicates the percent contribution at the outflow of a specified watershed source.

21.9. METHODS

21.9.1. Field Studies

A paired study will be used to determine the effect controlled drainage has on nitrogen reductions from agricultural fields. One set of fields will continue using uncontrolled drainage through the open ditch network already in place. The second set of fields will have controlled drainage implemented as previously described in the Best Management Practices section. ISCO auto-samplers will be used at the outflow point of each plot to sample flows for nutrient analysis. Flows from the uncontrolled plots will be measured using a stage discharge relationship developed for the site. The weir structure will be used to measure flow from the controlled drainage plots.

To determine the drainage water nitrate reduction due to in-stream denitrification, a 1,400 meter section of the drainage network will be isolated. All inputs and outputs of water and nitrogen species to the section will be measured. A mass balance for all the nitrogen species and the total flow will be conducted. Unaccounted-for nitrogen will be assumed to be denitrified in the sediments. Plant uptake is not a factor as the section chosen has no plant growth in the channel or along the immediate banks.

The same approach will be used to determine the drainage water nitrate reduction due to wetland. All inputs and outputs of water and nitrogen species to the section will be measured. A mass balance for all the nitrogen species and the total flow will be conducted. Unaccounted-for nitrogen will be assumed to be denitrified in the sediments of the wetland or taken up by the vegetation.

Flow will be calculated using a stage discharge relationships developed for each sampling site. The stage discharge relationship will be developed using a water level recorder and a manual flow meter. Flow will be measured at different flow depths allowing for the relationship between stage and flow to be developed. Water samples will be taken using ISCO auto-samplers as water passes into the treatment area and as it passes out of the treatment area. Samples will be taken on a flow proportional basis during flow events. Samples will be removed from the samples within 12 hours after the flow event, placed on ice, and taken to the lab for analysis. Samples will be analyzed for total nitrogen, nitrate, nitrite, ammonium, total phosphorous, phosphate, total suspended solids, and chlorophyll *a*.

A sampler will be located at the outlet of the sub-watershed, as well as the development of a stage discharge relationship to calibrate the AnnAGNPS model.

21.9.2. Modeling

The continuous simulation model AnnAGNPS will be used to model the Cabin-Teele sub-watershed. Different BMP type and placement scenarios will be simulated using 10 years of weather data to determine their optimal placement within the sub-watershed to maximize the benefit for improving water quality at the outlet as well as minimize cost. The model will be calibrated using flow and nutrient data collected from a series of gauging stations equipped with water samplers located within the sub-watershed. Previously collected water quality data and preliminary simulations will be used to aid in locating sampling equipment within the sub-watershed.

The Cabin-Teele sub-watershed was subdivided into fields based on the NRCS certified field delineation and assigned a field number using aerial photos and the ESRI software package ARC-Map. This allowed for the records kept by the Farm Service Agency (FSA) and farm chemical supply to be used for background crop and fertilization information as well as to have a constant field base from which to work. The drainage network was also delineated and subdivided into reaches using aerial photos and the ARC-Map software. Each reach is a section of the drainage network between individual field drainage input locations. This allowed for the drainage water with its associated nutrient and sediment loads from each field to enter the drainage network and be routed through the drainage network.

Soil maps and data, aerial photographs, and digital topographic maps of the sub-watershed were acquired from the USDA-NRCS Geospatial Data Gateway website (http://datagateway.nrcs.usda.gov).

An automated NOAA weather station exists at the Tallulah-Vicksburg Regional Airport located in the center of the sub-watershed at Mound, Louisiana. All the required weather input data for the AnnAGNPS model will be obtained from this weather station through the NOAA Southern Regional Climate Center, Louisiana State University. Additional tipping bucket rain gages may be co-located with flow sampling equipment.

21.10. COMPLETED WORK

The creation of a GIS inventory database including the past ten years of land use, fertilization, soil classification, and drainage information has been completed. Detailed certified field, drainage network, soil classification, and transportation maps have been completed and ground truthed. The field and drainage network discretization of the Cabin-Teele sub-watershed for the AnnAGNPS model has also been completed. Locations for the in-stream denitrification studies and wetland diversions have been identified.

21.11. EXPECTED RESULTS

The expected results of this study are to quantitatively document water quality improvements regarding nutrient and sediment loads within the sub-watershed using the implemented BMPs and determine the contribution that in-stream processes have on nutrient reduction in surface waters in northeastern Louisiana. The modeling aspect of the study will determine the effectiveness of different field targeting combinations of BMP Strategies within the sub-watershed. The economic analysis of the different BMP strategies will determine the cost effectiveness of each strategy.

ACKNOWLEDGEMENTS

The authors wish to acknowledge the contributions of Dr. John Day, Dr. Robert Lane, and Dr. John Westra at Louisiana State University, and Mr. Max Forbes at the Louisiana Department of Environmental Quality as cooperators on the project planning.

DISCLAIMER

Mention of a trade name, proprietary product, or specific equipment does not constitute a guarantee or warranty by the USDA and does not imply approval of the product to the exclusion of others that may be suitable.

REFERENCES

Bachand, P.A.M., and Horne, A.J., 2000. Denitrification in constructed free-water surface wetlands: II. Effects of vegetation and temperature. Ecol. Eng. 14:17-32.

Beman, J.M., Arrigo, K.R., and Matson, P.A., 2005. Agricultural runoff fuels large phytoplankton blooms in vulnerable areas of the ocean. Nature 434:211-214.

Binkley, D., Burnham, H., and Allen, H.L., 1999. Water quality impacts of forest fertilization with nitrogen and phosphorous. Forest Ecology and Management 121:191-213.

Boesch, D.F., Brinsfield, R.B., and Magnien, R.E., 2001. Chesapeake Bay eutrophication: scientific understanding, ecosystem restoration, and challenges in agriculture. Journal of Environmental Quality 30:303-320.

Chatarpaul, L., Robinson, J.B., and Kaushik, N.K., 1979. Role of tubificid worms on nitrogen transformations in stream sediment. J. Fish. Res. Board Can. 36:673-678.

Cooper, J.R., and Gilliam, J.W., 1987. Phosphorus redistribution from cultivated fields into riparian areas. Soil Science Society of America Journal 51:1600–1604.

Cronshey, R.G., and Theurer, F.G., 1998. AnnAGNPS non-point pollutant loading model. In Proc.1st Federal Interagency Hydrologic Modeling Conf.., 1-9 to 1-16. Las Vegas, Nev. April 19-23. Prepared by the Subcommittee on Hydrology of the Interagency Advisory Committee on Water Data.

Dahm, C.N., Grimm, N.B., Marmonier, P., Valett, H.M., and Vervier, P., 1998. Nutrient dynamics at the interface between surface water and groundwaters Freshwater Biol. 40:427-451.

Day, J.W., Ko, J.Y., Rybczyk, J., Sabins, D., Bean, R., Berthelot, G., Brantley, C.,

Cardoch, L., Conner, W., Day, J.N., Englande, A.J., Feagley, S., Hyfield, E., Lane, R., Lindsey, J., Mistich, J., Reyes, E., and Twilley, R., 2004. The use of wetlands in the Mississippi Delta for wastewater assimilation: a review. Ocean and Coastal Management. 47:671-691.

Dinnes, D.L., Karlen, D.L., Jaynes, D.B., Kasper, T.C., Hatfield, J.L., Colvin, T.S., and

Cambardella, C.A., 2002. Nitrogen management strategies to reduce nitrate leaching in tile-drained Midwest soils. Agron. J. 94:153-171.

Drury, C.F., Tann, C.S., Gaynor, J.D., Oloya, T.O., van Wesenbeeck, I. J., and

McKenney, D.J., 1997. Optimizing corn production and reducing nitrate losses with water table control-subirrigation. Soil Science Society of America Journal. 61:889-895.

EEA, 1995. Europe's Environment, the Dobris Assessment. European Environment Agency, Copenhagen.

EFOTG, 2004. National Resource and Conservation Service Field Office Technical Guide. Document can be found at the NRCS website: http://efotg.nrcs.usda.gov/technical/efotg/

Erisman, J.W., Brydges, T., Bull, K., Cowling, E., Grennfelt, P., Nordberg, L., Satake, K., Schneider, T., Smeulders, S., Hoek, K., Wisniewski, J.R., and Wisniewski, J., 1998. Nitrogen the Confer-N-s. First International Nitrogen Conference. March 23-27, 1998. Noordwijkerhout, Netherlands.

Evans, R.O., Cummings, J.R., andGilliam, J.W., 1989. Controlled drainage: A best management practice in North Carolina. ASAE Parer 892695. ASAE. St Joseph, MI.

Evans, R.O., Gilliam, J.W., and Skaggs, R.W., 1991. Controlled drainage management guidelines for improving water quality. Publication NO. AG-443. N.C. Coop. Extension Service.

Evans, R.O., Skaggs, R.W., and Gilliam, J.W., 1995. Controlled drainage versus conventional drainage effects on water quality. Journal of Irrigation and Drainage Engineering 121(4):271-276.

Fausey, N.R., King, K.W., Baker, B.J., and Cooper, R.L., 2004. Controlled drainage performance on Hoytville soil in Ohio. In: Drainage VIII Proceedings of the Eighth International Symposium, 21-24 (Sacramento, California USA). Ed. R. Cooke. Publication Date March 21, 2004. ASAE Publication Number 701P0304. ASAE, St Joseph, MI.Pp 84-88.

Gilliam, J.W., Skaggs, R.W., and Weed, S.B., 1979. Drainage control to diminish nitrate loss from agricultural fields. Journal of Environmental Quality 8(1):137-142.

Goolsby, D.A., Battaglin, W.A., Lawrence, G.B., Artz, R.S., Aulenbach, B.T., Hooper, R.P., Keeney, D. R., and Stensland, G.J., 1999. Flux and Sources of Nutrients in the Mississippi-Atchafalaya River Basin: Topic 3 Report for the Integrated Assessment on Hypoxia in the Gulf of Mexico. NOAA, Coastal Ocean Program Decision Analysis Series No. 17. NOAA Coastal Ocean Program, Silver Spring, MD. 130 pp.

Hill, A.R., and Sanmugadas, K., 1985. Denitrification rates in relation to stream sediment characteristics. Wat Res. 19: 1579-1586.

Hynes, H.B.N., 1983. Groundwater and stream ecology. Hydrobiologia. 100:93-99. Jansson, M., Leonardson, L., and Fejes, J., 1994. Denitrification and nitrogen retention in a farmland stream in southern Sweden. Am. Bio. 23(6): 326-331.

Jensen, K., Sloth, N.P., Rysgaard-Petersen, N., Rysgaard, S., and Revsbech, N.P., 1994. Estimation of nitrification and denitrification from microfiles of oxygen and nitrate model sediment systems. Appl. Environ. Microbial. 60(6):2094-2100.

Johnston, C.A. 1991. Sediment and nutrient retention by freshwater wetlands: effects on surface water quality. CRC Crit. Rev. Environ. Control. 21:491-565.

Matheson, F.E., Nguyen, M.L., Cooper, A.B., Burt, T.P., and Bull, D.C., 2002. Fate of ^{15}N-nitrate in unplanted, planted and harvested riparian wetland soil microcosms. Ecol. Eng. 19:249-264.

Mitsch, W.J., Day, J.W., Gilliam, J.W., Groffman, P.W., Hey, D.L., Randall, G.W., and Wang, N., 2001. Reducing nitrogen loading to the Gulf of Mexico from the Mississippi River Basin: Strategies to counter a persistent ecological problem. BioScience 51(5):373-388.

Mitsch, W.J., and Gosselink, J.C., 1993. Wetlands. 2nd Ed. Van Nostrand Reinhold, New York, 722p.

Power, J.F., Wiese, R., and Flowerday, D., 2001. Managing farming systems for nitrate control: A research review from management systems evaluation areas. Journal of Environmental Quality 30:1866-1880.

Rabalais, N.N., Turner, R.E., Justic, D., Dortch, Q., and Wiseman, W.J., Jr., 1999. Characterization of Hypoxia: Topic 1 Report for the Integrated Assessment on Hypoxia in the Gulf of Mexico. Coastal Ocean Program Decision Analysis Series No. 15. NOAA Coastal Ocean Program, Silver Spring, MD. 167 pp.

Rabalais, N.N., Turner, R.E., and Scavia, D., 2002. Beyond science into policy: Gulf of Mexico Hypoxia and the Mississippi River. BioScience. 52(2):129-142.

Renard, K.G., Foster, G.R., Weesies, G.A., McCool, D.K., and Yoder, D. C. coordinators, 1997. Predicting Soil Erosion by Water: A Guide to Conservation Planning with the Revised Universal Soil Loss Equation (RUSLE). USDA Agricultural Handbook No. 703. Washington, DC: USDA.

Seitzinger, S.P. 1988. Denitrification in freshwater and coastal marine ecosystems: ecological and geochemical significance. Limnol. Oceanogr. 33(4):702-704.

Skaggs, R.W. 1999. Chapter 20: Water Table Management: Subirrigation and Controlled drainage. In: R.W. Skaggs and J. van Schilfgaarde eds. Agricultural Drainage. Agron. Monogr. 38. American Society of Agronomy, Inc., Crop Science Society of America, Inc., Soil Science Society of America, Inc., Madison, Wisconsin, USA. Pp. 695-718.

Skaggs, R.W., Breve, M.A., and Gilliam, J.W., 1994. Hydrology and water quality impacts of agricultural drainage. Critical Reviews in Environmental Science and Technology 24:1-32.

Stanley, D.W., 1983. Nitrogen cycling and phytoplankton growth in the Nuese River, North Carolina. Tech. Rep. 204. Water Resources Research Institute of the University of North Carolina, Raleigh, N.C.

Thomas, D.L., Perry, C.D., Evans, R.O., Izuno, F.T., Stone, K.C., and Gilliam, J.W., 1995. Agricultural drainage effects on water quality in Southeastern U.S. Journal of Irrigation and Drainage Engineering 121(4):277-282.

Triska, F.J., Kennedy, V.C., Avanzino, R.J., Zellweger, G.W., and Bencalla, K.E., 1989. Retention and transport of nutrients in a third order stream: hyporheic processes. Ecology. 70(6):1893-1905.

USEPA, 2005. Fact Sheet, Nitrates/Nitrites. National Primary Drinking Water Regulations. http://www.epa.gov/safewater/contaminants/dw_contamfs/nitrates.html

Whitall, D.R., and Pearl, H.W., 2001. Atmospheric pollutants and trace gasses. Journal of Environmental Quality 30:1508-1515.

Xue, Y., Kovacic, D.A., David, M.B., Gentry, L.W., Mulvaney, R.L., and Lindau, C.W., 1999. In situ measurements of denitrification in constructed wetlands. Journal of Environmental Quality 28:263-269.

22

AUTONOMOUS VEHICLES TO MEASURE WATER QUALITY IN MULTIPLE DIMENSIONS IN COASTAL AREAS

Steven G. Hall[1], Randy R. Price[1], Amol Mudgundi[1], and Navneet Mandhani[1]
[1]Biological and Agricultural Engineering, Louisiana State University
AgCenter, 143 EB Doran, LSU, Baton Rouge, LA 70803, USA

ABSTRACT

Autonomous (self guided) vehicles can be programmed to perform a variety of tasks without further human input. They have been used for reduction of bird predation, monitoring water quality parameters, for agricultural, military and industrial tasks. A series of such devices have been developed and used in coastal Louisiana and the region. Vehicle types include small boats, tracked and wheeled vehicles and small airplanes. Guidance has been provided via local physical sensors, infrared, image processing, and GPS. Power for such devices can be provided from batteries, fossil fuel or solar radiation, and vehicles can be programmed to monitor not only their environments, but also their own condition, a primitive form of self awareness. This can be used for self-management of vehicles during field exercises. Water quality can be checked across a variety of areas with such vehicles, providing potentially better information than fixed systems. To test this, autonomous boats were equipped with three dissolved oxygen sensors (at depths of 30, 60 and 90 cm), four thermocouples and a datalogger. Data collected for various paths taken by the boat (horizontal, vertical and random) were used to analyze the variation of dissolved oxygen and temperature with depth across the water bodies. In some applications, feedback using a radio frequency transmitter and receiver could aid in management. In addition, such devices can send information, alarm signals or other data to fixed units, laboratories or managers, offering promise for better monitoring and potentially improved management of coastal water quality and environments.

22.1. INTRODUCTION

Water quality is of interest to managers of coastal waters, aquacultural ponds, drinking water reservoirs and related areas (Mizalkoff, 2003; Meyer and Brune, 1982; Losordo et al., 1986; Gobet et al., 2003). Changes in temperature and dissolved oxygen can significantly affect water quality in aquaculture ponds (Cathcart and Wheaton, 1983). Dissolved oxygen is one of the most important water quality parameters for fish survival and growth. As dissolved oxygen decreases, fish become stressed. Fish exposed to sublethal dissolved oxygen have significantly higher mortality rates than fish exposed to acceptable dissolved oxygen (Costa-Pierce and Laws, 1985). Automated systems have been developed for fixed water quality monitoring (Gobet et al., 2003; Krom et al., 1985; Losordo et al., 1987; Plaia, 1987), and various models have been developed to predict and assist in management

(Cathcart and Wheaton, 1987; Fang et al., 2003). In addition, a variety of systems have been developed to control water conditions in tanks and ponds (Hall et al., 2002; Landman and VandenHeuvel, 2003). Simultaneously, a variety of automated tractors and other robotic devices for use in agriculture have been developed (Toda et al., 1993; Ellis, 1995; Kim et al., 2000). Combining unique elements of each of these, our team has developed a series of autonomous vehicles which can move about performing a variety of operations in fish ponds, water quality reservoirs and other coastal areas (Hall et al., 2001; Hall and Price, 2003a, 2003b, 2003c; Hall et al., 2004). This report deals with use of an autonomous boat for data acquisition of water quality parameters such as dissolved oxygen and temperature. These publications detail the design and construction of the autonomous vehicle, its use and efficacy in reducing bird predation, and pointed to some extent to the possibility of using such vehicles to monitor and eventually control water quality parameters in aquacultural or urban reservoirs. The use of such vehicles in coastal areas may contribute to better understanding of coastal hydrodynamics, could be used to identify toxic plumes and potentially to aid in management or control of water quality in critical coastal areas.

22.2. OBJECTIVES

The goal of this project was to design and test an automated mobile water quality monitoring system and create a methodology of displaying and using this data for relevant management needs. The specific objectives were to:

1. Configure a GPS unit, dissolved oxygen (DO) sensors and thermocouples to acquire water quality data,
2. Capture real-time water quality data with reference to 3-D spatial coordinates and time,
3. Develop and show DO profiles by using three dimensional graphs, and
4. Develop a method for control of dissolved oxygen concentration in water bodies using wireless transmission and control.

22.3. METHODS AND MATERIALS

The system consisted of three dissolved oxygen sensors (Campbell Scientific Inc., Model no. CS511-L and CS510), an autonomous boat (Hall and Price, 2003a, 2003b, 2003c), T-type thermocouples, a 23X datalogger (Campbell Scientific Inc.), a GPS Unit (GARMIN 16-HVS – marketed by Campbell Scientific Inc.), RF transmitters and receivers (Abacomm Technologies), a Basic Stamp 2 (Parallax Inc.) and relevant associated hardware components.

In one configuration, an autonomous boat (Figure 1) was powered by two 12 volt (V) batteries along with solar panels which help in recharging the batteries. The solar panels provided approximately 2.0 amperes at 12V in bright sunlight. They also served to protect the electronic circuitry and the data logger that were placed under the solar panel on an aluminum chassis. The autonomous boat had two paddle wheels, one on each side of the

Fig. 1. Autonomous boat with instrumentation for capturing dissolved oxygen concentration and temperature in coastal water bodies.

boat. These were attached to two motors (Pittman, GM712-31). The motors used approximately 0.3A each during operation of the boat and a spike of 0.6A during switching of direction. The dual paddle wheel design allowed the boat to turn in a required direction including front, back, or turn (right or left). The turning of the boat is done by programming the chip to turn one motor on forward and the other backward or off.

Three oxygen sensors, four thermocouples, a GPS unit (GARMIN 16-HVS – marketed by Campbell Scientific Inc.), and a 23X datalogger were mounted on the autonomous boat. The data was captured every 15 seconds and stored on the datalogger. The electronics on the boat consisted of a single chip micro-controller. The micro-controller used was a Parallax Inc. BASIC II Stamp. The system had shore-feelers attached on the front end to prevent collision with objects. If the boat stopped in the water, then the GPS readings received by the Basic Stamp became constant. In this situation, the basic stamp was programmed to reverse the direction of the motion of the paddles so that the boat backed up for approximately 8 seconds to provide distance from shore and avoid obstacles. Then the boat was rotated by driving one paddlewheel forward and the other backward. Once the boat was turned the boat was again driven forward so that it could continue to move in a random fashion or according to a desired path.

A 23X Campbell scientific datalogger was powered with rechargeable lead acid cells was the central storage device on the experimental setup. This stored all the programming logic, dissolved oxygen data, the temperature data and the GPS data. It could also power the RF transmitter to transmit a wireless signal to the control station. At the control station the RF receiver may receive this signal and activate the aerator.

The GARMIN 16-HVS GPS unit included the GPS receiver and antenna in the same housing with one external cable for the power supply and communications. This unit was directly connected to the 23X datalogger and gives an output in the format:

$GPGGA,hhmmss,1111.111,a,nnnn.nnn,b,t,uu,v.v,w.w,M,x.x,M,y.y,zzzz*hh.

All DO sensors, thermocouples and the GPS unit were connected to the datalogger and the entire unit was placed on the boat (Figure 2).

Fig. 2. Instruments used for the design of the system.

The three DO sensors (manufactured by Campbell Scientific Inc, Model no. CS511-l and CS510) were deployed at depths of 30, 60 and 90cm. The supporting rope had a 2 kg mass tied at the bottom to minimize drift in the sensors due to the motion of the boat and water. The programming of the datalogger was done using the PC208W software.

Accuracy of the dissolved oxygen sensors was tested in air after calibration and was found to be ±0.5 mg/l. Initial results showed that three dimensional variations are common in reservoirs and other coastal waters and ongoing evaluation is continuing. In some situations, these devices could employ a control mechanism (Figure 3) to manage water quality parameters such as dissolved oxygen, or to provide information to a manager to assist in decision making.

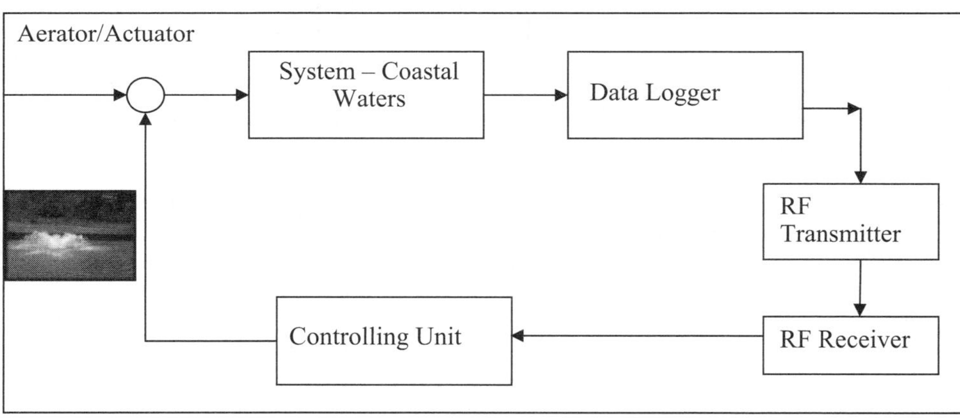

Fig. 3. Signal flow diagram explaining the working of the autonomous dissolved oxygen control system.

22.4. RESULTS

Variation of dissolved oxygen at depths of 30, 60 and 90 cm was plotted over time (Figure 4). The concentration of dissolved oxygen theoretically is supposed to decrease with depth and is seen to decrease in general, but local variations in animal or plant respiration, temperature or aeration could each alter expected trends.

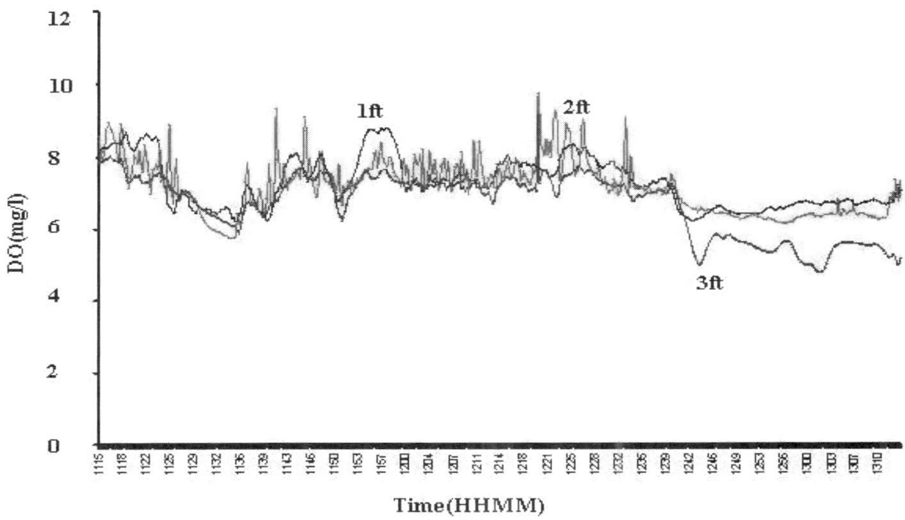

Fig. 4. Dissolved oxygen at depths 1ft (30cm), 2ft (60cm), and 3 ft (90cm) versus time over two hours shows a slow decrease in DO as respiration likely increases at warmer temperatures.

The plots for dissolved oxygen and temperature (Figure 5) are all two dimensional plots. They show the variation of dissolved oxygen and temperature as a function of depth and time. Figures 6 and 7 illustrate the variation of dissolved oxygen across the ponds. These contour (Figure 6) and surface (Figure 7) plots were constructed by capturing data a series of linear passes through the water. These maps give us an idea of concentration of dissolved oxygen across a water body which can be used as a tool to control the switching of aerators in a particular location of, for example, an aquaculture pond.

These preliminary results suggest that autonomous vehicles can be used to capture water quality data in multiple dimensions effectively across water bodies in coastal environments. Use of such devices may include deployment in difficult to access areas, capture of environmental, aquacultural, water reservoir or even off-shore data. The vehicles are stable, capture data effectively, and can be included in a feedback system to provide control, alarm or information to improve water quality management in coastal areas.

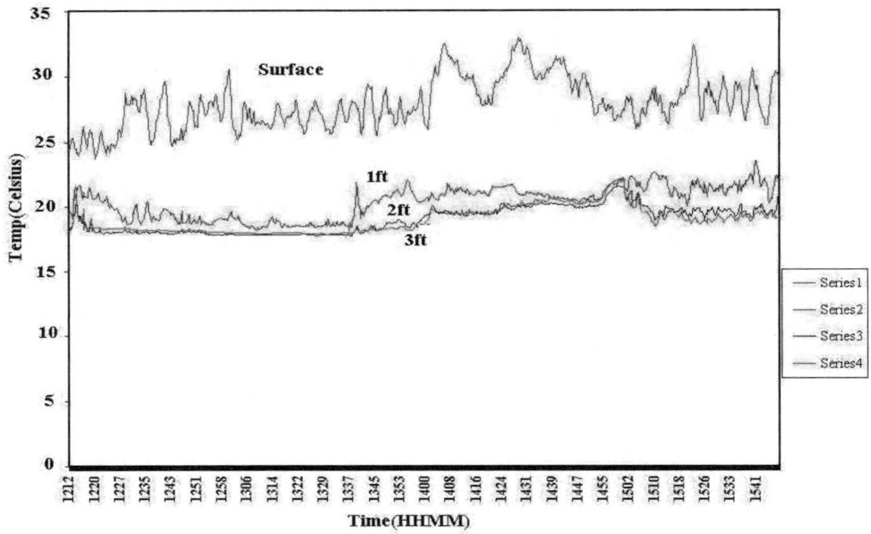

Fig. 5. Temperature at surface, 1ft (30cm), 2ft (60cm), and 3ft (90cm) versus time shows a slow increase in temperature at all depths later in the day.

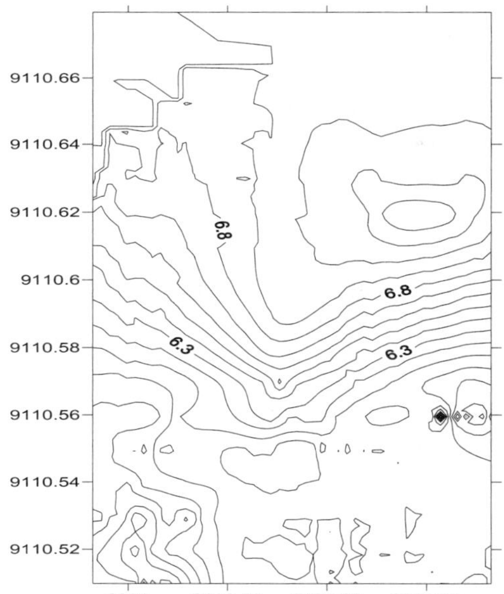

Fig. 6. A contour plot showing the concentration of dissolved oxygen in mg/l across the pond. Above map plots latitude, longitude and dissolved oxygen at one depth (30 cm).

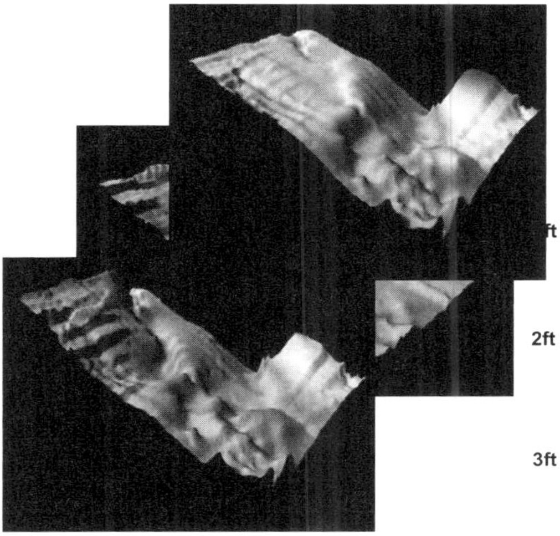

Fig. 7. These surface plots show a variation of concentration of dissolved oxygen at 1ft (30cm), 2ft (60cm) and 3ft (90cm) across a coastal water body with respect to the GPS co-ordinates (longitude and latitude).

These preliminary results suggest that autonomous vehicles can be used to capture water quality data in multiple dimensions effectively across water bodies in coastal environments. Use of such devices may include deployment in difficult to access areas, capture of environmental, aquacultural, water reservoir or even off-shore data. The vehicles are stable, capture data effectively, and can be included in a feedback system to provide control, alarm or information to improve water quality management in coastal areas.

22.5. CONCLUSIONS

This methodology of using an autonomous vehicle to capture water quality data has proven effective. Challenges included capturing and plotting data over time and space in three spatial dimensions and time. Use of similar systems for automated control, monitoring of other water quality parameters, or performance of needed tasks such as reduction of unwanted pests are all possible future developments.

ACKNOWLEDGEMENTS

The authors thank faculty, staff and students of the department of Biological and Agricultural Engineering for their assistance, the LSU College of Agriculture and the LSU AgCenter for funding. This manuscript has been approved by the LSU AgCenter as manuscript 06-22-0043.

REFERENCES

Cathcart, T.P., and Wheaton, F.W., 1987. Modeling temperature distribution in freshwater ponds. Aquaculture Engineering 6:237-257.

Costa-Pierce, B.A., and Laws, E.A., 1985. Effects of destratification on autotrophic and heterotrophic microplankton productivity in eutrophic aquaculture ponds. Aquaculture Engineering 50: 141-151.

Ellis, R., 1995. Robots are becoming used on farm. Macchine e Motori Agricoli 53(9):55-56.

Fang, X, Stefan, H. G., Eaton, J. G., and McCormick, J. H., and Alam, S.R., 2003. Simulation of thermal/ dissolved oxygen habitat for fishes in lakes under different climate scenarios. Ecological Modelling 172:13-37.

Gobet, J., Cardot, F., Santoll, E., and Rychen, P., 2003. Microelectrode array sensor for water quality monitoring. Water Science and Technology 47(2):127-134.

Hall, S., Price, R. R., and Mandhani, N., 2004. Use of Autonomous Vehicles for Drinking Water Monitoring and Management in an Urban Environment, Presented at ASAE/CSAE International Meeting, ASAE, St. Joseph, MI, Paper No. 047053.

Hall, S.G., and Price, R.P., 2003a. An Autonomous Scareboat to Reduce Bird Predation on Aquaculture Ponds. Louisiana Agriculture.

Hall, S., and Price, R.P., 2003b. Using Autonomous Robotic Vehicles for Aquacultural and Environmental Applications, CSAE Annual Meeting, Montreal, Canada, Paper no.03-202

Hall, S., and Price, R.P., 2003c. Mobile Semi-Autonomous Robotic Vehicles Improve Aquacultural Ventures by Reducing Bird Predation and Improving Water Quality Monitoring. Abstract, Proceedings of the World Aquaculture Society, Louisville, KY.

Hall, S.G., Finney, J., Lang, R.P., and Tiersch, T. R., 2002. Design and development of a geothermal temperature control system for brood stock management of channel catfish *Ictalurus punctatus*. Aquacultural Engineering 26: 277-289

Hall, S. G., Price, R.P., and Wei, L., 2001. Design of an Autonomous Bird Predation Reduction Device. ASAE International Meeting, ASAE St. Joseph, MI, Paper no. 01-3131.

Kim, S.C., Park, W.P., Jung, I.J., Chung, S.O., and Lee, W.Y., 2000. Development of an unmanned tractor with autonomous and remote control utility. Transactions in ASAE, Paper no. 001065

Krom, M.D., Porter, C., and Gordin, H., 1985. Description of water quality conditions in a semi-intensively cultured marine fish pond in Eilat, Israel. Aquaculture 49:141-157

Landman, M.J., and Van den Heuvel, M.R., 2003. An improved system for the control of dissolved oxygen in fresh water aquaria. Water Research 37:4337-4342

Losordo, T.M., Piedrahita, R. H., and Ebeling, J. M., 1986. Stratification measurement techniques in aquaculture ponds. Transactions in ASAE, Paper no. 86-5047

Losordo, T.M., Piedrahita, R. H., and Ebeling, J. M., 1988. An automated water quality data acquisition system for use in aquaculture ponds. Aquacultural Engineering, 7(4): 265-278.

Meyer, D.I , and Brune, D. E., 1982. Computer modeling of diurnal oxygen levels in a Stillwater aquaculture pond. Aquacultural Engineering 1: 245-261

Mizalkoff, B., 2003. Infrared optical sensors for water quality monitoring. Water Science and Technology 47(2):35-42

Plaia, W. C., 1987. A computerized environmental monitoring and control system for use in aquaculture. Aquacultural Engineering 6:27-37

Toda, M., Kitani, O., Okamoto, T., and Torii, T., 1993. Studies on autonomous vehicles for agricultural robotics. Transactions in ASAE, Paper no. 93-3091.

23

ASSESSMENT OF FORESTRY BEST MANAGEMENT PRACTICES IN A LOUISIANA WATERSHED

Adrienne Viosca, April Mason, Philip Saksa, and Y. Jun Xu
School of Renewable Natural Resources
Louisiana State University Agricultural Center, Baton Rouge, LA 70803, USA

Richard Stich
Plum Creek Timber Company
PO Box 717, Crossett, AR 71635, USA

ABSTRACT

Nonpoint source pollution is a major cause of poor water quality in many Louisiana streams. Traditional forestry practices can increase sediment and nutrient runoff, further impacting waterbodies that experience periods of ambient low oxygen levels. Best Management Practices (BMPs) have been developed to buffer runoff from forestry operations and maintain stream water quality. Many studies over the past twenty years have assessed and documented the usefulness of BMPs in reducing nonpoint source pollutants in streams adjacent to harvest areas. Our study aims to assess the effectiveness of forestry BMPs at the basin-wide scale in the Flat Creek Watershed, located in central Louisiana where forestry is the dominant land use. The objectives are to quantify changes in water quality characteristics with implementation of forestry BMPs, assess the magnitude of water quality changes, and develop a GIS model that will predict impacts of land use activities on water quality at the watershed scale. We use a paired-watershed design and employ two water quality monitoring approaches for a two year period. One approach assesses biological and physicochemical water quality characteristics near treatment sites, while the other assesses the same attributes throughout the watershed. We are using a GIS framework that integrates several spatial layers including land use, soil type, and drainage networks to develop the water quality prediction model. Presented are the major research concepts, methods, analytical techniques, and discussion of preliminary water quality results with hydrologic and GIS modeling system development.

23.1. INTRODUCTION

Forestry Best Management Practices (BMPs) are a combination of management tools with the purpose of reducing nonpoint source (NPS) pollution and have been developed to decrease sediment, nutrient, and organic matter runoff into nearby waterways. In the eastern United States, many studies have assessed the impacts of forest operations and concluded that operations which employ BMPs are effective in protecting water quality (Aust and Blinn, 2004). Numerous studies at the plot-scale have proven that forest operations with

BMP implementation can reduce NPS pollution to the adjacent waterways. However, there is a knowledge gap of how the small-scale benefits translate to protecting water quality at the larger watershed scale (Grace, 2005). As watersheds increasingly become the primary planning unit for natural resource management, there is an urgent need to understand the effectiveness of current forestry BMPs in protecting stream water quality from NPS pollution beyond an operational boundary.

Forestlands cover 49 percent of Louisiana and are owned by private forest industries, public agencies, and non-industrial private landowners (Louisiana Forestry Association, 2000). Since the amount of land under forest operation in Louisiana is so large, it is imperative that the Louisiana forestry community adopt effective sustainable techniques that aim to reduce NPS pollution. Louisiana's Department of Environmental Quality (LDEQ) manages the state's Nonpoint Source Pollution Program, which focuses on voluntary controls of NPS pollution and education efforts (U.S. EPA, 2002). Currently, Louisiana implements a voluntary program to reduce NPS pollution from forestry operations by distributing a manual to the forestry community entitled *Recommended Forestry Best Management Practices for Louisiana* (Louisiana Forestry Association, 2000). This manual outlines forestry BMP guidelines and is intended to educate the forestry community about practical and economic forest operations that are designed to reduce NPS pollution. The state does not have laws regarding compliance with this program, thus it is completely voluntary. However, it is also unknown how effective Louisiana's BMPs are in protecting water quality.

Traditionally, physical and chemical parameters have been used to determine water quality. However, an increasing amount of attention is now being directed to the biological characteristics of water, because they are indicative of long-term changes in stream health. Forest operations such as timber harvest can directly or indirectly affect stream channel morphology and the biological community. Adams (1993) found that forestry BMPs in South Carolina protected the benthic macroinvertebrate community and a similar study in Florida concluded that the effects of forestry operations with the implementation of BMPs did not adversely affect stream health, including the macroinvertebrate community (Vowell, 2001). Additionally, a study in Mississippi determined that the use of streamside management zones, a type of forestry BMP, was effective in maintaining macroinvertebrate density (Carroll et al., 2004). To date, research on the effectiveness of Louisiana's forestry BMPs is nonexistent. Currently, there is one other ongoing project, directed by Louisiana Tech University, that is researching forestry BMPs in Louisiana. Past studies in other states have only focused on the small plot-scale effect, which is a good foundation, but this research needs to be expanded to include impacts on a larger scale. In order to encourage the voluntary implementation of BMPs in Louisiana, evidence is needed of their effectiveness in protecting water quality.

This study aims to determine the magnitude of forestry BMP effectiveness within a medium size watershed in central Louisiana. The Flat Creek Watershed represents a typical forestry-dominated landscape in the Upper West Gulf Coastal Plain ecoregion, and has been identified by LDEQ as an impaired watershed. The main objectives of this study are to: 1) identify spatial and temporal water quality patterns and trends in the watershed; 2) quantify basin-wide magnitude of timber harvesting impacts on chemical, physical, and biological water quality parameters; and 3) develop a watershed model to predict land use impacts on water quality changes. The study is currently in its primary implementation phase, and this paper presents some preliminary results.

23.2. METHODS

23.2.1. Site Description

Located in Central Louisiana, the Flat Creek Watershed covers 365 square kilometers of the Ouachita River basin in Winn Parish (Figure 1). The watershed location coordinates are 32° 12' N 92° 36' W (upper left corner) and 31° 55' N 92° 16' W (lower right corner). Forest cover accounts for 61% of the total land area in this watershed in which various types of land use, ranging from agriculture to forestry, exist (LDEQ 2001). The forest structure is dominated by loblolly pines, while various hardwoods are common within the riparian zones. The soil composition is mainly hydrous over loamy comprising 78.6% of the entire area, and clayey soil covers about 16.5% of the area - usually found in riparian zones. The remaining 4.9% of the area, near the extreme upper and lower watershed boundaries, consists of other various soil types (Boyd et al., 1998).

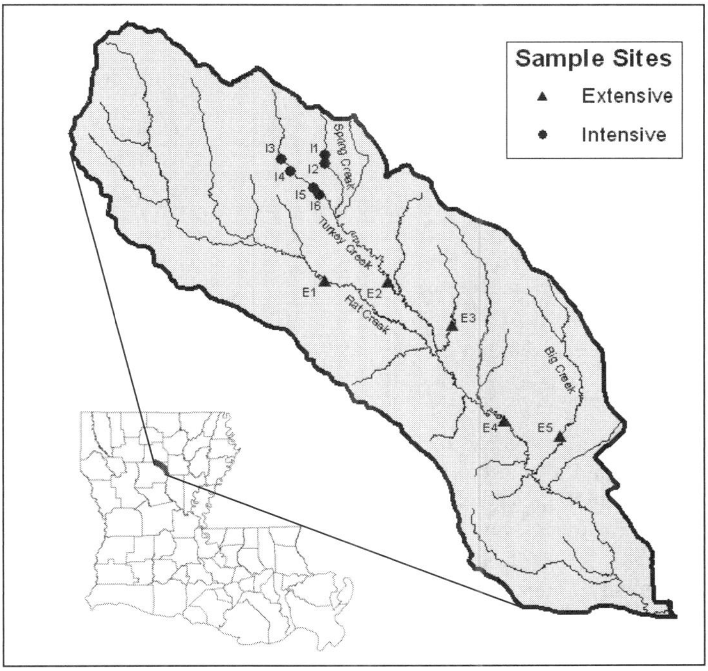

Fig. 1. Flat Creek Watershed monitoring locations.

23.2.2. Experimental Design and Data Collection

A paired-watershed approach is employed in this study. In the experimental sub-watershed, Turkey Creek, two forest tracts will be harvested with BMPs implemented. Changes in water quality parameters above and below each tract will be monitored. In an adjacent sub-watershed, Spring Creek, no forestry operations will be conducted and the same water quality measures will be recorded. Water quality will be monitored for one year prior to harvest activities, and one year after harvest activities. Pre- and post-harvest data will be compared in order to identify any changes in water quality related to forestry operations with BMPs.

NPS pollutants result from storm-generated runoff, thus a stormwater sampling approach is used in this study. Automated water samplers, ISCO 6712C (Teledyne Isco, Inc., Lincoln, Ne), were installed at six locations adjacent to the stream, above the bankfull channel, in order to collect storm-event water samples. At both harvest sites and the control site, one sampler was placed upstream of the treatment area and one was placed downstream below the treatment area. Five additional monitoring locations across the Flat Creek Watershed were identified, where monthly in-stream water quality and flow measurements are being conducted. Water samples at these five extensive monitoring locations are collected manually for laboratory analyses. The extensive monitoring design will allow a basin-wide characterization of stream physicochemical parameters (Table 1). The data from the intensive and extensive monitoring locations will be combined for assessment of spatial and temporal changes in water quality across the landscape.

Table 1. Water quality parameters to be analyzed at all monitoring locations.

Physical Parameters	Chemical Parameters
Total Suspended Solids	Total Phosphorous
Total Dissolved Solids	Dissolved Phosphorous
Total Solids	Nitrate
Temperature & pH	Nitrite
Specific Conductance	Total Kjeldahl Nitrogen
Dissolved Oxygen	Ammonia
Turbidity	Total Dissolved Carbon
Discharge	Total Organic Carbon

The instream water quality measurements include dissolved oxygen, temperature, conductivity, pH, and turbidity, using YSI 556 (YSI Inc., Yellow Springs, OH) and Hydrolab DataSonde (Hach Environmental, Loveland, CO) multi-probe meters. Water samples from storm-events and monthly grabs are analyzed for chemical and physical parameters (Table 1) by Louisiana State University (LSU) AgCenter Department of Agriculture Chemistry Laboratory.

At both intensive and extensive monitoring sites, the stream channels are surveyed. Stream stage and velocity are being measured across a range of storm events to develop a stage-discharge rating curve for each of the six intensive monitoring sites. A weather station was installed centrally within the watershed to collect relevant climatic variables including precipitation, air temperature, humidity, wind speed, and solar radiation. The weather data will be used for hydrologic modeling and hydrometeorological analyses.

Aquatic benthic macroinvertebrate will be collected at all eleven monitoring locations using a core sampling method. Specimens will be sorted and identified to the family level at LSU's School of Renewable Natural Resources Fisheries Laboratory. This data will be used to develop biological water quality indicators. All water quality, hydrological, and biotic data will be used interactively for the assessment of spatial, temporal, and forest operational effects on stream conditions across the Flat Creek Watershed.

23.2.3. Data Analysis

Water quality and biological data analysis will be based on the before-after control-impact paired (BACIP) design (Stewart-Oaten et al., 1986; Smith, 2002). Chemical and physical water quality parameters from each monitoring site will be assessed for spatial and temporal variations, utilizing monthly means of the measurements and mass loadings. Various indices, including species abundance and diversity, will be established from benthic macroinvertebrate samples. Analysis of variances will be applied to determine significant differences between treatment and control sites, as well as before and after effects. Regression techniques will also be used to investigate relationships among biological indices, water quality, and hydrology.

Stream discharge will be estimated with stage – discharge rating curves developed from storm event observations. Monthly discharge will be used to quantify monthly mass loadings of sediment and nutrients. A spatial hydrologic simulation will be performed using GIS-based modeling to analyze space-time variations of hydrological components, such as surface runoff, streamflow, and evapotranspiration. Hydrologic data from pre- and post-harvest activity will be used to determine spatial characteristics and impacts from forest operations on hydrological components. The relationship between hydrological responses of the watershed and the forest harvesting will then be investigated. An assessment of surface runoff in relationship with weather and land use variations will be completed from spatial analysis of hydrological and other environmental data.

In addition to data collected in the field, geo-referenced elevation, soil, and land cover data have been collected for spatial analyses of basin-wide forest BMPs effects on water quality. An initial hydrologic simulation for the Flat Creek Watershed was done using SWAT, the Soil and Water Assessment Tool (Arnold et al. 1998; Neitsch et al. 2002). A 1:250,000 scale DEM data (250K DEM) was used for watershed delineation and hydrologic simulation. Soil data, originally from STATSGO (State Soil Geographic Database), was used to determine parameters defining soil infiltration, evapotranspiration, and other processes in the hydrologic modeling. Land use and land cover data (LULC) was used to identify land cover status and management factors. Reach File Version 1.0 (RF1), developed by the U.S. EPA, was selected for automatic delineation as burnt-in river reaches, which allowed the GIS interface in BASINS 3.0 (U.S. EPA, 2001) to segment sub-watersheds accurately. Discharge data at the output location of the watershed was simulated for the years 1975 through 2005. This simulated data was used to plot seasonal variation in the discharge at the watershed outlet, which is located at the confluence of Flat Creek and Castor Creek.

Basic GIS layers such as watershed boundary, perennial streams, and ephemeral streams were acquired from the LDEQ. A landuse change assessment for the watershed was done using two satellite images from October 14, 1990 (Landsat TM 5) and August 30, 2000 (Landsat TM 7). The images were obtained from the Global Land Cover Facility at the University of Maryland. The downloaded data was preprocessed for geometric correction and atmospheric correction, and all seven bands were used to classify major land use types. The classification scheme was mainly developed to identify the forested and harvested areas. The initial classification was performed using a 12 class output, and these classes were further consolidated into 4 classes based on visual observation. The major classes identified are forest, pasture, harvested land, and waterbodies. The classified image was then post-processed to obtain uniform and distinct classification results with a minimum mapping unit of 28.50m by 28.50m.

23.3. RESULTS AND DISCUSSION

23.3.1. Land Use Change Assessment

A GIS framework integrating several spatial datasets, including land use and drainage network, has been established for the Flat Creek Watershed (Figure 2). This framework serves as a basis for future basin-wide assessment of forestry BMP effectiveness in water quality protection. An analysis of land cover changes in the watershed was conducted using LandSat images from 1990 and 2000. The results show a change in all three major land use classes - forest area, pasture land, and harvested area (Table 2). During that decade, about 30% (or 45.3 km^2) of the pasture land in this rural watershed was mostly converted into forest (Figures 2 and 3). Therefore, both an increase in forest area (13.9% or 19.4 km^2), and an increase in harvested area (35.8% or 25.7 km^2) are observed.

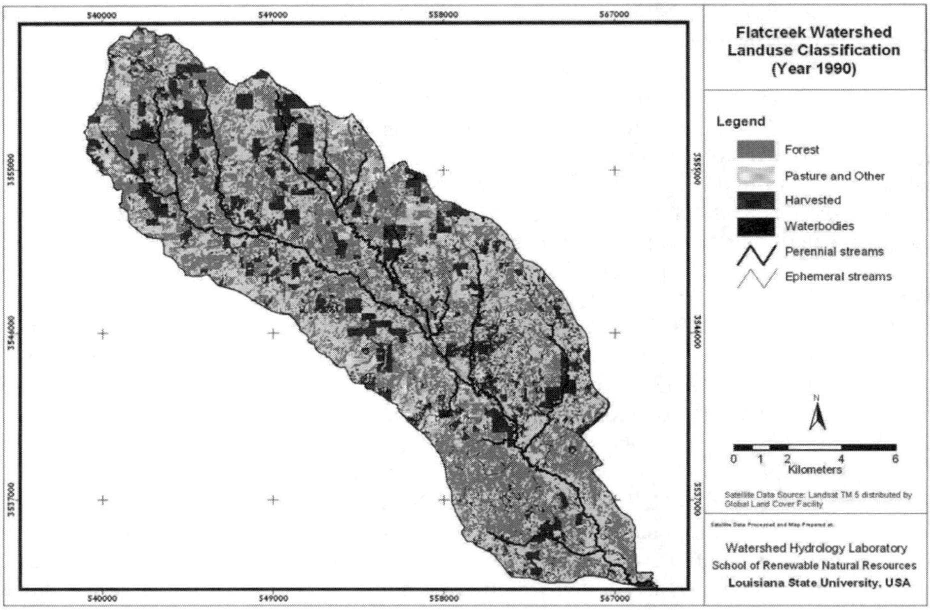

Fig. 2. Flat Creek Watershed land use classification in 1990.

Table 2. Land use changes in the Flat Creek Watershed.

Land Cover	Area in 1990 (km^2)	Area in 2000 (km^2)	Change in area from 1990-2000 (km^2)
Forest	139.6	159.0	19.4 (13.9%)
Pasture	153.6	108.3	-45.3 (-29.5%)
Harvested	71.8	97.5	25.7 (35.8%)
Water	0.1	0.3	0.2 (200%)
Total	365.1	365.1	

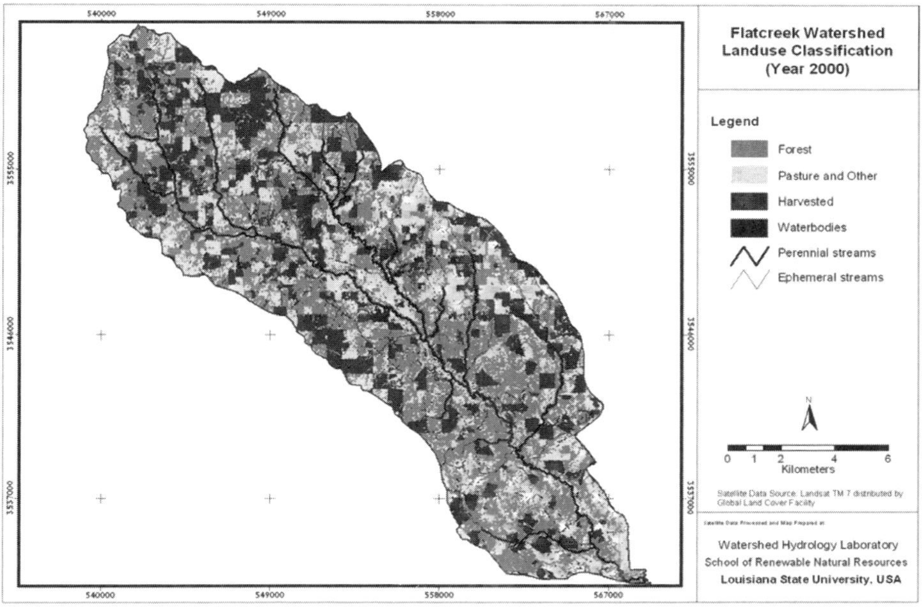

Fig. 3. Flat Creek Watershed land use classification in 2000.

It is not clear if regeneration of harvested land is the cause for the small increase of forest area. The observed increase in 2000 may have resulted because some areas were not harvested to the same extent as in 1990. The nearly 30% decrease in the pasture land, however, is a clear indication of land use change that may have affected, and will continue to affect, stream water characteristics in Flat Creek. Additional image analysis needs to be conducted to distinguish the agricultural, urban, and submerged areas from the pasture land area.

23.3.2. Hydrological Simulation

An initial hydrologic simulation for Flat Creek using the SWAT model was done for a 30-year period from 1975 to 2005. As an example, Figure 4 shows the seasonal trend of discharge in 1990 at the outlet of the watershed. The modeling results reveal that discharge from the Flat Creek Watershed is typically governed by precipitation, with the highest amounts from February through April, and the lowest amounts from July through September. Although monthly discharge has varied largely within each particular season over the last 30 years, the seasonality shown in Figure 4 represents the overall trend of the discharge pattern in this predominantly forested upland watershed.

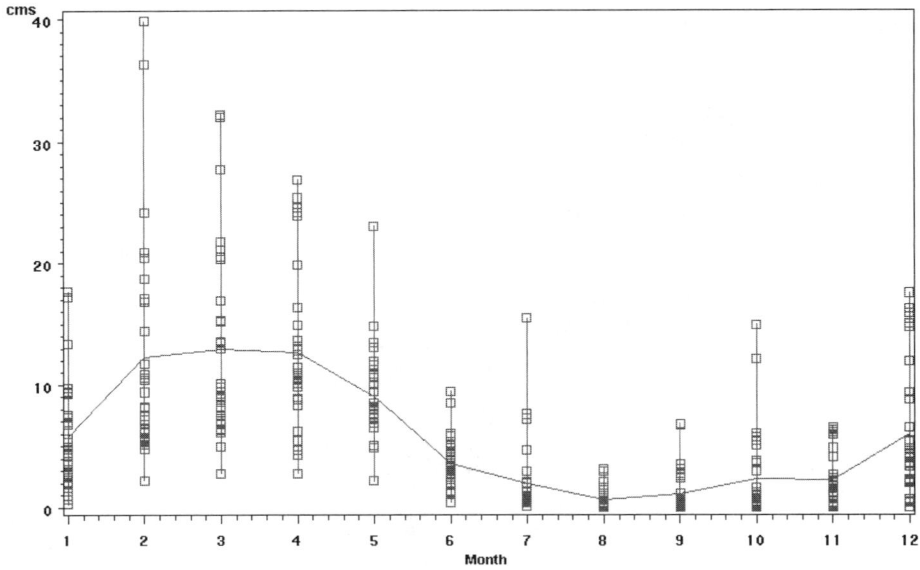

Fig. 4. Simulation of seasonal discharge variation in the Flat Creek Watershed.

23.3.3. Basin-Wide Variation of Dissolved Oxygen

An example of the differences in water quality values across the watershed is shown in Table 3. The average dissolved oxygen (DO) for each monitoring location ranges from 1.7 mg/L, at Turkey Creek Harvest I Downstream, to 7.9 mg/L at Spring Creek Control Upstream. Average temperature ranges from 6.6 ºC to 8.5 ºC at Fish Creek and Spring Creek Control Upstream, respectively. Similarly, there is a variation within conductivity values across the watershed. The average conductivity ranges from 70 μS to 180 μS at Spring Creek Control Upstream and both Turkey Creek Harvest I Downstream and Fish Creek, respectively. These preliminary results indicate a relatively large variation of stream water chemistry within this watershed, and even within the monitoring locations. Additional measurements are needed in order to validate this observation.

Previous studies addressing forestry BMP effectiveness on adjacent streams have concluded that water quality changes occurred. Therefore, we also expect a change in water quality to occur from pre- to post-harvest at the intensive sites. However, it is unknown whether water quality changes will be detected further downstream in the watershed, nor to what magnitude. Also uncertain is how the biological indicators will be affected. In a similar study, Carroll et al. (2004) found the number of macroinvertebrates decreased from pre- to post-harvest at both the control stream and the Streamside Management Zone (SMZ) treated stream.

This project will provide a comprehensive understanding of the effectiveness of forest BMPs on water quality at the watershed scale. Additionally, up-to-date water quality data for the Flat Creek Watershed and a geospatial model for predicting various land use effects on water quality will be produced. Importantly, these results will be available for use in future management planning for the Flat Creek Watershed.

TABLE 3: Temperature and dissolved oxygen levels across the Flat Creek Watershed. Sites I1 and I2 are located on Spring Creek. Sites I3 – I6 are located on Turkey Creek. E1 – E5 are located throughout the watershed.

Parameter	Date	Location										
		I1 Control US	I2 Control DS	I3 Harvest I US	I4 Harvest I DS	I5 Harvest II US	I6 Harvest II DS	E1 Upper Flat Creek	E2* Turkey Creek	E3 Fish Creek	E4 Lower Flat Creek	E5 Big Creek
DO (mg/L)	12/11/2005	8.7	4.6	1.8	0.3	2.4	0.8	2.5	NA	5.5	2.5	3.4
	1/11/2006	7.7	3.1	2.7	1.5	5.7	1.0	4.0	NA	1.8	3.5	4.3
	1/28/2006	7.3	3.6	4.1	3.3	3.1	3.5	4.7	7.5	6.1	7.8	4.7
	Average	7.9	3.8	2.9	1.7	3.7	1.8	3.8	7.5	4.5	4.6	4.1
Temperature (°C)	12/11/2005	6.2	5.8	7.6	5.8	5.7	5.3	4.9	NA	4.5	5.1	4.3
	1/11/2006	9.5	9.3	8.3	8.1	9.5	8.4	9.0	NA	7.6	9.0	8.0
	1/28/2006	9.7	8.9	9.0	9.1	8.8	9.4	8.3	8.8	7.8	9.4	9.8
	Average	8.5	8.0	8.3	7.7	8.0	7.7	7.4	8.8	6.6	7.9	7.3
Conductivity (µS)	12/11/2005	90	94	251	298	257	188	157	NA	386	232	164
	1/11/2006	57	64	133	134	138	108	130	NA	75	81	110
	1/28/2006	64	80	101	109	96	95	82	90	78	87	73
	Average	70	79	162	180	164	130	123	90	180	133	116

DS – Downstream; US – Upstream; * Site not included in summary statistics.

ACKNOWLEDEMENTS

The authors would like to recognize the Louisiana Department of Environmental Quality for providing financial support (Contract#: CFMS 595451) and Plum Creek Timber Company for assistance in the field and with logistical operations of this project.

REFERENCES

Adams, T. O., Hook, D. D., and Floyd, M. A., 1995. Effectiveness Monitoring of Silvicultural Best Management Practices in South Carolina. Southern Journal of Applied Forestry 19:170 - 176.

Arnold, J. G., Srinivasan, R., Muttiah, R. S., and Williams, J. R., 1998. Large Area Hydrologic Modeling and Assessment - Part 1: Model Development. Journal of the American Water Resources Association 34(1):73-89.

Aust, W. M., and Blinn, C. R., 2004. Forestry Best Management Practices for Timber Harvesting and Site Preparation in the Eastern United States: An Overview of Water Quality and Productivity Research During the Past 20 years (1982-2002). Water, Air, and Soil Pollution: Focus 4:5-36.

Boyd, W.H, Bowden, G.L., and Bordelon, M.J., 1998. Soil survey of Winn Parish, Louisiana. U.S. Dept. of Agriculture, Natural Resources Conservation Service, Washington, D.C.

Carroll, G. D., Shoenholtz, S. H., Young, B. W., and Dibble, E. D., 2004. Effectiveness of Forestry Streamside Management Zones in the Sand-Clay Hills of Mississippi: Early Indications. Water, Air, and Soil Pollution: Focus 4:275-296.

Grace, J. M., 2005. Forest Operations and Water Quality in the South. Transactions of the ASAE 48:871-880.

LDEQ, 2001. Flat Creek Watershed TMDL for Oxygen Demand: Surveyed July 18-20, 2000. Louisiana Department of Environmental Quality, http://www.deq.louisiana.gov/portal/portals/0/technology/tmdl/pdf/Flat-Creek-OD.pdf (July 10, 2005).

LFA/LDAF/LDEQ, 2000. Recommended Forestry Best Management Practices for Louisiana. Louisiana Forestry Association, Louisiana Department of Environmental Quality, Louisiana Department of Agriculture and Forestry, Baton Rouge, Louisiana, 81 pp.

Neitsch, S. L., Arnold, J. G., Kiniry, J. R., Srinivasan, R., and Williams, J. R., 2002. Soil and Water Assessment Tool User's Manual, Version 2000. GSWRL Report 02-02, BRC Report 02-06, Published by Texas Water Resources Institute, TR-192, College Station, Texas, 438 p.

Smith, E.P., 2002. BACI Design. Pages 141-148 *in* A.H. El-Shaarawi and W.W. Piegorsch (eds.) Encyclopedia of Environmetrics. John Wiley & Sons, Chichester, New York.146 p.

STATSGO (State Soil Geographic Database), Soil Spatial for Louisiana downloaded from National Cartography and Geospatial Data Center (http://www.ncgc.nrcs.usda.gov/products/datasets/statsgo/data/la.html) (10 March 2005).

Stewart-Oaten, A., Murdoch, W. W., and Parker, K. R., 1986 Environmental Impact Assessment: "Pseudoreplication" in Time? Ecology 67(4):929-940.

U.S. EPA. 2001. BASINS 3.0 Users' Manual. 337 p.

U.S. EPA. 2002. National Water Quality Inventory: 2000 Report. EPA-841-R-02-001, United States Environmental Protection Agency Office of Water, Washington DC.

Vowell, J. L., 2001. Using Stream Bioassessment to Monitor Best Management Practice Effectiveness. Forest Ecology and Management 143:237-244.

Section 5:

COASTAL WATER QUALITY MODELING

24

WATER QUALITY MODELING USING ARTIFICIAL NEURAL NETWORKS

Avinash Agarwal[1], Joydeep Dutta[2], S.K. Mishra[2], and M.K. Sharma[1]
[1] National Institute of Hydrology, Roorkee-247 667 (UA), India.
[2] Department of Water Resources Development and Management, Indian Institute of Technology Roorkee-247 667, UA, India

ABSTRACT

This paper attempts to develop Back Propagation Artificial Neural Network (BPANN) models for determination of sodium concentration in pre-monsoon and post-monsoon seasons using routinely monitored water quality parameters of ground water wells located in Jaipur district, Rajasthan (India). Identifying the input variables using the best subset procedure based on coefficient of determination (R^2) and F (Fisher's test) values, three-layer BPANN models were developed using the steepest descent optimization technique considering the fixed maximum 1000 iterations. Based on the statistical performance evaluation criteria of root mean square error (RMSE), correlation coefficient (CC), and coefficient of efficiency (CE), these models exhibited satisfactory performance in both calibration and validation.

24.1. INTRODUCTION

As a result of increasing industrialization, urbanization, civilization and other developmental activities, most of water bodies like ponds, lakes, streams, rivers as well as groundwater bodies are have become polluted. Once polluted, the purification of ground water is hopelessly difficult (Moly Kutty, 2000). Hence, the necessity of rapid monitoring of ground water quality is imperative. In developing countries, like India, it is however very difficult to analyze routinely the ground water for all water quality parameters where laboratory facilities and available funds are inadequate (Mariappan et al., 2001). It is, therefore, in order if indirect approaches are developed and used to assess, within desired precision, the quality of water using some easily measurable chemical constituents. The use of artificial neural network (ANN) models can be advantageous and, during the last decade, these have been popularly applied to a wide range of problems (Agarwal, 2002). In the water sector, ANNs have found applications in a range of problems dealing with surface water, groundwater, management of water resources systems, water quality, and so on (Agarwal, 2002).

ANN refers to computing systems whose central theme is borrowed from the analogy of biological neural networks. ANNs represent highly simplified mathematical models of the available understanding (Jain, 2004). They have the ability to (a) learn and generalize from examples to produce meaningful solutions to problems even when input data contain error or are incomplete, (b) change over time to compensate for changing circumstances, and (c) process information rapidly as a new model (Jain, 2004).

Many of the problems that engineers must deal with are exactly the types of problems for which ANN models appear to be most applicable. Engineers are interpreters of incomplete, noisy data, such as interpreting sensor information to determine the existence and location of damage in a structural system. Engineers are designers and controllers of complex systems for which there is no exact model of behaviour and whose expected performance is unknown, requiring that it to be estimated-such as a heating ventilation and air-conditioning (HVAC) system within a large multifunction building

Bowers and Shedrow (2000) attempted to predict water quality in small streams of Savannah River Site in USA using ANN. The selected input variables were local precipitation, stream flow rates, and turbidity for the initial prediction of suspended solids in the stream. A single hidden-layer feed forward neural network using back propagation learning algorithms was developed with a detailed analysis of model design of those factors affecting successful implementation of the model. All features of neural model were investigated including training set creation, number and layers of neurons, neural activation functions, and back propagation algorithms. Least-squares regression was used to compare model predictions with test data sets. Most of the model configurations offered excellent predictive capabilities. Using either the logistic or the hyperbolic tangent neural activation function did not significantly affect the predicted results. This was also true for the two learning algorithms, the Levenberg-Marquardt and Polak-Ribiere conjugate-gradient descent methods. The most important step in model development and training was the representative selection of data records for training of the model. Gibbs et al. (2004) attempted two data-driven techniques, namely linear regression models and multiplayer perceptron artificial neural networks, to predict chlorine concentrations at two key locations in the Hope Valley water distribution system located in the north of Adelaide, South Australia. A 5-year data set containing routinely measured parameters was used for model development and validation. When compared with the measured data, both the techniques predicted chlorine concentrations in the distribution system successfully. Overall, the performance of the multiplayer perceptron was slightly better than that of the regression model, suggesting the presence of some non-linearities in the underlying physical processes governing chlorine decay. In this study, an attempt is made to develop Back Propagation ANN (BPANN) models for prediction of sodium concentration in pre-monsoon and post-monsoon seasons using usually monitored water quality parameters of ground water wells located in Jaipur district, Rajasthan (India).

24.2. STUDY AREA AND DATA USED

The study area (Figs. 1 and 2) falls in the Jaipur district (in the north-eastern part) of Rajasthan State (India). It is located between 26^0 25' and 27^0 51' North latitude and 74^0 55' and 76^0 10' East longitude covering an area of 10878 sq. km. The minimum and maximum temperatures are 3^0C and 45^0C, respectively, while the mean temperature is 24^0C. In the distinct rainy season usually from June to September, the normal annual rainfall is 548.2 mm. The oldest groups of rock, in the district are schist, gneisses, migmatite and quartzite of Pre-Aravalli, which are considered to be nearly 2,500 million years old. The soils of the district are greyish brown to brown and yellowish brown, light to medium textured, and deep to very deep. These soils can be classified in Entisols order by 7th approximation classification; some soils belong to Aridisols order. During the pre- and post monsoon

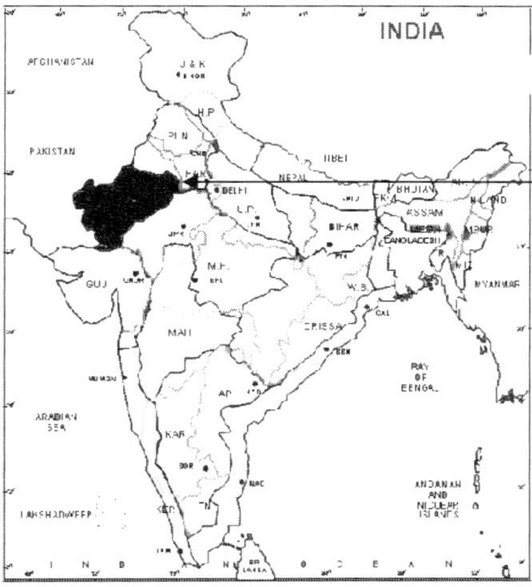

Fig. 1. Location of Rajasthan in India.

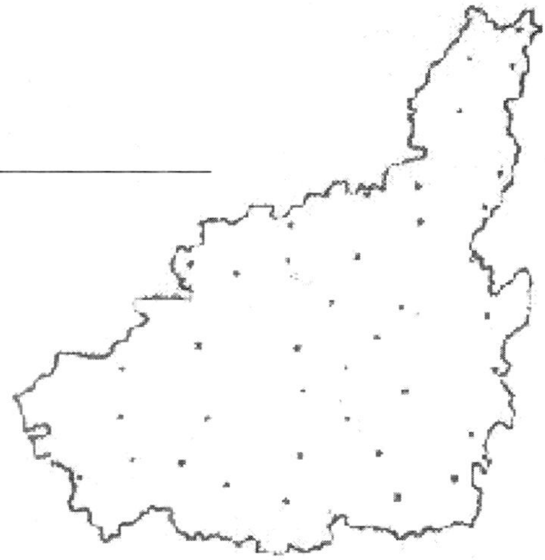

Fig. 2. Study area Jaipur district.

seasons of May 2002 and November 2003, thirty-eight samples were collected from the study area from different sources, viz, hand pumps, open wells, and tube-wells, which are extensively used for drinking and other domestic purposes. The physico-chemical analysis was performed following standard methods.

24.3. SELECTION OF INPUT VARIABLES FOR ANN MODELS

For selection of the most appropriate input variables for development of ANN-based models, the best subset procedure that helps decide the best sub-set (containing specific input variables) was employed using criteria of F (Fisher's Test), Coefficient of determination (R^2), and Residual sum of squares (SSE).

F-criterion

F can be computed as:

$$F = \left(\frac{N - K - 1}{K}\right)\left(\frac{R^2}{1\text{-}R^2}\right) \tag{1}$$

where R^2 = explained variation of y_i, $(1\text{-}R^2)$ = unexplained variation of y, N= number of data points, and K= number of independent variables. From Eq. 1, it is evident that F value, which is the ratio of the explained to the unexplained variation in y, will be large when the proportion of explained variation in y_i will be large, and vice versa. The regression will be significant when the F-value is large.

R^2-criterion

The square of the multiple correlation coefficients R^2 is defined as

$$R^2 = \frac{SSR}{SS_Y} = 1 - \left(\frac{SSE}{SS_Y} \right) \tag{2}$$

where SS_y = sum of squares about the mean $= \sum (y_i - \bar{y})^2$, SSE = sum of squares about regression = $= \sum (y_i - \hat{y}_i)^2$, and SSR = sum of squares due to regression $= \sum (\hat{y}_i - \bar{y})^2$, $\bar{y}$ is the average value of dependent variable, and $\hat{y}_i$ is the model computed values of the dependent variable. The stronger the linear association between y_i and $\hat{y}_i$, it will yield a large value of R^2 and vice-versa. Unfortunately, wherever comparing a subset model to a large model including the subset, R^2 provides an inadequate criterion for subset model selection because large model always have an R^2-value larger than that for the subset model. However, for a fixed number of independent variables (= k), R^2 can be used to compare different models with a large value of R^2 indicating the preferred model.

In the next step, taking Na as a dependent variable, different combinations were investigated for performance using R^2-, F- and SSE- values as criteria. Tables 2 and 3 shows selected sets/subsets candidate for possible model independent variables with sodium using Best Subset Procedure based on highest R^2 and F values and lowest SSE values for pre-monsoon and post-monsoon season respectively. Each set of model variables is represented by model numbers from ANN1 to ANN8 for pre-monsoon and ANN9 to ANN16 for post-monsoon respectively.

Table 1. R^2 of water quality parameters with Na.

Water quality parameters	Pre-monsoon	Post-monsoon
pH	0.044	0.029
EC	0.933	0.939
ALK	0.535	0.604
Hard	0.199	0.137
Cl	0.731	0.729
SO_4	0.724	0.703
NO_3	0.048	0.084
PO_4	0.035	0.001
F	0.413	0.349
K	0.068	0.057
Ca	0.129	0.122
Mg	0.243	0.127

Table 2. Selected Sets/ subsets candidate for possible model independent variables for pre-monsoon season with Sodium.

Model No.	Variables	Set of independent variables	N	R^2	F-Value	SSE
ANN8	8	EC, Hard, Cl, SO_4,ALK, F, Mg, Ca	25	0.998	1250.850	15585.703
ANN7	7	EC, Hard, Cl, SO_4, ALK, F, Mg	25	0.998	1503.632	15743.596
ANN6	6	EC, Hard, Cl, SO_4, ALK, F	25	0.998	1807.624	16176.648
ANN5	5	EC, Hard, Cl, SO_4, ALK	25	0.998	1935.822	19127.669
ANN4	4	EC, Hard, Cl, SO_4	25	0.998	2166.428	22481.231
ANN3	3	EC, Hard, Cl	25	0.995	1421.265	47850.323
ANN2	2	EC, Hard	25	0.990	1115.338	95349.739
ANN1	1	EC	25	0.933	319.913	654847.045

Table 3. Selected Sets/ subsets candidate for possible model independent variables for post-monsoon season with Sodium.

Model NO.	Variables	Set of independent variables	N	R^2	F-Value	SSE
ANN16	8	EC, Hard, Cl, SO_4, ALK,Mg,Ca,F	25	0.999	1804	7.145×10^3
ANN15	7	EC, Hard, Cl, SO_4, ALK,Mg,Ca	25	0.999	2146	7.295×10^3
ANN14	6	EC, Hard, Cl, SO_4, ALK ,Mg	25	0.999	2572	7.519×10^3
ANN13	5	EC, Hard, Cl, SO_4, ALK	25	0.999	3195	7.666×10^3
ANN12	4	EC, Hard, Cl, SO_4	25	0.999	3615	8.914×10^3
ANN11	3	EC, Hard, Cl	25	0.998	4082	1.105×10^4
ANN10	2	EC, Hard	25	0.993	1525	4.622×10^4
ANN9	1	EC	25	0.939	355	3.920×10^5

For model testing, out of thirty-eight water quality data values for both pre-monsoon and post-monsoon seasons, the first twenty-five were used for model formulation, and the remaining thirteen for model validation. To form various appropriate sub-sets, the Pearson correlation coefficient (R) between water quality parameters was computed. To enhance the visualization of the correlation matrix, the square of correlation coefficient (R^2) or coefficient of determination was calculated for showing the contribution of individual water quality parameter in explaining the variation in the dependent variable as shown in Table 1. Since pH, nitrate, phosphate, and potassium had no significant correlation with sodium for pre-monsoon and post-monsoon data sets, respectively, these parameters were excluded from model formulation. Using the R^2 explaining the proportion of variation in the dependent variable, different best subsets of independent variables were identified, and their performance assessed from resulting R^2- and F-values, and the number of observations used in model development. Notably, models with high R^2- and F- values were preferred.

24.4. ANN MODEL DEVELOPMENT

As above, ANNs are highly distributed interconnections of adaptive nonlinear processing-elements (PEs) (Fig. 3). When implemented in digital hardware, the processing-element is a simple sum of products followed by non-linearity. The connection strengths, also called the network weights, can be adapted such that the network's output matches the desired response. Errors are commonly converged using the steepest descent optimization technique.

In multi-layered perceptron, hidden layer means third layer of processing elements or units in between the input and output layers that increases computational power. In principle, the hidden layer can be more than one layer. In practice, the number of layers and neurons in these layers is evaluated by trial and error. Hornik et al. (1989) proved that a single hidden layer containing a sufficient number of neurons could be used to approximate any measurable functional relationship between input and output variable to any desired accuracy. In addition, De Villars and Barnard (1993) showed that an ANN comprising of two hidden layers tends to be less accurate than its single hidden layer counterpart.

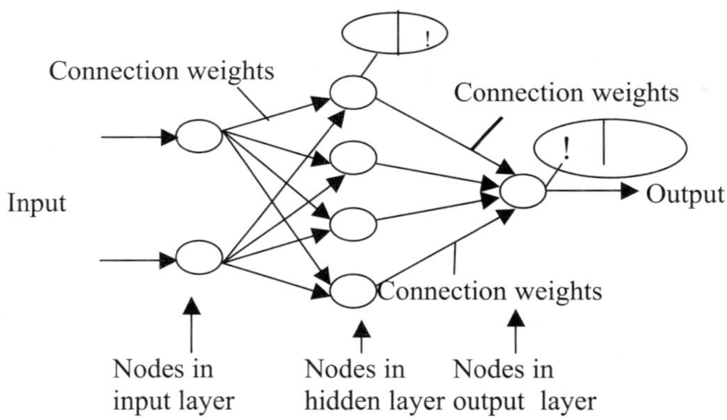

Fig. 3. The Building Blocks of ANN.

Accordingly, a three-layered BPANN structure with a non-linear sigmoid function between the layers, hidden nodes twice as input nodes, and numbers of input variables equal to the number of input nodes was adopted for all models. The number of nodes in the output layer was taken as one for single output reasons, values were normalized by their maximum value, and the number of iterations fixed at 1000. Here, with initial value of 0.5 for both, the learning rate (!) and momentum factor (!) varied with iteration. Their succeeding was estimated as Initial value minus (Initial value/No. of Iterations) to avoid oscillation problem. The model performance was evaluated using the following criteria: root mean square error (RMSE) and correlation coefficient (CC), and coefficient of efficiency (CE), expressed, respectively, as:

$$\text{RMSE} = \sqrt{\frac{\text{residual variance}}{\text{n}}} = \left[\sum_{j=1}^{n} \left(Y_j \,!\, \hat{Y}_j\right)^2 / n \right]^{1/2},$$

$$CC = \frac{\sum_{j=1}^{n} \left\{ \left(Y_j - \bar{Y}\right)\left(\hat{Y}_j - \bar{\hat{Y}}\right) \right\}}{\left\{ \sum_{j=1}^{n} \left(Y_j - \bar{Y}\right)^2 \sum_{j=1}^{n} \left(\hat{Y}_j - \bar{\hat{Y}}\right)^2 \right\}^{\frac{1}{2}}} \times 100,$$

and

$$CE = \left\{ 1 - \frac{residual\ variance}{Initial\ variance} \right\} x\ 100\% = \left\{ 1 - \frac{\sum_{j=1}^{n} (Y_j - \hat{Y}_j)^2}{\sum_{j=1}^{n} (Y_j - \bar{Y})^2} \right\} x\ 100\%$$

where Y and $\hat{Y}$ are the observed and estimated values of sodium, respectively, $\bar{Y}$ *and* $\bar{\hat{Y}}$ are mean of observed and estimated values, respectively, and n is the number of observations.

24.5. DISCUSSION OF RESULTS

Tables 4 and 5 show the resulting values of the above performance indicators for pre-monsoon and post-monsoon, respectively. For pre-monsoon season, from Table 4, the resulting RMSE is minimum for ANN4 in calibration, whereas it is more than that for ANN3 in verification. In calibration, the values of CE for both the models ANN3 and ANN4 are the same as 99%. However, in verification, CE is 97% for ANN3 and it is 95% for ANN4. Thus, both these criteria lead to prefer ANN3 to ANN4. A similar inference can also be drawn if CC is considered. Similarly, with CE values equal to 99% in both

Table 4. Comparative performance of selected models for pre-monsoon season for sodium.

Model No.	Nodes-input, hidden output	Performance evaluation of models					
		Calibration (Training)			Verification (testing)		
		RMSE	CC%	CE%	RMSE	CC%	CE%
ANN 1	1,2,1	363.02	97.07	66	133.64	91.86	83
ANN 2	2,4,1	73.89	99.37	99	72.62	98.55	95
ANN 3*	3,6,1	62.11	99.56	99	58.55	99.08	97
ANN 4	4,8,1	55.41	99.64	99	73.57	98.09	95
ANN 5	5,10,1	47.51	99.74	99	110.92	94.79	88
ANN 6	6,12,1	44.25	99.77	99	107.74	94.95	89
ANN 7	7,14,1	46.20	99.75	99	103.33	95.55	90
ANN 8	8,16,1	44.38	99.51	99	105.85	94.90	89

Table 5. Comparative performance of selected models for post-monsoon for sodium.

Model No.	Nodes-input, hidden output	Performance evaluation of models					
		Calibration (Training)			Verification (testing)		
		RMSE	CC%	CE%	RMSE	CC%	CE%
ANN 9	1,2,1	273.83	93.51	71	134.86	91.86	83
ANN 10	2,4,1	67.65	99.17	98	47.95	99.05	98
ANN 11*	3,6,1	46.01	99.62	99	33.05	99.51	99
ANN 12	4,8,1	45.05	99.63	99	38.04	99.37	99
ANN 13	5,10,1	42.77	99.67	99	47.38	99.03	98
ANN 14	6,12,1	43.89	99.65	99	50.41	98.86	98
ANN 15	7,14,1	41.74	99.68	99	47.00	99.28	98
ANN 16	8,16,1	40.76	99.70	99	58.26	99.09	97

calibration and validation for post-monsoon season, ANN11 (Table 5) can be considered to be the most appropriate model. Since CE values are quite high, the proposed models can be described to have performed satisfactorily. Figures 4-5 compare the computed (validated) sodium concentration values with the observed values for pre- and post-monsoon seasons, respectively, indicating a satisfactory agreement.

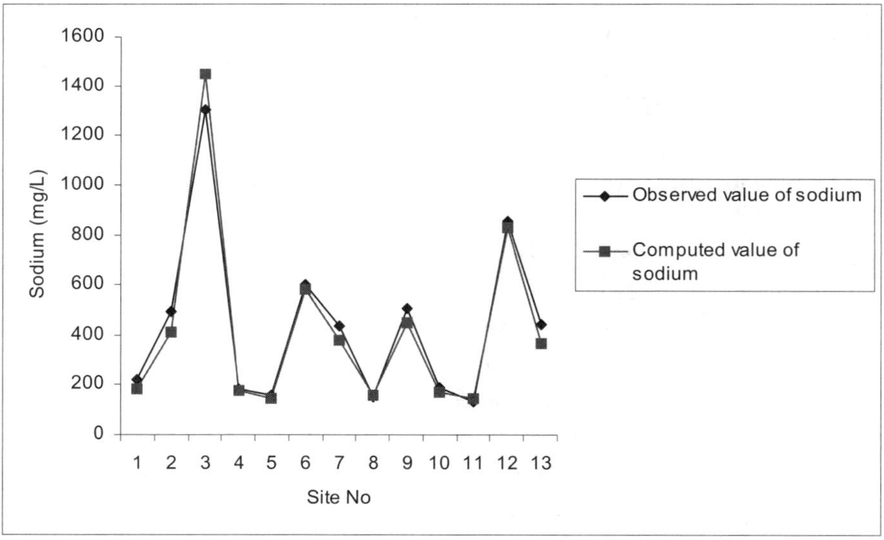

Fig. 4. Comparison of observed and computed values of sodium for Pre-monsoon.

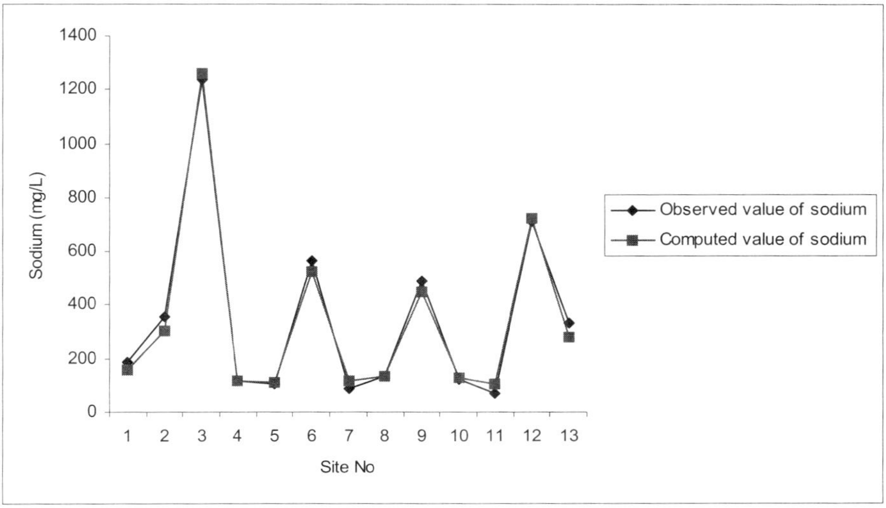

Fig 5. Comparison of observed and computed values of sodium for Post-monsoon.

24.6. CONCLUSIONS

The following conclusions can be drawn from the study:

1. ANN models could be useful in prediction of sodium concentration in ground water wells of Jaipur district, Rajasthan, in pre- and post- monsoon seasons, from the easily measurable constituents of water quality.
2. Based on RMSE, ANN4 performed the best in calibration for pre-monsoon season, and ANN3 in verification. Based on CE, both ANN3 and ANN4 perform equally well in calibration with CE = 99%. In verification, CE = 97% for ANN3, CE = 95% for ANN4. Thus, ANN3 is preferable to ANN4. A similar inference was drawn when CC was considered.
3. For the post-monsoon, ANN11 performed the best with CE = 99% in both calibration and validation, indicating a satisfactory performance.

ACKNOWLEDGEMENT

The authors gratefully owe to the National Institute of Hydrology, Roorkee (India), for providing laboratory facilities for water quality analysis.

REFERENCES

Agarwal, A., 2002. Artificial Neural Networks and their application for simulation and prediction of runoff and sediment yield, PhD. Thesis ,G.B. Pant University of Agriculture and Technology, Pantnagar-263145, Uttaranchal (India)

Bowers, J.A.,.and Shedrow, C.B., 2000. Predicting Stream Water Quality using Artificial Neural Networks,WSRC-MS-2000-00112,Westinghouse Savannah River Company, Aiken, South Carolina.

De Villars, J., and Barnard, E., 1993. Back Propagation Neural Nets with one and two Hidden layers, IEEE Trans, Neural News, 4(1), 136-141.

Draper, N.R., and Smith, H., 1981. Applied Regression Analysis(second edition).John Wiley and Sons, New York.

Gibbs, M.S, Morgan, N., Maier, H.R., Dandy, G.C., and Nixon, J.B., 2004. Use of Artificial Neural Networks for Modelling Chlorine Residuals in Water Distribution Systems, Center for Applied Modelling in Water Engineering, The University of Adelaide.

Hornik, K., Stinchcombe, M., and White, H., 1989. Multilayer Feed Forward Networks are Universal Approximators, Neural Networks, 2(5), 259-366.

Jain, S. K., 2004. Overview of Artificial Neural Networks, In: Training Workshop on Artificial Neural Networks and its applications in Water Resources-Organized by National Institute of Hydrology, Roorkee-247667 at Central Water Commission, Sewa Bhawan, R. K. Puram, New Delhi-110066,June 09, 2004.

Jain,C.K., Ali, Imran, and Sharma, M.K., 1998. Salinity Modelling of Ground water of a Coastal Region Using Best Subset Procedure. Indian Journal Environmental Protection 18(10): 762-768.

Mariappan, P., Yegnaraman, V., and Vasudevan, T., 2001. Estimation of Groundwater Quality Using Rainfall - A Case Study. Indian Journal Environmental Protection 21(11): 1012-1017.

Molykutty, M.V., 2000. Groundwater Quality and Pollutant Transport Modelling – A Discussion. Indian Journal Environmental Protection 20(3): 200-205.

Sarkar,A., Agarwal, A., Singh, R.D., and Jain, S.K., 2003. Artificial Neural Networks for daily rainfall runoff modeling in Sutlej basin, India. Proceedings of the International conference on Hydrological perspectives for sustainable development (HYPESD-2005), 23-25 February, 2005, Roorkee (India),196-205.

25

A COMPUTATIONAL FRAMEWORK FOR EVALUATION OF NPS MANAGEMENT SCENARIOS: ROLE OF PARAMETER UNCERTAINTY

Mazdak Arabi[1], Rao S. Govindaraju[2], and Mohamed M. Hantush[3]

[1] Department of Agricultural and Biological Engineering, Purdue University, 225 South University Street, West Lafayette, IN 47907, USA

[2] School of Civil Engineering, Purdue University, 550 Stadium Mall Drive, West Lafayette, IN 47907, USA

[3] National Risk Management Research Laboratory US Environmental Protection Agency, Cincinnati, OH 45268, USA

ABSTRACT

Utility of complex distributed-parameter watershed models for evaluation of the effectiveness of non-point source sediment and nutrient abatement scenarios such as Best Management Practices (BMPs) often follows the traditional {calibrate → validate → predict} procedure. Despite its simplicity, this approach is subject to non-uniqueness of the calibrated parameter set. In this study, a computational framework is developed, in which investigation of uncertainty provides complementary quantitative and qualitative information in support of BMP evaluation. The Generalized Likelihood Uncertainty Estimation (GLUE) method is employed to generate a cumulative likelihood for sediment and nutrient outputs of the Soil and Water Assessment Tool (SWAT) for two scenarios representing outputs with and without representation of BMPs. Quantile analysis of the cumulative likelihoods yields expected sediment and nutrient loads as well as their corresponding uncertainty bounds. While comparison of expected values determines the effectiveness of BMPs, uncertainty bounds could be used to obtain a Margin Of Safety (MOS) for such evaluations. The methodology was applied over the Dreisbach watershed within the Maumee River basin. The Maumee River is the longest river in the Great Lake system, where elevated levels of phosphorus have been a major concern. Results indicated that effectiveness of BMPs evaluated through the traditional method fell well between the estimated uncertainty bounds. It was concluded that parameter uncertainty accounted for nearly 15% of the variation in the estimated effectives of BMPs.

25.1. INTRODUCTION

The analysis of uncertainty associated with utility of simulation models is an important consideration in development of watershed management plans. Modeling uncertainty should be rigorously addressed in development and application of models, especially when stakeholders are affected by the decisions contingent upon model-supported analyses (NRC, 2001). Watershed models are commonly utilized to investigate rainfall-runoff generation, and fate and transport of contaminants resulting from nonpoint source activities. These are perceived to be the most important source of pollution in the United States (Ice, 2004). The evaluation of the success of best management practices (BMPs) in meeting their original goals has also been facilitated by these models (Santhi et al., 2001; Kirsch et al., 2002; Santhi et al., 2003; Arabi et al., 2005). Uncertainty associated with absolute estimates of design quantities tends to be very high because of data sparsity and model limitations (Benaman and Shoemaker, 2004). Thus, models are found to be more useful when making relative comparisons rather than making absolute predictions. It would be more meaningful to implement the uncertainty associated with effectiveness of BMPs in the planning process.

Although the literature is replete with sensitivity analysis and uncertainty analysis methods (Saltelli et al., 2000), implications of uncertainty associated with model predictions have not been widely endorsed in the decision making process mainly as a result of large uncertainty estimates. The magnitude of uncertainty itself is a key factor in its acceptance as the cost of implementation of management actions such as the Total Maximum Daily Load (TMDL) program may significantly increase with larger uncertainty estimates (Dilks and Freedman, 2004). In a case study in the Cannonsville Reservoir system watershed (1178 km^2) located in upstate New York, Benaman and Shoemaker (2004) concluded that even in the presence of observed data it was not possible to reduce the uncertainty of absolute sediment predictions in their study to practical values for the TMDL program. The argument, however, is that if the goal of a modeling study is to examine the impact of management scenarios on water quality of a study area, it may be neither practical nor necessary to incorporate large uncertainty of absolute predictions in the decision making process. It would be perhaps more feasible (and more desirable) to communicate and implement uncertainty of estimated effectiveness of management scenarios rather than uncertainty of absolute predictions. Moreover, the importance of such a formulation would be particularly appreciated when the uncertainty of estimated reduction of a variable of concern (sediment, nutrients, etc) due to implementation of an abatement action is less than estimated uncertainty of absolute predictions. In such cases, evaluation of impact of management scenarios would not be inhibited by uncertainty of model outputs.

In this paper, a Monte Carlo-based probabilistic approach is utilized (i) to examine the effect of modeling uncertainties on evaluation of long-term water quality impacts of BMPs using a distributed watershed model, SWAT; and (ii) to provide a comparison between magnitudes of uncertainties associated with absolute predictions versus effectiveness of BMPs. The analysis is demonstrated for a small watershed in Indiana where water quality data were collected and several structural BMPs were implemented.

25.2. METHODOLOGY

The computational framework utilized in this study is comprised of (1) a process-based watershed model (SWAT; Soil and Water Assessment Tool, Arnold et al., 1998) employed for simulating the fate and transport of sediment and nutrients in the study watersheds under two scenarios- before and after implementation of BMPs, (2) a BMP representation method (adapted from Arabi et al., 2005) for incorporation of water quality impacts of BMPs, (3) an uncertainty analysis methodology based on Generalized Likelihood Uncertainty Estimation (GLUE; Beven and Binely, 1992). The methodology was tested on a watershed located in northeast Indiana.

25.2.1. Watershed Model

Soil and Water Assessment Tool (SWAT) (Neitsch et al., 2001) is a process-based distributed-parameter simulation model, operating on a daily time step. The model was originally developed to quantify the impact of land management practices in large, complex watersheds with varying soils, land use, and management conditions over a long period of time. SWAT uses readily available inputs and has the capability of routing runoff and chemicals through streams and reservoirs, and allows for addition of flows and inclusion of measured data from point sources. Moreover, SWAT has the capability to evaluate the relative effects of different management scenarios on water quality, sediment, and agricultural chemical yield in large, ungaged basins. Major components of the model include weather, surface runoff, return flow, percolation, evapotranspiration (ET), transmission losses, pond and reservoir storage, crop growth and irrigation, groundwater flow, reach routing, nutrient and pesticide loads, and water transfer. Table 1 provides a listing of important SWAT input parameters corresponding to the abovementioned components.

Table 1. Listing of the parameters used in uncertainty analysis: LB and UB refer to lower and upper bounds of parameter vector, parameters identified by * were altered as a percentage of the default value.

Parameter	Description	Units	Range	
			LB	UB
CN2*	SCS runoff curve number	%	30	98
SOL-AWC	available soil water capacity	mm/mm	0.01	0.4
ESCO	soil evaporation compensation factor		0	1
OV-N	Manning's "n" value for overland flow		0.1	0.3
SLOPE*	average slope steepness	m/m	0	0.6
SLSUBBSN	average slope length	m	0.15	1.2
GWQMN	minimum threshold depth of water in the shallow aquifer for return flow to occur	mm	0	5000
REVAPMN	minimum threshold depth of water in the shallow aquifer for "revap" to occur	mm	0	500
GW-REVAP	groundwater "revap" coefficient		0.02	2
GW-DELAY	groundwater delay	days	0	500
ALPHA-BF	baseflow alpha factor for recession constant	days	0	1
SURLAG	surface runoff lag time		1	12
SFTMP	snowfall temperature	oC	-5	5
CH-S1*	average slope for tributary channels	m/m	0	10
CH-N1	Manning's "n" value for tributary channels		0.008	0.065
CH-K1	effective hydraulic conductivity in tributary channels	mm/hr	0	150
CH-S2*	average slope for the main channels	m/m	0	10
CH-N2	Manning's "n" value for the main channel		0.01	0.3
CH-K2	effective hydraulic conductivity in the main channel	mm/hr	0	150
CH-EROD	channel erodibility factor	cm/hr/Pa	0	0.6
CH-COV	channel cover factor		0	0.6

Parameter	Description	Units	Range	
			LB	UB
CN2*	SCS runoff curve number	%	30	98
SOL-AWC	available soil water capacity	mm/mm	0.01	0.4
ESCO	soil evaporation compensation factor		0	1
OV-N	Manning's "n" value for overland flow		0.1	0.3
SLOPE*	average slope steepness	m/m	0	0.6
SLSUBBSN	average slope length	m	0.15	1.2
GWQMN	minimum threshold depth of water in the shallow aquifer for return flow to occur	mm	0	5000
REVAPMN	minimum threshold depth of water in the shallow aquifer for "revap" to occur	mm	0	500
GW-REVAP	groundwater "revap" coefficient		0.02	2
GW-DELAY	groundwater delay	days	0	500
ALPHA-BF	baseflow alpha factor for recession constant	days	0	1
SURLAG	surface runoff lag time		1	12
SFTMP	snowfall temperature	oC	-5	5
CH-S1*	average slope for tributary channels	m/m	0	10
CH-N1	Manning's "n" value for tributary channels		0.008	0.065
CH-K1	effective hydraulic conductivity in tributary channels	mm/hr	0	150
CH-S2*	average slope for the main channels	m/m	0	10
CH-N2	Manning's "n" value for the main channel		0.01	0.3
CH-K2	effective hydraulic conductivity in the main channel	mm/hr	0	150
CH-EROD	channel erodibility factor	cm/hr/Pa	0	0.6
CH-COV	channel cover factor		0	0.6
PRF	peak rate adjustment factor for in-stream channel routing		0	2
SPCON	linear coefficient for in-stream channel routing		0	0.001
SPEXP	exponent coefficient for in-stream channel routing		1	1.5
BIOMIX	biological mixing efficiency		0.01	1
BIOMIN	minimum plant biomass for grazing	Kg/ha	0	1
USLE-P	USLE equation support practice factor		0.1	1
USLE-C*	minimum value of USLE equation cover factor	%	0.001	0.5
USLE-K	USLE equation soil erodibility factor		0.01	0.65
ORGP-AG	initial organic P in soils for agriculture land use	mg/kg	1	1000
ORGP-PAST	initial organic P in soils for pasture land use	mg/kg	1	500
LABP-AG	initial soluble P in soils for agriculture land use	mg/kg	1	100
LABP-PAST	initial soluble P in soils for pasture land use	mg/kg	1	50
ORGN-AG	initial organic N in soils for agriculture land use	mg/kg	1	10000
ORGN-PAST	initial organic N in soils for pasture land use	mg/kg	1	5000
SOLN-AG	initial NO_3 in soils for agriculture land use	mg/kg	0.1	5
SOLN-PAST	initial NO_3 in soils for pasture land use	mg/kg	0.1	3

25.2.2. Generalized Likelihood Uncertainty Estimation (GLUE) Method

The GLUE methodology is based on recognition of the importance of the set of parameters to produce the behavior of the system, not individual parameters. The acceptable model realizations (i.e. behavior set) obtained from Monte Carlo simulations are given likelihood weights according to observed data and a likelihood function. Several choices for an appropriate likelihood function can be obtained from Beven and Freer (2001). A likelihood measure based on Nash-Sutcliff efficiency criterion with shaping factor N is defined as (Freer and Beven, 1996):

$$L(\alpha \mid y) = (1 - \frac{\sigma_\varepsilon^2}{\sigma_o^2})^N; \quad \sigma_\varepsilon^2 < \sigma_o^2 \tag{1}$$

where $L(\alpha \mid y)$ is the likelihood of parameter set (α), given the observed data (y). The quantities σ_ε^2 and σ_o^2 refer to the error variance between model simulations and observed data, and the variance of the observed data, respectively. For N=1, L in Eq 1 is the well-known Nash-Sutcliff efficiency coefficient that is often used for calibration of hydrologic and water quality models. A negative L value indicates that the corresponding model output is dissimilar to the behavior of the system under study, and the likelihood of such a simulation in mimicking the system behavior is zero. Therefore, the likelihood measure is rewritten as:

$$L(\alpha \mid y) = \max[(1 - \frac{\sigma_\varepsilon^2}{\sigma_o^2}), 0] \tag{2}$$

For each simulation from a random parameter set, a likelihood weight is obtained from Eq 2. Then, these weights are rescaled by dividing each of them by their total sum. The rescaled likelihood weights are used to construct a cumulative distribution for the output of interest, which can be used for estimation of uncertainty bounds associated with the output by computing its quantiles. The GLUE method is subjective not only to the choice of likelihood function, but also to the cutoff criterion used for classification of Monte-Carlo simulations to behavior and non-behavior sets.

25.2.3. Case Study Watershed

The Black Creek watershed (Fig. 1) located in northeast Indiana is a typical watershed in the upper Maumee River basin. In mid 1970's and early 1980's several BMPs were implemented in the watershed and detailed water quality monitoring was carried out at various locations within the watershed to examine short-term water quality impacts of soil and water conservation techniques (Lake and Morrison, 1977; Lake and Morrison, 1978;

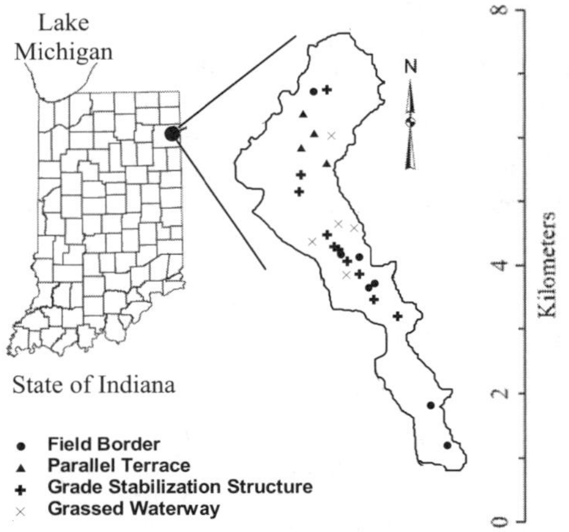

Fig. 1. Location map of study watersheds and implemented BMPs.

Morrison and Lake, 1983). Data collected from automated samplers at the outlet of Dreisbach (6.23 km^2) were the most complete and were used in this study. The dominant hydrological soil group in the study watersheds is type C, representative of medium to poorly drained soils. Land use in Dreisbach is mostly pasture in the upper portion, while row crops are wide spread in the remainder of the watershed.

25.2.4. Representation of BMPs

For this study, a method presented by Arabi et al. (2005) was utilized to evaluate the water quality impacts of grassed waterways, grade stabilization structures, field borders and parallel terraces. The method was developed based on published literature pertaining to BMP simulation in hydrological models and considering the hydrologic and water quality processes simulated in SWAT. Based on the function of the BMPs and hydrologic and water quality processes that are modified by their implementation, corresponding SWAT parameters were selected and altered. Table 2 summarizes SWAT parameters and their corresponding values for representation of the BMPs.

Table 2. Representation of field borders, parallel terraces, grassed waterways, and grade stabilization structures in SWAT.

BMP	Function	Representing SWAT Parameter		
		Parameter (input file)	Range	Value when BMP implemented
Field Border	increase sediment trapping	FILTERW (.hru)	0-5 (m)	5 (m)
Parallel Terrace	reduce overland flow	CN(2) (.mgt)	0-100	*
	reduce sheet erosion	USLE_P (.mgt)	0-1	0.2 (terraced)
	reduce slope length	SLSUBBSN (.hru)	10-150	**
Grassed Waterway	increase channel cover	CH_COV (.rch)	0-1	0.0 (fully protected)
	reduce channel erodibility	CH_EROD (.rch)	0-1	0.0 (non-erosive)
	increase channel roughness	CH_N(2) (.rch)	0-0.3	0.24
Grade Stabilization Structure	reduce gully erosion	CH_EROD (.rch)	0-1	0.0 (non-erosive)
	reduce slope steepness	CH_S(2) (.rch)	-	***

*estimated based on land use and hydrologic soil group of the HRU where it is installed for terraced condition.

** estimated for each parallel terrace based on its features and SWAT assigned overland slope of the HRU where it is installed:

$$SLSUBBSN = (A \times S + B) \times 100/S$$

where S is average slope of the HRU; A=0.21 and B=0.9 (ASAE, 2003).

*** estimated for each grade stabilization structure based on its features and SWAT assigned slope, $CH\_S(2)_{old}$, and length of the channel segment where it is installed:

$$CH\_S(2)_{new} = CH\_S(2)_{old} - D/CH\_L(2)$$

where D is height of the structure (1.2 m) and $CH\_L(2)$ is length of the channel segment.

25.2.5. Analysis of Uncertainty

In the present uncertainty analysis, 5000 model simulations were performed for the 1974-1978 period when water quality data were collected. All model parameters in Table 1 were included in the analysis. For each realization of the parameter space, the uncertainty analysis entailed the following steps:

1. Select parameter values randomly assuming that they vary uniformly within their specified ranges;
2. Perform model simulation with the selected values to compute model outputs for the scenario without BMPs;
3. Compute likelihood of the simulation for monthly output variables by applying Eq 2;
4. Represent BMPs by changing appropriate model parameters from Table 2;
5. Compute model outputs for the scenario with BMPs in place;
6. Using GLUE-likelihood weights computed at step (3), generate a cumulative density function for outputs of interest for both model simulations with and without BMPs. The same likelihood computed in step 3 applies.

It is worthwhile to mention that if data have been collected after implementation of BMPs, the likelihood of parameter sets in generating behavior of the system should be computed based on model simulations with BMPs in place. Therefore, step 3 should be performed after step 5, based on model outputs from the scenario with BMPs.

25.3. RESULTS AND DISCUSSION

The computational procedure described previously was performed to quantify the uncertainty associated with both model simulations and estimated effectiveness of BMPs. Figure 2 depicts the cumulative likelihood distribution of average monthly values for various output constituents at the outlet of the Dreisbach watershed. A likelihood weight from Eq 2 was assigned to each of 5000 model realizations resulting from Monte Carlo simulations. Each panel in the figures contains two cumulative probability distributions for the two scenarios. The solid lines correspond to results from the GLUE method for the scenario without BMPs (scenario A), while the dashed lines demonstrate the results for the scenario with BMPs represented in the model (scenario B). It is evident that the difference between GLUE-likelihoods for scenarios A and B was marginal for streamflow, but substantial for sediment, total P, and total N computations. Following Figure 2, the results are summarized in in Table 3 for further discussion. Lower and upper bounds in Table 3 refer to 5th and 95th percentiles of the cumulative likelihoods, respectively, while the median represents the 50th percentile. The top panel on the right in Figure 2 demonstrates how various percentiles were determined. For scenario A, comparison of median values to the average of monthly observed data, also presented in Table 3, indicate that averages of monthly observed values for streamflow were within ±15% of the median in the study watershed, and fell well within the uncertainty ranges from GLUE. Sediment yield measurements at the outlet of the Dreisbach watershed were within ±20% of the median values, also well covered by the uncertainty bounds. The median values for total P appeared to adequately match the average of monthly observations as they underpredicted by only 27%. Moreover, the uncertainty bounds covered the data. The median for total N differed from the average of monthly observed values by only 7%, indicating satisfactory results from GLUE.

These values may now be compared to corresponding quantities for scenario B listed in Table 3 Comparison of the expected values for scenarios A and B revealed the estimated effectiveness of BMPs, shown as a percent reduction. The percent reduction of each constituent as a result of implementation of BMPs was determined as:

$$r_i = \frac{\hat{y}_{i,a} - \hat{y}_{i,b}}{\hat{y}_{i,a}} \times 100 \tag{3}$$

where i is the constituent of interest, i.e., streamflow, sediment yield, total P load, or total N load, r_i is percent reduction of constituent i, $\hat{y}_{i,a}$ is the median of constituent i for scenario A, and $\hat{y}_{i,b}$ is the median of constituent i for scenario B. The results indicated that implementation of BMPs as shown in Figure 1 would not reduce streamflows significantly. This did not come as a surprise because the BMPs implemented in the watersheds were mostly sediment control BMPs. Implementation of the 26 BMPs in the Dreisbach watershed would lower median sediment yield by nearly 57%, from 0.035 t/ha/month to 0.015 t/ha/month. The estimated reduction rates for total P and total N at the outlet were 33% and 31%, respectively.

The uncertainty bounds associated with absolute predictions of the SWAT model were much larger than the ones corresponding to the estimated effectiveness of BMPs. For example, the uncertainty bound for monthly sediment yield at the outlet of Dreisbach for the simulation period under scenario A (scenario with no BMP represented) was determined to be [0.015,0.083] with a median value of 0.035 t/ha/month (Table 3). Incorporation of this large uncertainty in decision making and management may be extremely costly and not feasible. However, as it is evident in Table 3, the estimated 5th and 95th uncertainty bounds for estimated effectiveness of BMPs in reducing sediment yield at its outlet were [53%,60%] with a median (50th percentile) of 57% Similar results were observed for streamflow, total P, and total N computations for the study watershed.

Table 3. Summary of the results from analysis of uncertainty: $\hat{y}$ refers to 50th percentile (median) of output variables for 5000 Monte Carlo simulations of the SWAT model, $\hat{r}_i$ is the 50th percentile (median) of estimated reduction of output variables due to implementation of BMPs computed by Eq 3 for 5000 Monte Carlo simulations of the SWAT model, and "range" refers to the 5th and 95th percentile of the Monte Carlo simulations.

Variable	Units	Average Monthly Observed Value	Scenario A		Scenario B		Reduction (%)	
			$\hat{y}_a$	range	$\hat{y}_b$	range	$\hat{r}_i$	range
Streamflow	m^3/s	0.039	0.041	[0.026,0.062]	0.037	[0.023,0.056]	9.8	[9.70,11.5]
Sediment	t/ha/month	0.031	0.035	[0.015,0.083]	0.015	[0.007,0.033]	57.1	[53.3,60.2]
Total P	kg/ha/month	0.095	0.069	[0.030,0.125]	0.046	[0.015,0.086]	33.3	[31.2.4,50]
Total N	kg/ha/month	1.228	1.145	[0.487,2.385]	0.788	[0.360,1.502]	31.2	[26.1,37.0]

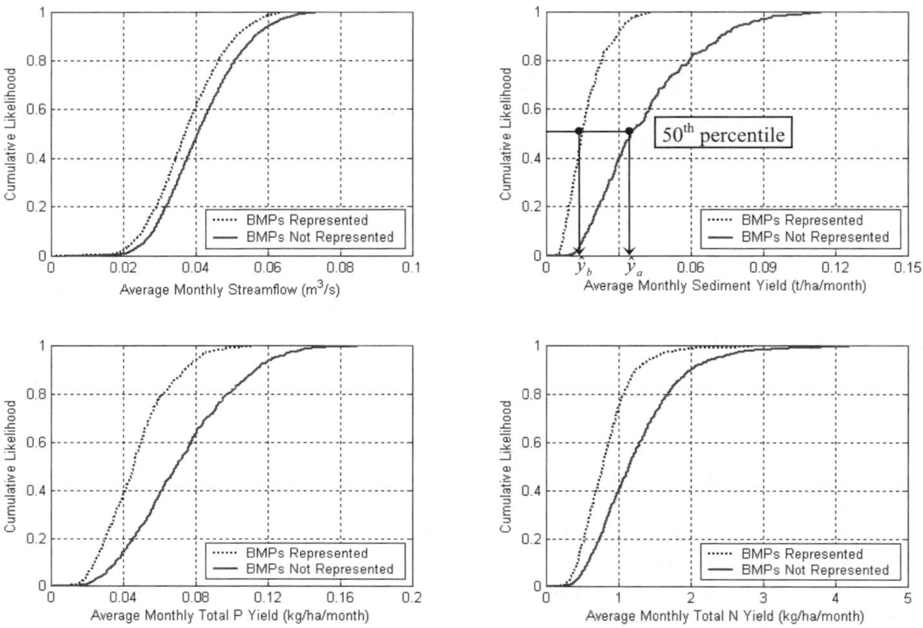

Fig. 2. Cumulative GLUE-likelihood for variables simulated at the outlet of the Dreisbach watershed, Indiana. Each panel represents 5000 model realizations. The top panel in the right demonstrates how various percentiles were determined. $\hat{y}_a$ is the expected value for scenario A (BMPs not represented), and $\hat{y}_b$ is the expected value for scenario B (BMPs represented).

25.4. DISCUSSION

The conceptual simplicity is an attractive feature of the computational analysis developed in this study. The utilized likelihood measure, Nash-Sutcliff coefficient, is widely used by modelers for calibration and validation of watershed models. Also, GLUE methodology will, more than likely, produce various sets of model parameters that adequately satisfy calibration criteria that are usually set based on Nash-Sutcliff coefficient. This helps modelers avoid the cumbersome practice of manual calibration. That the uncertainty bounds of various constituents encompassed the corresponding observed values strongly supports this suggestion. It appears that for the Dreisbach watershed, sediment and nutrient outputs of the SWAT model bear more uncertainty than streamflow simulations.

Consolidation of results in Table 3 indicates that reducing the uncertainty associated with absolute sediment and nutrient outputs of the SWAT model to practical ranges may not be feasible. Some of the parameters that are used to determine the transport capacity of channel segments (SPCON, SPEXP, and PRF in Table 1) cannot be measured in the field and are usually calibrated from a broad "suggested" range. With availability of more data, more narrow uncertainty bounds for absolute predictions may be determined. Nevertheless, it may still not be adequate for reducing the uncertainty associated with these absolute model predictions to small enough ranges to be useful for practical management decisions. However, comparison of sediment and nutrient outputs from model simulations with and without representation of BMPs would not suffer from such limitations as demonstrated here. It is evident that incorporation of modeling uncertainty in informed decision making is more practical through communicating the uncertainty associated with evaluation of effectiveness of management actions.

The foregoing results indicate that the computational framework developed in this study could be endorsed by watershed management programs such as the TMDL as advocated in Zhang and Yu (2004). Once the level of protection to be provided by the Margin of Safety (MOS) is specified by decision makers, the probabilistic framework presented in this study could be applied to verify the success of a particular management action in achieving its designated goals. Implementation of the estimated MOS is more feasible in this context, as opposed to the MOS computed from uncertainty analysis of absolute model predictions where the additional cost of implementation may render its consideration impractical. Finally, the methodology can be easily incorporated in watershed models such as SWAT that are commonly used for TMDL development.

CONCLUSIONS

A computational framework was developed, in which investigation of uncertainty provides complementary quantitative and qualitative information to support management and decision making. The procedure was applied to a small Indiana watershed. It was demonstrated that the uncertainty associated with estimated effectiveness of BMPs was substantially smaller than the uncertainty associated with absolute predictions. Future work will focus on investigation of the utility of the developed methodology in the TMDL process. Especially, the means for coping with the Margin of Safety (MOS) in the TMDL formula should be explored.

REFERENCES

ASAE, 2003. Design, Layout, Construction and Maintenance of Terrace Systems. ASAE Standards S268.4 FEB03, St. Joseph, MI.

Arabi, M., Govindaraju, R.S., and Hantush, M.M., 2005. Role of watershed subdivision on evaluation of long-term impact of best management practices on water quality. Journal of the American Water Resources Association, in press, 40 pp.

Arabi, M., 2005. A modeling framework for evaluation of watershed management practices for sediment and nutrient control. Ph.D. Dissertation, Purdue University, West Lafayette, IN 47907.

Arnold, J.G., Srinivasan, R., Muttiah, R.S., and Williams, J.R., 1998. Large area hydrologic modeling and assessment part I: model development. Journal of the American Water Resources Association 34(1): 73-89.

Benaman, J., and Shoemaker, C.A., 2004. Methodology for analyzing range of uncertain model parameters and their impact on total maximum daily load process. Journal of Environmental Engineering - ASCE 130(6): 648-656.

Beven, K.J., and Binely, A.M., 1992. The future of distributed models: model calibration and uncertainty prediction. Hydrological Processes 6: 279-298.

Beven, K.J., and Freer, J., 2001. Equifinality, data assimilation, and uncertainty estimation in mechanistic modeling of complex environmental systems using the GLUE methodology. Journal of Hydrology 249(1-4): 11-29.

Dilks, D.W. and Freedman, P.L., 2004. Improved consideration of the Margin of Safety in Total Maximum Daily Load development. Journal of Environmental Engineering – ASCE 130(6):690-694.

Freer, J. and Beven, K., 1996. Bayesian estimation of uncertainty in runoff prediction and the value of data: An application of the GLUE approach. Water Resources Research 32(7): 2161-2173.

Ice, G. 2004. History of innovative best management practice development and its role in addressing water quality limited waterbodies. Journal of Environmental Engineering-ASCE 130(6): 684-689.

Kirsch, K., Kirsch, A., and Arnold, J.G., 2002. Predicting sediment and phosphorous loads in the Rock River basin using SWAT. The Transactions of ASAE 45(6): 1757-1769.

Lake, J., and Morrison, J., 1977. "Environmental impacts of land use on water quality, Final report on the Black Creek project-project data." EPA-905/9-77-007-C, U.S. Environmental Protection Agency, Chicago, IL.

Lake, J., and Morrison, J., 1978. Environmental impact of land use on water quality- Final report on the Black Creek project- Technical report. EPA-905/9-77-007-B, U.S. Environmental Protection Agency, Chicago, IL.

Morrison, J., and Lake, J., 1983. Environmental impacts of land use on water quality, Black Creek Project- Final report. Allen County Soil and Water Conservation District, Fort Wayne, IN.

NRC, 2001. Assessing the TMDL approach to water quality management. Committee to Assess the Scientific Basis of the Total Maximum Daily Load Approach to Water Pollution Reduction, Water Science and Technology Board, Division on Earth and Life Studies, National Research Council, Washington, D.C.

Neitsch, S.L., Arnold, J.G., Kiniry, J.R., Williams, J.R., and King ,K.W., 2001. Soil and Water Assessment Tool theoretical documentation, version 2000. Grassland, Soil and Water Research Laboratory, Agricultural Research Service, Temple. TX.

Saltelli, A., Chan, K., and Scott, E.M., 2000. Sensitivity Analysis, Wiley, Chichester, England.

Santhi, C., Arnold, J.G., Williams, J.R., Dugas, W.A., Srinivasan, R., and Hauck, L.M., 2001. Validation of the SWAT model on a large river basin with point and nonpoint sources. Journal of the American Water Resources Association, 37(5): 1169-1188.

Santhi, C., Srinivasan, R., Arnold, J.G., and Williams, J.R., 2003. A modeling approach to evaluate the impacts of water quality management plans implemented in the Big Cypress Creek Watershed. Second conference on Watershed Management to Meet Emerging TMDL Environmental Regulations, Albuquerque, NM, pp. 384-394.

Zhang, H.X., and Yu, S.L., 2004. Applying the first-order error analysis in determining the margin of safety for total maximum daily load computations. Journal of Environmental Engineering 130(6): 664-673.

26

COMPUTING STREAM RE-AERATION COEFFICIENT FOR KALI RIVER USING ANN AND EMPIRICAL EQUATIONS

Sharad Kumar Jain and Ramakar Jha
National Institute of Hydrology
Roorkee - 247667, India

ABSTRACT

Dissolved oxygen (DO) is one of the most useful indices of river's health and the stream reaeration coefficient is an important input to computations related to DO. Normally, this coefficient is expressed as a function of several variables, such as mean stream velocity, shear stress velocity, bed slope, flow depth, and Froude number. In recent years, different functional forms have been advanced to represent the reaeration coefficient for different data sets. In the present study, the artificial neural network (ANN) technique has been applied to estimate the reaeration coefficient (K_2) using data sets measured in different reaches of Kali River in India. Observed values of stream velocity, bed slope, flow depth, cross-section area, and reaeration coefficient were used for the analysis. Different combinations of variables were tested to compute the reaeration coefficient. The performance of ANN was compared with other estimation methods. It was found that K_2 estimations using ANN was much closer to the observed values as compared with the empirical equations. Among the ANN structures, RBF network was found to give somewhat better results compared to multi-layer perceptron (MLP).

26.1. INTRODUCTION

Reaeration is the physical transfer of oxygen from the atmosphere to water at the air-water interface. A number of Biochemical Oxygen Demand and Dissolved Oxygen (BOD-DO) models have been proposed (Thomman and Muller, 1987; Jha et al., 2004) in which reaeration coefficient is an essential input parameter. The rate of transfer of oxygen from atmosphere to the liquid body is proportional to the difference between the existing concentration C and the equilibrium concentration C^* of oxygen in the liquid. Mathematically:

$$\frac{dC}{dt} = K_L\left(\frac{A}{V_1}\right)(C^* - C) \tag{1}$$

The parameter $K_L(A/V_1)$, generally denoted as $(K_2)_T$, is termed as the reaeration coefficient at temperature T (°C). The value of $(K_2)_T$ is related to the value at 20 °C, $(K_2)_{20}$, as:

$$(K_2)_{20} = (K_2)_T / \theta^{(T-20)} \tag{2}$$

where, $\theta = 1.024$ for pure water. In this paper, the term K_2 has been used to denote $(K_2)_{20}$.

The measurement of reaeration coefficient requires considerable effort. Therefore, reaeration coefficients are often predicted from channel characteristics and flow data. Besides the theoretical approaches, many empirical equations have been developed to relate K_2 with mean flow velocity (V), shear stress velocity (V_*), depth of flow (H), Froude number (F_r), and channel bed slope (S). Moog and Jirka (1998) proposed two empirical equations based on slope criteria. Thackston and Dawson (2001) claimed that the equation provides better results.

During the last decade, ANNs have emerged as a powerful tool for pattern recognition and modeling output of a system using the input data. This paper compares the performance of ANN with some commonly used equations to estimate the reaeration coefficient using channel and flow characteristics.

Before proceeding further, we briefly discuss the concepts behind ANN and review some relevant applications in water resources.

26.2. ARTIFICIAL NEURAL NETWORKS

An ANN is a simplified mathematical representation of the human biological neural network. The application of ANNs to water resources problems has become very popular due to their immense power and potential in modeling non-linear systems. Among the many ANN structures that have been studied, the most widely used structure in hydrology is the multi-layer, feed-forward network.

An ANN consists of a number of data processing elements called neurons or nodes; in a multi-layer ANN, these are grouped in layers. The neurons in the input layer receive the input vector and transmit the values to the next layer of processing elements across connections. This process is continued until the output layer is reached. This type of network in which data flows in only forward direction is known as a feed forward network.

A three-layer feed forward ANN, shown in Figure 1, has an input layer, an output layer and a hidden middle layer. In this multi-layer perceptron (MLP) each neuron in a layer is

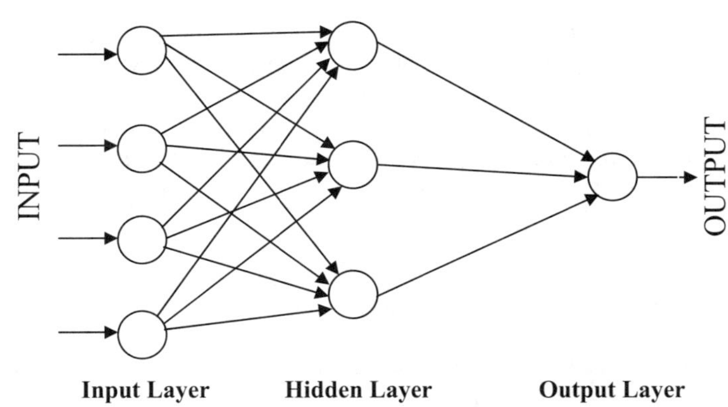

Fig. 1. A typical three-layer feed forward artificial neural network.

connected to all the neurons of the next layer and the neurons in one layer are not interconnected. All the nodes within a layer act synchronously. The data passing through the connections from one neuron to another are multiplied by the interconnection weights. The signal emanating from the output node(s) is the network's solution to the input problem. When the interconnection weights are modified, the ANN output changes.

To produce output, each neuron multiplies every input by its inter-connection weight, sums the product, and then passes the sum through a transfer function. This transfer function is usually a steadily increasing S-shaped curve, called a sigmoid function. This function is continuous, differentiable everywhere, and is monotonically increasing. The input to the function can vary between $\pm\infty$ and the output y_j is always bounded between 0 and 1. Under this function, the output y_j from the jth neuron in a layer is

$$y_j = f(\Sigma\ w_{ji}\ x_i) = \frac{1}{1 + e^{-(\Sigma\ w_{ji}\ x_i)}} \tag{3}$$

where w_{ji} is the weight of the connection joining the jth neuron in a layer with i^{th} neuron in the previous layer and x_i is the value of the ith neuron in the previous layer.

An ANN stores the knowledge about the problem in terms of weights of inter-connections. The process of determining ANN weights is called learning or training. The ANNs are trained with a training set of input and known output data. Error back propagation algorithm is the most population learning algorithm. At the beginning of training, the initial value of weights can be assigned randomly or based on experience. The weights are systematically changed by the learning algorithm such that for a given input, the difference between the ANN output and the actual output is small. Many learning examples are repeatedly presented to the network, and the process is terminated when this difference is less than a specified value. At this stage, the ANN is considered trained. After training is complete, the weights are frozen and the ANN performance is to be validated. A data set which the ANN has not encountered before is presented to validate its performance. Depending on the outcome, either the ANN has to be re-trained or it can be implemented for its designated use. ANN theory has been described in many books such as Rumelhart et al. (1986), Vemuri (1992) and Yegnanarayana (1999), Jain and Singh (2003). A comprehensive review of the underlying concepts and applications was carried out by the ASCE task committee on *Application of Artificial Neural Networks in Hydrology* (ASCE 2000a and 2000b). Maier and Dandy (2000) also presented a review of ANNs applications.

26.3. RADIAL BASIS FUNCTION NETWORK

Although multi-layer perceptron (MLP) trained with the BP algorithm is the most popular architecture in hydrologic modeling, the BP algorithm may not lead to good generalization when the data are limited (Sudheer, 2000). Fernando and Jayawardena (1998) reported that a radial basis function (RBF) type network performs better than an MLP trained with BP algorithm. Further, in RBF the non-linearity is embedded in the transfer functions of its hidden-layer neurons, making the optimization of tunable parameters a linear search. The RBF network is briefly described in the following sections.

Similar to the MLP network, RBF is also a feed-forward ANN (Fernando and Jayawardena, 1998) having input, hidden, and output layers of neurons. Figure 2 shows a

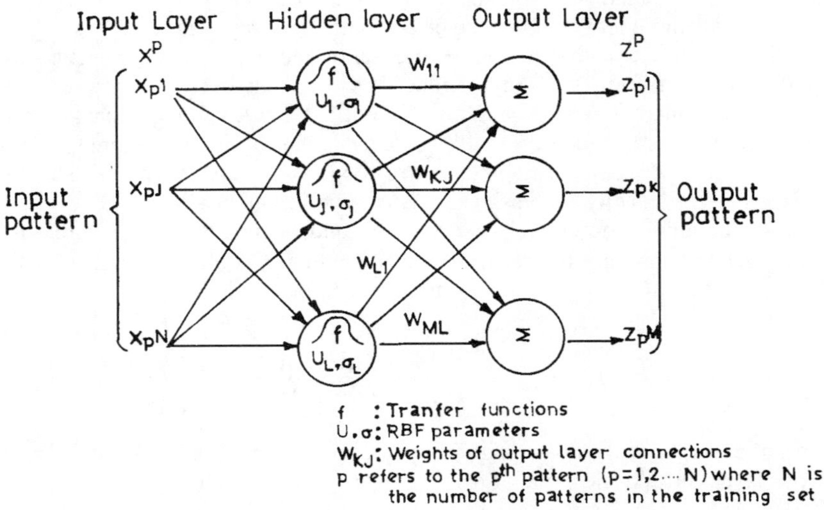

Fig. 2. A typical radian basis function network.

RBF network with N, L, and M nodes, in the input, hidden, and output layers respectively. Here the N-dimensional input patterns [X] is being mapped to M-dimensional outputs [Z]. The nodes in the adjacent layers are exhaustively connected. While the value of N and M are problem dependent, the value of L is to be determined by the network designer.

The input enters the network at the input layer such that one variable is fed into one node. The input layer sends copies of the variables to each node in the hidden layer. Each neuron in the hidden layer has a transfer function *f*. These transfer functions are radially symmetric basis functions. The choice of the radial basis function is not crucial to the network performance (Stinchcombe and White, 1989) as long as localized activations are near the center and function decays rapidly as values move away from the center.

The most commonly used radial basis functions is the Gaussian. The Gaussian RBF center of the jth hidden neuron is specified by the mean U_j and the deviation σ_j. The RBF centers are also points on the N-dimensional input space. An input pattern falling within $U_j \pm \sigma_j$ will cause that node to produce a significant response. For each input-pattern, the hidden layer nodes compute the distance between their centers and the point on the input space that corresponds to the input pattern. The responses of the hidden nodes depend on the distance, which takes a maximum of unity if the distance is zero and rapidly diminishes to zero if it is greater than σ_j. For the *p*th input pattern X^p, the response (y_j) of the *j*th hidden-layer node is

$$y_j = f\left\{\frac{\| X^p - U_j \|}{2\sigma_j^2}\right\} \tag{4}$$

where $\|.\|$ denotes the Euclidian distance norm. The network response (Z_{kp}) at the kth node in the output layer is given by the linear summation of the weighted responses as:

$$Z_{kp} = \sum_{j=1}^{L} j_j w_{kj} \tag{5}$$

The parameters that need to be optimized are the RBF functions on the hidden layer nodes and the output layer weights. Determining the RBF function involves finding suitable RBF centers and spreads. A variety of techniques have been evolved to optimize the number of RBF centers.

26.4. APPLICATIONS OF ANN IN WATER RESOURCES

By far, ANNs have been extensively applied to the problems of water resources, such as rainfall-runoff modeling, flow predictions, flow/pollution simulation, and modeling non-linear / input-output time series. Hsu et al. (1995) reported that multi-layer feed-forward networks are the best tools for input-output function approximation. Sandhu and Finch (1996) used ANNs to relate flow conditions and gate positions in the Sacramento San Joaquin Delta to salinity levels in the interior and along the boundary of the delta. Hutton et al. (1996) also used ANNs for trihalomethane (THM) formation and transport in delta waters. Maier and Dandy (1996) illustrated the utility of ANNs to estimate salinity at the Murray Bridge on the Murray River in Australia.

A neural network model was applied by Roehl and Conrads (1999) to simulate the hydrodynamics and water quality of two rivers in USA. An ANN model was developed to predict river flow and quality by Imrie et al. (2000). They concluded that ANNs can perform as well as existing process and statistical models. Neural network models for an urban watershed and a rural sub-urban watershed were developed by Cannon and Whitfield (2001) to detect changes in system response resulting from landscape changes. Jain (2001) applied ANN to establish an integrated rating between river stage, discharge, and sediment discharge. Huang and Foo (2002) presented an application of ANN to assess salinity variations in Apalachicola River, Florida.

A few past applications using RBF networks for rainfall-runoff modeling (Fernando and Jayawardena, 1998; Sudheer et. al, 2002; Sudheer and Jain, 2003) indicate that the RBF performs better than a feed forward network with sigmoid transfer function. This provided an impetus to examine RBF for estimation of K_2.

26.5. DATA COLLECTION

The data used in this study came from Kali River, India. Many visits were made during March 1999 to February 2000 to this river and data were collected under various conditions of flow. A total of 576 data sets on flow depth, velocity, cross-section, slopes, and reaeration coefficient were measured/ analyzed from 22 locations (16 river points and 6 effluent points) on Kali River. The water quality variables estimated by Standard Methods (APHA, 1985) on different reaches of the River Kali are: dissolved oxygen (DO) of upstream, downstream and incoming tributaries; biochemical oxygen demand (BOD) of upstream, downstream, and incoming tributaries; photosynthesis and respiration by plants, and water temperature. Details are given in Jha (2001), Jha (2004), and Jain and Jha (2005).

26.6. PERFORMANCE EVALUATION

A large number of statistical criteria are available to compare the goodness/adequacy of a given model. The frequently used performance evaluation statistics are the root mean square error (RMSE), coefficient of determination (CoD), mean multiplicative errors (MME), etc. The RMSE is computed by:

$$RMSE = \left(\sum_{i=1}^{N} \frac{\left(K_P - K_M\right)^2}{N} \right)^{0.5} \tag{6}$$

where K_P and K_M are predicted and measured values of reaeration coefficient and N is the number of data points. If the model is good, the predicted and measured values will be close and this will result in a small value of RMSE. The coefficient of determination (CoD), denoted by R^2, is computed by

$$R^2 = 1 - S_{ee}/S_{yy} \tag{7}$$

where, S_{ee} is the sum of squares of difference between the observed and computed values and S_{yy} is the sum of squares of departures of observed values of K_2 from the mean of observed values. CoD gives the fraction of variance that is explained by the model. Clearly, the closer this value is to unity, the better the model is.

The MME was considered by Moog and Jirka (1998) as another measure to represent the inaccuracies in predicting the reaeration coefficient. It is defined as:

$$MME = \exp\left[\frac{\sum_{i=1}^{N} \left| \ln\left({K_P}/{K_M} \right)_i \right|}{N} \right] \tag{8}$$

If the model under use produces good results, the values of K_P and K_M will be close and therefore, MME will be close to unity. Note that the statistical criteria described above are independent of each other. The RMSE uses the difference in magnitude of the observed and predicted values while MME uses their ratios.

26.7. ANALYSIS AND RESULTS

Pre-processing of input and output data is an important step in ANN application. In the present study, the data for a variable x were scaled so as to fall in the range [0, 0.9] by the following formula:

$$x_{si} = x_i / x_{max} \tag{9}$$

where x_{max} a value that is 1.1 times the maximum of the x values, and x_i and x_{si} are the original and scaled values of the ith data item of x.

In estimation of parameters of a hydrologic model, the available data are divided in two parts. The first part is used to calibrate the model and the second, to validate it. This practice is known as 'split-sample' test. The general practice is to use about two-third of the data for calibration and the remaining for validation.

The data of Kali River consisted of 576 patterns of depth, velocity, slope, and reaeration coefficient. Using these, the Froude number was calculated. Out of the data sets, 400 randomly selected patterns were used for training and 176 patterns for testing. To estimate the reaeration coefficient, empirical equations by Thackston and Dawson (2001), and Moog and Jirka (1998) and ANN were used.

For ANN, several combinations of inputs were tried: (a) depth, velocity, slope; (b) depth, velocity, slope, Froude number; (c) depth, velocity, slope, shear velocity; and (d) Froude number and shear velocity. Combination (b) gave the best results for both training and test data. Note that the Froude number is calculated from velocity and depth and therefore by including it in inputs, no additional information is really being provided. Its inclusion, however, gives additional flexibility to the network. Thus, the ANN had 4 inputs and 1 output, namely K_2.

During training of BP network, the number of nodes in the hidden layer was systematically changed to find the value that gave the best result. For the hidden layer, tan-sigmoid transfer function was used and for the output layer, linear transfer function was chosen. The Levenberg-Marquardt technique was used foe training. The ANN training epoch consisted of 500 cycles or a tolerance limit of 0.01, whichever occurs earlier.

The results for the Kali River analysis giving the RMSE, MME and CoD for K_2 estimation are given in Table 1. The RMSE for MLP & RBFANN is substantially smaller than the other approaches indicating that ANN estimates are closer to the observed values. MME is another indicator of goodness of fit that was used by Moog and Jirka (1998). It is seen from Table 1 that for both ANNs, MME is very close to unity while for the other approaches, MME is quite far from unity. Recall that closer MME value is to unity, the better the model is. For ANN, the CoD is also closer to unity compared to the other techniques, both for training and validation data. This indicates that a larger fraction of variance is explained by the ANN models compared to the other techniques. Thus, among the four approaches whose results have been presented in Table 1, the estimations by ANN are the best based on all the three statistical indices.

Table 1. Performance evaluation of reaeration equations for data sets of Kali River, India.

K_2 estimation method	FOR CALIBRATION / TRAINING DATA (400 PATTERNS)			For Validation / Test data (176 patterns)		
	RMSE	MME	CoD	RMSE	MME	CoD
Moog and Jirka (1998)	6.636	6.37	0.18	4.24	2.23	0.17
Thackston and Dawson (2001)	1.095	6.16	0.63	0.75	2.22	0.74
MLP ANN	0.293	1.16	0.72	0.303	1.16	0.76
RBF ANN	0.2578	1.15	0.944	0.3661	1.17	0.89

In Figure 3, the scatter diagram between the observed K2 and those computed using RBF ANN for the test data is presented. An examination of this figure shows that the scatter between the observed and modeled values is quite small. Similar results for MLP ANN were obtained by Jain and Jha (2005).

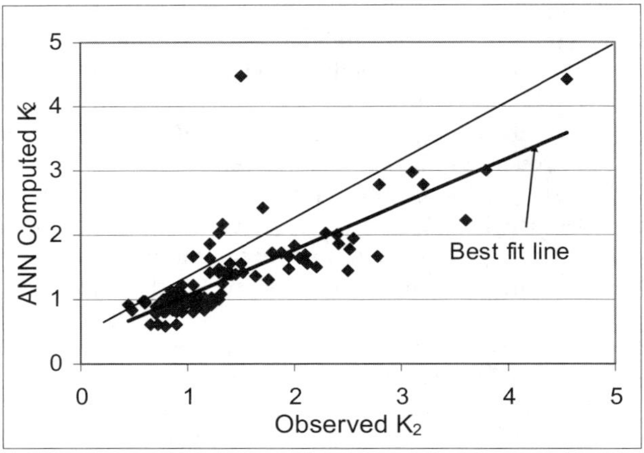

Fig. 3. Scatter plot observed and K2 computed by RBF –test data.

26.8. CONCLUSIONS

A large number of empirical equations have been developed to estimate reaeration coefficient from flow data and channel characteristics. The ANNs has been successfully used in many hydrological studies and this was the motivating factor for its application to the present study to examine if improved estimates can be obtained using ANNs. The performance of ANN and reaeration equations was tested using RMSE, MME and CoD. It was found that on all the three counts, the ANN approach turned out to be better than the other methods for the available data. Among the ANN models tried, RBF results were slightly better for the training data while the results were comparable for the test data.

REFERENCES

APHA, 1985. Standard methods for the examination of water and wastewater (sixteenth edn.). American Public Health Association, Washington DC, 1268 pp.

ASCE Task Committee, 2000a. Artificial neural networks in hydrology. I: Preliminary concepts. Journal of Hydrologic Engineering - ASCE 5(2): 115-123.

ASCE Task Committee, 2000b. Artificial neural networks in hydrology. II: Hydrologic applications. Journal of Hydrologic Engineering - ASCE 5(2): 124-137.

Cannon, A.J., and Whitfield, P.H., 2001. Modelling transient pH depressions in Coastal streams of British Columbia using neural networks. Journal of American Water Resources Association 37(1): 73-89.

Dawson, C.W., and Wilby, R., 1998. An artificial neural network approach to rainfall-runoff modelling. Hydrological Sciences Journal 43(1): 47-66.

Fernando, D. A. K., and Jayawardena, A. W., 1998. Runoff forecasting using RBF networks with OLS algorithm. Journal of Hydrologic Engineering - ASCE 3(3): 203-209.

Haykin, S., 1994. Neural networks: a comprehensive foundation. MacMillan, New York.

Hsu, K-L., Gupta, H.V., and Sorooshian, S., 1995. Artificial neural network modeling of the rainfall-runoff process. Water Resources Research 31(10): 2517-2530.

Huang, W., and Foo, S., 2002. Neural network modelling of salinity variation in Apalanchicola River. Water Research 36: 356-362.

Hutton, P.H., Sandhu, N., and Chung, F.I., 1996. Predicting THM formation with artificial neural networks. Proc. North Am. Water and Envir. Conf., ASCE, 3547-3556.

Imrie, C.E., Durucan, S., and Korre, A., 2000. River flow prediction using artificial neural networks: generalization beyond the calibration range. Journal of Hydrology 233: 138-153.

Jain, S. K., 2001. Development of Integrated Sediment Rating Curves using Artificial Neural Networks. Journal of Hydraulic Engineering - ASCE 127(1): 30-37.

Jain, S. K., and Singh, V. P., 2003. Water Resources Systems Planning and Management. Elsevier, Amsterdam, The Netherlands.

Jain, S.K., and Jha, R., 2005. Comparing stream re-aeration coefficient estimated from ANN and empirical models. Hydrological Sciences Journal, IAHS, 50(6): 1037-1052.

Jha, R., Ojha, C.S.P., and Bhatia, K.K.S., 2001. Refinement of predictive reaeration equations for a typical Indian river. Hydrological Processes 15(6): 1047-1060.

Jha, R., Ojha, C.S.P., and Bhatia, K.K.S., 2004. A supplementary approach for estimating reaeration rate coefficients. Hydrological Processes 18: 65-79.

Maier, H. R., and Dandy, G. C., 1996. The use of artificial neural networks for the prediction of water quality parameters. Water Resources Research 32(4): 1013-1022.

Maier, H. R., and Dandy, G. C., 2000. Neural networks for the prediction and forecasting of water resources variables: a review of modeling issues and applications. Environmental Modeling and Software 15: 101-124.

Moog, D. B., and Jirka, G. H., 1998. Analysis of reaeration equations using mean multiplicative error. Journal of environmental engineering - ASCE 112(2): 104-110.

Roehl, E.A., and Conrads, P.A., 1999. Real-time control for matching wastewater discharges to the assimilative capacity of a complex, tidally affected river basin. in Proceedings of the 1999 South Carolina Environmental Conference, March 15-16, Myrtle Beach, SC.

Rumelhart, D.E., McLelland, J.L., and the PDP Research Group, 1986. Parallel Distributed Processing, Explorations in the Micro Structure of Cognition, Vol. I: Foundations, The MIT Press, Cambridge.

Sandhu, N., and Finch, R., 1996. Emulation of DWRDSM using artificial neural networks and estimation of Sacramento River flow from salinity. Proc., North Am. Water and Env. Conf, ASCE, New York, 4335-4340.

Stinchcombe, M., and White, H., 1989. Universal approximation using feed forward networks with non-sigmoidal hidden layer activation functions. Proceedings of the International Conference on Neural Networks, IEEE Neural Network Committee, IEEE, N. Y., 1: 607-611.

Sudheer, K. P., and Jain, S. K., 2003. Radial Basis Function Neural Network for Modeling Rating Curves. Journal of Hydrologic Engineering - ASCE 8(3): 161-164.

Thackston, E. L., and Dawson, J.W., 2001. Recalibration of a reaeration equation. Journal of environmental engineering - ASCE 127(4): 317-321.

Vemuri, V.R., 1992. Artificial Neural Networks: Concepts and Control Applications. IEEE Computer Society Press, California.

Yegnanarayana, B., 1999. Artificial Neural Networks. Prentice-Hall of India Pvt. Ltd., New Delhi.

27

SEDIMENT TMDL CALCULATIONS FOR THE UPPER AMITE RIVER

Zhiqiang Deng and Pradeep Mishra
Department of Civil and Environmental Engineering , Louisiana State University
Baton Rouge, LA 70803, USA

ABSTRACT

The Amite River basin is situated in Southeast of Louisiana and Southwest of Mississippi and comprises of 2200 sq miles of area. 11,000 acres of the basin have experienced degradation with significant loss of shore life and aquatic habitat. The primary cause is due to erosion and it is thus necessary to make a BMP plan for erosion control based on the sediment TMDLs. Sediment TMDL calculations have been done for the Upper Amite River. The calculations involved (1) combined application of USLE model and ARCVIEW GIS for soil erosion estimation in the Upper Amite River, (2) HEC-RAS modeling for flow computation, and (3) numerical solution of a new advection-dispersion equation for suspended sediment transport. The entire Amite River basin is divided into 15 sub-basins. Digital elevation data was imported into the GIS. The GIS database provides inputs for land use, soil type, slopes and elevation. Meteorological data from 1987 to 2004 from the USGS stations were used. Surface erosion from land catchments, settling, scouring, and bank erosion were considered. The calculated soil erosion rate for the Upper Amite River Basin is 5.42 ton/acre/year, producing sediment load of 0.103 kg/m^3 to the Amite River. It is found that sediment criteria for the Amite River can be met by adopting practices such as terraces on the steep slopes, creation of buffer zones along the river. Results indicate that the new model can be an effective tool for sediment TMDL calculations.

27.1. INTRODUCTION

TMDL is defined as "a calculation of the maximum amount of a single pollutant - from all contributing point and nonpoint sources - that a waterbody can receive and still meet water quality standards [without adverse impact to fish, wildlife, recreation, or other public uses]...." That calculation must include a margin of safety and account for seasonal variations in water quality. An overview of current TMDLs by the EPA shows that over 40% of the United States assessed waters still do not meet the water quality standards which the states, territories, and authorized tribes have set for them. This amounts to over 20,000 individual river segments, lakes, and estuaries. These impaired waters include approximately 300,000 miles of rivers and shorelines and approximately 5 million acres of lakes -- polluted mostly by sediments, excess nutrients, and harmful microorganisms from nonpoint sources. In fact, the largest water pollutants in the United States, by volume, are instream suspended sediment (Fowler and Heady, 1981). The Amite River is identified as one of the 15 water bodies impaired by sediments in Louisiana and sediment TMDL calculation is required by EPA for the Amite River. The overall goal of this paper is thus to present the sediment TMDL calculation for the Amite River.

The sediment TMDL calculation is composed of four parts: (1) Development of a new sediment transport and dispersion model for the Amite River, (2) Flow computation, (3) Estimation of sediment loads (sources) produced by watershed erosion, and (4) Determination of sediment TMDL for the Amite River.

27.2. DERIVATION OF 1-D COHESIVE SEDIMENT DISPERSION AND TRANSPORT MODEL

In order to calculate the TMDLs in Louisiana waterways it is essential to determine the transport and fate of the suspended sediment in the waterways. To that end, a sediment transport model is necessary. Using the mass conservation principle and Reynolds transport theorem a new model for suspended cohesive sediments has been developed for computation of sediment transport in the Amite River.

$$\frac{\partial S}{\partial t}+V\frac{\partial S}{\partial x}=\frac{1}{A_f}\frac{\partial}{\partial x}\left(A_f K_x \frac{\partial S}{\partial x}\right)+\frac{\alpha(u_*-u_{*c})(S_*-S)}{h}-\frac{\beta\omega_S S}{h}+\frac{q_L}{A_f}(S_*-S) \quad (1)$$

in which A_f = channel cross-sectional area (L^2), V = flow velocity, h = flow depth, S = sediment concentration (M/L^3), S_L = sediment concentration of lateral inflow (M/L^3), S_* = suspended sediment concentration under equilibrium conditions or suspended-load carrying capacity (M/L^3) which is determined using the formula proposed by WIHEE (Chien and Wan, 1999), Q = volumetric flow rate (L^3/T), q_L = lateral inflow rate (L^3/T-L), t = time [T], x = distance [L], K_x = longitudinal dispersion coefficient (L^2/T) which is calculated using the method presented by Deng et al. (2001), u_* = shear velocity (L/T), u_{*c} = critical shear velocity (L/T), ω_s = settling velocity of sediment particles (L/T) which is calculated using the equation presented by Cheng (1997), α = constant , β = constant.

The main advantage of Eq. (1) over existing 1-D sediment transport models is that sediment erosion (described by the second term on the right hand side of the equation) and sediment settling (represented by the third term on the RHS) are treated as two different processes and thus modeled by two separate terms. The last term on the RHS of Eq. (1) stands for the influence of tributaries on sediment transport in the Amite River.

27.3. FLOW COMPUTATION

The area A_f, flow velocity, and flow depth h in Eq. (1) are calculated using the HEC-RAS software. HEC-2 input files for the Amite River above Denham Springs labeled "UAR.DAT" were obtained from US Army Corps of Engineers New Orleans district. The Schematic representation of the Amite River is shown in Fig. 1. The river is divided into numerous cross sections. Each cross–section represents a location and the corresponding river mile from the downstream end is labeled. The river mile 97 is the starting point and river mile station 50.1 is the most downstream location for the river reach. The direction of the flow is from the higher river mile station towards the lower river mile stations. Other required flow parameters like Manning's coefficient, cross-section profile are appropriately defined at each cross section.

The flow computation includes steady flow analysis and unsteady flow analysis. The steady flow analysis is intended for calculating water surface profiles for steady gradually

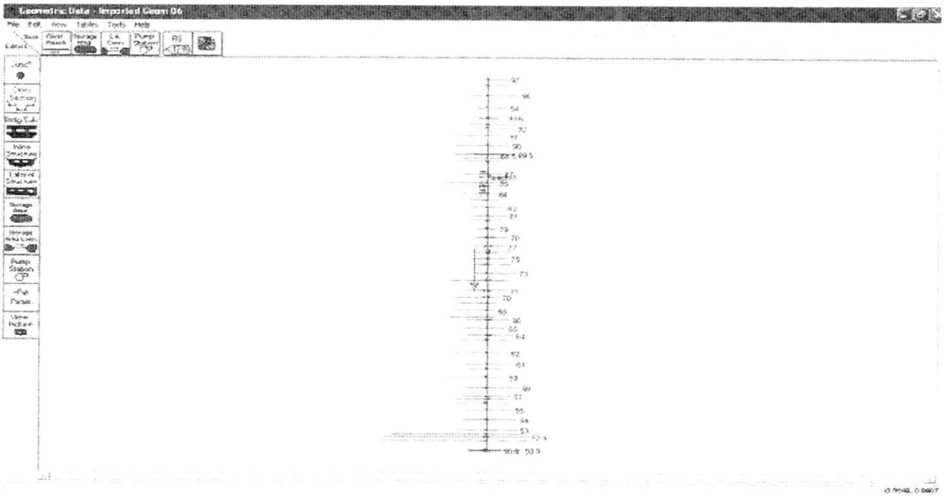

Fig. 1. Schematic representation of the Amite River with various River Mile Stations.

varied flow. The basic computational procedure is based on the solution of the energy equation. Effects of various obstructions such as bridges, culverts, weirs, spillways and other structures in the flood plain have been considered in the computations. Once all of the geometric data were entered, it was possible then to enter any steady flow data that were required. The steady flow data editor was brought up, and "Steady Flow Data" was selected from the "Edit" menu on the HEC-RAS main window. The last step in developing the steady flow data was to save the information to a file. The outputs from HEC- RAS gave the necessary flow information in the Amite River. Figures 2 and 3 show the computed variations of flow depth in the longitudinal and lateral directions, respectively.

Sediment concentration is very sensitive to flow variations. Therefore, to study the impact of unsteady flow regime on sediment transport in the Amite River arising due to individual storm events, it is important to get flow parameters under unsteady flow conditions. The objective of unsteady flow simulation is to get the flow hydrograph at each of the river mile stations of the Amite River. The hydraulic calculations for cross-sections, bridges, culverts, and other hydraulic structures that were used for the steady flow analysis were also incorporated into the unsteady flow module. For the current problem, a lateral inflow hydrograph was taken as the boundary condition. In addition to the boundary conditions, the initial conditions of the system at the beginning of the unsteady flow simulation were also established. Initial conditions consisted of flow and stage information at each of the cross sections. Once all of the geometry and unsteady flow data was entered in the HEC-RAS main window "Unsteady Flow Analysis" was selected from the "Run" menu to perform unsteady flow simulation. The calculated flow velocity and discharge of the Amite River vary in the range of 0.34 – 2.4 m/s and 285 – 771 m^3/sec, respectively.

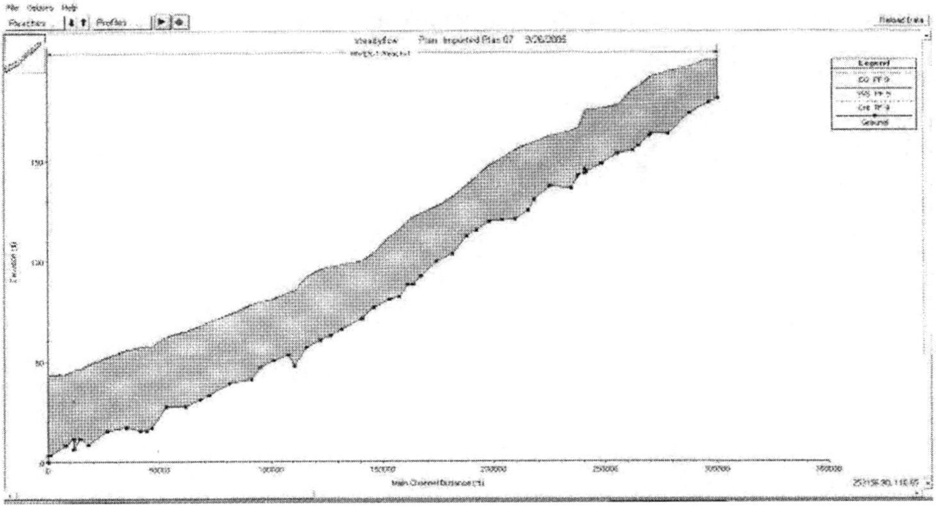

Fig. 2. Flow depth profile along the Amite River.

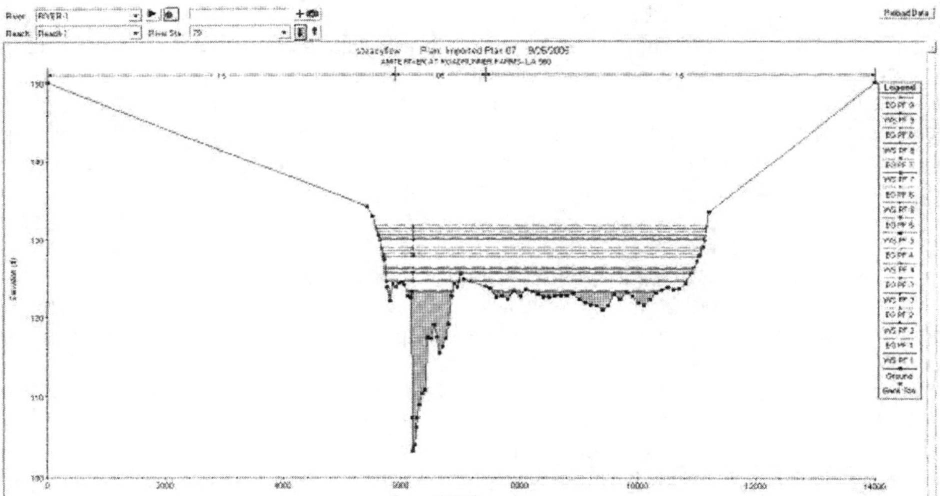

Fig. 3. Cross-section 18 of the Amite River.

27.4. NUMERICAL SOLUTION OF 1-D SEDIMENT DISPERSION AND TRANSPORT MODEL

To solve the new sediment transport equation 1, a standard split approach by Sobey (1983), is used. Such an approach requires solving the advection and diffusion parts separately at each time step. The advection-dispersion equation is decomposed into the hyperbolic (pure advection) and the parabolic (pure dispersion with sink and source terms) partial differential equations. The two sub-equations are then solved separately in two or three consecutive fractional steps by the corresponding numerical approaches that best fit the features of each PDE for one time step. Based on the split-operator algorithm it is commonly assumed that the pure advection process and the pure dispersion process alternate with time: the advection process occurs in the first sub-time step, the dispersion takes place in the second sub-time step, and the reaction is considered in the final sub-time step (Holly and Preissmann, 1977). In the current case, there is no reaction term.

27.5. SOIL EROSION CALCULATION FOR AMITE RIVER BASIN

The USLE model developed by Wischmeier and Smith (1978) has been the most widely utilized for estimation of sediment loss in pollution control programs. The USLE for estimating average annual soil erosion is expressed as

$$A = RKLSCP \tag{2}$$

where A = average annual soil loss in t/a (tons per acre), R = rainfall erosivity index, K = soil erodibility factor, LS = topographic factor, L = slope length, S = slope, C = cropping factor, P = conservation practice factor. Values for each of the six factors are determined from research data. It is important to note that values generally differ from one location to another. Predicted soil loss is nothing but the calculated annual soil loss expressed in tons per acre. In most of the nonpoint source models, USLE or one of its other versions is used for predicting soil erosion. Although originally developed for agricultural purpose, its use has been extended to watershed with other land uses. It is therefore suitable, to apply this model for predicting soil erosion in Amite River Basin. For this work, the ArcGIS spatial analyst was used to derive the erosion map from all the input maps, as shown in Figure 4.

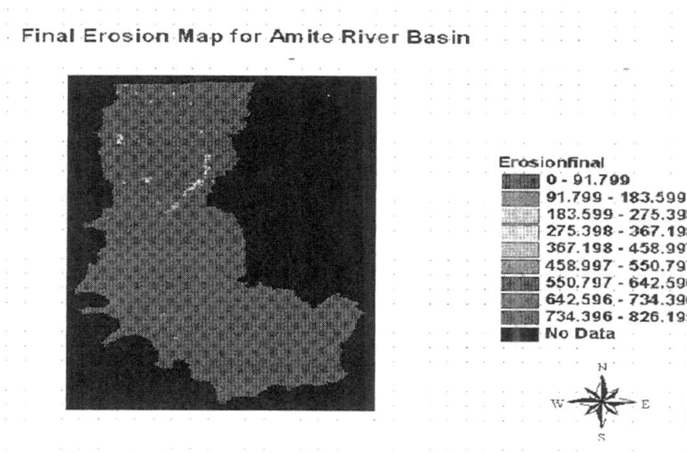

Fig. 4. Soil erosion map for the Amite River.

Sediment yield of the Amite River Basin is estimated using the model developed by the USDA (1972). The calculated soil erosion rate for the Upper Amite River Basin is 5.42 ton/acre/year, producing sediment load of 0.103 kg/m^3 to the Amite River. Even though the soil erosion value was very high at some points in the basin, as shown in Figure 4, this value is in consonance with the soil erosion values between 5-10 tons/acre/year in the Amite River basin, which is cited in literature (US Army Corps of Engineers, 2000).

27.6. CALCULATION OF SEDIMENT TMDL

To develop a TMDL it is necessary to establish quantitative measures, or indicators, that can be used to establish the relationship between pollutant sources and their impact on water quality. Once an indicator has been selected, a target value for that indicator which distinguishes between the impaired and unimpaired state of the water body must be established (US EPA, 1999). One such indicator is turbidity which is measured in terms of Nephelometric turbidity units (NTU). Turbidity is the measure of the optical property of water that causes light to be either scattered or absorbed. Turbidity may be influenced by a number of factors but is primarily affected by suspended matter such as clay, silt, plankton, or microscopic organisms. Thus, the criteria that was taken as benchmark for establishing TMDL was 50 NTU or 64 mg/L. This numeric target is prescribed by the EPA for the Amite River Basin. Based on this numeric target and sediment concentration calculated from the numerical solution of Eq. (1) the load reduction is made.

The estimate of the actual average load capacity from the Amite River watershed for 1990 was calculated as 1.35×10^6 tons/year. The 1-D model predicts a maximum sediment concentration of 114 mg/L, ranging from 3 mg/l to 114 mg/l. The average concentration of SSC in the Amite River is 25 mg/L. The difference between the standard load capacity permissible for sustenance of aquatic life and actual load capacity derived from suspended sediment concentration (SSC) would be used for determination of load reduction. Due to variations in SSC concentration along the river, there would exists different load capacities. For determining the load reduction, the largest difference between the two load capacities was taken into account. The largest value of the difference was found to be 219.456 tons/day. This value (i.e., the load reduction) is apportioned among the various sources of the sediments produced by human activities. In establishing TMDLs, EPA generally apportions the loading capacity among: (1) the background loading; (2) the waste load allocations for point sources; and (3) the load allocations for non-point sources. For this TMDL, there are no point sources, so the waste load allocations equal zero. The background loading is not considered in this case and the TMDL is expressed as the sums of three major components, namely, load allocations, margin of safety and future growth. Even though there might be some point sources and NPDES dischargers in the watershed, it is assumed that the effect on water quality due to point sources is negligible. The existing loads contributed by these facilities are unknown and will need to be determined in the future through monitoring of effluent and ambient receiving stream flow. Therefore, the allocation for point sources is to be considered as future work. At this stage of the TMDL, the assumed condition would be effect of non point source pollution.

The composition of the watershed indicates a mixture of rural non-point sources and wetland which may contribute to the downstream impairment. These sources tend to become dominant under higher flow conditions. Since load reductions are primarily attributed to nonpoint sources, then the MOS should be relatively large. Using a MOS of

25% for the Phase I TMDL and taking the population growth in the Amite River basin to be around 1.3 % (US Army Corps of Engineers, 2000), the future growth values is taken to be the same for TMDL calculations. Taking into all these values the final TMDL can be easily calculated. The final calculations show that TMDL for the Amite River (= 3693.28 tons/day + 0.25*3693.28tons/day + 0.013*3693.28 tons/day) is 4665.235 tons/day.

In order to seek the impact of unsteady flow on sediment concentration and thus on the sediment TMDL, the erosion was calculated using the MUSLE model. The unsteady lateral inflow induces an unsteady pattern in the discharge at the particular cross section upstream. Due to this variation in discharge at the cross section, the suspended sediment concentration was found to vary and reach a peak value close to 140 mg/L. The calculated sediment TMDL is 5418.357 tons/day. This value is higher compared to the TMDL for steady flow. Therefore, the incorporation of unsteady flow factor into future TMDL development would give a better picture of the Best Management Practices that are to be followed for controlling non point source pollution.

27.7. CONCLUSIONS

(1) Using mass conservation and Reynolds transport theorem, a new 1-D sediment dispersion and transport model has been developed for the Amite River sediment TMDL calculations. The new model is capable of predicting suspended sediment transport in the Amite River.

(2) The combination of USLE model and GIS technology is an efficient and effective approach for estimation of watershed sediment erosion. GIS is very useful compared to traditional methods by breaking up the land surface into many small cells which enables an analysis to be performed on both large regions as well as discrete areas. GIS not only generates inputs for USLE, but also displays outputs. Spatial variation of soil loss correlated with land use can be observed. Based on USLE and the average intensity rainfall of 1990, erosion rate of the Amite River Basin is found to be 5.41 tons/acre/year. This erosion rate represents the average annual erosion rate for the entire basin. This erosion value can be used for sediment TMDL calculations under steady flow conditions. The MUSLE model which is a single event model can be employed to determine the soil erosion for the TMDL calculations under unsteady flow conditions.

(3) The flow parameters for both the steady and unsteady flow conditions can be efficiently computed using the HEC-RAS. The calculated flow velocity and discharge of the Amite River vary in the range of 0.34 – 2.4 m/s and 285 – 771 m^3/sec, respectively.

(4) The 1-D model predicts a maximum sediment concentration of 114 mg/L and the average concentration of 25 mg/L, ranging from 3 mg/l to 114 mg/l in the Amite River.

(5) Based on the EPA's and LDEQ's water quality standard of 50 NTU, the calculated sediment TMDL for the Amite River is 4665.235 tons/day. The daily reduction in this case is found to be approximately 220 tons. The TMDL accounted not only for waste load allocation, but also for margin of safety (MOS) and future growth.

(6) Unsteady flow is found to have a significant effect on TMDL calculations. Event based rainfalls produce unsteady flows in the river and a higher erosion rate and

thereby sediment concentration in the river, resulting in a higher TMDL for the Amite River. The TMDL value for unsteady flow is found to be 5418.357 tons/day. The median daily reduction is found to be 2653.29 tons/day for the Amite River.

REFERENCES

Cheng, N.-S., 1997. Simplified settling velocity formula for sediment particle, Journal of Hydraulic Engineering 123(2): 149-152.

Chien, N., and Wan, Z., 1999. Mechanics of Sediment Transport, ASCE Press, Reston, Virginia.

Deng, Z.-Q., Singh, V. P., and Bengtsson, L., 2001. Longitudinal dispersion coefficient in straight rivers. Journal of Hydraulic Engineering - ASCE 127(11): 919-927.

Fowler, J. M., and Heady, E.O., 1981. Suspended sediment production potential on undisturbed forest land. Journal of Soil Water Conservation 36: 47–49.

Holly, F. M., and Preissmann, A., 1977. Accurate calculation of transport in two dimensions. Journal of Hydraulic Diversion - ASCE 103(HY11): 1259-77.

Sobey, R. J., 1983. Fraction step algorithm for estuarine mass transport. International Journal for Numerical Methods in Fluids 3: 749-772.

USDA, 1972. Sediment sources, yields, and delivery ratios. National Engineering Handbook, Section 3 Sedimentation.

US EPA (Environmental Protection Agency), 1999. Noyo River total maximum daily load for sediment; comment responsiveness summary. Region IX Water Division. San Francisco, CA. 16 December 1999.

U.S. Environmental Protection Agency, 1999. Protocol for Developing Sediment TMDLs. EPA, 841-B-99-004.

US Army Corps of Engineers, 2000. Amite River and Tributaries, Louisiana, Ecosystem Restoration and Reconnaissance Study.

Wischmeier, W.H., and Smith, D.D., 1978. Predicting rainfall erosion losses - a guide to conservation planning. The USDA Agriculture Handbook No. 537.

28

NUMERICAL MODELLING OF POINT AND NON-POINT SOURCE POLLUTANT DISCHARGE FROM COASTAL WATERSHED TO MARINE / LAKE COAST

A. Ghosh Bobba
National Water Research Institute
Environment Canada
Burlington, ON, Canada, L7R 4A6

ABSTRACT

Human activities on coastal watersheds provide the major sources of nutrients entering shallow coastal ecosystems. Contamination loadings from watersheds are the most widespread factor that alters structure and function of receiving aquatic ecosystems. Subsurface water discharges are recognized as potentially significant contaminant pathway from the watershed to the coast by human activities. This interaction affects the water quality and quantity in both surface and subsurface waters. The diversity, in combination with the range of disciplines and of time and space scales involved; complicate the use of data for purposes other than those envisioned by the original investigators. As land and water resource development increases in the coastal watersheds, it is becoming readily apparent that subsurface and surface waters interaction must be considered in establishing water management policies. All these interactions occur along coastal watersheds of Great Lakes and other coast. In this paper, some example watersheds shows the roles of geo-hydrology and hydro-geochemical processes are investigated by application of numerical models for water quantity and quality modelling of receiving waters. This model also predicts the movements of contaminants in subsurface water to coast.

28.1. INTRODUCTION

Subsurface water discharge from watersheds to coastal and estuarine waters has been a topic of theoretical and practical interest for at least a century (Bobba, 1993a). Most work has stressed controls on saltwater intrusion to freshwater aquifers and it has been only recently that field measurements of groundwater discharge have shown the importance of subsurface flow on water and nutrient budgets in coast and estuaries. Groundwater discharge has shown to be a source of nitrogen, typically as nitrate, in shallow sediments of lakes and coastlines. For example, upland aquifers contribute to direct groundwater discharge greater than 20% of the freshwater and 75% of the nitrogen that enters the Great South Bay, new York (Bokuniewicz, 1980). Excess nitrogen in groundwater derived from sewage and fertilizer has drastically affected water quality in other estuaries and coastal lagoons as well (Valiela et. al., 1992).

Ecosystem of the world's coastlines are receiving extraordinary amounts of nutrients as a consequences of human activities such as fertilizers, industrial emissions to the atmosphere, and disposal of waste water in watersheds adjoining coastal waters. The loadings of nitrogen and phosphorous to coastal aquatic environments even exceed those to fertilized agro-ecosystem. Increased nutrient loading from anthropogenic sources is pervasive and function of shallow coastal ecosystems during coming decades.

Although increased nutrient loading by precipitation has been documented, most research has focused on deeper estuaries in which flow from rivers and streams dominates water budgets and contributes the major of nutrients. Rivers and direct precipitation, however, are not the sole source of freshwater - borne nutrients to coastal environments. Even in places without rivers, salinity is often depleted in coastal waters due to groundwater input. Groundwater flow is especially important where underlying coastal sediments are coarse, unconsolidated sands of glacial or marine origin. In such situations flow of groundwater may be the major source of nutrients to coastal waters.

In unconsolidated sediments, groundwater moves through the watershed shoreward in paths that have downward vertical as well as horizontal vector. Downward flow is caused by additional water infiltrating along the path of the water. Freshwater eventually moves close enough to shore to meet the denser saltwater that saturates interstitial space in sediments beneath the sea. The presence of sea water in the pore space acts together with lower head pressures in the near shore zone compared with offshore to deflect the path of fresh groundwater sharply upward (Bobba, 1993). As a result, most of the groundwater flow occurs very near the shore.

During the last decade, it has become apparent that groundwater flow and transport of nutrients into shallow coastal water are far more significant and widespread than had been realized. The importance of ground water is not so much because of the magnitude of flow rates, but rather because of the high nutrient concentrations in groundwater compared to those in receiving coastal water. Although highly variable, the nutrient content of groundwater discharging onto coastal water may be up to five orders of magnitude larger than concentrations in receiving seawater.

28.2. CONSEQUENCES OF GROUNDWATER NUTRIENT TRANSPORT

There is enormous freshwater literature that deals with the consequences of eutrophication. The response of primary producers to nutrient loading within an estuary must be dependent on the balance between increased growth due to elevated inputs of the limiting nutrient and losses related to the flushing rate. This is the point of the well documented relation between phosphorous loading and phytoplankton chlorophyll. Additional evidence, based on correlations and inference from N:P values, on enrichments within small containers, and on whole - system enrichments in fresh waters, is consistent with the phosphorus loading results, and shows that in fresh water phosphorus limits potential primary production.

In coastal systems the evidence of the role of nitrogen in limiting phytoplankton growth is based on inferences from loading calculations, on ambient nutrient concentrations, and more importantly, on nutrient enrichment experiments.

Groundwater entering the coastal areas flows from watersheds where urbanization has taken place. Human activities such as disposal of wastewater via septic tanks, and use of fertilizers increase nutrient concentrations in groundwater. We presently lack of data groundwater along coastal areas, but can make preliminary comparisons using data from nearby locations, where watersheds include some urbanized areas. This paper focuses on application of numerical model to predict subsurface water and pollution discharge to coast.

28.3. SUBSURFACE HYDROLOGY OF A COASTAL WATERSHED

The two main sources of pollutants are point sources (Table 1) and non-point sources (Table 2). Pollutants from the two sources may be released continuously or at discrete intervals (Figure 1). Point sources of pollution can be geometrically defined and the dimensions are amenable to mathematical analysis in assessing pollution loads and rates of discharge determined. Distributed sources of pollutants are much more widespread and can rarely be geometrically defined as precisely as a point source. Hence, it is more difficult to subject the input/output source to precise mathematical analysis. Rather a measured and intelligent assumption of the affected area is made for use in modelling and analysis. In heavily polluted areas, both point sources and distributed sources may occur together or may be independent of one another.

Table 1. Point sources of pollution in a coastal watershed.

Type of Pollution	Examples
Sewage disposal systems	Sewage lagoons, septic systems, cesspools, barnyards/feed lots
Surface waste disposal sites	Landfills, garbage dumps, surface waste dumps.
Underground waste disposal sites	Storage tanks (low, medium, high level wastes)
Spills, washings, and intrusions	Oil, gas, waste spills: Auto workshop washings, Research laboratory washings, seawater or saltwater intrusions.
Mining sources	Acid mine drainage: Mine waste dumps, seepages gas explosions.
Natural mineral / ore deposits	Saline springs, hot spring waters, anhydrite, pyrite deposits etc.

Table 2. Non-point sources of pollution in a coastal watershed.

Source	Examples
Agriculture	Cropland, irrigated land, woodland, feed lots
Silviculture	Growing stock, logging, road building
Construction	Urban development, highway construction
Mining	Surface, underground
Utility maintenance	Highways, streets and de-icing
Urban runoff	Floods and snowmelt

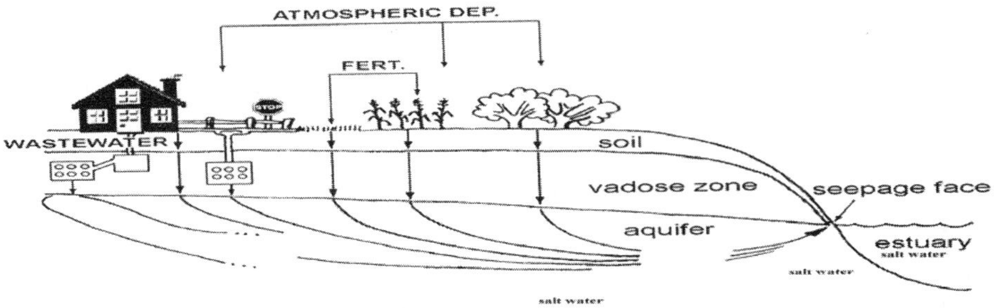

Fig. 1. Different sources of pollutants in a coastal watershed.

A watershed is a topographically defined as an area that water enters through precipitation and leaves as evapotranspiration, surface runoff, and subsurface water discharge. In the case of a coastal watershed, runoff and subsurface water discharge enter the sea (Figures 1 & 2). Rainfall and potential rate of evapotranspiration are determined by climate. Actual evapotranspiration is limited by the climatically controlled potential rate, vegetation and the wetness of the soil. Soil properties, topography, and the history of rainfall and evapotranspiration determine runoff and subsurface water discharge. For the purpose of this discussion, surface runoff includes direct runoff which occurs as stream flow immediately following a rainstorm, and drainage of subsurface water into creeks which accounts for stream flow between streams. The amount of direct runoff generated by a storm depends on the amount of rainfall and on the moisture condition of the soil. In general, more runoff occurs when the soil is initially wet. The subsurface water discharge can be calculated if rainfall, evapotranspiration, runoff, and the change in the amount of water stored on the watershed are known.

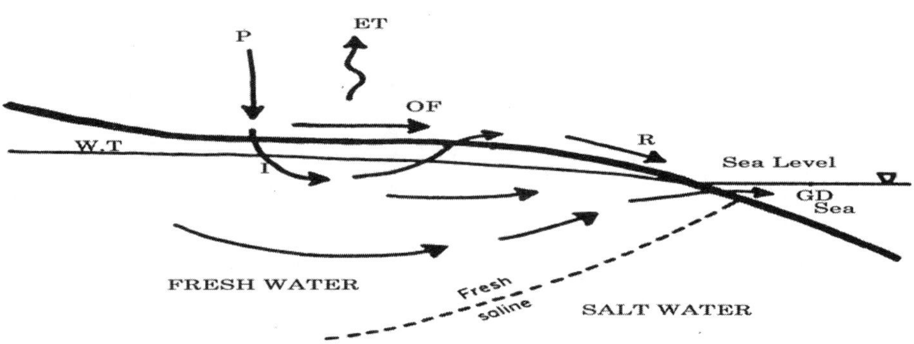

Fig. 2. Hydrological pathways in a coastal watershed.

A link between rising sea level and changes in the water balance is suggested by the general description of the hydraulics of subsurface water discharge at the coast. Fresh subsurface water rides up over denser salt water in the subsurface system on its way to the sea (Figure 2), and subsurface water discharge is focused into a narrow zone that overlaps with the inter tidal zone. The width of the zone of subsurface water discharge measured perpendicular to the coast is indirectly proportional to the discharge rate. The shape of the water table and the depth to the fresh/saline interface are controlled by the difference in density between freshwater and saltwater, the rate of freshwater discharge and the hydraulic properties of the subsurface system. The elevation of the water table is controlled by mean sea level through hydrostatic equilibrium at the shore.

Because pollutants are transported in large part by the bulk motion of subsurface water, the parameters of subsurface water flow are of major importance in the understanding of pollution processes. The various aspects of the subsurface water environments, as well as stratigraphic factors that control or could influence subsurface water motion are also of major consideration. The subsurface hydrology environment is shown schematically in Figure 2. It consists mainly of saturated and unsaturated zones. The unsaturated zone occurs above the capillary fringe where the soil pores are partially saturated with water. This zone is important in waste management because in most cases, it is the burial zones for wastes. Consequently, a thick unsaturated zone may sometimes be preferred for waste disposal since it would take a much longer time for pollutants to reach the water table. In the saturated zone, the pores are saturated with water. When this zone is capable of transmitting significant quantities of water for economic use it is referred to as an aquifer. In most field situations, two or more aquifers occur, separated by impermeable strata or aquitard. In the situation illustrated in Figure 2, the upper or unconfined aquifer is much more prone to pollution than the lower confined aquifer.

28.4. NUMERICAL SIMULATION OF SUBSURFACE AND CONTAMINATION DISCHARGE TO COAST

Bobba and Singh (1995) presented a detailed description of available model characteristics and recent trends in subsurface water flow and contamination transport modelling. Flow models are used to determine the quantitative aspects of subsurface water motion, such as direction, rate, changes in water table or potentiometric head, stream-aquifer interaction, etc. Transient 2-D models for simulating groundwater flow are widely available.

One common goal of these models is to predict and characterize the movement of the transition zone in the aquifer where fresh water and salt water meet in coast. Another purpose of modeling is to predict the degree and extent of mixing that occurs in this transition region. In this way, models allow problems to be defined before they actually occur. The details of the models presented earlier by Bobba (1996), Bobba and Singh (1995), Bobba et. al. (2000).

In this paper, a finite element model, SUTRA (Voss, 1984) is applied to three different types of cases. This model simulates water movement and the transport of either dissolved substances or energy in the subsurface system. The model can be applied areally or in cross section. It uses a two-dimensional, combined finite-element and integrated-finite difference method to approximate the equations that describe the two interdependent processes being simulated. When used to simulate saltwater movement in the subsurface system in cross section, the two interdependent processes are the density-dependent saturated subsurface-water flow and the transport of dissolved solids in the subsurface water. Either local - or regional - scale sections having dispersed or relatively sharp transition zones between saltwater and freshwater may be simulated. The results of numerical simulation of saltwater movement show distributions of fluid pressures and dissolved - solids concentrations as they vary with time and also show the magnitude and direction of fluid velocities as they vary with time. Almost subsurface properties that are entered into the model may vary in value throughout the simulated section. Sources and boundary conditions may vary with time. The finite - element method using quadrilateral elements allows the simulation of irregular areas with irregular mesh spacing. The model has been applied to real field data and observed to give favorable are explained as follows (Voss, 1984; Bobba, 1993b; 1998; Bobba et al., 2000).

28.4.1. Case Study 1: Lambton County, Ontario, Canada

Lambton County is located in the province of Ontario, Canada (Figure 3). The study region occupies most of Lambton County is located along the Saint Clair River between Lakes Huron and Saint Clair. The region is boundary to the by Lake Huron and to the west by the Saint Clair River. Numerous chemical and petrochemical companies operate in this region

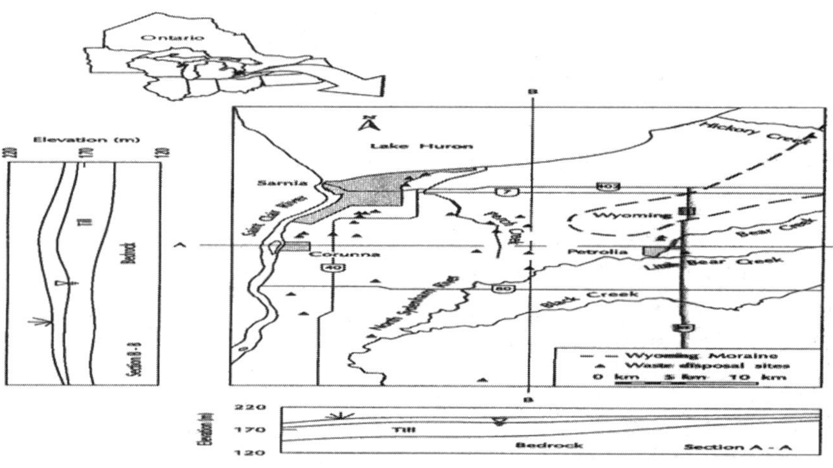

Fig. 3. Location of Lambton County where Sections A-A and B-B shows the surface geology and hydrogeology of region in East-West and North-South cross sections (Piggott, Bobba and Novakowski, 1996).

and the wastes generated by these facilities have historically been managed using near surface burial and deep well injection. Sporadic discharge of formation fluids, and possibly industrial waste, to the surface has caused concern over the contamination of Lake Huron, the Saint Clair River, and the regional aquifer that serves as a water supply for rural Lambton County. Migration of wastes injected at depth to the near surface via discontinuities in the confining strata and abandoned deep wells is a plausible mechanism for the contamination of surface and subsurface waters. The numerical models are used to determine rates and directions of groundwater flow for contamination transport studies, predict the impact of varying climatic conditions and groundwater consumption, guide sampling programs for natural and anthropogenic contamination, and determine groundwater discharge to surface water bodies. The geology and hydrogeology of the site are described earlier (Bobba, 1993). The groundwater is obtained from sand and gravel deposits at depths below 60 m. The aquifer is called as Fresh Water Aquifer is floating on top of high saline water due to density difference. Flow in this aquifer is likely be strongly influenced by precipitation events and results in either discharge to shallow ditches, creeks and rivers of recharge to the deeper formations.

The application of the numerical model explained earlier (Bobba, 1993). Figure 4 shows the computed hydraulic head map which best approximates the observed water table. The model results indicate that virtually all of the water flowing through the aquifer discharges to the St. Clair River and Lake Huron. The water level contours in the bedrock valley indicate a generally westward groundwater flow system. Since hydraulic gradient varies across the bedrock valley, the rates of groundwater flow also vary. Calculations indicate that groundwater flux into and out of the valley are about 0.9 and 0.5 m^3 / sec, respectively, for a 100 m strip of aquifer. The amount which flows westward out of the (0.5 m^3 /sec) probably discharges to the St. Clair River. Discharge of groundwater from the Canadian side of the Fresh Water Aquifer is calculated to be between 0.45 and 0.50 m^3 / sec for that portion of the river between Lake Huron and Stag Island.

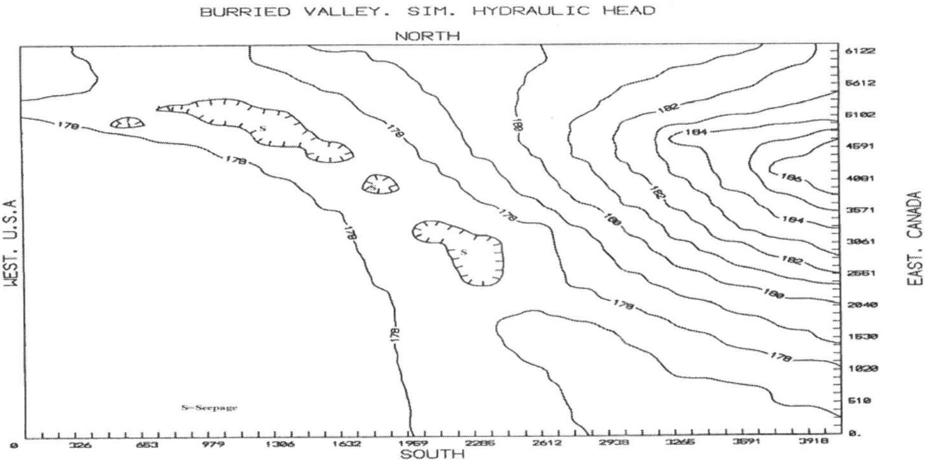

Fig. 4. Simulated hydraulic head map of the buried valley, and groundwater seepage locations in Saint Clair River, Canada and USA boundary (Bobba, 1993).

28.4.2. Case Study 2: Port Granby radioactive disposal site

Canada has a uranium refinery, at Port Hope, Ontario. The waste from refinery was disposed at Port Granby waste management site located on the north shore of Lake Ontario (Figure 5). In recognition of concern over the possible contamination of surface lake waters, the concentrations of radium and uranium were measured in water samples collected of Lake Ontario coastal zone near waste site. These data showed that the leachate infiltrating and seeping to coastal zone of Lake Ontario. The plume, moving parallel to the shoreline in the direction of the prevailing wind direction (Figure5). The finite element model applied to calculate hydraulic head and contamination discharge to lakeshore. The predicted Ra-226 contamination concentration from waste disposal site to beach is shown in Figure 6. The groundwater discharge from the waste disposal site between the east and west creeks is computed by applying Darcy's law.

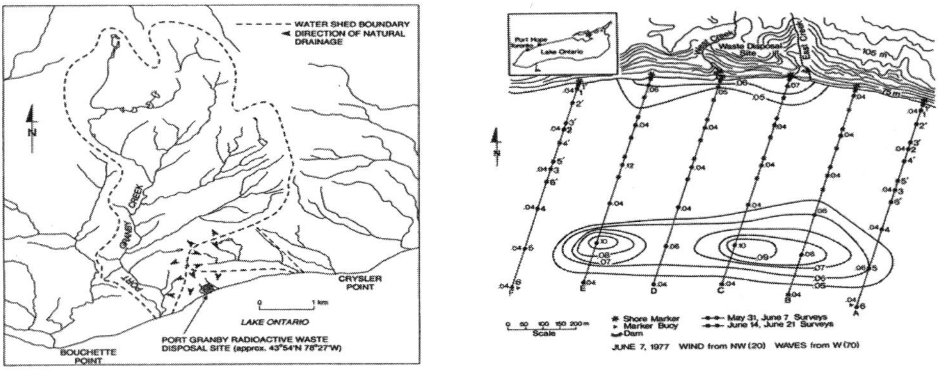

Fig. 5. Location of waste disposal site and Ra-226 concentration in Lake Ontario, Canada.

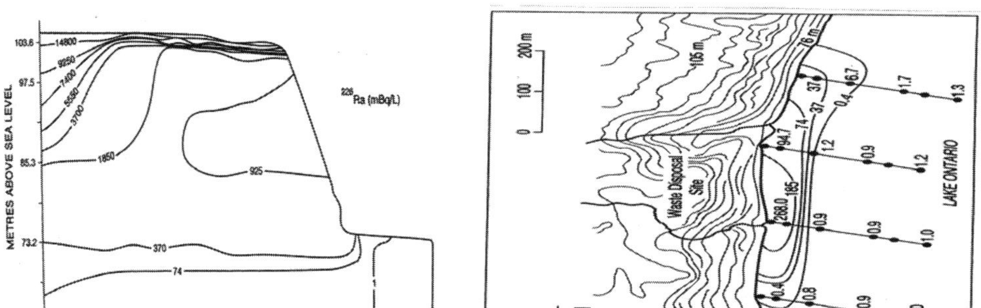

Fig. 6. Computed Ra -226 concentration in waste site, beach and observed concentration in coast (Bobba and Joshi, 1988).

The groundwater flow to Lake Ontario from the site is 2.5 x 10^5 m^3/yr. This annual volume of water is about 3 - 4% of the total solid volume of the site, assuming a depth of about 35 m. If the ^{226}Ra activity in the shore piezometers averages about 100 Bq m^{-3}, about 2.5 x 10^{7} Bq y^{-1} is carried into the lake. But the total amount of ^{226}Ra which has been disposed of at the site is 2.3 x 10^{13} Bq, so that only about 10^{-6} of this is lost from the site per year. A comparable calculation for uranium indicates that about 25 kg of this element reaches the lake. The details of the application of model and interpretation of results were presented earlier (Bobba and Joshi, 1988, 1989).

28.4.3. Case Study 3: Godavari Delta, India

The Godavari delta is located in East Coast of India (Figure 7). The details of geology and environmental problems have been explained earlier (Bobba 2002). The Godavari delta lies between the sea level and 12-m contour. The delta has a projection of about 35-km into the sea from the adjoining coast.

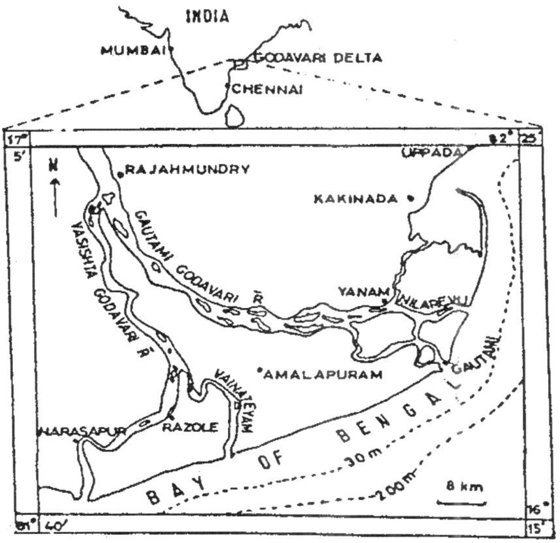

Fig. 7. Location of the Godavari delta.

The Godavari delta consists of alluvial plain. It has a very gentle land slope of about 1-m per km. The coastal line along the study area measures to about 40 km and the general elevation varies from about 2m near the sea to about 13m at the upper reach. Texturally, a major part of the study area consists of sandy loams and sandy clay loams. The silty soils, which are very deep, medium textured with fine loamy soils is located all along the Godavari River as a recent river deposits. The very deep, coarse textured soils with sandy sub-soils representing the coastal sand are also found along the sea.

The delta has a network of canal systems and increased use of new crops along with chemical fertilizers and pesticides have brought about rapid growth in the agricultural output. The crops irrigated with river water throughout the year except between the last week of April and the second week of June. As ample surface water is made available to irrigate the delta, there has been no effort to use groundwater. Potassium fertilizers are extensively used for increasing the crop yield. The variation in potassium concentrations in groundwater in the study area is shown in Figure 8. The peak values are generally observed in November. Higher concentrations are found due to the application of potash fertilizer besides contribution from soils. The temporal variation of chloride and bicarbonate from groundwater samples is shown in Figure 9.

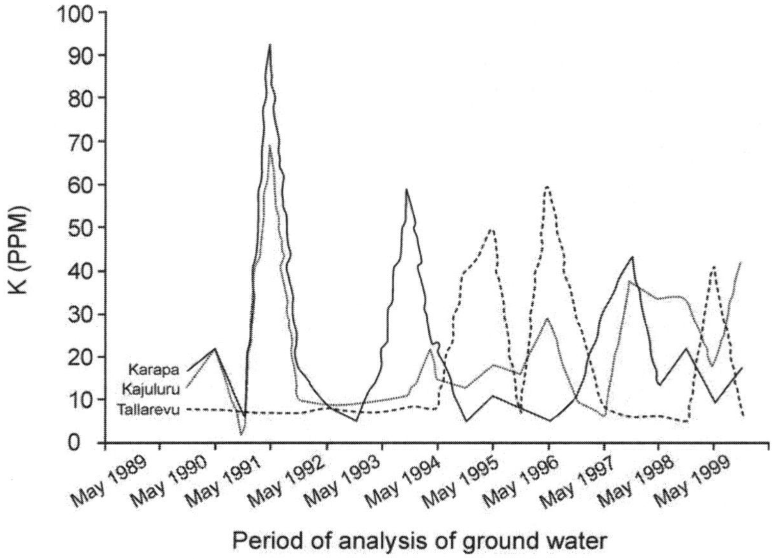

Fig. 8. Variation of potassium concentration in groundwater in the Godavari Delta (Chachadi and Teresa, 2002).

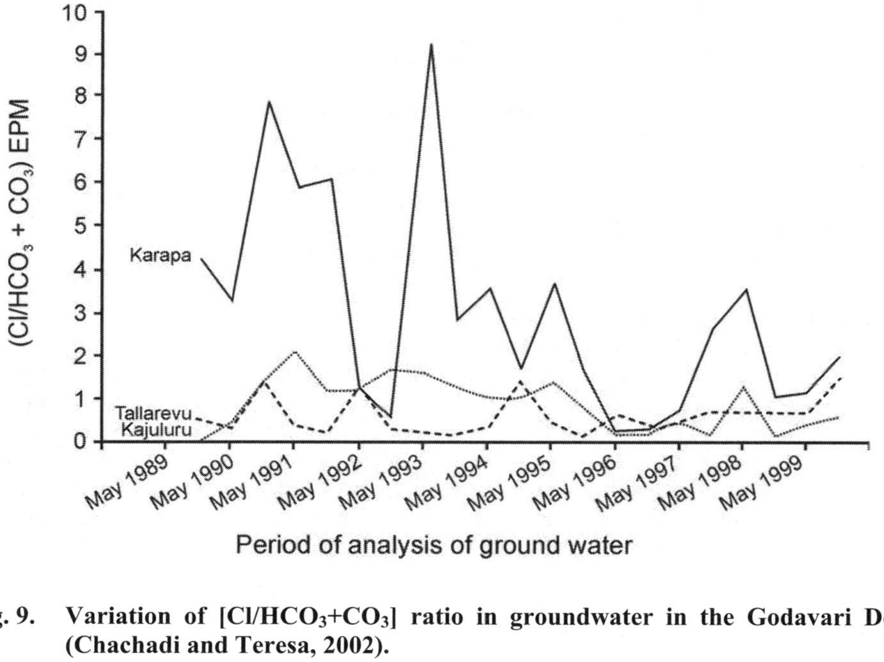

Fig. 9. Variation of $[Cl/HCO_3+CO_3]$ ratio in groundwater in the Godavari Delta (Chachadi and Teresa, 2002).

The ratio of chloride/bicarbonate + carbonate can be used as criteria to evaluate sea water intrusion. Chloride is the dominant ion in seawater and it is only available in small quantities in groundwater while bicarbonate, which is available in large quantities in groundwater, occurs only in very small quantities in seawater. The details of the model and application to the Godavari delta basin have been explained earlier by Bobba (2000, 2002). The prediction of the water table depth due to irrigation and saltwater intrusion reported earlier (Bobba, 2000, 2002a). Figures 10b, and 11b depict the influence of irrigation (rainy) and non- irrigation seasons. During the irrigation season periods, the water table is raised due to irrigation water recharged to subsurface. The distance between surface and water table in the coastal area is very small, and the material is generally composed of sands, which do not retain significant amounts of moisture under unsaturated conditions. Hence,

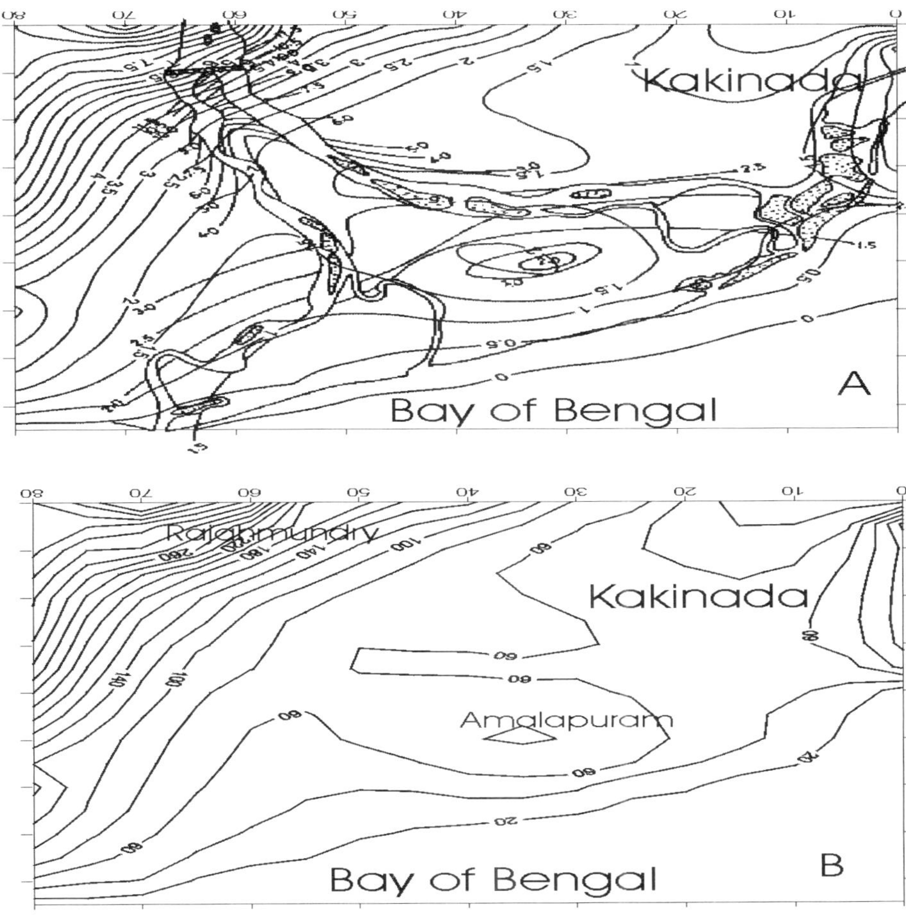

Fig. 10. Simulated hydraulic heads (A) and freshwater depth (B) of Godavari delta in non-irrigation months (Bobba, 2002).

the irrigated water overflows and directly recharges the unsaturated zone. The distance between the water table and surface is at a minimum in the central portion of the delta. It has been observed that areas of minimum depth from the ground level to the water table have high freshwater potential whereas lowering of the water table from the ground surface reduces the freshwater potential substantially. The water table elevation varies from 0.5 m to 1 m from MSL and decreases gradually towards the coastal side. Patches of freshwater zones are also present along coastal areas (Figures 10b and 11b).

The distance between the water table and surface is at a minimum in the central portion of the delta. It has been observed that areas of minimum depth from the ground level to the water table have high freshwater potential whereas lowering of the water table from the ground surface reduces the freshwater potential substantially. The water table elevation varies from 0.5 m to 1 m from MSL and decreases gradually towards the coastal side. Patches of freshwater zones are also present along coastal areas (Figures 10b and 11b).

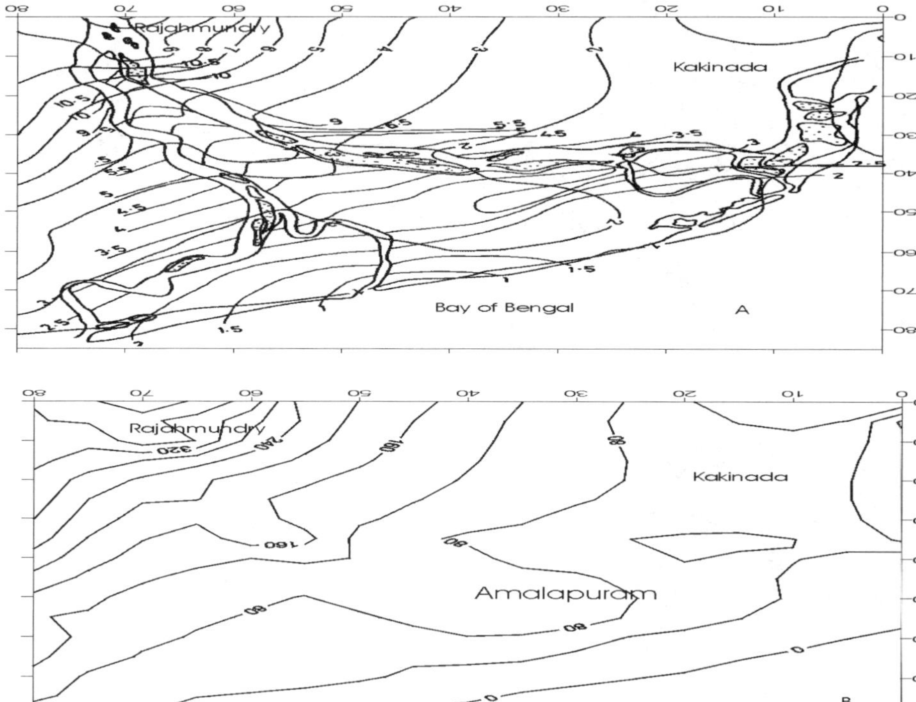

Fig. 11. Simulated hydraulic heads (A) and freshwater depth (B) of Godavari delta in irrigation season months (Bobba, 2002).

Figure 12 shows the results in different environmental conditions due to heavy and long rainy season and drought conditions due to high temperature and less rainfall (climate change). Higher water table conditions are observed due to heavy rain and irrigated water is recharged to the aquifer. The salt water was flushed out or stopped saltwater intrusion to the aquifer. However, if the severe drought conditions (higher temperature, lesser rainfall) occur in the delta, the water table is reduced due to higher evapotranspiration and over pumping the ground water for irrigation and domestic purposes. The salt water intruded to the aquifer and freshwater thickness reduced in the delta.

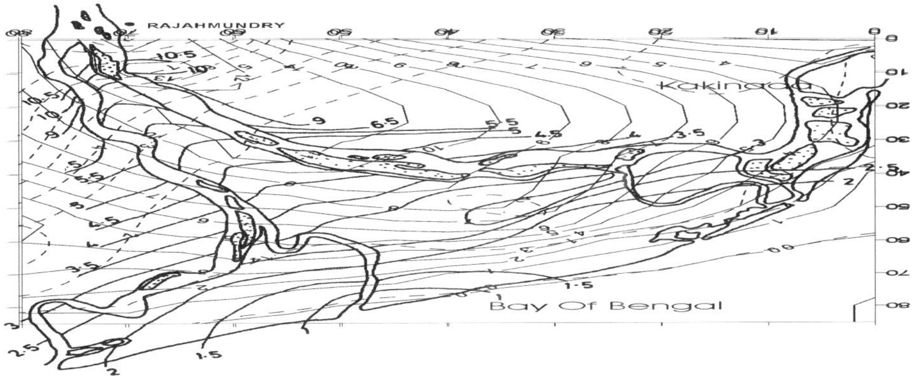

Fig. 12. Simulated hydraulic heads of Godavari delta in different seasons (solid line in heavy rainy season, ---- drought conditions) (Bobba, 2002).

28.5. SUMMARY AND CONCLUSIONS

Mathematical and numerical models are useful to estimate subsurface water and contamination discharge to coastal areas from point and non-point source areas. Applications of two types of numerical models were presented in this paper. As an example, a finite element model, considering open boundary conditions for coasts and a sharp interface between freshwater and salt water was applied under steady-state conditions to the phreatic aquifer for fresh water surplus and deficits at the coastline due to El Nino effect. When recharges of saltwater occur at the coastline, essentially of freshwater deficits, a hypothesis of mixing for the freshwater - saltwater transition zone allows the model to calculate the resulting seawater intrusion in the aquifer. Hence, an adequate treatment and interpretation of the hydrogeological data that are available for a coastal aquifer are of main concern in satisfactorily applying the numerical model.

The results of the steady state simulations showed reasonable calculations of the water table levels and the freshwater and saltwater thicknesses, as well as, the extent of the interface and seawater intrusion into the aquifer for the total discharges or recharges along the coastline. As a result of the present hydrogeological simulations on the subsurface system, a considerable advance in seawater intrusion would be expected in the coastal watershed if the sea level rises due to climate change and El Nino effects.

REFERENCES

Bobba, A.G., 2002. Numerical modelling of saltwater intrusion due to human activities and sea level change in the Godavari delta. Hydrological Sciences Journal 47(S): S67-S80.

Bobba, A.G., 2000. Numerical simulation of saltwater intrusion into coastal basin of Indian sub-continent due to anthropogenic effects. ICIWRM - 2000. Proceedings of International Conference on Integrated Water Resources Management for Sustainable Development. Published by National Institute of Hydrology, Roorkee, India. V. 1. pp. 323-340.

Bobba, A.G., 1998. Application of numerical model to predict freshwater depth in islands due to climate change effect: Agati Island, India. Journal of Environmental Geology 6: 1-13.

Bobba, A.G., 1993a. Mathematical models for saltwater intrusion in coastal aquifers-Literature review. Water Resources Management 7: 3-37.

Bobba, A.G., 1993b. Field validation of "SUTRA" groundwater flow model to Lambton County, Ontario, Canada. Water Resources Management 7: 289-310.

Bobba, A.G., Singh, V.P., and Bengtsson, L., 2000. Application of environmental models to different hydrological systems. Journal of Ecological Modelling 125: 15-49.

Bobba, A.G., and Singh, V.P., 1995. Groundwater contamination modelling. Chapter#8. Environmental Hydrology. Edited by V.P. Singh. Kluwer Academic Publishers. pp. 225-319.

Bobba, A.G., and Joshi, S.R., 1988. Groundwater transport of Radium-226 and uranium from Port Granby Waste management site to Lake Ontario. Nuclear and Chemical Waste Management 8: 199-209.

Bokuniewicz, H.J., 1980. Groundwater seepage into Great South Bay, New York. Estuarine and Coastal Marine Science 10: 437-444.

Chachadi, A.G., and Teresa, L., 2002. Health of the groundwater regime in a coastal delta of East Godavari, Andhra Pradesh. Coastin, A Coastal Policy Research Newsletter, Published by TERI, 9607/13, Multani Dhanda, Paharganj, New Delhi-110 055, India. Pp.5-8.

Valiela, I., Foreman, K., LaMontagne, M., Hersh, D., Costa, J., Peckol., DeMeo-Anderson., D'Avanzo, C., Babione, M., Sham, C., Brawley, J., and Lajyha, K., 1992. Coupling of watersheds and coastal waters: Sources and consequences of nutrient enrichment in Waquoit Bay, Massachusetts. Estuaries 15: 443-457.

Voss, C.I., 1984. SUTRA-A finite element simulation model for saturated – unsaturated fluid density dependent groundwater flow with energy transport or chemically reactive single species solute transport, USGS Water Resource. Investigation Report 84-4269.

Section 6:

COASTAL ECOLOGY, ENVIRONMENT AND RESTORATION

FISH ASSEMBLAGE INSTABILITY AND HYDROLOGIC INFLUENCES IN LAKE PONTCHARTRAIN, LOUISIANA (USA), A DEGRADED OLIGOHALINE ESTUARY

Martin T. O'Connell
Pontchartrain Institute for Environmental Sciences
349 CERM Building - Research and Technology Park
University of New Orleans, New Orleans, LA 70148, USA

Robert C. Cashner
Department of Biological Sciences
University of New Orleans, New Orleans, LA 70148, USA

Christopher S. Schieble
Pontchartrain Institute for Environmental Sciences, 2079 Science Building
University of New Orleans, New Orleans, LA 70148, USA

ABSTRACT

Lake Pontchartrain, an oligohaline estuary in southeastern Louisiana, has been subject to numerous anthropogenic impacts over the last half century including urban and agricultural runoff, shell dredging, over-fishing, artificial saltwater and freshwater inputs, shoreline alteration, and industrial discharges. Previous analysis of long-term fish data showed that significant changes in fish assemblages have occurred between 1954 and 2000. To assess the current condition of fish assemblages in Lake Pontchartrain relative to multiple ongoing and past environmental stressors, we conducted a three-year (2000-2003) monthly survey of fishes using trawls (7 sites), gillnets (5 sites), and beach seines (5 sites). Because increased fish assemblage variability has been used as an identifiable symptom of perturbed situations, we compared relative multivariate dispersion indices for assemblages representing different sites to determine which region appeared most affected by anthropogenic impacts. We also examined possible correlations between fish assemblage stability and five hydrologic factors: salinity, temperature, Secchi disk depth, dissolved oxygen, and specific conductivity. Beach seine and gillnet data from the southeastern region exhibited the most assemblage dispersal over the three-year period. This region is hydrologically associated with an artificial connection to saltwater, the Mississippi River Gulf Outlet. Compared to the other regions, trawl data from the southwestern region had the highest dispersal readings and also showed the second highest dispersal in beach seine and gillnet data. The northeastern region of Lake Pontchartrain had the least assemblage dispersal over time for all three gear types. This area has the most submersed aquatic vegetation in the estuary and is considered the least impacted. Of the hydrologic factors examined, temperature was most commonly correlated to fish assemblage changes over the three years. These results suggest that the greatest ecological instability occurred in areas associated with artificial waterways.

29.1. INTRODUCTION

Habitat quality is the primary factor determining stability in ecological assemblages of species (Ross et al., 1985; Kutka and Richards, 1996). For example, in ecosystems across the planet habitat loss is recognized as the leading threat to biodiversity (Dobson et al., 1997). In aquatic ecosystems hydrologic factors are the key components affecting habitat quality and these abiotic influences are especially important in estuaries (Day et al., 1989; O'Connell et al., 2005). Changes in factors such as salinity, temperature, and water clarity often lead to changes in the composition of local aquatic organisms as species abandon habitats that are no longer suitable (Hastings et al., 1987; Wagner, 1999; Matern et al., 2002; O'Connell et al., 2004). As a rule, where habitat quality is consistently high over time the composition of ecological assemblages remains stable (Araujo et al., 2000; Clarke and Warwick, 2001; Chavez-Lopez et al., 2005).

The aquatic habitats and fish assemblages of Lake Pontchartrain, an oligohaline estuary in southeastern Louisiana, have been subject to numerous anthropogenic impacts over the last half century (Thompson and Fitzhugh, 1985; O'Connell et al., 2004). Urban and agricultural runoff, shell dredging, over-fishing, artificial saltwater and freshwater inputs, shoreline alteration, and industrial discharges have all contributed to the degradation of local estuarine aquatic habitats (Penland et al., 2002). In turn, the fisheries and fish assemblages of the estuary have also been negatively affected (Thompson and Fitzhugh, 1985; O'Connell et al., 2004). While recent interest and research funding has focused on habitat restoration within Lake Pontchartrain, recovery to a fully functioning estuarine ecosystem will require extensive management efforts (Lopez, 2005). A key to these recovery efforts will be determining which regions and habitats of the estuary are the most degraded and, therefore, require the most attention.

The objective of our research was to examine the relative stability of Lake Pontchartrain fish assemblages to determine which regions and habitats of the estuary appear the most degraded. Once the most degraded areas are identified we can establish which anthropogenic impacts and hydrologic conditions are associated with these localities. This information could then be used to better manage habitat restoration in the most affected regions and habitats. Specifically, we asked: 1. which regions of Lake Pontchartrain exhibited the least stable fish assemblages over time based on three years' of monthly fish survey data from three habitat types (demersal, pelagic, and inshore), and 2. what were the hydrologic conditions most influencing any changes in fish assemblages.

29.2. MATERIALS AND METHODS

Lake Pontchartrain has a surface area of 1,630 km^2 and a mean depth of 3.7 m (Sikora and Kjerfve, 1985). The hydrology of this ecosystem is influenced by a combination of natural and anthropogenic influences. The northwest region has the highest natural freshwater input (mostly from Lake Maurepas to the west and the Tangipahoa River to the north) and the least modified hydrology. Increases of urbanization and agricultural run-off from dairy farms have recently begun to affect this region (Penland et al., 2002). In the southwest region, the largest environmental influence is the Bonnet Carre Spillway (BCS). This structure was built in 1931 as a means of flood control and forms an artificial connection between the Mississippi River and the Lake (Sikora and Kjerfve, 1985). During periods of exceptionally high discharge in the Mississippi River (e.g., 1937, 1945, 1950, 1973, 1975, 1979, 1983, and 1997) fresh water is diverted from the river, through the BCS, and into southwestern Lake Pontchartrain. Other factors that impact this region are industrial discharge from oil refineries and urban runoff from canals draining New Orleans. The majority of submersed aquatic vegetation (SAV) in Lake Pontchartrain occurs in the northeast region. Like the northwest region, this area receives natural freshwater input from rivers, but exhibits more degradation from agricultural and urban runoff. The area of the estuary that is most impacted by anthropogenic influences is southeastern Lake Pontchartrain. An extensive artificial shoreline (i.e., armored seawalls, rip-rap) and high urban runoff are the characteristics that best describe this region. The influence of greatest magnitude on this region is the artificial connection between Lake Pontchartrain and the Gulf of Mexico created by the construction of the Mississippi River-Gulf Outlet (MR-GO), a deep-water canal completed in 1963 (Sikora and Kjerfve, 1985). Through its connections with the Inner Harbor Navigation Canal and the Intra-coastal Waterway, the MR-GO has allowed unnaturally high salinity water to enter this region of the estuary (Poirrier, 1978). Lake Pontchartrain's natural connection to marine waters occurs in the eastern-most region. This region receives saltwater input from Lake Borgne to the east and is consistently the region of highest salinity in Lake Pontchartrain.

To assess possible fish assemblage changes within Lake Pontchartrain, we conducted a three-year (2000-2003) monthly survey of fishes using trawls (7 sites), gillnets (5 sites), and beach seines (5 sites). These sites correspond to the different natural and anthropogenic influences discussed above while the three sampling gear types (trawls, gillnets, and beach seines) represent different habitat types: demersal, pelagic, and inshore habitats, respectively. The regions represented by these sites are as follows: P1 (east region), P2 (southeast region), P4 (northeast region), P5 (southwest region), P6 (mid-lake west region), P7 (mid-lake central region), and P14 (northwest region; Fig. 1). The two mid-lake sites (P6 and P7) were only sampled with trawls because gillnets and beach seines cannot be used in these deeper habitats. Because increased fish assemblage variability has been used as an identifiable symptom of perturbed situations (Warwick and Clarke, 1993), our first goal was to compare relative multivariate dispersion indices for assemblages representing different sites to determine which regions and habitats appeared most affected by anthropogenic impacts. More plainly, we wanted to determine which sites in Lake Pontchartrain showed the most fish assemblage variability over our three year sampling period. Calculating relative multivariate dispersion indices begins with the construction of a triangular non-metric multidimensional scaling (MDS) similarity matrix as described in Clarke and Warwick (2001). The values in this matrix are measures of Bray-Curtis similarity for all pair-wise comparisons of assemblage data. For example, if the fish assemblage collected at Time 1 from Site A is very similar to that collected at Time 1 from Site B, the Bray-Curtis similarity would be relatively large. A pair of fish assemblages that were dissimilar would have a lower Bray-Curtis similarity index and so on. All pair-wise comparisons are included in the MDS similarity matrix and are the basis for calculating a relative multivariate dispersion index for each sampling site. From the matrix, only those intra-site similarities (i.e., changes over three years at each site in the present study) are averaged to calculate relative dispersion for the fish assemblage of a given site. Greater values indicate greater dispersion or increased fish assemblage instability. For the present study, monthly samples were taken from each site between July 2000 and June 2003. Monthly sampling for gillnets and beach seines consisted of one collection per month per site while trawl samples were taken twice a month at each site. To retain equal sample sizes for statistical purposes, if a monthly sample was missed at any site (due to gear failure, etc.) then data for all sites were omitted from analyses for that month. Final total sample sizes over the three years for trawl, gillnet, and beach seine samples were 465, 119, and 184 samples, respectively. Details of sampling methods are discussed in O'Connell et al. (2004). All analyses were conducted using PRIMER v5 software (Clarke and Warwick, 2001).

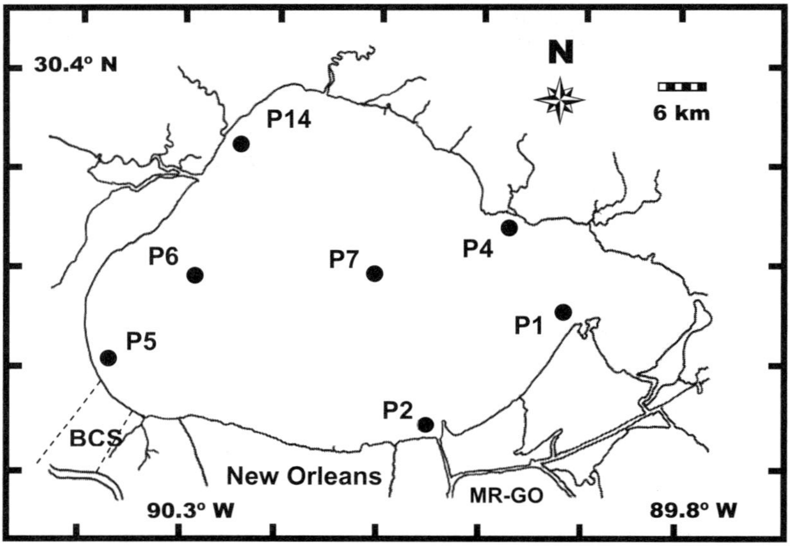

Fig. 1. Location map of Lake Pontchartrain, Louisiana showing the seven fish assemblage sites that were sampled monthly between July 2000 and June 2003. Note that only trawl samples were collected at mid-lake sites P6 and P7. Sources of anthropogenic disturbances shown are the Bonnet Carre Spillway (BCS), the City of New Orleans, and the Mississippi River Gulf Outlet (MR-GO).

Our secondary goal was to determine which hydrologic conditions could be influencing changes in fish assemblages over time. Previous research with estuarine fish assemblages has shown that assemblages from impacted habitats are not only more variable over time but also respond less to natural changes in hydrologic factors (Araujo et al., 2000; Peterson et al., 2000; Clarke and Warwick, 2001; O'Connell et al., 2004). For example, in less-impacted estuaries the fish assemblages may change seasonally as more marine fishes enter the system during periods of increased salinity. Such correlations between species composition and local hydrologic conditions are weaker in impacted estuaries (Araujo et al., 2000; Matern et al., 2002; O'Connell et al., 2004). To test for correlations between fish assemblage change over time and local hydrologic conditions we used the BIO-ENV and RELATE procedures of Clarke and Warwick (2001). As with calculating relative multivariate dispersion indices, these procedures are based on an MDS of changes in fish assemblages over time for each sampling site. The BIO-ENV procedure compares measurements of hydrologic conditions from these sites with fish assemblage changes to test for agreement in pattern (i.e., correlation). For the present study we measured salinity (ppt), temperature (°C), Secchi disk depth (m), dissolved oxygen (ppm), and specific conductivity (mS/cm) at each fish sampling event to determine which hydrologic factor was the most correlated with fish assemblage change at each site. When the factor with the highest correlation to assemblage change was identified by the BIO-ENV procedure, we then used the RELATE procedure to test for the significance of the relationship. The RELATE procedure first calculates the actual correlation (ρ) between the factor and fish assemblage change based on rank correlation. This value is then compared to 1000 randomly generated correlation values based on the original data set to determine if the actual value is significantly different from expected. For these tests we chose a critical value of $p = 0.025$ and the following analytical options in PRIMER v5 were used: weighted Spearman ranking, Euclidean distances were compared, and data were neither transformed nor standardized. The outcome of using these two procedures was to determine which hydrologic factors were most correlated with fish assemblage change at each site and to then determine if this correlation was significant.

29.3. RESULTS

For all three gear types (trawl, gillnets, and beach seines), sites P2 and P5 consistently exhibited the greatest relative dispersion (i.e., assemblage instability) over the three year sampling period (Table 1). Out of all sites, P2 had the highest RMD indices for both gillnet and beach seine samples (RMD = 1.186 and 1.204, respectively) while exhibiting only relatively moderate dispersion (RMD = 0.939) in trawl samples (Table 1). Site P5 had the highest RMD index for trawl samples (RMD = 1.485) and was second only to P2 in high dispersion values for gillnet and beach seine samples (RMD = 1.033 and 1.092, respectively). The order of RMD indices was identical for both gillnet and beach seine samples: P2 (greatest dispersion) > P5 > P1 > P14 > P4 (least dispersion). This exact pattern was not observed for the trawl samples. Site P4 had the lowest RMD indices across all three gear types (Table 1).

Table 1. Relative multivariate dispersion (RMD) indices for fish assemblages collected over three years from seven sites in Lake Pontchartrain.

Trawl Samples		Gillnet Samples		Beach Seine Samples	
site (region)	RMD index	Site (region)	RMD index	site (region)	RMD index
P5 (SW)	1.485	P2 (SE)	1.186	P2 (SE)	1.204
P14 (NW)	1.424	P5 (SW)	1.033	P5 (SW)	1.092
P7 (mid.)	1.061	P1 (E)	0.985	P1 (E)	1.026
P2 (SE)	0.939	P14 (NW)	0.954	P14 (NW)	0.975
P6 (mid)	0.939	P4 (NE)	0.915	P4 (NE)	0.702
P1 (E)	0.818				
P4 (NE)	0.333				

Significant correlations between fish assemblage changes and hydrologic factors were identified in 10 of the 17 site/gear type combinations (Table 2). For 6 of these 10 significant correlations, temperature was identified as the most significant correlate with fish assemblage change: trawl samples at P5; gillnet samples at P1, P2, and P5; and beach seine samples at P1 and P14 (Table 2). The remaining significant correlates were specific conductivity (trawl samples at P4), salinity (trawl samples at P7), Secchi depth (gillnet samples at P4), and dissolved oxygen (beach seine samples at P5).

Table 2. Hydrologic factors most often correlated with fish assemblagechange at seven sites in Lake Pontchartrain over three years.

Sampling Gear	Site (Region)	Best Correlate	Correlation (ρ)	p Value (critical value = 0.025)
trawl				
	P1 (E)	Secchi depth	0.004	0.095
	P2 (SE)	salinity	-0.037	0.223
	P4 (NE)	**specific conductivity**	**0.161**	**0.005**
	P5 (SW)	**temperature**	**0.06**	**0.018**
	P6 (mid.)	Secchi depth	0.072	0.051
	P7 (mid.)	**salinity**	**0.134**	**0.004**
	P14 (NW)	Secchi depth	0.043	0.094
gillnet				
	P1 (E)	**temperature**	**0.341**	**0.001**
	P2 (SE)	**temperature**	**0.279**	**0.001**
	P4 (NE)	**Secchi depth**	**0.231**	**0.004**
	P5 (SW)	**temperature**	**0.413**	**0.001**
	P14 (NW)	specific conductivity	0.098	0.046
beach seine				
	P1 (E)	**temperature**	**0.052**	**0.011**
	P2 (SE)	Secchi depth	0.074	0.207
	P4 (NE)	salinity	0.049	0.074
	P5 (SW)	**dissolved oxygen**	**0.228**	**0.001**
	P14 (NW)	**temperature**	**0.203**	**0.001**

29.4. DISCUSSION

Between July 2000 and June 2003, fish assemblages in the southern portion of Lake Pontchartrain were less stable than fish assemblages in other parts of the estuary. While a casual observer could have predicted this outcome based solely on the obvious and extensive habitat degradation found along the south shore of the Lake (i.e., armored seawalls, urban development, etc.), our quantification of this ecological instability further verifies the connection between habitat destruction and negative impacts on local fishes. While sites P2 and P5 both have numerous habitat limitations associated with their locations, it is their proximity to two artificial waterways (the MR-GO and the BCS, respectively) that offers the most obvious explanation for the instability exhibited by their fish assemblages (Fig. 1). For example, site P2 is located adjacent to a "dead zone" that has been created by the construction of the MR-GO (Poirrier, 1978; Abadie and Poirrier, 2000; Lopez, 2005). Higher salinity waters entering Lake Pontchartrain through this artificial connection has led to increased stratification in the water column of the southeastern region. Stratification, in turn, has led to consistent anoxic or hypoxic conditions in local demersal habitats which has negatively impacted local populations of the Rangia clam (*Rangia cuneata*), a keystone estuarine species (Abadie and Poirrier, 2000; O'Connell et al., 2005). Likewise, the area where overflow water from the Mississippi River enters Lake Pontchartrain through the BCS is just south of site P5 (Fig. 1). Unlike the MR-GO connection with the Lake, this river-lake link only exists during rare overflow periods when high Mississippi River discharge poses a potential flooding threat to New Orleans (Sikora and Kjerfve, 1985). Prior to the construction of levees along this portion of the River, a natural crevasse existed in the area where the BCS was built. This crevasse allowed seasonal flooding to inundate and nourish the thriving cypress swamp that once surrounded the southwest portion of Lake Pontchartrain. Without this regular hydrological connection, though, the swamp ecosystem that once supported the animals and plants of this region of the Lake has largely died-off. Irregular openings of the BCS (only 8 openings since it was built in 1931) are inadequate replacements for a natural flooding regime because during these events river water is quickly flushed into Lake Pontchartrain and not allowed to slowly spread out into the swamp. One result is that river sediments are not given time to deposit in swamp areas but rather accumulate in the estuary. These sediments have created a substrate in the southwest portion of Lake Pontchartrain that is an unstable combination of loose detritus and mud which is sub-standard habitat for most local fish species (MTOC, personal observation). The negative ecological effects caused by both these artificial waterways could be reduced or eliminated with proper management of the hydrological connections.

It would also not be surprising to the casual observer familiar with Lake Pontchartrain that we measured the greatest fish assemblage stability for all three gear types at site P4 in the northeast region of the estuary (Table 1). This area retains the most SAV in the Lake and these beds of vegetation provide valuable habitat for fishes and other aquatic organisms (Day et al., 1989; Zimmerman et al., 2000; O'Connell et al., 2005). For example, juveniles of important nekton species such as brown shrimp (*Farfantepenaeus aztecus*) are more commonly associated with vegetated habitats than non-vegetated areas (Zimmerman et al., 2000). The benefits of this habitat association are twofold in that these vegetated habitats offer both increased resources for growth (e.g., invertebrate prey items for fishes) and protection from potential predators (Turner and Brody, 1983; Zimmerman et al., 2000). Unfortunately most of Lake Pontchartrain's SAV has disappeared due to habitat degradation (Cho and Poirrier, 2002). As recently as 50 years ago researchers described SAV as being abundant on both the north and south shores of the Lake (Suttkus et al., 1954). While the cessation of shell-dredging in 1990 has led to some recent SAV recovery (Cho and Poirrier, 2002), the northeastern region of Lake Pontchartrain remains the only portion of this oligohaline estuary to exhibit both habitat and fish assemblage stability. Natural resource managers and conservationists concerned with the recovery of Lake Pontchartrain would do well to recognize the association between SAV and the stability of estuarine fish assemblages.

Temperature was the hydrologic factor most associated with fish assemblage change at the majority of Lake Pontchartrain sites we sampled and this may also reflect recent steps toward environmental recovery. Under natural conditions, estuarine fishes move in and out of oligohaline habitats on a seasonal basis and temperature serves as a proxy measurement of this process. A seasonal response by fishes to natural fluctuations in temperature occurs in other estuaries (Hastings et al. 1987; Wagner 1999; Matern et al. 2002) and is likely to become more prevalent in degraded systems when anthropogenic factors are reduced (Araujo et al., 2000; Matern et al., 2002; O'Connell et al., 2004). There were no direct associations between fish assemblage stability and any of the hydrologic factors. For example we might have expected the two least stable sites (P2 and P5) to have no significant correlations to hydrologic factors while more stable sites such as P4 would have seen fish assemblages changing in close response to seasonal shifts in environmental conditions. In fact, site P5 showed significant correlations with hydrologic factors for all three gear types: temperature for trawl and gillnet samples and dissolved oxygen for beach seine samples (Table 2). It is likely that factors beyond local environmental conditions need to be considered when interpreting these results. For instance, gear avoidance has been recognized as playing a role in fish assemblage measurements in Lake Pontchartrain (Schieble et al., 2002). This might explain the significant correlation between fish assemblage change and Secchi depth (a measurement of water clarity) at site P4 for gillnet samples because some fishes detect and avoid gillnets more under clear water conditions.

There are some trends in the hydrologic results that suggest inter-habitat differences and that agree with other long-term research we have conducted on Lake Pontchartrain fisheries (O'Connell et al., 2004). The three gear types sample different habitats and it is interesting to note that a significant seasonal (i.e., temperature) response was only observed at 1 of 7 trawl sites (P5). A significant seasonal response occurred at 2 of 5 beach seine sites (P1 and P14) and 3 of 5 gillnet sites (P1, P2, and P5). The broad interpretation of these findings is that the lack of seasonal response in demersal habitats (sampled by trawls) means these sites are less healthy than the other habitats. In turn, Lake Pontchartrain pelagic or mid-water habitats (sampled by gillnets) appear the least impacted and inshore habitats (sampled by beach seines) are midway in condition between the other two habitats. This result for 3 years' worth of data matched what we found when comparing fish assemblage stability in Lake Pontchartrain over 50 years (O'Connell et al., 2004). In these long-term data we observed that changes in fish assemblages from inshore (beach seine) and pelagic (gill net) habitats were more closely related to seasonal environmental fluctuations suggesting some level of relative health. Conversely, the long-term trawl data showed a reduced assemblage role of benthic fishes and a marked assemblage change. These results suggested that over the last half century demersal habitats in Lake Pontchartrain have been significantly impacted (O'Connell et al., 2004). The consistency in results between these two studies using different temporal scales (50 years versus 3 years) and different methods of analysis (Canonical Correspondence Analysis versus Non-metric Multidimensional Scaling) implies two possibilities. First, even after recent efforts to restore the estuarine habitats of Lake Pontchartrain, the same habitats that were most disturbed over the last half century (i.e., demersal habitats) are still the least healthy today. Second, while analysis of long-term data over a large spatial scale allowed us to discriminate natural effects from other, possibly anthropogenic effects in a complex and highly variable estuarine system in the earlier study, we reached a similar conclusion regarding fish assemblage stability and habitat health using a shorter time frame (i.e., 3 years) and much less data in the current study. While long-term data are essential to identifying negative impacts in estuaries throughout the world (Nordstrom and Roman, 1996; Araujo et al., 1998; Marshall and Elliott, 1998; Power et al., 2002), perhaps the methods used in the present study could be used to identify possible regions of ecological instability based solely on a few years' worth of data. Finally, as with other degraded estuaries (Matern et al., 2002), it is unlikely that fish assemblages and habitats in Lake Pontchartrain will stabilize until more natural processes return and the effects of human disturbance seriously reduced.

ACKNOWLEDGMENTS

This research was supported through grants from the United States Fish and Wildlife Service and the National Marine Fisheries Service (NOAA Grant # NA96FW0380). Significant additional support was provided by the Louisiana Department of Wildlife and Fisheries. We are appreciative of the efforts of M. Catallo, H. Finley, K. Foote, D. Norriss, R. Pausina, and M. Schexnayder for supporting this and other ongoing research conducted by our laboratory. We thank K.G. Blanke, G.N. Fuentes, C.M. Kemp, N.E, Rios, and A.M.U. O'Connell for their provision and collection of data. We also thank J.M. King and S.L. Penland for their continued encouragement and support of these endeavors. This manuscript represents publication No. 3 for the Nekton Research Laboratory, Pontchartrain Institute for Environmental Sciences.

REFERENCES

Abadie, S. W., and Poirrier, M. A., 2000. Increased density of large Rangia clams in Lake Pontchartrain after the cessation of shell dredging. Journal of Shellfish Research 19:481-485.

Araujo, F. G., Bailey, R. G., and Williams, W. P., 1998. Seasonal and between-year variations of fish populations in the middle Thames estuary: 1980-1989. Fisheries Management and Ecology 5(1): 1-21.

Araujo, F. G., Williams, W. P., and Bailey, R. G., 2000. Fish assemblages as indicators of water quality in the middle Thames estuary, England (1980-1989). Estuaries 23(3): 305-317.

Chavez-Lopez, R., Franco-Lopez, J., Moran-Silva, A., and O'connell, M.T., 2005. Long-term fish assemblage dynamics of the Alvarado Lagoon Estuary, Veracruz, Mexico. Gulf and Caribbean Research 17: 145-156.

Cho, J., and Poirrier, M.A., 2002. Biological Resources - Submersed Aquatic Vegetation in S. Penland, A. Beall and J. Kindinger (editors), Environmental Atlas of the Lake Pontchartrain Basin: Lake Pontchartrain Basin Foundation, New Orleans, Louisiana, U.S. Geological Survey Open File Report 2 (pp. 2-6), CD-ROM (available at http://coastal.er.usgs.gov/pontchartrain/atlas).

Clarke, K. R., and Warwick, R. M., 2001. Change in marine communities: an approach to statistical analysis and interpretation, 2nd Edition. PRIMER-E: Plymouth, United Kingdom.

Day, J. W., Jr., Hall, C.A.S., Kemp, W.M., and Yanez-Arancibia, A., 1989 Estuarine Ecology. John Wiley & Sons, New York. 576 pp.

Dobson, A.P., Bradshaw, A. D., and Baker, J. M., 1997. Hopes for the future: restoration ecology and conservation biology (in human-dominated ecosystems). Science 277 (5325): 515-522.

Hastings, R. W., Turner, D. A., and Thomas, R. G., 1987. The fish fauna of Lake Maurepas, an oligohaline part of the Lake Pontchartrain Estuary. Northeast Gulf Science 9(2): 89-98.

Kutka, F.J., and Richards, C., 1996. Relating diatom assemblage structure to stream habitat quality. Journal of the North American Benthological Society 15(4): 469-480.

Lopez, J.A. 2005. Comprehensive plan for the Lake Pontchartrain Basin. Plan submitted to the Lake Pontchartrain Basin Foundation. Available online at saveourlake.org. 125 pp.

Marshall, S., and Elliot, M., 1998. Environmental Influences on the Fish Assemblage of the Humber Estuary, United Kingdom. Estuarine, Coastal, and Shelf Science 46: 175-184.

Matern, S.A., Moyle, P. B., and Pierce, L. C., 2002. Native and Alien Fishes in a California Estuarine Marsh: Twenty-One Years of Changing Assemblages. Transactions of the American Fisheries Society 131(5): 797–816.

Nordstrom, K. F., and Roman, C. T., 1996. Estuarine shores: evolution, environments, and human alterations. John Wiley and Sons Ltd., Chichester. 486 pp.

O'connell, M.T., Cashner, R.C., and Schieble, C.S., 2004. Fish assemblage stability over fifty years in the Lake Pontchartrain Estuary; comparisons among habitats using Canonical Correspondence Analysis. Estuaries 27(5):807-817.

O'connell, M.T., Franze, C.D., Spalding, E.A., and Poirrier, M.A., 2005. Biological resources of the Louisiana coast: Part 2. coastal animals and habitat associations. Journal of Coastal Research 44:146-161.

Penland, S., Mccarty, P., Beall, A., and Maygarden, D., 2002. Environmental overview of the Lake Pontchartrain Basin in S. Penland, A. Beall and J. Kindinger (editors), Environmental Atlas of the Lake Pontchartrain Basin: Lake Pontchartrain Basin Foundation, New Orleans, Louisiana, U.S. Geological Survey Open File Report 2 (pp. 2-6), CD-ROM (available at http://coastal.er.usgs.gov/pontchartrain/atlas).

Peterson, M. S., Comyns, B. H., Hendon, J. R., Bond, P. A., and Duff, G. A., 2000. Habitat use by early life-history stages of fishes and crustaceans along a changing estuarine landscape: differences between natural and altered shoreline sites. Wetlands Ecology and Management 8(2/3): 209-219.

Poirrier, M. A. 1978. Studies of salinity stratification in southern Lake Pontchartrain near the Inner Harbor Navigational Canal. Proceedings of the Louisiana Academy of Sciences. 61:26-35.

Power, M., Attrill, M. J., and Thomas, R. M., 2002. Environmental influences on the long-term fluctuations in the abundance of gadoid species during estuarine residence. Journal of Sea Research 47(2): 185-194.

Ross, S.T., Matthews, W.J., and Echelle, A.A., 1985. Persistence of stream fish assemblages: effects of environmental change. The American Naturalist 126(1): 24-40.

Schieble, C. S., O'connell, M. T., and Cashner, R. C., 2002. Species occurrence in nocturnal and diurnal trawl samples of an oligohaline estuary: behavioral differences or gear avoidance. Abstract in Program Book and Abstracts for the Joint Meeting of Ichthyologists and Herpetologists - July 3-8, 2002. p. 269.

Sikora, W. B., and Kjerfve, B., 1985. Factors influencing the salinity regime of Lake Pontchartrain, Louisiana, a shallow coastal lagoon: analysis of a long-term data set. Estuaries. 8 (2A): 170-180.

Suttkus, R. D., Darnell, R. M., and Darnell, J. H., 1954. Biological study of Lake Pontchartrain. Tulane University Zoology Department. Annual Report submitted to the Louisiana Department of Wildlife and Fisheries.

Thompson, B. A., and Fitzhugh, G. R., 1985. Synthesis and analysis of Lake Pontchartrain environments, influencing factors and trends. Report submitted to the Office of Water Resources, Louisiana Department of Environmental Quality.

Turner, R.E., and Brody, M.S., 1983. Habitat suitability index models: northern Gulf of Mexico brown shrimp and white shrimp. US Department of Interior, US Fish and Wildlife Service Biological Services Program, FWS/OBS-82/10.54 24.

Wagner, C. M. 1999. Expression of the estuarine species minimum in littoral fish assemblages of the Lower Chesapeake Bay Tributaries. Estuaries 22(2A): 304-312.

Zimmerman, R.J., Minello, T.J., and Rozas, L.P., 2000. Salt marsh linkages to productivity of penaeid shrimps and blue crabs in the northern Gulf of Mexico. In: WEINSTEIN, M.P. and KRIEGER, D.A. (eds.), Concepts and Controversies in Tidal Marsh Ecology. Dordrecht, Netherlands: Kluwer Academic Press. 293–314 pp.

30

CONSTRAINING RATES AND TRENDS OF HISTORICAL WETLAND LOSS, MISSISSIPPI RIVER DELTA PLAIN, SOUTH-CENTRAL LOUISIANA

Julie C. Bernier
ETI Professionals, Inc.
600 4th Street South, St. Petersburg, FL 33781, USA

Robert A. Morton
U.S. Geological Survey
Center for Coastal and Watershed Studies
600 4th Street South, St. Petersburg, FL 33781, USA

John A. Barras
U.S. Geological Survey
National Wetlands Research Center
Coastal Restoration Field Station
P.O. Box 25098, Baton Rouge, LA 70894, USA

ABSTRACT

The timing, magnitude, and rate of wetland loss were described for five wetland-loss hotspots in the Terrebonne Basin of the Mississippi River delta plain. Land and water areas were mapped for 34 dates between 1956 and 2004 from historical National Wetlands Inventory (NWI) datasets, aerial photographs, and Landsat Thematic Mapper (TM) satellite images. Since 1956, the emergent land area at the five study areas in south-central Louisiana has decreased by about 50%. Comparison of the water-area curve derived from the 29 TM images with water-level records from the nearby Grand Isle, Louisiana tide gauge (NOS #8761724) clearly shows that changes in land and water areas fluctuate in response to variations in regional water levels. The magnitude of water-area fluctuations decreased from the 1980s to the 1990s as former areas of wet marsh within and immediately adjacent to the wetland-loss hotspots became permanently submerged. The most rapid wetland loss occurred during the late 1960s and 1970s. Peak wetland-loss rates during this period were two to four times greater than both the pre-1970s background rates and the most recent wetland-loss rates. These results provide constraints on predicting future delta-plain wetland losses and identify Landsat TM imagery as an important source for analyzing land- and water-area changes across the entire delta plain.

30.1. INTRODUCTION

Determining the magnitude, rate, and timing of wetland loss in south Louisiana and identifying the underlying processes that cause historical wetland loss have been high-priority topics of scientific investigation throughout the 1990s (e.g., Britsch and Dunbar, 1993; Barras et al., 1994, 2003; Boesch et al., 1994; Williams and Cichon, 1994; Day et al., 2000; Morton et al., 2002, 2003, 2005; Penland et al., 2002; Gagliano et al., 2003). These issues take on even greater importance and urgency considering that the state is seeking substantial federal funding to restore parts of coastal Louisiana and to compensate for some of the historical wetland loss (LCWCRTF, 1998; USACE, 2004).

Approximately 4000 km^2 of low-lying coastal lands on Louisiana's delta plain have become submerged since the 1930s. Previous studies (Britsch and Dunbar, 1993; Barras et al., 1994, 2003) reported peak delta-plain land-loss rates of 60 to 75 km^2/yr from the 1960s to 1980s, with continued land loss to present at rates that are significantly higher than the 1930s to 1950s background rates (Fig. 1). Projections based on these studies, which show some decrease in land-loss rates since the 1980s, indicate future delta-plain land losses of approximately 1000 km^2 by 2050 (Barras et al., 2003). However, the reported rates are averaged over 10- to 20-year time periods. Improving the resolution of the land-loss curve will better constrain the timing and duration of peak land-loss rates, and incorporating more recent data will provide more accurate constraints on present land-loss trends.

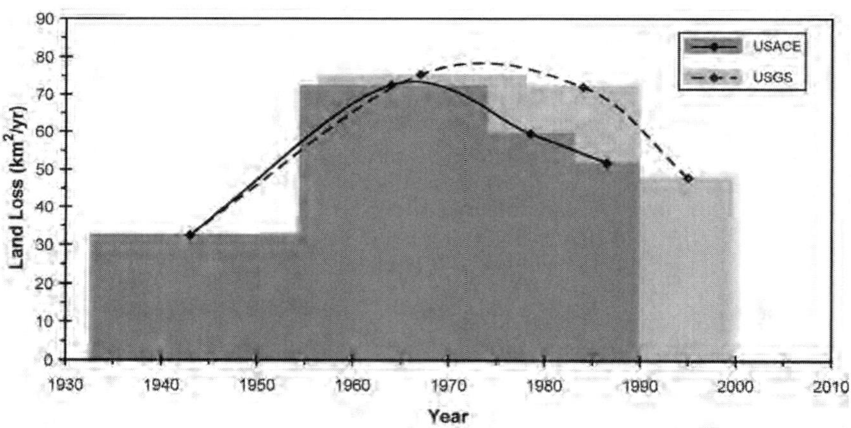

Fig. 1. Land-loss rates, Louisiana delta plain, 1932-2000. Data from the U.S. Army Corps of Engineers (USACE; Britsch and Dunbar, 1993) and the U.S. Geological Survey (USGS; Barras et al., 2003).

The purpose of this study is to further examine the timing, magnitude, and historical trends of wetland losses by establishing detailed patterns of land- and water-area changes at five representative wetland-loss hotspots (Fig. 2) in the Terrebonne Basin of the Mississippi River delta plain in south-central Louisiana. The delta-plain ecosystem is dominated by wetlands consisting of saline, brackish, intermediate, and freshwater marshes. Other delta-plain land areas include barrier islands, beaches, natural and artificial levees, fastlands, and

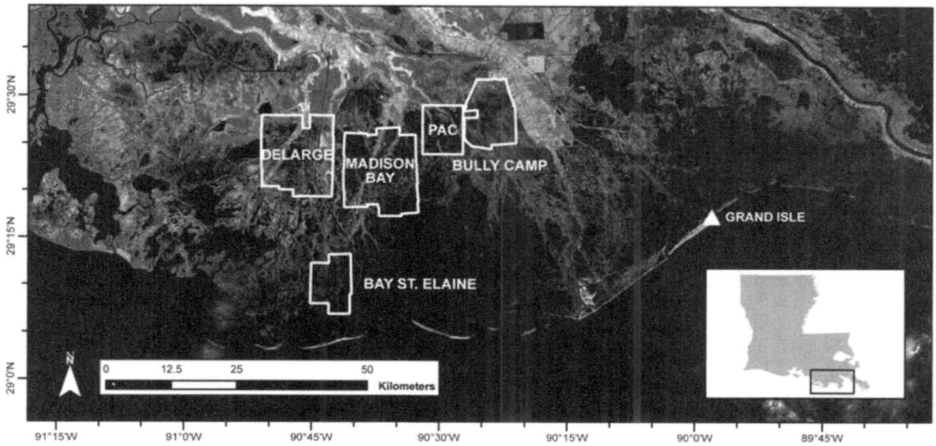

Fig. 2. Regional map of the Mississippi River delta plain in south-central Louisiana showing study area locations and the Grand Isle, Louisiana tide gauge (NOS #8761724). Landsat TM image acquired 4/16/05; visual display uses band 5 (mid-infrared). PAC = Pointe au Chien.

freshwater swamps (LCWCRTF, 1998; USACE, 2004). Because the selected study sites form only a small subset of the delta plain, land areas within the combined study area consist primarily of wetlands and levees, and essentially all land losses are wetland losses.

30.2. METHODS

30.2.1. Imagery Sources and Classification Methods

Modified National Wetlands Inventory (NWI) habitat data (Wicker, 1980; Barras and others, 1994), aerial photographs, and Landsat Thematic Mapper (TM) satellite images were used to quantify land and water areas for 34 dates between 1956 and 2005 (Table 1) for the five study areas (Fig. 2). The 1956, 1978, and 1988-1989 NWI data and the 11/1/90, 11/18/99, and 2/27/02 TM images were previously used to assess recent historical trends of land loss and land gain for coastal Louisiana (Barras et al., 1994, 2003). The 1956 land-water classification was used to establish the initial conditions for this study. Additional aerial photographs and TM images were acquired to provide better definition of temporal trends for the study areas. The 1968-1969 panchromatic (PAN) photographs and the 1974 color-infrared (CIR) photographs provided two additional datasets for assessing land- and water-area changes before 1978. Multiple TM images allowed for high-frequency (semi-annual to annual) temporal analysis of land- and water-area changes between 1983 and 2005.

The PAN and CIR photographs were rectified and the land-water distribution was classified using the methods described by Morton et al. (2005). These images were scanned from prints and often needed to be enhanced to discriminate between land and water in areas of poor contrast or sun flare. The 1:20,000 (1x1-m) spatial resolution of the PAN

Table 1. Land-water classification by date for the combined study area (Bay St. Elaine, Madison Bay, DeLarge, Pointe au Chien, and Bully Camp wetland-loss hotspots) and daily average water levels recorded at the Grand Isle, Louisiana tide gauge (NOS #8761724) for the corresponding Landsat Thematic Mapper (TM) image acquisition dates. MWL = mean water level; NWI = National Wetlands Inventory; PAN = panchromatic; CIR = color-infrared.

Date	Data	Land Area (km^2)	Water Area (km^2)	Total Area (km^2)	% Land	% Water	Water Level (m NAVD88)	Predicted MWL (m NAVD88)
1956	NWI	538.80	132.03	670.83	80.32%	19.68%		
1968-1969[1]	PAN	468.52	201.84	670.36	69.89%	30.11%		
1974[2]	CIR	349.19	317.47	666.66	52.38%	47.62%		
Oct. 1978	NWI	386.53	284.21	670.74	57.63%	42.37%		
1/6/1983[3]	TM	394.20	276.63	670.83	58.76%	41.24%	-0.157	-0.105
4/6/1984	TM	399.49	271.34	670.83	59.55%	40.45%	-0.289	-0.097
9/29/1984	TM	358.26	312.57	670.83	53.40%	46.60%	-0.060	-0.093
1/19/1985[3]	TM	379.62	291.21	670.83	56.59%	43.41%	-0.178	-0.091
8/31/1985	TM	311.85	358.98	670.83	46.49%	53.51%	0.386	-0.087
3/27/1986	TM	374.04	296.79	670.83	55.76%	44.24%	-0.235	-0.083
10/8/1987[3]	TM	368.90	301.93	670.83	54.99%	45.01%	-0.161	-0.073
1/28/1988	TM	399.96	270.87	670.83	59.62%	40.38%	-0.384	-0.071
1988-1989[4]	NWI	396.09	274.61	670.71	59.06%	40.94%		
12/16/1989	TM	398.36	272.47	670.83	59.38%	40.62%	-0.290	-0.058
11/1/1990	TM	328.52	342.31	670.83	48.97%	51.03%	0.052	-0.052
10/11/1991[3]	TM	329.32	341.51	670.83	49.09%	50.91%	-0.019	-0.045
10/5/1992	TM	314.22	356.61	670.83	46.84%	53.16%	0.094	-0.039
1/25/1993[3]	TM	337.41	333.42	670.83	50.30%	49.70%	0.007	-0.037
3/17/1994	TM	329.23	341.60	670.83	49.08%	50.92%	-0.101	-0.029
11/15/1995[3]	TM	317.37	353.46	670.83	47.31%	52.69%	-0.070	-0.018
4/7/1996	TM	327.66	343.17	670.83	48.84%	51.16%	-0.092	-0.015
10/3/1997[3]	TM	311.70	359.13	670.83	46.47%	53.53%	-0.067	-0.005
2/24/1998	TM	325.87	344.96	670.83	48.58%	51.42%	-0.225	-0.002
1/26/1999	TM	323.42	347.41	670.83	48.21%	51.79%	-0.148	0.004
11/18/1999[3]	TM	289.91	380.92	670.83	43.22%	56.78%	0.020	0.010
10/11/2000	TM	295.89	374.94	670.83	44.11%	55.89%	-0.098	0.016
10/30/2001[3]	TM	296.22	374.61	670.83	44.16%	55.84%	0.021	0.023
2/27/2002	TM	325.95	344.88	670.83	48.59%	51.41%	-0.243	0.025
12/28/2002	TM	299.80	371.03	670.83	44.69%	55.31%	-0.184	0.031
10/20/2003	TM	284.05	386.78	670.83	42.34%	57.66%	0.155	0.037
11/7/2004	TM	290.24	380.59	670.83	43.27%	56.73%	0.024	0.044
4/16/2005[3]	TM	285.37	385.46	670.83	42.54%	57.46%	-0.037	0.047
10/9/2005	TM	274.23	396.60	670.83	40.88%	59.12%	0.137	0.050
10/25/2005	TM	282.10	388.73	670.83	42.05%	57.95%	-0.062	0.050

[1] 1968-1969 images acquired 12/15/68, 3/9/69, 3/16/69, 3/17/69, 3/19/69, 3/26/69, and 3/27/69

[2] 1974 images acquired 9/17/74 and 10/18/74

[3] Image-acquisition dates for which the daily, previous day, and 3- and 7-day running average water levels at the Grand Isle, Louisiana tide gauge (NOS #8761724) were within one standard deviation (0.0910 m) of the mean water-level trend line

[4] 1988-1999 NWI source imagery acquired 11/21/88 and 1/17/89

photography revealed detailed land-water patterns not visibly evident in either the small-scale (1:119,000 to 1:128,000) 1974 CIR photography or the satellite imagery (28.5x28.5-m spatial resolution).

The Landsat TM data consisted of cloud-free to nearly cloud-free TM 4, 5, and 7 scenes. All scenes were converted from radiance to reflectance values prior to land-water classification. The scenes were then classified applying the same methods used to develop the original TM land-water data sets (Barras et al., 2003). The multi-band TM imagery provided better spectral discrimination of marsh "wetness" than either the PAN or CIR photography. Wet marsh is a sub-category of land that appears wetter than the surrounding marsh in the TM imagery due to the absorption of mid-infrared and near-infrared wavelengths but has a higher spectral reflectance of the infrared bands than water. A RGB visual display combination of TM spectral bands 4 (near-infrared), 5 (mid-infrared), and 3 (visible red) can be used to discriminate between water, wet marsh, and emergent marsh (land) areas. Wet marsh is usually located adjacent to ponds and in areas where the emergent marsh is fragmenting into networks of broken marsh and open water. The extent of wet marsh varies depending on water level, marsh elevation, spectral-reflectance variations linked to the seasonal marsh-vegetation growth cycle, and density of vegetative cover.

All classified land-water datasets were resampled to a standard 25x25-m cell size to provide a consistent spatial resolution for assessing changes in land and water area over time. A thematic raster data set was created containing the minimum shared area for each study area encompassed by all land-water datasets from 1956 to 2005. The maximum extent of the compared area for each site was controlled by the availability of the 1968-1969 PAN frames. A composite land-water summary was created for the combined study area for each image-acquisition date.

30.3. ANALYTICAL METHODS

Historical land-water changes are assessed by comparing changes in the distribution and total size of land and water areas between selected image dates. Average annual rates of change are derived for each period by dividing the total change in water or land area by the number of years between observations. Ideally, environmental conditions effecting the land-water distribution at the time of image acquisition should be similar, and there should be no major differences in classification methods that might increase or decrease the interpreted land or water area.

The interpretation of land- and water-area trends derived from the historic NWI habitat data and aerial photographs between 1956 and 1978 were based on decadal or multi-decadal comparisons due to the high cost of obtaining and interpreting coast-wide photography. The aerial photographs were often acquired over a period of several days due to weather conditions and acquisition-platform capabilities. As a result, water levels likely fluctuated between frames. These effects could not be quantitatively assessed because accurate daily water-level data and multiple photographs bracketing known high- and low-water conditions were not available for this time period.

Determining the present average land and water areas and current land-loss rates requires developing classification methods that account for natural variations in land and water areas. One approach to obtaining recent trend estimates is to acquire and classify multiple images separated by short periods using a standard acquisition platform. Ranges and averages of land and water areas can be compared over time and general trends can be described, regardless of short-term fluctuations in land and water areas (Fig. 3). An overall increase in average water area through time indicates continued land loss as low-lying wetlands become permanently submerged.

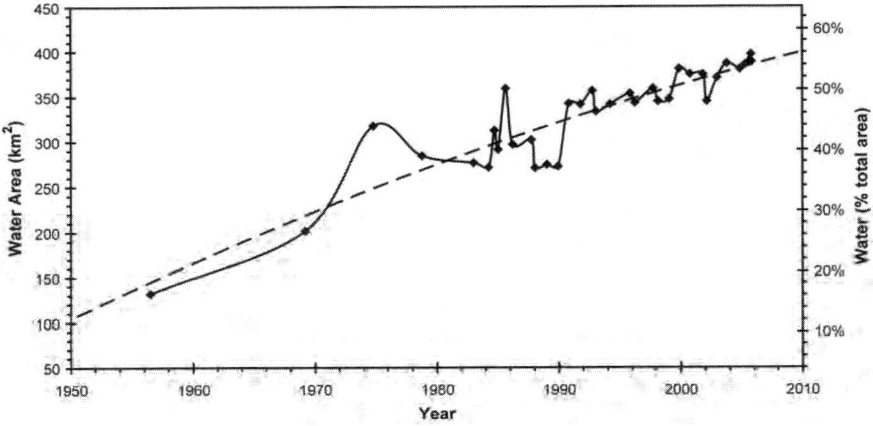

Fig. 3. Composite historical water area from 1956 to 2005 for the Bay St. Elaine, Madison Bay, DeLarge, Pointe au Chien, and Bully Camp wetland-loss hotspots. Although water-area fluctuations are evident, the trend line shows a general increase in open water area through time, indicating continued wetland loss.

A second approach is to identify those data points that are most representative of "average" water-level conditions, and use this subset to quantify trends in land- and water-area changes. Analysis of the 29 TM data points, representing a 22-year period, provides a clearer understanding of the effects of water-level variations on land-water classification over annual and decadal time scales. Variations in water levels recorded at the National Ocean Service (NOS) Grand Isle, Louisiana tide gauge (NOS #8761724) correlate well with the shape of the classified water-area curve (Fig. 4). Image-acquisition dates for which the classified TM data forms peaks on the water-area curve (e.g., 8/31/85, 10/5/92, and 10/9/05) correspond to dates on which peak water levels were recorded at NOS #8761724. Conversely, image-acquisition dates for which the classified TM data forms lows on the water-area curve (e.g., 4/6/84, 1/28/88, 12/16/89, and 2/27/02) correspond to recorded low water levels at NOS #8761724. These relationships may reflect some seasonal effects: the high water area/high water level images were acquired during the late summer and early fall whereas the low water area/low water level images were acquired during the winter and early spring.

Water-level readings from NOS #8761724 were used to identify Landsat TM acquisition dates that represented "average" water-level conditions. Daily and monthly average water levels at NOS #8761724 from 1981 to 2005 were computed from the raw hourly records. A linear regression of the monthly averages represents the best estimate of

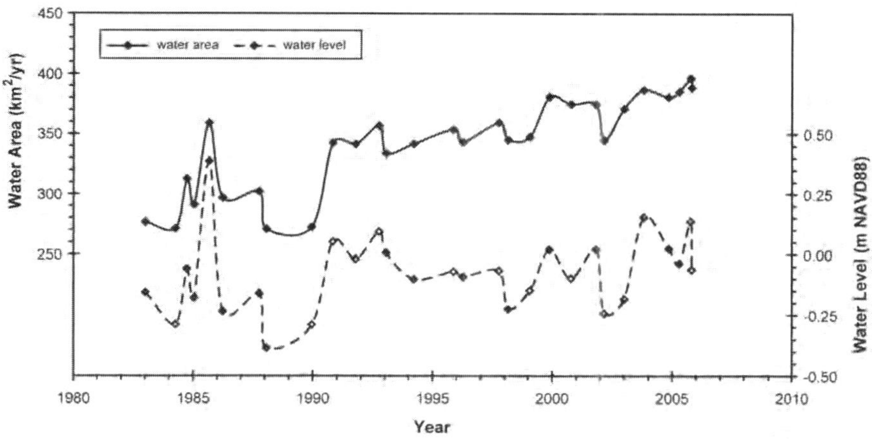

Fig. 4. Composite historical water area from 1983 to 2005 for the Bay St. Elaine, Madison Bay, DeLarge, Pointe au Chien, and Bully Camp wetland-loss hotspots compared to daily average water levels recorded at the Grand Isle, Louisiana tide gauge (NOS #8761724) for the corresponding Landsat TM image acquisition dates.

"average," or expected, water levels through time. "Average" water-level data points were defined as those TM acquisition dates for which the daily, previous day, and 3- and 7-day running averages all were within one standard deviation (0.0910 m) of this mean water-level trend line (Table 1, Fig. 5). The classified land and water areas from these dates were

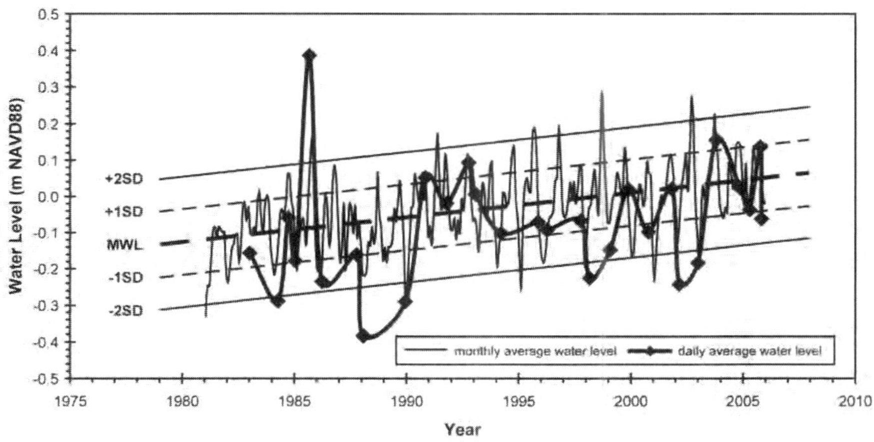

Fig. 5. Monthly mean water level from 1981 to 2005 compared to daily average water levels for Landsat TM image acquisition dates at the Grand Isle, Louisiana tide gauge (NOS #8761724). Similar plots were generated using daily average water levels from one day prior to image acquisition and 3- and 7-day water-level averages. MWL = monthly mean water level.

used to calculate the post-1983 land-loss rates averaged over approximately 4-year time periods (Fig. 6). By using those images that were acquired on "average" water-level days, the effects of extreme seasonal or meteorological conditions on the land-water classification are minimized.

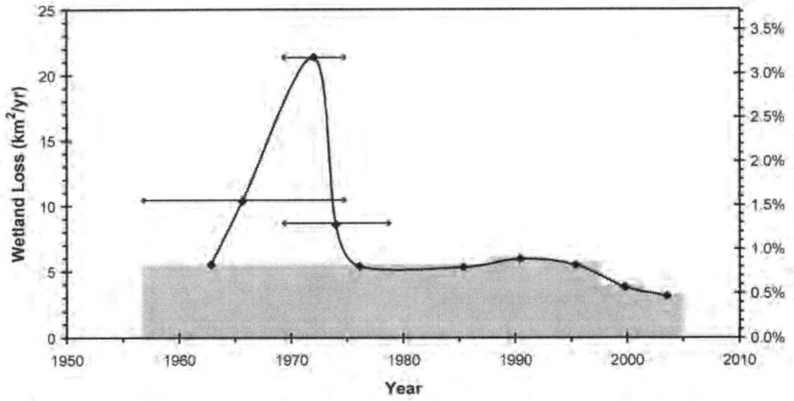

Fig. 6. Rate of wetland loss for the combined study area from 1956 to 2005. The post-1983 rates were calculated from Landsat TM imagery acquired on dates for which the daily, previous day, and 3- and 7-day running average water levels at the Grand Isle, Louisiana tide gauge (NOS #8761724) were within one standard deviation of the mean water-level trend line. Peak wetland-loss rates are two to four times higher than the pre- and post-1970s background rates.

30.4. RESULTS

Total wetland loss in the combined study area is about 250 km^2, or about 50% of the 1956 land area. Analysis of land- and water-area changes for the combined study area (Fig. 3) reveals a pattern of increasing land-area loss from the 1950s through the late 1960s, rapidly accelerating land loss in the 1970s, and short-term fluctuations between land loss and land gain from the 1980s to present. Net water area varied by as much as 10% between observed high- and low-water conditions since 1983 based on the TM data. The derived wetland-loss curve (Fig. 6) shows a rapid increase in wetland-loss rates from the late 1960s to 1970s. Wetland-loss rates during this time period were two to four times higher than the pre-1970s rates. In contrast, average wetland-loss rates since the 1980s are of similar magnitude to the pre-1970s rates.

30.5. DISCUSSION

30.5.1. Historical Land-Water Changes, 1956-2005

The ability to interpret the effects of water-level variations during the 1960s and 1970s was hampered by the lack of a standard image acquisition platform with multiple data points, the necessity of combining data from multiple image-acquisition dates to cover the entire study area, and the unavailability of daily water-level data. The 1974 data were likely acquired under high-water conditions based on the total water area and visual comparison with later images. Comparison of the 1974 CIR imagery and the original 1978 CIR photographs used to develop the NWI habitat data suggests that the 1978 data were acquired under lower water-level conditions; however, visual review of the original imagery suggests that the NWI classification methodology also appears to have underestimated the total water area.

Between 1983 and 1989, only the 8/31/85 water area exceeded the 1974 water area for the combined study area (Fig. 3). The 8/31/85 peak water area and the corresponding extreme elevated water levels recorded at the Grand Isle tide gauge (Figs. 4 and 5) were likely effected by Hurricane Elena, which was located offshore of southeastern Louisiana in the Gulf of Mexico. The counter-clockwise easterly winds would have pushed water onshore along the Louisiana coastline. Excluding this data point, the average water area for 1983 to 1989, including the NWI habitat data, was about 5% less than the 1974 water area. The TM data may underestimate water area compared to the aerial photography (Ramsey and Laine, 1995), and/or the 1974 photography may reflect anomalous high-water conditions. High-water conditions in 1974 may have resulted from incomplete recovery of the delta-plain wetlands after severe flooding from the Mississippi River in 1973 and/or remnant elevated water levels following landfall of Hurricane Carmen near Morgan City, Louisiana on 9/8/74.

After 1990, the water area from the TM data consistently exceeded the 1974 water area, indicating that some marsh areas that were previously alternately exposed or submerged under varying water-level conditions had become permanently submerged. At the same time, the average water area increased to over 50% of the total study area, corresponding to the submergence of remnant wet marsh that was exposed during low water events in the 1980s. Permanent submergence of previously emergent wet marsh is also suggested by comparing the magnitude of water-area fluctuations between 1983 to 1990 and 1991 to present. From 1983 to 1990, the total water area varied by 5 to 10%; since 1991, the total water area has varied by only 2 to 5% (Fig. 3). Between these same time periods, however, the average range of water-level variations recorded at the Grand Isle tide gauge did not change significantly. Thus, by the 1990s, some areas of wet marsh were completely submerged and no longer exposed under low-water conditions.

30.6. HISTORICAL SUBSIDENCE OF DELTA-PLAIN MARSHES AND FORMATION OF WETLAND-LOSS HOTSPOTS

The image analysis enables quantification of land- and water-area changes and visual depiction of the resultant temporal trends, but interpretation of the classified land-water data and trends also requires examination and comparison of source photography and satellite imagery at local and regional scales to visually verify changes. The five study sites were previously identified (Morton et al., 2005) as areas of extensive wetland losses ("hotspots") in south-central Louisiana for which visual review of historical imagery indicated that emergent marsh areas converted to open water very rapidly during the late 1960s to mid 1970s. Morton et al. (2003, 2005) described the shallow subsurface stratigraphy at each of the five study areas from sediment vibracores taken near the center of each wetland-loss hotspot. Correlation of recent peat deposits showed that subsidence was the primary mechanism of wetland loss at these locations. The uppermost peats from recently-submerged cores are similar in thickness to the uppermost peats from adjacent emergent-marsh core locations, indicating that subsequent erosion of the delta-plain marsh was relatively minor. This is consistent with the observed pattern of wet marsh submergence, with permanent submergence occurring only after the land has subsided beyond some critical threshold. Although some shoreline erosion occurs within the study areas, subsidence and water-level effects appear to have a greater impact on observed water area and wetland loss in the interior marshes.

The timing and patterns of regional land- and water-area changes provide insight into how local wetland-loss histories fit into the overall patterns of wetland loss. Visual examination of selected (unclassified) aerial photographs from 1940, 1953, and 1957, as well as the original 1978 CIR photography, showed that between 1940 and 1957, the regional landscape was dominated by unbroken interior marsh between distributary levees, and the 1953 and 1957 landscape closely resembled the 1956 land-water conditions. Some marsh-surface fragmentation was apparent within future hotspot areas by 1968-1969. Large historic wetland-loss hotspots characterized by widespread marsh fragmentation and well-developed ponds with large tracts of adjacent wet marsh were well-defined by 1974. The 1974 and 1978 landscape more closely resembled the mid-1980s landscape than it did the 1968-1969 landscape, indicating that that much of the historic wetland loss occurred before 1978. Some areas identified as water in 1974 were exposed as marsh in 1978, indicating that water-level fluctuations effected land-loss interpretations. Areas that were identified as water in 1974 and land in 1978 were generally permanently submerged by the late 1980s or early 1990s, often after repeated cycles of exposure and submergence. Remnant emergent-marsh islands, natural levees, and spoil mounds are often the only remaining areas of exposed land within the historic wetland-loss hotspots.

The study sites fall within scene 2240 (path/row) of the Landsat TM World Reference System. Visual analysis of multiple 180x185-km scenes under varying meteorological and water-level conditions between 1983 and 2005 shows that persistent wet marshes were commonly located next to former or developing wetland-loss hotspots. Wet marshes located within and adjacent to historic (1970s) delta-plain wetland-loss hotspots typically became permanently submerged and converted to open water between 1983 and 2005. Identification of persistent wet marsh over multiple periods indicated that new wetland-loss hotspots also formed after 1983 but were not as large or as numerous as those that formed between 1956 and 1978.

30.7. CONCLUSIONS AND IMPLICATIONS

Analysis of historical photographs and satellite images from five wetland-loss hotspots in south-central Louisiana show that delta-plain wetland losses had initiated by the late 1960s. Rapid acceleration of wetland loss and the development of large wetland-loss hotspots with associated wet marsh occurred during the 1970s. The late 1970s through 1980s were characterized by continued wetland loss. During the 1980s and 1990s, land and water areas fluctuated significantly, reflecting variations in regional water levels. The magnitude of the land- and water-area fluctuations decreased in the 1990s relative to the 1980s, corresponding to complete submergence of former wet-marsh areas that occurred within and immediately adjacent to historic wetland-loss hotspots. The most recent wetland-loss rates are significantly lower than the 1970s peak rates and are similar in magnitude to the pre-1970s background rates.

Although wetland-loss rates for discrete periods are probably not linear, average annual wetland-loss rates based on historical data provide constraints for estimating future wetland losses and are important for assessing the feasibility of wetland-restoration projects in coastal Louisiana. The Landsat TM satellite imagery does not have the high spatial resolution of most aerial photographs; however, it offers a relatively inexpensive way of monitoring both short- and long-term changes in land and water areas across a broad geographic area. In addition, the standardized filtering and classification process produces rapid and reproducible results. The identification of wet marsh areas as potentially high-risk areas for future wetland loss offers a previously-unrecognized predictive tool that can be incorporated into marsh-management and wetland-loss mitigation plans.

REFERENCES

Barras, J.A., Bourgeois, P.E., and Handley, L.R., 1994. Land loss in coastal Louisiana 1956-90. National Biological Survey, National Wetlands Research Center Open-File Report 94-01, 4 p.

Barras, J., Beville, S., Britsch, D., Hartley, S., Hawes, S., Johnston, J., Kemp, P., Kinler, Q., Martucci, A., Porthouse, J., Reed, D., Roy, K., Sapkota, S., and Suhayda, J., 2003. Historical and projected coastal Louisiana land changes: 1978-2050. U.S. Geological Survey Open-File Report 03-334, 39 p.

Boesch, D.F., Josselyn, M.N., Mehta, A.J., Morris, J.T., Nuttle, W.K., Simenstad, C.A., and Swift, D.J.P., 1994. Scientific assessment of coastal wetland loss, restoration and management in Louisiana. Journal of Coastal Research, Special Issue 20, 103p.

Britsch, L.D., and Dunbar, J.B., 1993. Land-loss rates: Louisiana coastal plain. Journal of Coastal Research 9:324-338.

Day, J.W. Jr., Shaffer, D.P., Britsch, L.D., Reed, D.J., Hawes, S.R., and Cahoon, D.R., 2000. Pattern and process of land loss in the Mississippi Delta, a spatial and temporal analysis of wetland habitat change. Estuaries 23:425-438.

Gagliano, S.M., Kemp, E.B., Wicker, K.M., Wiltenmuth, K.S., and Sabate, R.W., 2003. Neo-tectonic framework of southeast Louisiana and applications to coastal restoration: Transactions – Gulf Coast Association of Geological Societies 53: 262-276.

Louisiana Coastal Wetlands Conservation and Restoration Task Force and the Wetlands Conservation and Restoration Authority (LCWCRTF), 1998. Coast 2050: Toward a sustainable coastal Louisiana. Louisiana Department of Natural Resources, Baton Rouge, LA, 161p.

Morton, R.A., Bernier, J.C., Barras, J.A., and Ferina, N.F., 2005. Rapid subsidence and historical wetland loss in the Mississippi delta plain: likely causes and future implications. U. S. Geological Survey Open-file Report 2005-1216, 116p.

Morton, R.A., Tiling, G., and Ferina, N.F., 2003. Causes of hotspot wetland loss in the Mississippi delta plain. Environmental Geosciences 10:71-80.

Morton, R.A., Buster, N.A., and Krohn, M.D., 2002. Subsurface controls on historical subsidence rates and associated wetland loss in south-central Louisiana. Transactions – Gulf Coast Association of Geological Societies 52:767-778.

Penland, S., Beall, A.D., Britsch, L.D., and Williams, S.J., 2002. Geologic classification of coastal land loss between 1932 and 1990 in the Mississippi River delta plain, southeastern Louisiana. Transactions – Gulf Coast Association of Geological Societies 52:799-807.

Ramsey, E.W. III and Laine, S.C., 1997. Comparison of Landsat Thematic Mapper and high resolution photography to identify change in complex coastal wetlands. Journal of Coastal Research 13:281-292.

USACE (U.S. Army Corp of Engineers), 2004. Louisiana Coastal Area (LCA) Final Report–Ecosystem Restoration Study, 296p.

Wicker, K.M., 1980. The Mississippi deltaic plain habitat mapping study. U.S. Fish and Wildlife Service, Office of Biological Services, FWS/PBS-79/07.

Williams, S.J. and Cichon, H.A., 1994. Processes of coastal wetland loss in Louisiana. U.S. Geological Survey Open-File Report, no. 94-0275, 226p.

31

ECOLOGICAL CONSEQUENCES OF CHANGING HYDROLOGICAL CONDITIONS IN WETLAND FORESTS OF COASTAL LOUISIANA

Richard F. Keim, Jim L. Chambers, Melinda S. Hughes, J. Andrew Nyman, Craig A. Miller, and J. Blake Amos
School of Renewable Natural Resources, Louisiana State University Agricultural Center
Renewable Natural Resources Building, Baton Rouge, LA 70803, USA

William H. Conner
Department of Forestry and Natural Resources,
Baruch Institute of Coastal Ecology & Forest
Science, Clemson University P.O. Box 596
Georgetown, SC 29442, USA

John W. Day, Jr.
School of the Coast and Environment, Coastal
Ecology Institute, Louisiana State University
Energy Coast and Environment Building,
Baton Rouge, LA 70803, USA

Stephen P. Faulkner
National Wetlands Research Center, USGS
700 Cajundome Blvd., Lafayette, LA 70506, USA

Emile S. Gardiner
Center for Bottomland Hardwoods Research, USDA
Forest Service, Southern Hardwoods Laboratory
P.O. Box 227, Stoneville, MS 38776, USA

Sammy L. King
LSU Agricultural Center
School of Renewable Natural Resources
USGS Louisiana Coop. Fish & Wildlife Rsch Unit
Renewable Natural Resources Building
Baton Rouge, LA 70803, USA

Kenneth W. McLeod
Savannah River Ecology Laboratory
University of Georgia
P.O. Drawer E, Aiken, SC 29802, USA

Gary P. Shaffer
Department of Biological Sciences, Southeastern Louisiana University
Meade Hall 107, Hammond, LA 70402, USA

ABSTRACT

Large-scale and localized alterations of processes affecting deltaic coastal wetlands have caused the complete loss of some coastal wetland forests and reduced the productivity and vigor of many areas in coastal Louisiana. This loss and degradation threatens ecosystem functions and the services they provide. This paper summarizes ecological relationships controlled by hydrological processes in coastal wetland forests of the Mississippi River delta and presents two case studies that illustrate the complexity of assessing hydrological control on swamp forest establishment and growth. Productivity of overstory trees has been affected by these changes, but the first case study illustrates that the relationship between flooding and growth may be site-specific. An important effect of increased flooding has been to reduce regeneration of swamp forest trees. The second case study is an outline of the kind of hydrological analysis required to assess probability of regeneration success.

31.1. INTRODUCTION

Wetland hydrology and vegetation are closely linked, and most wetland functions rely on their interaction. However, understanding how hydrology controls wetland ecosystems is often hampered by the complex nature of this interaction and sparse data (NRC, 1995). Interactions between hydrological processes and ecosystems are critically important in coastal wetlands of Louisiana, where extensive hydrological engineering works coincide with globally significant wetlands composed of diverse ecosystems vulnerable to loss by coastal erosion, land subsidence, sea-level rise, and saltwater intrusion (Day et al., 2000). Of these coastal ecosystems, coastal marshes have received the most attention because it is there that land loss is prominent. However, the same processes place coastal wetland forest ecosystems at risk of loss or conversion to other ecosystems (Conner and Brody, 1989; Conner and Toliver, 1990).

Louisiana's coastal wetland forests are of tremendous economic, ecological, cultural, and recreational value. Large-scale and localized alterations of processes affecting coastal wetlands have caused the complete loss of some coastal wetland forests and reduced the productivity and vigor of remaining areas. This loss and degradation threatens ecosystem functions and the services they provide. An example of the societal importance of coastal wetland forests is the commission by the Governor of Louisiana of an *ad hoc* group of scientists, the Coastal Wetland Forest Conservation and Use Science Working Group (SWG), to assess management issues in coastal forests (Chambers et al., 2005). This paper builds upon the results of that collaborative effort.

In the deltaic plain of the Mississippi River, coastal forested swamp wetlands occupy an ecotone topographically below the bottomland hardwoods that occupy natural levees of distributary channels, and topographically above treeless freshwater and saltwater marshes that are at sea level. Of the 800,000 ha of forested wetlands throughout Louisiana, over half are in the coastal parishes (Figure 1). Both swamps and bottomland hardwoods are formally classified as palustrine wetlands in the Cowardin classification of the National Wetlands Inventory (Cowardin et al., 1979), but swamps are flooded for most if not all of the growing season and are dominated by baldcypress, pondcypress and water tupelo (Conner and Buford, 1998). Bottomland hardwoods are seasonally inundated for varying lengths of time and contain as many as 70 commercial tree species depending on the hydroperiod (Hodges, 1997). The SWG concentrated on swamps in the deltaic plain, and referred to them as "coastal wetland forests".

The objectives of this paper are to (1) summarize ecological relationships controlled by hydrological processes in coastal wetland forests of the Mississippi River delta, and (2) compare hydrological regimes of two pairs of typical sites to illustrate the complexity of assessing hydrological control on swamp forest establishment and growth.

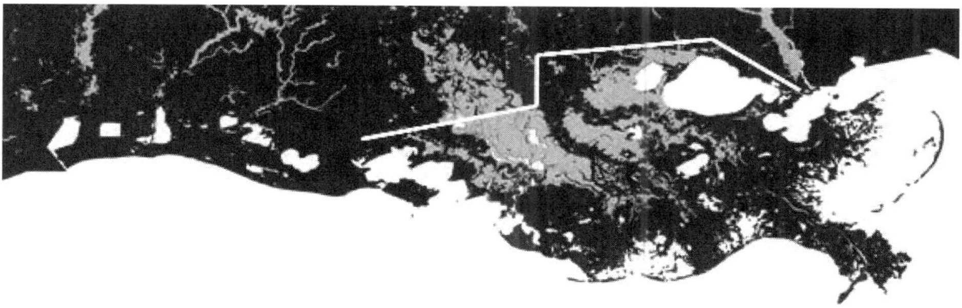

Fig. 1. Wetland forests of the delta of the Mississippi River and coastal Louisiana. The white line indicates the approximate ecological boundary between riverine-dominated forests to the north and deltaic, coastal forests to the south, although deltaic geological processes encompass all of the Atchafalaya Basin.

31.2. HYDROLOGICAL CHANGES

The geological processes controlling the physical development of the Mississippi River delta are aggradational (accumulation of riverine sediment) and degradational (consolidation of freshly deposited sediments). Aggradational processes in the Mississippi River delta have been on balance larger than degradational processes over the past 5,000 years, with the result being seaward growth of the deltaic plain as delta lobes were progressively formed and abandoned (Roberts, 1997; Coleman et al., 1998). Net growth of the delta occurs where the local rate of accretion is greater than consolidation, but subsidence can result in land loss where local additions are less than consolidation rates.

Management of the wetlands and the river has altered the stochastic patterns of sediment accumulation and contributed to subsidence land loss. The foremost cause is the flood-control levees along the Mississippi River that eliminated diffuse riverine input to most of the delta and contributed to wetland loss. Hydrological disruption via control of rivers has reduced freshwater and sediment inputs, while canal construction has led to much greater saltwater intrusion into coastal wetlands (Turner, 1997; Day et al., 2000). Increasing water levels resulting from eustatic sea-level rise and subsidence are also major degradation factors. Without the annual flood of new sediments, subsidence exceeds sedimentation in many areas (Baumann et al., 1984), and most of coastal Louisiana is presently experiencing an apparent water level rise of up to 1 m per century (Gagliano, 1998). Large-scale processes have been exacerbated by management practices and societal infrastructure that have also altered and degraded ecosystems. Local alterations to hydrological patterns include impoundments by highways, railroads, and dredge spoil; river diversions; and water level stabilizations by flood control structures.

Evidence of long-term, slow changes in hydrological processes can be found in water level records maintained by the USGS and US Army Corps of Engineers. For example, subsidence and diversion of Mississippi River water into the Atchafalaya River have combined to raise the average water level in Morgan City, Louisiana, by nearly 1 m from

1936-2004 (Figure 2). These gauge records can be extrapolated to infer historical water levels in wetland forests. Doing this, Conner and Day (1988) found that one swamp forest changed from being flooded about half the year to being flooded nearly continuously in the 1970s. Another swamp forest changed from infrequent inundation to flooding for more than half of the year, and a nearby hardwood ridge became flooded for up to 250 days per year (Figure 3). Since the 1950s, flood water levels in the swamps of the Pontchartrain Basin have doubled (Thomson et al., 2002).

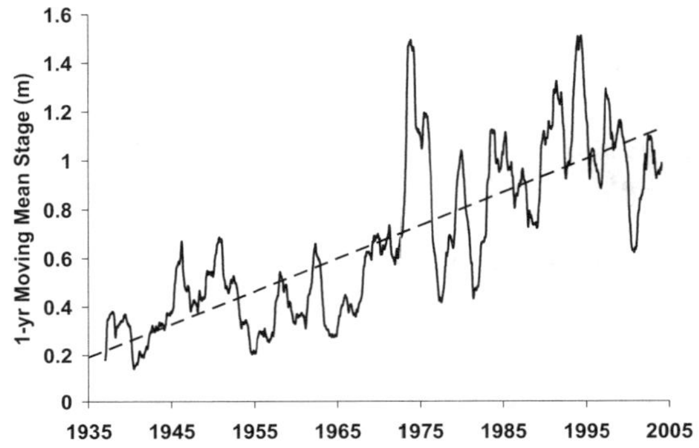

Fig. 2. 365-day moving mean water level on the Atchafalaya River in Morgan City, Louisiana, from 1936-2004. The dashed line is the best-fit 68-year linear trend.

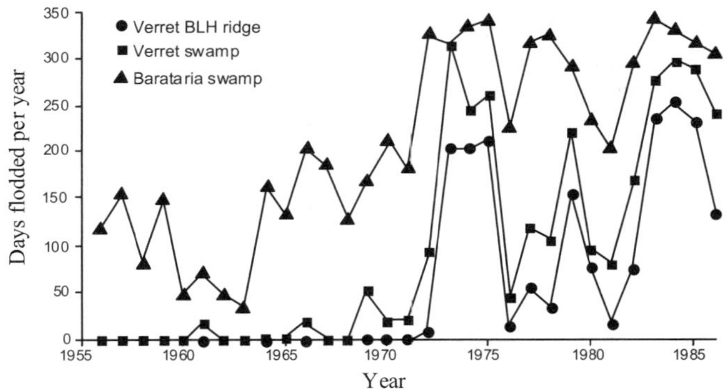

Fig. 3. The number of days flooded per year in the Barataria and Verret swamp forests (Conner and Day, 1988). Reprinted by permission from Allen Press.

31.3. ECOLOGICAL CONSEQUENCES

As water levels continue to rise, coastal forests will be subjected to more prolonged and deeper flood events. Even though many of the forest species growing in these areas are adapted to prolonged inundation (Kozlowski, 1984), extended flooding during the growing season can cause mortality of these tree species. Already, many of the trees in these areas are showing evidence of severe stress (Conner et al., 1981). Even baldcypress and water tupelo, two of the dominant species in Louisiana's coastal forests, slowly die when exposed to prolonged, deep flooding (Harms et al., 1980; Brown, 1981).

Pre-settlement baldcypress–water tupelo swamps in the delta, which may have exceeded 1,000,000 ha in extent, were mostly clearcut from about 1890–1930 and subsequently either converted to agriculture or other land uses or naturally regenerated to second-growth, even-aged stands (Conner and Toliver, 1990). Some data to assess historical growth of these stands comes from the US Forest Service Forest Inventory and Analysis (FIA) program, which has measured tree growth on permanent plots since the 1930s. Total volume of baldcypress in the region has not increased since the 1974 FIA measurement (Figure 4), but diameter of the average-sized tree increased from 22.3 cm in 1974 to 28.1 cm in 2003. The fact that baldcypress trees are continuing to grow in diameter but little additional wood volume is accumulating indicates environmental stresses are causing mortality.

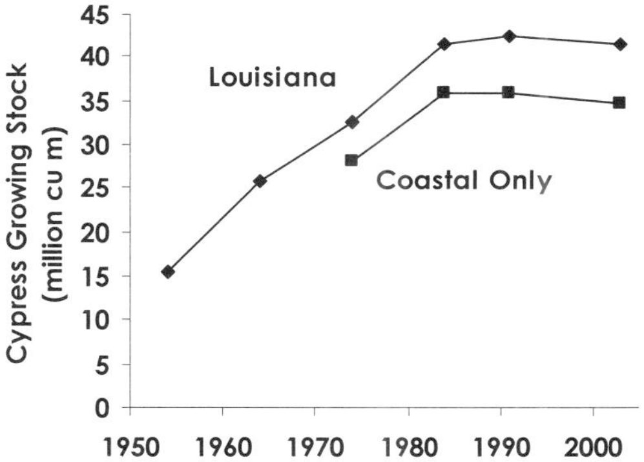

Fig. 4. Baldcypress growing stock volume in Louisiana from US Forest Service data.

Another important factor to be considered in these coastal forests is the recruitment of new individuals into the forest. Baldcypress and water tupelo must have dry periods for the seed to germinate and establish (DeBell and Naylor, 1972; Hook, 1984; Kozlowski, 1997). In many cases, dry periods have not been sufficient for seedling establishment (Conner et al., 1986) and if water levels continue to rise, coastal forested areas will eventually be replaced by scrub-shrub stands, marsh, or open water. Young seedlings in a wetland environment must grow rapidly to reduce the risk of canopy submersion by future floods during the growing season (Conner et al., 1986). Baldcypress seedlings can withstand complete inundation for up to 45 days in the laboratory (Souther and Shaffer, 2000), but long-term flooding above the foliage results in high mortality and longevity in turbid water is likely less. Natural regeneration of swamplands after the widespread logging of 1890–1930 occurred when hydrological conditions were less affected by levees and subsidence.

The SWG developed a hydrological classification for coastal wetland forests to reflect the hydrological conditions controlling tree regeneration (Chambers et al., 2005). Class I (best) sites have seasonal flooding with sufficient dry-down during the growing season for seedlings to grow tall enough to withstand flooding during the next growing season. Class II (artificial regeneration only) sites experience sufficient flooding during the growing season to prevent establishment of seedlings, but are not too deeply flooded (maximum of about 0.6 m) to prevent artificial regeneration during the winter months. Class III (no regeneration) sites experience sufficient flooding during the growing season to prevent establishment of seedlings, and are too deeply flooded to allow practical artificial regeneration.

Changes to ecological structure affect wildlife habitat in important ways. Songbirds, wading birds, waterfowl, raptors, reptiles, amphibians, mammals, crawfish, and fish are all common inhabitants of Louisiana's coastal forests. Millions of songbirds migrate through Louisiana's coastal forests, including virtually all of the eastern landbird species in the United States and numerous species from the western United States (Lowery, 1974; Barrow et al., 2005). While several dozen songbird species breed in Louisiana's coastal forests, Zoller (2004) found that the number of species of breeding migrant songbirds was less in forests degraded by hydrologic changes than in relatively undegraded or moderately degraded forests. Louisiana's coastal forests are also important habitat for two bat species listed as federal species of concern (the southeastern bat and Rafinesque's big-eared bat) and the threatened Louisiana black bear. Bats utilize hollow trees for roost and hibernation sites (Gooding and Langford, 2004), and black bears commonly use tree dens in frequently flooded areas (Alt, 1984; White et al., 2001; Hightower et al., 2002). Though other tree species are used, Rafinesque's big-eared bat frequently uses hollow water tupelo trees > 90 cm dbh that are characteristic of older baldcypress–water tupelo forests (Lance et al., 2001; Gooding and Langford, 2004). The U.S. Fish and Wildlife Service listing rule for the Louisiana black bear defines candidate den trees as baldcypress or water tupelo > 91 cm dbh with a visible cavity, occurring along rivers, lakes, streams, bayous, sloughs, or other water bodies. Thus, forest structure is critical for the maintenance of wildlife populations in these forests.

31.3.1. Case Study I: Inundation and Productivity

To better understand how hydrological changes are affecting the coastal forest ecosystem, Amos et al. (2005) used dendrochronology of baldcypress at two contrasting sites to compare the historical relationship of hydrology and productivity of overstory trees. Both sites are adjacent to long-term records of water levels, and are in similar topographic positions near lakes in backswamps of the historical Atchafalaya River distributary basin, but have strongly contrasting hydrological conditions. The first site is near Grand Lake, where diversion of Mississippi River water into the Atchafalaya Basin Floodway has drastically increased flooding and sediment and nutrient additions. The second site is near Lake Verret, where flood control levees have eliminated seasonal riverine flooding but inundation has increased because of subsidence, impeded drainage, and backwater flooding from the outlet of the Atchafalaya Basin.

Growth rates of overstory baldcypress trees in the two stands have decreased over the past 30 years (Figure 5). Qualitatively, the growth decrease coincides with increased flooding, especially at the Lake Verret (non-riverine) site (Figure 6). A principle component ordination of stage variables resulted in significant correlation between the first PCA axis and growth at each site. At the Grand Lake (riverine) site, the first axis was positively correlated with annual growth (r = 0.77), and was most related to average annual stage in the current and previous years. This suggests that deeper flooding in the riverine wetland may increase growth. At the Lake Verret (non-riverine) site, the first axis was negatively correlated with annual growth (r = -0.56), and was most related to 5- and 10-year average stage. This suggests that long-term increases in flooding in the non-riverine wetland may decrease growth. At both sites, the flooding variables were much more correlated to growth rates than were climate variables such as temperature and precipitation.

The mechanisms by which flooding affects growth rates in these two stands are not clear; two possible explanations are that (1) flooding in the riverine site brings nutrients that increase growth, whereas flooding in the non-riverine site only produces hypoxia and stresses trees; or (2) the reduction in growth at the non-riverine site may be caused by the change in hydrological conditions, which has been more severe than in the riverine site. We are adding new plots to the analysis to determine whether the differences in these sites are consistent.

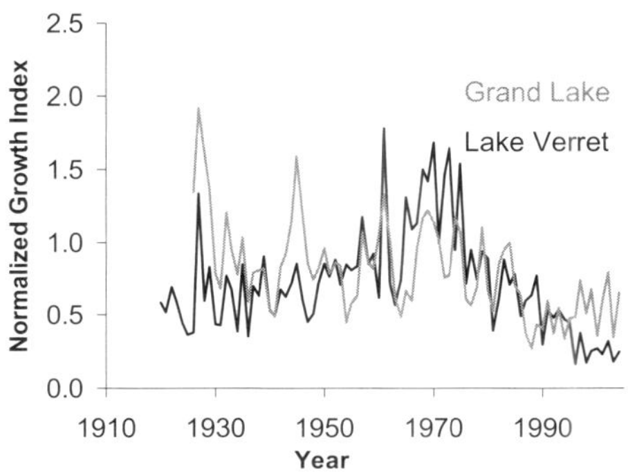

Fig. 5. Tree ring growth indices (normalized to remove biological trends) for two baldcypress stands in contrasting hydrological conditions in Louisiana.

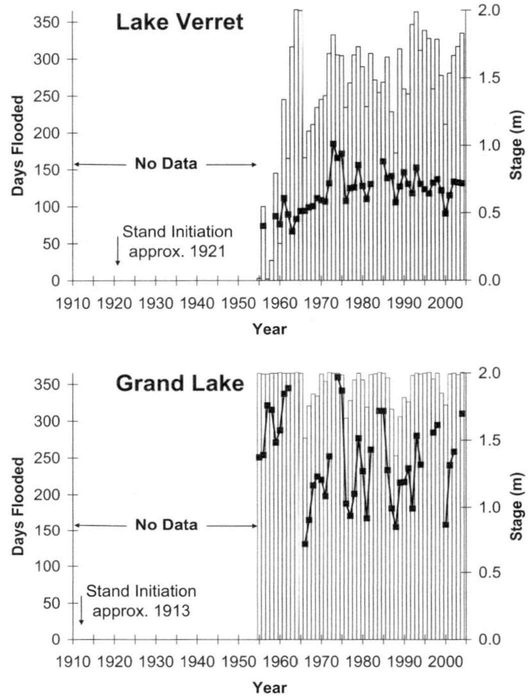

Fig. 6. Inundation histories for two baldcypress stands with contrasting hydrological conditions in Louisiana. Yearly average stage (symbols/lines) was obtained from a nearby, hydrologically connected water level gauge, and number of days flooded per year (bars) were estimated from site elevations.

31.3.2. Case Study II: Inundation and Regeneration

Complexity of hydrological interactions with regeneration arises because seasonality and timing of inundation are not easily quantifiable in ecological terms. Here we attempt to classify two swamps into the SWG hydrological condition classes (Chambers et al., 2005). The sites are both in the Atchafalaya Basin Floodway (ABF), Louisiana, and both are in the same swamp bounded on the north by Bayou Pigeon. One site is near the upper topographic limit of baldcypress-tupelo forest, near the natural levee of the brownwater Bayou Pigeon. The other site is lower, and occupies the low natural levee of the blackwater Bayou Mallet about 5.7 km south of the Bayou Pigeon site. Tree growth and response to thinning at these sites have been previously reported by Dicke and Toliver (1988, 1990). The annual hydrograph of the ABF mostly follows that of the Mississippi River, because most ABF flow originates as diverted river water. Thus, seasonally high flows in the spring following snowmelt often last into the summer, and are followed by low flows in the fall.

To obtain flooding variables relevant to regeneration, we used the US Army Corps of Engineers water level gauge at Bayou Sorrell, which is 6.3 km north of the Bayou Pigeon site. This gauge has a daily record beginning in 1955, and shows no long-term changes in mean water level. Thus, it is a good candidate for analysis of flooding probabilities. Field experience is that water levels in the area do not vary appreciably in space in this part of the ABF, so we assumed historical water depth at the sites would be fully recoverable from the gauge record by simple subtraction based on spot field observations of water depth at the sites. Based on these assumptions, Bayou Pigeon floods at stage 1.6 m MSL, and Bayou Mallet floods at stage 0.8 m MSL. We assumed the limiting hydrological factor for regeneration is consecutive unflooded days during the growing season (April 1 – October 31: 214 days), in which time a seedling must grow tall enough to withstand flooding during the next growing season.

The average length of seedling growing time at Bayou Pigeon is 116 days, 25% of years have at least 137 days, and 75% of years have at least 97 days (Figure 7). The average

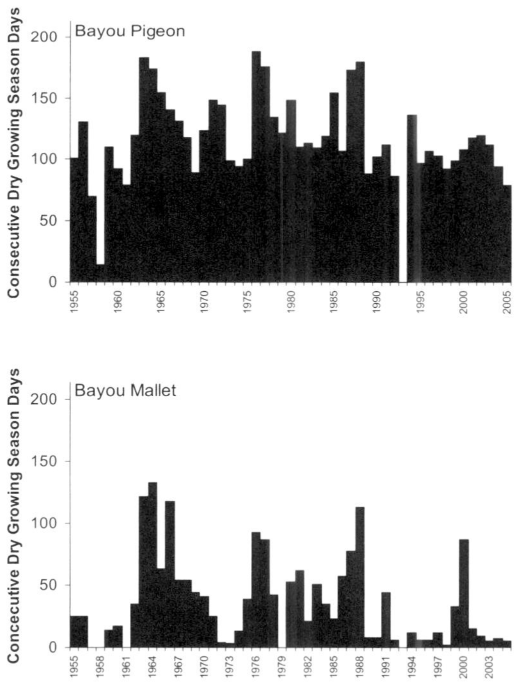

Fig. 7. Maximum consecutive non-inundated days during the growing season at two sites in the Atchafalaya Basin Floodway, Louisiana.

length of seedling growing time at Bayou Mallet is 35 days, 25% of years have at least 54 days, and 75% of years have at least 6 days. Assuming a baldcypress seedling is able to germinate on the first dry day and grow at maximum rate of 0.9 m yr^{-1} (Williston et al., 1980; Keeland and Conner, 1999), these flooding regimes dictate that seedlings can grow to be 49 cm in an average year at Bayou Pigeon, but just 15 cm at Bayou Mallet. In dry years (25% probability), seedlings can grow 58 cm at Bayou Pigeon and 23 cm at Bayou Mallet. In wet years (75% probability), seedlings can grow 41 cm at Bayou Pigeon and 3 cm at Bayou Mallet.

These estimates clearly indicate that Bayou Mallet cannot be classified as Class I (best), because seedlings 15–23 cm tall are unlikely to withstand the summer flooding of the next year, and regeneration will likely be successful only in extraordinary circumstances outside the range of the observed flooding. In contrast, the expected seedling height of about 50 cm at Bayou Pigeon may be sufficient to withstand flooding the next year. More information is needed about the specifics of seedling flood tolerance to confidently classify Bayou Pigeon as Class I (best). For example, is the early-growing-season flooding that is typical of these sites more tolerable than late season flooding?

The operational possibility of artificial regeneration at the two ABF sites is defined by dormant-season flooding of less than 0.6 m. Bayou Pigeon is flooded to this depth on average 28% of the dormant season and Bayou Mallet is flooded to this depth on average 66% of the dormant season (Figure 8). Although it might be difficult to schedule a planting

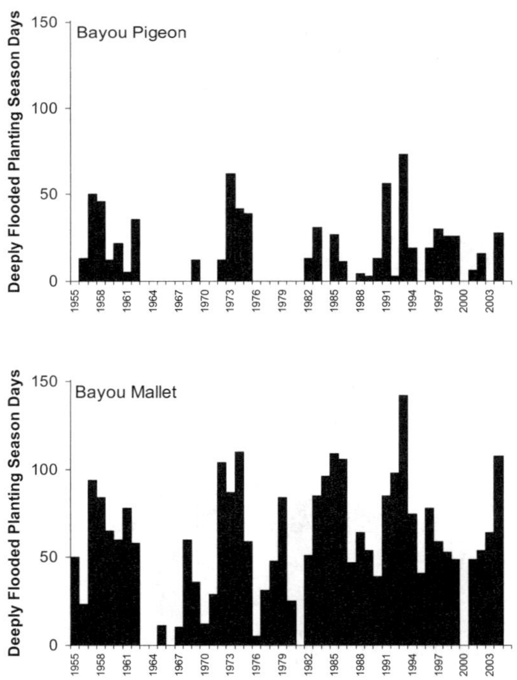

Fig. 8. Number of days flooded to depth >0.6 m during the dormant, planting season at two sites in the Atchafalaya Basin Floodway, Louisiana.

time at Bayou Mallet, it would not likely be impossible, and planting at Bayou Pigeon would usually not be a problem. Thus, neither site seems to warrant status as Class III (no regeneration), with the main caveat being that seedlings would need to be large enough to withstand flooding the next few years.

31.4. CONCLUSIONS

Widespread changes have been made to hydrological and delta-building processes in the Mississippi River delta. These changes have mainly exacerbated subsidence and increased flooding. Productivity of coastal wetland forests has been reduced by this increased flooding, but the mechanisms responsible for changes in site-specific productivity are not clear. Regeneration of coastal wetland forest tree species has also been reduced by the increased flooding. The Coastal Wetland Forest Conservation and Use Science Working Group, convened by the Governor of Louisiana, has developed a hydrological classification to aid in assessing sustainability of coastal wetland forests based on their potential to regenerate. A detailed analysis of the hydrologically-mediated regeneration capacity of two sites in the Atchafalaya Basin Floodway illustrates that more research is needed before confident assessment of wetland forest sustainability can be made.

REFERENCES

Alt, G.L., 1984. Black bear cub mortality due to flooding of natal dens. Journal of Wildlife Management 48:1432–1434.

Amos J.B., Keim, R.F., and Doyle, T.W., 2005. Coastal forest productivity and wetland hydrology: A dendrochronological study. Eos, Transactions, American Geophysical Union, 86(18), Joint Assembly Supplement, Abstract NB33C–06.

Barrow, W.C., Jr., Johnson, L.A., Randall, L., Woodrey, M.S., Cox, J., Ruelas, E., Riley, C.M., Hamilton, R.B., and Eberly, C., 2005. Coastal forests of the Gulf of Mexico: a description and some thoughts on their conservation. In: USDA Forest Service General Technical Report PSW–GTR–191.

Baumann, R.H., Day, J.W., Jr., and Miller, C.A., 1984. Mississippi deltaic wetland survival: sedimentation versus coastal submergence. Science 224:1093–1095.

Brown, S., 1981. A comparison of the structure, primary productivity, and transpiration of cypress ecosystems in Florida. Ecological Monographs 51:403–427.

Chambers, J.L., Conner, W.H., Day, J.W., Faulkner, S.P., Gardiner, E.S., Hughes, M.S., Keim, R.F., King, S.L., McLeod, K.W., Miller, C.A., Nyman, J.A., and Shaffer, G.P., 2005. Conservation, Protection and Utilization of Louisiana's Coastal Wetland Forests, Final Report to the Governor of Louisiana from the Science Working Group on Coastal Wetland Forest Conservation and Use. 102 p.

Coleman, J.M., Roberts, H.H., and Stone, G.W., 1998. Mississippi River delta: an overview. Journal of Coastal Research 14:698–716.

Conner, W.H., and Brody, M., 1989. Rising water levels and the future of southeastern Louisiana swamp forests. Estuaries 12:318–323.

Conner, W.H., and Buford, M.A., 1998. Southern deepwater swamps. Pages 261–287 in M.G. Messina and W.H. Conner, editors. Southern Forested Wetlands, Ecology and Management. Lewis, Boca Raton, Florida, USA.

Conner, W.H., and Day, J.W., Jr., 1988. Rising water levels in coastal Louisiana: implications for two forested wetland areas in Louisiana. Journal of Coastal Research 4:589–596.

Conner, W.H., and Toliver, J.R., 1990. Long-term trends in the bald-cypress (*Taxodium distichum*) resource in Louisiana (U.S.A.). Forest Ecology and Management 33/34:543–557.

Conner, W.H., Gosselink, J.G., and Parrondo, R.T., 1981. Comparison of the vegetation of three Louisiana swamp sites with different flooding regimes. American Journal of Botany 68:320–331.

Conner, W.H., Toliver, J.R., and Sklar, F.H., 1986. Natural regeneration of baldcypress (*Taxodium distichum* (L.) Rich.) in a Louisiana swamp. Forest Ecology and Management 14:305–317.

Cowardin, L.M., Carter, V., Golet, F.C., and LaRoe, E.T., 1979. Classification of wetlands and deepwater habitats of the United States. U.S. Fish and Wildlife Service, Office of Biological Services, FWS/OBS–79/31, Washington, DC.

DeBell, D.S., and Naylor, A.W., 1972. Some factors affecting germination of swamp tupelo seeds. Ecology 53:504–506.

Day, J.W., Shaffer, G.P., Britsch, L.D., Reed, D.J., Hawes, S.R., and Cahoon, D., 2000. Pattern and process of land loss in the Mississippi delta: A spatial and temporal analysis of wetland habitat change. Estuaries 23:425–438.

Dicke, S.G., and Toliver, J.R., 1988. Effects of crown thinning on baldcypress height, diameter, and volume growth. Southern Journal of Applied Forestry 12:252–256.

Dicke, S.G., and Toliver, J.R., 1990. Growth and development of bald-cypress water-tupelo stands under continuous versus seasonal flooding. Forest Ecology and Management 33:523–530.

Gagliano, S.M., 1998. Faulting, Subsidence, and Land Loss in Coastal Louisiana. Coastal Environments Inc., Baton Rouge, LA.

Gooding, G., and Langford, J.R., 2004. Characteristics of tree roosts of Rafinesque's big-eared bat and southeastern bat in Northeastern Louisiana. The Southwestern Naturalist 49:61–67.

Harms, W.R., Schreuder, H.T., Hook, D.D., and Brown, C.L., 1980. The effects of flooding on the swamp forest in Lake Ocklawah, Florida. Ecology 61:1412–1421.

Hightower, D.A., Wagner, R.O., and Pace, R.M. III., 2002. Denning ecology of female American black bears in south central Louisiana. Ursus 13:11–17.

Hodges, J.D., 1997. Development and ecology of bottomland hardwood sites. Forest Ecology and Management 90:117–125.

Hook, D.D., 1984. Waterlogging tolerance of lowland tree species of the South. Southern Journal of Applied Forestry 8:36–149.

Keeland, B.D., and Conner, W.H., 1999. Natural regeneration and growth of *Taxodium distichum* (L.) Rich. In Lake Chicot, Louisiana after 44 years of flooding. Wetlands 19:149-155.

Kozlowski, T.T., 1984. Responses of woody plants to flooding. Pages 129–163. In T.T. Kozlowski (ed.). Flooding and Plant Growth. New York: Academic Press.

Kozlowski, T.T., 1997. Responses of woody plants to flooding and salinity. Tree Physiology Monograph No. 1. 29 pp.

Lance, R.F., Hardcastle, B.T., Talley, A., and Leberg, P.L., 2001. Day-roost selection by Rafinesque's big-eared bats (*Corynorhinus rafinesquii*) in Louisiana forests. Journal of Mammalogy 82:166–172.

Lowery, G.H. Jr., 1974. Louisiana Birds. Louisiana State University Press, Baton Rouge.

National Research Council, 1995. Wetlands, Characteristics and Boundaries. National Academy Press, Washington.

Roberts, H.H., 1997. Dynamic changes of the holocene Mississippi river delta plain: the delta cycle. Journal of Coastal Research 13:605–627.

Souther, R.F., and Shaffer, G.P., 2000. The effects of submergence and light on two age classes of baldcypress (*Taxodium distichum* (L.) Richard) seedlings. Wetlands 20:697–706.

Thomson, D.A., Shaffer, G.P., and McCorquodale, J.A., 2002. A potential interaction between sea-level rise and global warming: implications for coastal stability on the Mississippi River deltaic plain. Global Planetary Change 32:49–59.

Turner, R.E., 1997. Wetland loss in the northern Gulf of Mexico: multiple working hypotheses. Estuaries 20:1–13.

White, T.H. Jr., Bowman, J.L., Jacobson, H.A., Leopold, B.D., and Smith, W.P., 2001. Forest management and female black bear denning. Journal of Wildlife Management 65:34–40.

Williston, H.L., Shropshire, F.W., and Balmer, W.E., 1980. Cypress management: a forgotten opportunity. Forestry Report SA-FR 8. USDA Forest Service, Southeastern Area, Atlanta, GA.

Zoller, J.A., 2004. Seasonal differences in bird communities of a Louisiana swamp and manipulation of the breeding density of Prothonotary Warblers. M.S. Thesis, Southeastern Louisiana University, Hammond.

32

EVALUATION OF THE MARSHLAND UPWELLING SYSTEM FOR THE REMOVAL OF $CBOD_5$ FROM COASTAL DWELLING WASTEWATER

Lorna A. Putnam[1], Robert E. Watson Jr.[2], Robert P. Gambrell[1], and Kelly A. Rusch[2]

[1]Wetland Biogeochemistry Institute
Energy, Coast and Environment Bldg.
Louisiana State University, Baton Rouge, LA 70803, USA

[2]Institute for Ecological Infrastructure Engineering
Engineering Lab Annex Bldg.
Louisiana State University, Baton Rouge, LA 70803, USA

ABSTRACT

Five-day carbonaceous biochemical oxygen demand ($CBOD_5$) removal efficacy was evaluated in the marshland upwelling system (MUS) that was installed on private camps in the Grand Bay National Estuarine Research Reserve, Moss Point, Mississippi, and Bayou Segnette, Louisiana in the Barataria Terrebonne National Estuary Program. The system is used for decentralized wastewater treatment of coastal dwellings. Raw wastewater was injected into the surrounding saltmarsh subsurface in Moss Point at flow rates of 1.9 and 2.8 L min^{-1} with an injection frequency of 30 min every 3 h. An additional Moss Point study with a flow rate of 2.8 L min^{-1} with an injection frequency of 15 min every 1 h was performed. The Bayou Segnette studies, located in an intermediate marsh, had flow rates of 1.0 and 1.9 L min^{-1} with injection frequencies of 15 min every 1 h. The overall (unfiltered) $CBOD_5$ influent mean was 301 ± 168 and 318 ± 115 mg L^{-1}, which was reduced to an overall effluent mean of 13 ± 13 and 9 ± 19 mg L^{-1} by a vector distance of 7 and 5 meters for Moss Point and Bayou Segnette, respectively. $CBOD_5$ removal followed a first-order decay rate with respect to vector distance for all studies. This resulted in removal rate constants for Moss Point ranging from 0.47-1.00 m^{-1} and predicted surface concentrations from 9.5-30.6 mg L^{-1}, while Bayou Segnette ranged from 0.92-1.14 m^{-1} and 2.9-5.4 mg L^{-1}, respectively. All studies, except one, achieved effluent $CBOD_5$ levels below NPDES standard levels of 25 mg L^{-1}.

32.1. INTRODUCTION

Wetlands, once considered to have very little or no value to human society, are now recognized for the many functions and services they provide. In addition to serving as habitat and protection for a number of fish and wildlife species, they are invaluable as sources, sinks and transformers of many biological and chemical resources (Mitsch and

Gosselink, 2000). The latter enables wetlands to remove contaminants from waters, which has become important as pollutants from anthropological sources increase. Because of these properties, natural wetlands have long been used for waste treatment purposes. Ancient China and Egypt filtered their waste though natural wetlands. More recently, in the 1960s and 70s, researchers have realized the potential benefits both natural and constructed wetlands provide as waste treatment systems (Kadlec and Knight, 1996; Mitsch and Gosselink, 2000). In contrast to constructed wetlands, natural wetlands used as waste treatment systems have the benefit of being a cheap and effective alternative to many other waste treatment systems (Kadlec and Knight, 1996).

Onsite domestic waste treatment is typically performed by septic systems. Approximately 30% of American households utilize the septic system for onsite waste treatment (USEPA, 1999). For most areas, septic systems prove to be an efficient and long-term solution for onsite waste treatment. However, septic systems are unsuitable for areas that have impermeable soils, shallow bedrock and/or high water tables (USEPA, 1999). The poor soil absorption that occurs in these situations can result in system failure and, ultimately, lead to surface ponding of wastewater (USEPA, 1999; USEPA, 2002). Ponding can then result in contamination of nearby waterbodies.

Five-day carbonaceous biochemical oxygen demand ($CBOD_5$) is one of the most common measurements used in determining the effectiveness of many waste treatment systems, including subsurface wetlands (Kadlec and Knight, 1996; Vymazal, 2002; Al-Omari and Fayyad, 2003; Mantovi, et al., 2003; Metcalf and Eddy, 2003). Several physical, chemical and biological operations and processes combine to create the reduction of BOD in wastewater, including filtration and decomposition (Gopal, 1999; Meuleman, et al., 2003). Aerobic decomposition of organic matter is much more efficient than the anaerobic decomposition that occurs in the subsurface of wetlands. However, the slow movement of water through wetlands, and subsequently organic matter, creates a longer retention time than would be found in open water. This allows for a longer time period to decompose the organic matter being presented as wastewater in subsurface wetlands. Many studies have shown that using subsurface wetlands for wastewater treatment significantly reduces the $CBOD_5$ present in the wastewater (Nguyen, 2000; Vymazal, 2002; Al-Omari and Fayyad, 2003; Mantovi et al., 2003; Meuleman et al., 2003).

To address the special issues that typify coastal environments; i.e., sparsely populated areas, high water tables, poor hydraulic soil conditions, anaerobic soils and saline groundwater, the marshland upwelling system (MUS) was developed as an alternative wastewater treatment system (Stremlau, 1994; Watson Jr., 2000; Watson Jr. and Rusch, 2001; Richardson, 2002; Watson Jr. and Rusch, 2002; Fotenot, 2003; Addo, 2004; Richardson, et al., 2004; Evans, 2005; Richardson and Rusch, 2005; Turriciano, 2005). The MUS consists of a collection/distribution tank, injection pump, programmable timer, injection well and saturated subsurface soils (Figure 1). Effective system operation depends upon several factors: 1) groundwater salinity, 2) injection flow rate, 3) injection frequency and 4) filtration properties of the native soil matrix.

Black and grey wastewater from coastal dwellings is collected in the collection/distribution tank and, after solids have settled, is injected intermittently into the marsh subsurface via an injection well. The intermittent injection of the wastewater allows for subsurface pressure dissipation between injection cycles. The energy created by the injection pump causes wastewater to disperse radially from the injection point. The freshwater waste plume that is injected into brackish or salt groundwater moves by

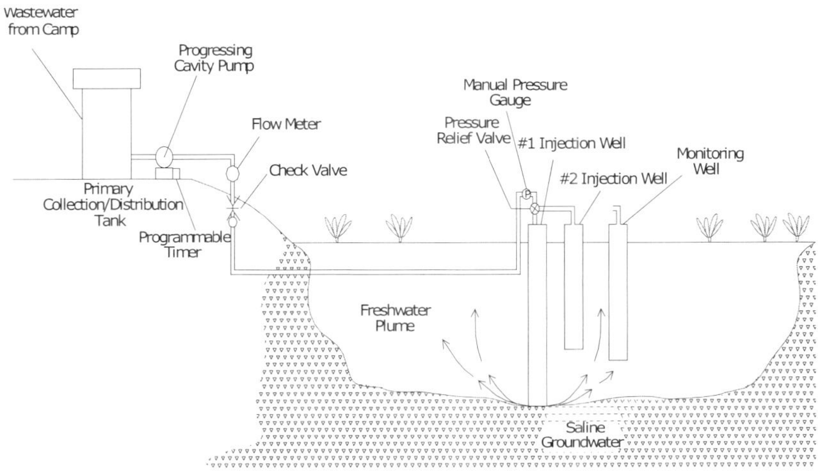

Fig. 1. Schematic of the marshland upwelling system.

buoyancy forces and natural groundwater flow between injection cycles. As buoyancy effects decline, lateral dispersion takes over plume movement. Wastewater treatment is accomplished by physical filtration, microbial decomposition, nutrient adsorption by mineral and organic soil solids, die off (for microorganisms), carbon and nitrogen transformation to gaseous forms, plant uptake and aerobic biofilm processes in the upper soil and root zones.

Before a wastewater treatment can be certified for public use, design and operational criteria must be developed that are applicable to a wide range of environmental conditions. The main objectives of these studies were to: 1) evaluate the $CBOD_5$ treatment efficacy of the MUS, 2) determine loading rates and 3) determine appropriate design and operational criteria for saline and intermediate marsh environments.

32.2. METHODOLOGY

The research sites were located within the Grand Bay National Estuarine Research Reserve, Moss Point, Mississippi, and along Bayou Segnette, Louisiana, in the Barataria Terrebonne National Estuary. The Moss Point research site was located in close vicinity to Bayou Cumbest. Bayou Cumbest has approximately 45 residences located directly on its banks, while Bayou Segnette has approximately 150 residences. Both regions have suffered from poor water quality, which has led to problems for shellfish harvesting. Bayou Cumbest was listed as the most impaired water body in the Mississippi List of Most Impaired Waterbodies (MDEQ, 1996), while Bayou Segnette was listed on the 1998 EPA 303(d) list of impaired waterbodies.

Soil from the Moss Point research site was characterized using a continuous intrusion cone penetrometer (CICP) in combination with separate hand augured soil borings. Soil immediately adjacent the experimental area was analyzed by standard methods (ASTM, 1995; Bardet, 1997). The marsh surface was composed of dark, highly organic soils that are consistent with Scatlake series soils, which are clayey, semifluid, mineral soils that are generally poorly drained (USDA, 1984). Soil from the Bayou Segnette research site was characterized by hand-augered soil samples taken from soil near the experimental area and analyzed using standard methods (ASTM, 1995). Sand, silt and clay content were identified using sieve and hydrometer analyses (ASTM, 1995; Bardet, 1997). Soils are consistent with the Kenner muck series, which exhibit high permeability within the organic layers and low permeability in the clay layers (USDA, 1983). For both sites, as the depth increased, the organic substrate was gradually replaced with a more compact sand/clay mixture, with the sand being predominate at increasing depth (Table 1).

Table 1. Selected soil properties at field sites from various depths.

	Moss Point		Bayou Segnette	
	Depth Interval (m)		Depth Interval (m)	
Property	**0-2.4**	**2.4-3.8**	**0-2.7**	**2.7-4.6**
Sand (%)	40.5	74.0	80.0	78.9
Silt (%)	42.0	16.0	16.5	14.8
Clay (%)	17.5	10.0	3.5	6.3
Median grain size diameter (d_{50}) (mm)	0.07	0.13	0.26	0.10
Fraction of organic content (f_{oc}) (%)	9.0 ± 0.5	0.5 ± 0.1	35.7 ± 1.2	11.6 ± 2.5

The Moss Point MUS treated wastewater from two privately owned camps and a public restroom facility, while the Bayou Segnette MUS treated wastewater from a permanent single-family residence. Wastewater for both sites was a combination of grey and black water. The Moss Point experimental MUS consisted of a 3.8 m deep injection well surrounded by 38 monitoring wells at depths of 1.5, 2.3 and 3.0 m (Figure 2). The monitoring wells allowed for examination of the wastewater plume in three dimensions. Three wells were installed away from the injection site to monitor background conditions during the course of the study. A main holding tank of 1,325 L served to store the wastewater until injection into the subsurface. The Bayou Segnette experimental MUS consisted of two injection wells at 6.1 and 4.3 m surrounded by 38 monitoring wells at depths of 2.7, 4.0, 4.3, 4.6, 6.1 and 7.6 m. One well was installed away from the injection site to monitor background conditions. For this study, only the 4.3 m injection well was used. Large solids were allowed to settle out prior to injection and to further prevent large solids from entering the injection line, a standpipe was installed at the effluent end of the tank. Wastewater was pumped out of the holding tank by a 0.093 kW low flow high pressure (276 kPa) cavity pump (Cole Palmer, Inc.) and transported to the injection site by a 1.9 cm diameter PVC line. A programmable timer and float switch were used to control wastewater injection. A water meter (Aquatic Ecosystems, Inc.) installed on the injection line allowed for documentation of cumulative wastewater volume and confirmed flowrates. Continuous monitoring of subsurface pressure was performed using an automated pressure transducer and data logger. For further details on the experimental system and construction and installation procedures, refer to Watson Jr. (2000), Richardson (2002), Fotenot (2003), Addo (2004) and Turriciano (2005).

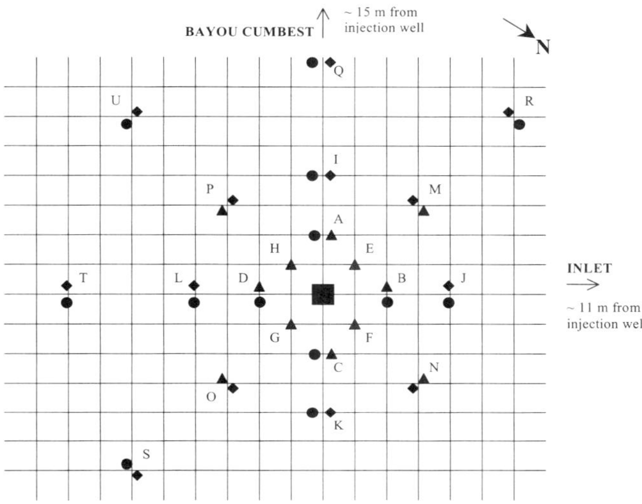

Fig. 2. A map representing the MUS layout consisting of one injection well (■) and 38 monitoring wells at depths of 1.5 m (◆), 2.3 m (▲), and 3.0 m (●). Each grid square is equal to 0.61 m X 0.61 m.

Native groundwater conditions were sampled before system activation. This allowed for evaluation of system impacts on pH, salinity, temperature (Table 2) and native parameter concentrations. The largest difference between the two sites was in the salinity regime. Moss Point had a mean background salinity of 31 ppt and Bayou Segnette of 9 ppt. A surface water sample from each site was also collected during each sampling event for comparison of the system effluent to ambient conditions. Moss Point system operation began in June 2001 and Bayou Segnette in November 2002. Various injection flow rates and frequencies were examined to determine the impact on the system (Table 3). Samples were collected once to twice a month to monitor the effect of the wastewater plume within the marsh subsurface. Each monitoring well contained its own permanently installed neoprene tubing to prevent cross-contamination between wells during sampling. All water samples were collected, stored and transported following standard protocol (APHA, 1998). In situ water quality parameters, including salinity, pH and temperature, were sampled for each monitoring well and the influent collection tank. In addition, dissolved oxygen was measured for the influent collection tank. Unfiltered and filtered samples from the influent tank and filtered samples from selected monitoring wells were analyzed for fecal coliforms (9222 D), $CBOD_5$ (5210 B), total and volatile suspended solids (influent only, 2540 D and E/G) and nutrients including, total phosphorus (4500-P B and E), orthophosphate (4500-P E), total ammonia nitrogen (4500-NH_3 D), total Kjeldahl nitrogen (4500-N org C and 4500-NH_3 D) and nitrate (4500-NO^-_3 E) and nitrite (4500-NO^-_2 B) nitrogen (APHA, 1998). With the exception of $CBOD_5$, only the influent results of the above parameters are presented (Table 4).

Table 2. Native properties from Moss Point and Bayou Segnette MUS sites over the course of study.

	Moss Point				Bayou Segnette			
	Surface	Subsurface Water			Surface	Subsurface Water		
Property	Water	1.5 m	2.3 m	3.0 m	Water	4.6 m	6.1 m	7.6 m
Salinity (ppt)	17.1 ± 7.3 (n=32)	31.7 ± 3.4 (n=15)	30.3 ± 3.7 (n=15)	30.9 ± 3.6 (n=15)	1.2 ± 1.5 (n=31)	7.3 ± 1.6 (n=16)	9.6 ± 2.0 (n=14)	11.5 ± 1.9 (n=10)
pH	7.5 ± 0.4 (n=28)	6.5 ± 0.2 (n=14)	6.6 ± 0.1 (n=14)	6.8 ± 0.3 (n=14)	7.2 ± 0.5 (n=35)	6.3 ± 0.1 (n=8)	6.6 ± 0.1 (n=7)	6.7 ± 0.1 (n=5)
Temperature (°C)	24.1 ± 6.8 (n=27)	21.8 ± 2.7 (n=14)	22.1 ± 2.6 (n=14)	22.9 ± 2.7 (n=14)	23.0 ± 8.0 (n=36)	20.9 ± 0.3 (n=8)	21.4 ± 1.0 (n=7)	21.0 ± 1.6 (n=5)

Table 3. Injection flow rates and frequencies employed over course of the studies.

Study period (month/day/year)	Elapsed time (days)	Injection flow rate ($L\ min^{-1}$)	Injection frequency	Volume of wastewater injected (L)	Theoretical hydraulic loading rate[a] ($L\ day^{-1}$)
06/21/01-09/17/01[c]	87	1.9	30 min every 3 h	13,215	456
09/17/01-11/05/01c	49	5.5	30 min every 3 h	7,737	1,320
11/05/01-01/21/02[c]	77	2.8	30 min every 3 h	9,236	672
01/21/02-04/02/02[b,c]	71	2.8	30 min every 3 h	32,274	672
04/02/02-06/03/02[c]	62	2.8	15 min every h	11,553	1,008
03/10/03-11/03/03[d]	238	1.0	15 min every h	59,977	342
11/03/03-03/22/04[d]	207	1.9	15 min every h	52,902	684

[a]Assumes an adequate volume of wastewater is present in holding tank to trigger injection.
[b]This study was using a synthetic waste.
[c]This study is from Moss Point.
[d]This study is from Bayou Segnette.

Influent data were used to calculate hydraulic, $CBOD_5$ and solids loading rates. Data from the monitoring wells were analyzed using *SAS* (version 9.1) and *SIGMA PLOT* (version 8.0). All statistical tests were performed at α=0.05. Mean in situ parameters were calculated by averaging over the number of sampling events within a given study. $CBOD_5$ data are presented by vector distance from the point of injection for each study. Removal rate coefficients were calculated from vector distance versus $CBOD_5$ concentration curves. Predicted surface concentrations were calculated assuming the wastewater traveled directly to the surface after injection, thus the distance used is the length of the injection well. Predicted travel distances were calculated using the NPDES standard limit of 25 mg L^{-1} for the concentration. Removal efficiencies were calculated by dividing the difference of the influent and the furthest vector distance $CBOD_5$ value by the influent $CBOD_5$ value.

Table 4. Influent wastewater parameters.

Parameters	Units	Moss Point		Bayou Segnette		Typical Wastewater[a]
		n	Results	n	Results	
Fecal coliform[b]	colonies/ 100mL	23	$5.53 \times 10^4 \pm 2.2 \times 10^6$	5	$2.77 \times 10^5 \pm 4.9 \times 10^5$	$10^4 - 10^5$
TSS	mg L^{-1}	22	208 ± 182	18	184 ± 40	100 – 350
VSS	mg L^{-1}	28	139 ± 163	18	129 ± 24	80 – 275
NO_2 –N filtered	mg L^{-1} – N	19	0.1 ± 0.1	22	BDL[c]	0
NO_3 – N filtered	mg L^{-1} – N	13	BDL	22	BDL	0
TAN unfiltered	mg L^{-1} – N	14	141 ± 53	26	100 ± 34	12 – 50
TAN filtered		21	115 ± 59	28	98 ± 35	
TKN unfiltered	mg L^{-1} – N	16	155 ± 72	26	116 ± 28	8 – 35
TKN filtered		21	124 ± 71	28	106 ± 27	
TP unfiltered	mg L^{-1} – P	17	16.2 ± 10.0	22	14.1 ± 3.2	4 – 15
TP filtered		18	13.5 ± 9.1	24	12.4 ± 3.3	
PO_4–P unfiltered	mg L^{-1} – P	15	11.7 ± 7.0		N/A	1 – 5
PO_4–P filtered		17	11.6 ± 6.9	31	11.3 ± 2.9	
$CBOD_5$ unfiltered	mg L^{-1}	24	282 ± 173		N/A	110 – 400
$CBOD_5$ filtered		21	217 ± 104	25	214 ± 14	

[a]Metcalf and Eddy (2003).
[b]Geometric mean
[c]Below detection limit

32.3. RESULTS AND DISCUSSION

Overall, the MUS for both sites performed well, with over 94% removal by the furthest vector distance from the point of injection, excluding the first Moss Point study. A total of 74,015 L of wastewater was injected at Moss Point over the entire study length and 112,879 L for Bayou Segnette (Table 3). The range of theoretical loading rates for Moss Point was 456-1,320 L day^{-1}, and for Bayou Segnette the theoretical loading rates ranged from 342-684 L day^{-1}. These rates varied depending on the flow rate and injection frequency being studied. $CBOD_5$ influent concentrations ranged from 26 to 676 mg L^{-1} with an overall average of 301 mg L^{-1} for the Moss Point site over the entire length of the study. For Bayou Segnette, concentrations ranged from 170 to 606 mg L^{-1} with an overall average of 318 mg L^{-1} for the length of the study presented. No seasonal trends appear to be present in the influent concentrations, though this is hard to determine from the short length of the studies and the varying flow rate and injection frequencies.

The first Moss Point study, at 1.9 L min^{-1} with an injection frequency of 30 min every 3 h, had a removal rate constant of 0.63 m^{-1}, a predicted surface concentration of 16.2 mg L^{-1} and travel distance of 3.11 m (Table 5). The second study, at 5.5 L min^{-1} with an injection frequency of 30 min every 3 h, was performed to determine if the system was capable of performing under maximum conditions. It is not discussed in this paper because it represents conditions to which the system would not generally be subjected. The third study, 2.8 L min^{-1} at 30 min every three h, had a removal rate constant of 0.95 m^{-1} and a predicted surface concentration of 12.6 mg L^{-1} and travel distance of 3.07 m. The fourth study was only different from the third study in that it was performed using a synthetic wastewater. The removal rate constant, predicted surface concentration and travel distances were 0.47 m^{-1}, 30.6 mg L^{-1} and 4.23 m, respectively. The final Moss Point study was performed at 2.8 L min^{-1} with an injection frequency of 15 min every h. The removal rate constant, predicted surface concentration and travel distances were 1.00 m^{-1}, 9.5 mg L^{-1} and 2.83 m, respectively. The first Bayou Segnette study, 1.0 L min^{-1} with an injection frequency of 15 min every h, had a removal rate constant of 1.14 m^{-1}, a predicted surface concentration of 2.9 mg L^{-1} and travel distance of 2.40 m, while the respective values for the second study, 1.9 L min^{-1} 15 min every h, were 0.92 m^{-1}, 5.4 mg L^{-1} and 2.64 m.

Table 5. Predicted $CBOD_5$ removal as a function of distance from injection point.

Injection flow rate ($L\ min^{-1}$)	First-order rate constant (m^{-1})	Predicted surface concentration ranges ($mg\ L^{-1}$)	Predicted travel distance ranges to achieve effluent standard (m)
1.9[c]	0.63	16.2	3.11
2.8[c]	0.95	12.6	3.07
2.8[a,c]	0.47	30.6	4.23
2.8[b,c]	1.00	9.5	2.83
1.0[d]	1.14	2.9	2.40
1.9[d]	0.92	5.4	2.64

[a]Synthetic wastewater study. [b]Injection frequency of 15 min every h. [c]Data from Moss Point. [d]Data from Bayou Segnette.

Overall, for Moss Point, excluding the synthetic wastewater study, it can be seen that the removal rate constants increase over time. All studies, except the synthetic wastewater study, had predicted surface concentrations that were below the NPDES standard of 25 mg L^{-1} and travel distances that were below the 3.8 m length of the injection well. Increasing the flow rate increased the rate constant while the predicted surface concentrations and travel distances decreased. This indicates the Moss Point site acclimated to the wastewater being injected over time and that the rate constant would continue to increase to a maximum value if the studies were continued. Bayou Segnette has rate constants and predicted surface concentrations and travel distances similar to Moss Point. In comparing these studies, Bayou Segnette does not perform significantly different ($p>0.8$) from Moss Point, indicating that the different salinity regiments between the two sites do not have a significant impact on performance of the MUS for $CBOD_5$.

Regardless of flow rate or injection frequency, $CBOD_5$ concentration decreases with increasing vector distance from injection point for all studies. The first Moss Point study reached the NPDES standard limit of 25 mg L^{-1} by a vector distance of 3 m. The third study was not consistently below 25 mg L^{-1} until a vector distance of 6.94 m, while the fourth and fifth studies were below 25 mg L^{-1} by 6.14 m. Because the first study took place during the initial acclimation period of the system, the standard deviations for each individual vector distance are high (Figure 3). Initially, the influent $CBOD_5$ value started off low and then increased as the study progressed. An overall mean of these values led to a lower $CBOD_5$

concentration for each vector distance, which explains why the first study reached the NPDES limit of 25 mg L^{-1} within a distance less than half of the subsequent studies. For both Bayou Segnette studies, $CBOD_5$ concentration fell below the 25 mg L^{-1} standard by a vector distance of 3.05 m. This study was performed after an initial acclimation period, thus the Bayou Segnette site achieves a drop below 25 mg L^{-1} within a lower travel distance than the Moss Point site. Salinity is often a stressor for microbes and can have negative impacts on microbial decomposition (Rietz and Haynes, 2003; Mendelssohn and Slocum, 2004). Therefore, the lower salinity regime at Bayou Segnette may account for the faster degradation of BOD.

All flow rates and injection frequencies performed well, achieving $CBOD_5$ removal to below 25 mg L^{-1} within a distance less than that of the injection well length, except the third Moss Point study. For Moss Point, doubling the flow rate increased the removal of $CBOD_5$ and increasing the injection frequency from 30 min every 3 h to 15 min every h increased the rate constant. However, the increase in the removal rate constant from 0.95 to 1.00 m^{-1} may not be a significant enough increase to account for the increase in the amount of pumping required between the injection frequencies. Though this does indicate the system is able to operate effectively at a higher injection frequency. In comparison to Moss Point, the first Bayou Segnette study has the lowest flow rate and an injection frequency (15 min every h) higher than all but one of Moss Point. The second Bayou Segnette study had a flow rate and injection frequency more comparable to the Moss Point studies. The second study has a removal rate constant that is within the range of the Moss Point studies. This implies the Bayou Segnette system operates as efficiently as the Moss Point system. Overall, the Bayou Segnette system with an injection flow rate of 1.0 L min^{-1} and frequency of 15 min every h performed the best.

Daily hydraulic, $CBOD_5$ and solids loading rates were calculated for each study (Table 6). Mean hydraulic loading rates (HLR) varied from 113 – 482 L day^{-1} for Moss Point, while Bayou Segnette varied from 314 – 407 L day^{-1}. All HLRs were less than the theoretical HLR (Table 3) calculated assuming a continuously adequate amount of wastewater. The highest HLR occurred during the fourth Moss Point study, performed at 2.8 L min^{-1} with an injection frequency of 30 min every h, which used a synthetic wastewater source. This study also had the highest $CBOD_5$, total solids and volatile solids loading rates (CLR, TLR and VLR, respectively) for all Moss Point studies. This indicates the synthetic wastewater had a higher strength wastewater than the domestic waste used in all other studies, which explains why it performed the worst in terms of removal constants. It also shows that the system is capable of treating a higher strength wastewater, but not within the bounds of other Moss Point studies. Mean CLRs ranged from 30 – 71 g day^{-1} for Moss Point and 80 – 148 g day^{-1} for Bayou Segnette. Mean TLRs ranged from 16 – 163 g day^{-1} and mean VLRs from 1 – 15 g day^{-1} for Moss Point. The last Moss Point study did not have enough data points available for an overall mean; however, the data points available do not indicate the overall mean for this study would have had a higher mean than the fourth study. Bayou Segnette TLR and VLR ranged from 43-86 and 36-72 g day^{-1}, respectively. Bayou Segnette solid loading rates indicate that the wastewater from this site has a higher organic matter component than at Moss Point. The higher organic matter content may explain why the CLRs for Bayou Segnette were higher than all of the Moss Point CLRs.

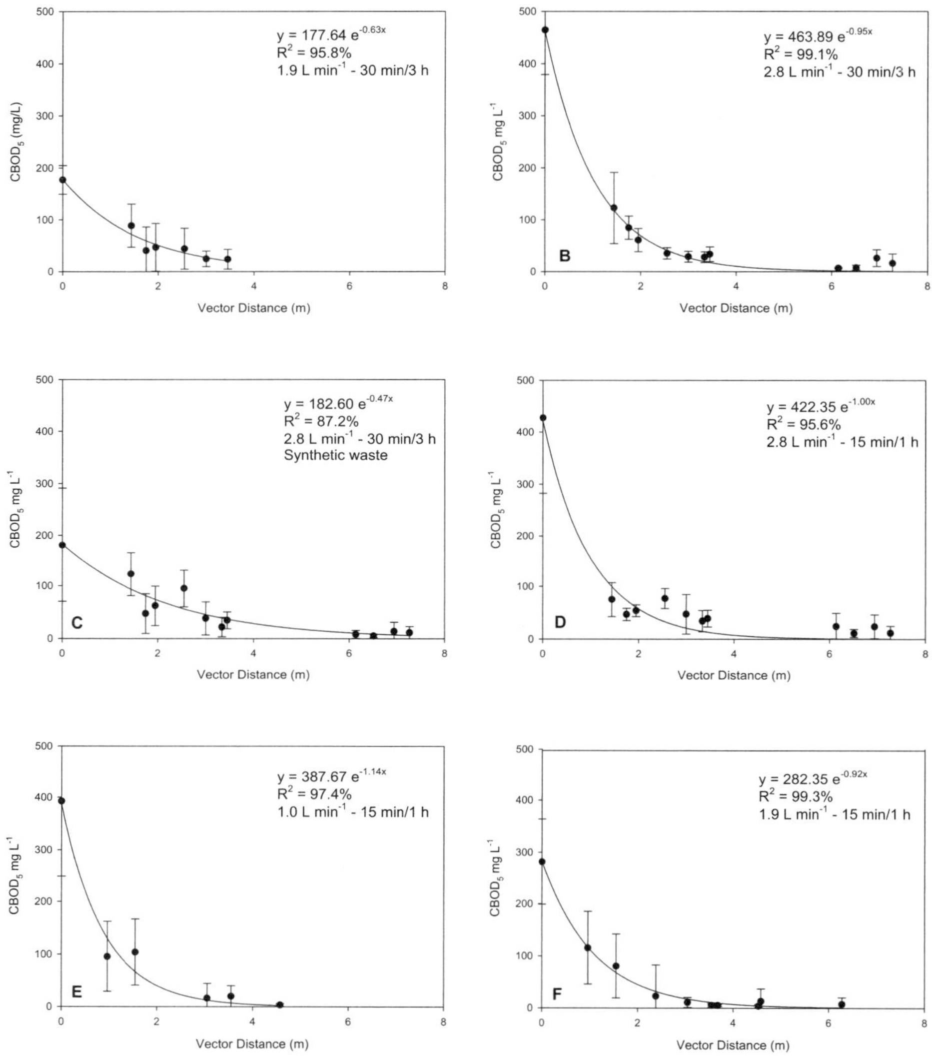

Fig. 3. **$CBOD_5$ removal with respect to vector distance from injection well for Moss Point, studies 1, 3, 4 and 5, (a – d) and Bayou Segnette (e – f). Each data point represents overall mean $CBOD_5$ concentration at given vector distance. Error bars represent standard deviation for each data point.**

Table 6. Daily hydraulic, $CBOD_5$ and solids loading rates for influent wastewater. [ ] represent overall mean values.

Injection flow rate (L min^{-1})	Hydraulic loading rate (L day^{-1})	$CBOD_5$ loading rate (g day^{-1})	Total suspended solids loading rate (g day^{-1})	Volatile suspended solids loading rate (g day^{-1})
1.9[d]	18-273 [177]	23-35 [30]	13-20 [16]	0-2 [1]
5.5[d]	44-226 [158]	33-54 [44]	5-31 [18]	0-7 [3]
2.8[d]	64-183 [113]	18-57 [31]	9-40 [21]	2-7 [4]
2.8[a,d]	125-826 [482]	8-124 [71]	82-221 [163]	0-27 [15]
2.8[b,d]	23-272 [155]	8-49 [40]	3[c]	1[c]
1.0[e]	5-1653 [314]	2-316 [80]	1-122 [43]	0-112 [36]
1.9[e]	279-477 [407]	71-254 [148]	[86][f]	[72][f]

[a]Synthetic wastewater study. [b]Injection frequency of 15 min every h.
[c]Only two data points available. [d]Data from Moss Point. [e]Data from Bayou Segnette. [f]Only mean values are available for these dates.

The first Moss Point study achieved an 87% removal efficiency of $CBOD_5$ by a vector distance of 3.45 m and the third, fourth and fifth Moss Point studies achieved 97, 94 and 97% removal efficiency by a vector distance of 7.27 m. The first study occurred while the subsurface was acclimating to system conditions, thus accounting for the lower removal efficiency. The Bayou Segnette studies achieved 97 and 99% removal efficiencies by a vector distance of 6.28 m. These removal efficiencies are comparable to other studies of constructed subsurface wetlands (Vymazal, 2002; Al-Omari and Fayyad, 2003; Mantovi et al., 2003; Meuleman et al., 2003). These studies achieved removal efficiencies from 84 – 96%. In contrast, a study performed by Al-Omari and Fayyad (2003) had much lower removal efficiencies, ranging from 33 – 65%. Overall, the Moss Point and Bayou Segnette studies performed well and much better than some studies.

32.4. CONCLUSIONS

Overall, the MUS system performs well in the treatment of organic matter in wastewater. $CBOD_5$ analysis of the domestic wastewater tested shows that treatment below the NPDES standard limit of 25 mg L^{-1} is achievable. For Moss Point, the standard can be achieved within a vector distance of 7 m, while at Bayou Segnette it can be achieved by 3 m. Overall, an injection flow rate of 2.8 L min^{-1} (15 min every h) provided the best treatment scenario for Moss Point, while for Bayou Segnette a flow rate of 1.0 L min^{-1} (15 min every h) was the best. High strength wastewater requires a longer travel distance to reach NPDES limits. Therefore, a MUS installed under these conditions should consider the area required for proper treatment during installation of the system. Flow rates appeared to have a higher impact on the removal constants than injection frequency, though this may depend on the environment in which the system is installed. Efficiency increased over time, suggesting an acclimation period is essential for adequate wastewater treatment. Both salt and intermediate marshes serve well as treatment areas, however further studies should be done at Bayou Segnette to fully understand how an intermediate marsh performs under MUS conditions.

ACKNOWLEDGEMENTS

Robert E. Watson, Jr. was the contributor of the Moss Point data; however, he passed away (May 3, 2002) before he was able to publish his data.

REFERENCES

Addo, B. K., 2004. Bacterial plume dynamics in the marshland upwelling system employed in the near freshwater conditions. MS Thesis. Louisiana State University, Baton Rouge, LA.

Al-Omari, A., and Fayyad, M., 2003. Treatment of domestic wastewater by subsurface flow constructed wetlands in Jordan. Desalination 155:27-39.

(APHA) American Public Health Association, 1998. Standard methods for the examination of water and wastewater, 20th ed. American Water Works Association, Water Pollution Control Federation, and the American Public Health Association, Washington, D.C.

(ASTM) American Society for Testing and Materials, 1995. Annual book of ASTM standards, Soil and rock (I). ASTM, Philadelphia, PA.

Bardet, J., 1997. Experimental soil mechanics. Prentice-Hall, Upper Saddle River, NJ.

Evans, D. A., 2005. Assimilation and removal of phosphorus within the marshland upwelling system. MS Thesis. Louisiana State University, Baton Rouge, LA.

Fontenot, J., 2003. The fate of nitrogen in the marshland upwelling system. MS Thesis. Louisiana State University, Baton Rouge, LA.

Gopal, B., 1999. Natural and constructed wetlands for wastewater treatment: potentials and problems. Water Science and Technology 40:27-35.

Kadlec, R. H., and Knight., R. L., 1996. Treatment wetlands. Lewis Publishers, Boca Raton.

Mantovi, P., Marmiroli, M., Maestri, E., Tagliavini, S., Piccinini, S., and Marmiroli, N., 2003. Application of a horizontal subsurface flow constructed wetland on treatment of dairy parlor wastewater. Bioresource Technology 88:85-94.

Mendelssohn, I. A., and Slocum, M. G., 2004. Relationship between soil cellulose decomposition and oil contamination after an oil spill at Swanson Creek, Maryland. Marine Pollution Bulletin 48:359-370.

MDEQ (Mississippi Department of Environmental Quality), 1996. State of Mississippi list of impaired waterbodies 1996. Office of Pollution Control, Jackson, MS.

Metcalf and Eddy, Inc., 2003. Wastewater engineering: treatment and reuse, 4th edition. McGraw-Hill, Boston, MA.

Meuleman, A. F. M., van Logtestijn, R., Rijs, G. B. J., and Verhoeven, J. T. A., 2003. Water and mass budgets of a vertical-flow constructed wetland used for wastewater treatment. Ecological Engineering 20:31-44.

Mitsch, W. J., and Gosselink, J. G., 2000. Wetlands, 3rd edition. John Wiley, New York, NY.

Nguyen, L. M., 2000. Organic matter composition, microbial biomass and microbial activity in gravel-bed constructed wetlands treating farm dairy wastewaters. Ecological Engineering 16:199-221.

Richardson, S. D., 2002. Wastewater plume dynamics and bacteria transport within the marshland upwelling system. MS Thesis. Louisiana State University, Baton Rouge, LA.

Richardson, S. D., and Rusch, K. A., 2005. Fecal coliform removal within a marshland upwelling system consisting of scatlake soils. Journal of Environmental Engineering-Asce 131:60-70.

Richardson, S. D., Willson, C. S., and Rusch, K. A., 2004. Use of rhodamine water tracer in the marshland upwelling system. Ground Water 42:678-688.

Rietz, D. N., and Haynes, R. J., 2003. Effects of irrigation-induced salinity and sodicity on soil microbial activity. Soil Biology & Biochemistry 35:845-854.

Stremlau, H. T., 1994. Feasibility study on the use of shallow upwelling systems in coastal areas as a polishing treatment to remove bacterial contamination from wastewater. MS Thesis. Louisiana State University, Baton Rouge, LA.

Turriciano, A. R., 2005. Evaluation of the spatial removal of nitrogen in the marshland upwelling system. MS Thesis. Louisiana State University, Baton Rouge, LA.

USDA (United States Department of Agriculture), 1983. Soil survey of Jefferson Parish, Louisiana. Louisiana Soil Conservation Service, Baton Rouge, LA.

USDA (United States Department of Agriculture),1984. Soil survey of Lafourche Parish, Louisiana. Louisiana Soil Conservation Service, Baton Rouge, LA.

USDA (United States Department of Agriculture), 1999. The class V underground injection control study, Volume 5: large-capacity septic systems. EPA/816-R-899-014e.

USEPA (United States Environmental Protection Agency), 2002. Onsite wastewater treatment systems manual. EPA/625-R-600-008.

Vymazal, J., 2002. The use of sub-surface constructed wetlands for wastewater treatment in the Czech Republic: 10 years experience. Ecological Engineering 18:633-646.

Watson Jr., R. E., 2000. Evaluation of a marshland upwelling system for the treatment of raw domestic wastewater. MS Thesis. Louisiana State University, Baton Rouge, LA.

Watson Jr., R. E., and Rusch, K. A., 2001. Performance evaluation of a marshland upwelling system for the removal of fecal coliform bacteria from domestic wastewater. Water Environment Research 73:339-350.

Watson Jr., R. E., and Rusch, K. A., 2002. *Escherichia coli* removal efficacy of a marshland upwelling system. Journal of Environmental Engineering-Asce 128:643-652.

33

IMPACTS OF SALINITY AND SOLUBLE ORGANIC MATTER ON MS-2 ADSORPTION AND TRANSPORT IN SEDIMENTS

Haibo Cao, Frank T-C. Tsai, and Kelly A. Rusch
Department of Civil and Environmental Engineering,
Louisiana State University, Baton Rouge, LA 70803, USA

ABSTRACT

In this research, the adsorption and transport of MS-2 in saturated sediments under the impacts of salinity and soluble organic matter (SOM) are studied through one-dimensional column experiments. The experiments are conducted with the prepared fine sand (d_{50}=0.13mm) from industry and the mixture of sand and field soil (d_{50}=0.16mm) obtained from Bayou Segnette State Park, Louisiana. The controlled conditions in all experiments include constant flow rates (0.2 mm/s), neutral condition (pH=7±0.5), and constant temperature (25±0.2^{0}C). Bromide is adopted as the conservative tracer to evaluate the retardation factors of saltwater, MS-2, and SOM in the column. In the first sorption and transport experiment with MS-2 mixing with deionized water, the MS-2 showed slight adsorption effect onto sediments. When the SOM added, the adsorption of MS-2 was decreased because of the presence of SOM, the competitor of MS-2 on the sorption sites of sediments. The MS-2 with salinity experiment showed dramatically enhanced adsorption effect. It can be explained that the salinity compressed double layers, thereby increasing attachment rates.

33.1. INTRODUCTION

Coastal regains are an invaluable source of recreational pleasure, biological diversity, and natural resources. However, rapidly increasing population and the corresponding disposal of poorly treated domestic wastewater in this area has led to a decline in coastal water quality. Septic systems, which are the most widely used on-site wastewater system, usually don't function well as the result of high water table, poor soil conditions, or influent flows exceeding the adsorptive capacity of the soil (USEPA, 2002). The consequential threats are generally linked to public health concerns and economic loss. It is reported that about 71% of all waterborne illness outbreaks in the United States since 1990 resulted from contaminated ground water (Kramer et al., 1996). Septic systems were also identified as a contributing factor to the increased number of shellfish harvesting sites that have been closed due to fecal coliform contamination (NOAA, 1997).

One of the efforts in coastal remediation is the development of the Marshland Upwelling System (MUS) to replace conventional septic systems in coastal areas (Richardson et al., 2005; Richardson et al., 2004; Watson et al., 2002). Wastewater from coastal dwellings is injected into the subsurface sediments, which function as the treatment media for physical/chemical operations and biological processes. Researchers (Richardson et al, 2004; Watson et al, 2002) have demonstrated the ability of the MUS to treat bacteria and nutrients. Corbin (Corbin, 2004) found that the bacteriophage, MS-2, was removed effectively in a MUS system located in Louisiana. While the bacteriophage removal mechanism is discussed as adsorption, inactivation, and transport in porous media, it is difficult to be investigated at the field scale.

In the past two decades, numerous batch, column, and field scales studies have been performed to investigate adsorption, inactivation, and transport of various microorganisms in different porous media (Bales et al., 1991; Chu et al., 2003; Gannon, 1991; Grant et al., 1993; Guan et al., 2003; Penrod et al., 1996). Research on virus transport has been of particular interest because of its long transport distance and potential threat to public health. It has been shown that virus type, soil type, pH, ionic strength and multivalent cations, and organic matter can affect the adsorption of the virus, whereas temperature and unsaturated conditions are crucial for virus inactivation (Schijven et al., 2000). However, most of the reported studies have been performed under the low soluble organic matter and ionic strength conditions that mimic the characteristics of ground water. Little research has been conducted to investigate the fate and transport of viruses in wastewater moving through high salinity groundwater, which is typical for coastal areas.

The purpose of the work described in this paper was to investigate the impact of high soluble organic matter (SOM) and high salinity, which represent the wastewater moving in coastal subsurface, to the adsorption and transport of virus in saturated sediments in lab scale. Preliminary results of recent column studies in both prepared fine sand from industry and the mixture of the sand and the soil which was obtained from Bayou Segnette State Park, Louisiana, are presented.

33.2. MATERIALS AND METHODS

Column Studies. Four continuous-flow column studies were conducted with bacteriophage MS-2 at pH 7.0±0.5, which is typical for ground water in coastal areas. The temperature was held constant at 25±0.2 °C in a low temperature incubator (Thermolyne, Type 4200). Experiments were performed in 45.3cm (length) × 3 cm (diameter) acrylic columns using the prepared sand and a sand/soil mixing representing of sediments in coastal areas. The sand was TECO-sil®Fused Silica, while the soil was from the Bayou Segnette State Park, Louisiana. The particle distribution analysis showed that the sand sample contains 24% coarse sand (0.5-2 mm), 11% medium sand (0.25-0.5 mm), 55% fine sand (0.05-0.25 mm), and 10% silica flint (<0.05 mm). The soil was mixed with the sand (sand/soil mixture) to obtain the similar particle distribution and the mixture has 30% coarse sand, 5% medium sand , 56.5% fine sand, 4.5% silt (0.002-0.005 mm) and 4% clay (<0.00 2mm). The mean diameters (d_{50}) of the sand and sand/soil mixture are 0.13 mm and 0.16 mm, respectively. The porosities for both are 37%. The sand was washed with 10% HNO_3, 0.5M NaOH and deionized (DI) water successively, then oven-dried at 90°C for 20-24 hours prior to packing. The MS-2 solution was added to each column continuously using a peristaltic pump. The velocity was controlled at 0.2 mm/s (1.57 pore volumes/hour). To avoid MS-2 adsorption, Teflon tubing was used everywhere in the system except for a length of PharMed® tubing (MasterFlex® 06485-13) in the pump. Before the experiment, each column was flushed with about 2 pore volumes of DI water to establish a steady-state flow condition. Samples were collected at the top of column using glass vials and stored in an ice chest. Analyses were performed within 2 hours.

Materials and Chemicals. Bromide (potassium bromide at 6-10 mg/L) was used as a conservative chemical tracer to evaluate flow dispersion in the column and analyzed using Ion Chromotography. The soluble organic matter (SOM) was prepared with 0.27g Dextrose, 0.27g glutamic acid, 1.18mL Algal Culture Formula Part A (Kent®), 0.018g sodium phosphate, 0.072g ammonium chloride per liter DI water to obtained a COD of 700 mg/L. 0.5M Na_2CO_3 was added to adjust the pH of the solution. The soluble organic matter was analyzed as COD by spectrophotometry (HACH DR/4000U). The salt water solutions were prepared by dissolving sea salts (Instant Ocean®) into DI water. Salinity was measured using a refractometer.

Bacteriophage and Assays. Baceriophage MS-2 is an icosahedral phage with a diameter of 26.0-26.6 nm and pH_{iep} of 3.9 (Bales, 1993). It was widely used as the model organism for human pathogenic enteric viruses as its morphology and survival characteristics closely resemble those of some of the important human enteric viruses (Sundram et al, 2002). MS-2 (ATCC 15597-B1) was grown and purified in *Escherichia coli* (ATCC 700891). The MS-2 concentration in solution used for column test was between 30-300 plague forming units (PFU). Samples were assayed for MS-2 within 2 hours after collection using the plaque-forming-unit method (Adams, 1959). All samples assays were performed duplicate and the mean was reported.

Data Analysis. This research focused on the change of MS-2 retardation factor with the effect of SOM and salinity. Under the assumption of linear isotherm, the retardation factor is the ratio of the time of the breakthrough curve of a solute to that of a conservative tracer. The mean retention time (MRT) was considered as the following (Levenspiel, 1999),

$$T = \int_0^1 \bar{t}\, d\bar{C} \Big/ \int_0^1 d\bar{C} \qquad (1)$$

where $\bar{C}$ is the normalized concentration (dimensionless) by the initial concentration ($\bar{C} = C / C_0$); and $\bar{t}$ is normalized time (dimensionless) by the time of one pore volume. The solutes in this study is MS-2, SOM and saltwater; and the conservative tracer is bromide. The retardation factor RF was calculated by (Richardson et al, 2004)

$$RF = \frac{T}{T_{Br^-}} \qquad (2)$$

where T_{Br^-} is the MRT of bromide.

33.3. RESULTS

A series of experiments were performed to study the adsorption of MS-2 onto sediments in the DI water, DI water with SOM, and saline water. The breakthrough curves of MS-2, SOM, salinity and bromide under different experiment conditions are illustrated in Figure 1 to Figure 4. The calculated MRT and RF values are listed in Table 1.

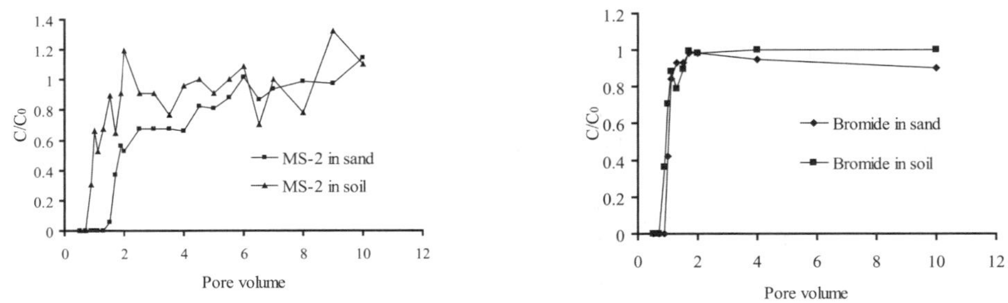

(a) **(b)**

Fig. 1. The breakthrough curves of MS-2 with bromide in deionized water. (a) MS-2 and (b) bromide.

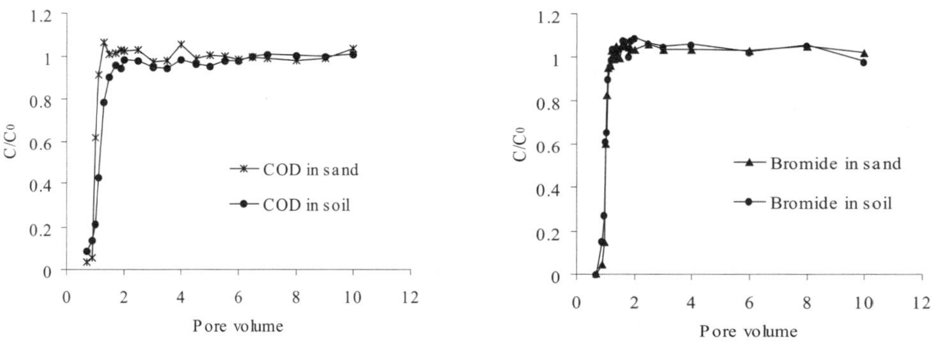

(a) **(b)**

Fig. 2. The breakthrough curves of SOM with bromide in deionized water. (a) SOM and (b) bromide.

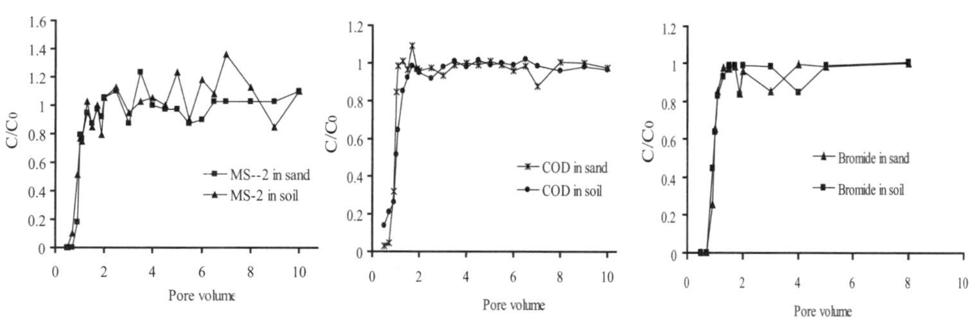

(a) **(b)** **(c)**

Fig. 3. The breakthrough curves of MS-2, SOM with bromide in deionized water. (a) MS-2, (b) SOM, and (c) bromide.

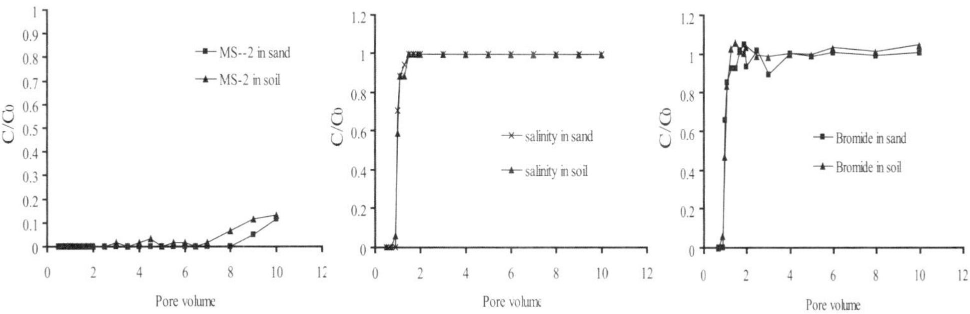

(a) (b) (c)

Fig. 4. The breakthrough curves of MS-2, salt with bromide in deionized water. (a) MS-2 (b) salinity and (c) bromide

Table 1. Mean retention times (MRT) in pore volume and retardation factors (RF) for column studies.

Experiments		Sand				Sand/soil mixture			
		Bromide	MS-2	SOM	Salinity	Bromide	MS-2	SOM	Salinity
(1) MS-2 and bromide	MRT	1.02	4.03	-	-	1.06	2.24	-	-
	RF	1	3.64	-	-	1	2.10	-	-
(2) SOM and bromide	MRT	1.03	-	1.05	-	1.07	-	1.51	-
	RF	1	-	1.02	-	1	-	1.41	-
(3) MS-2, SOM and bromide	MRT	1.03	1.86	0.99	-	1.05	1.31	1.21	-
	RF	1	1.80	0.96	-	1	1.25	1.15	-
(4) MS-2, salinity and bromide	MRT	1.02	9.57	-	1.08	1.09	8.13	-	1.08
	RF	1	9.37	-	1.06	1	7.45	-	0.99

Experiments were first conducted with the MS-2 and bromide suspended in DI water. From the breakthrough curves of MS-2 (Figure 1(a)), at about 2.0 pore volumes the ration of C/C_0 reached to 1.0 in the sand/soil mixture column. It showed that the adsorption of MS-2 does exist, but not strong. The RF calculated was 2.10. In the sand column, the steady state ($C/C_0=1$) happened around 5.5 pore volumes and RF was calculated to be 3.64. The MS-2 adsorption effect in the sand was greater than that in the sand/soil mixture. For the bromide tracer, Figure 1(b) shows that it moved faster in the sand/soil mixture column than the sand before 1.1 pore volumes. Between 1.1 to 1.7 pore volumes, bromide transported slower in the soil. After 1.7 pore volumes, bromide reached the steady stage in both columns. The sand/soil mixture has larger dispersion than that in the sand for MS-2 transport.

The second experiment was performed with SOM and bromide solution. According to Figure 2(b), the bromide showed similar behavior as with MS-2 in Figure 1(b). This indicates the bromide independence of MS-2 and SOM. SOM in Figure 2(a) reached the steady state around 1.3 pore volumes for sand and 1.7 pore volumes for sand/soil mixture. The RF was 1.02 and 1.40 for sand and sand/soil mixture, respectively. The sand/soil mixture column showed higher retention effect for SOM than sand column. Moreover, the SOM appeared earlier in sand/soil mixture column than that in sand, which is consistent with the results in Figure 1(a) that the sand/soil mixture has high dispersion than that in the soil column.

When soluble organic matter (COD=700) was added into the solution with MS-2 and bromide, the adsorption effects of MS-2 in both sand/soil mixture and sand columns decreased dramatically. The steady state of MS-2 appeared at 1.3-1.8 pore volumes for both of them and the RF dropped to 1.80 and 1.25 for sand and sand/soil mixture, respectively. Again, the bromide had the same behavior as in previous two experiments. SOM reached the steady state at 1.6 pore volumes in the sand/soil mixture column and 1.3 pore volumes in the sand column. The SOM adsorption decreased slightly with the existence of MS-2.

The last experiment was conducted to test the MS-2 adsorption effect under high salinity condition (20 ppt) by adding salt (Instant Ocean®) into the solution of MS-2 and bromide. From the breakthrough curve (Figure 4(a)), it can be seen that the adsorption of MS-2 onto sediments was enhanced dramatically. The RF increased from 3.64 to 9.37 in the sand column and from 2.10 to 7.45 in sand/soil mixture column. At 10 pore volumes, the MS-2 concentration only reached to 12% and 13% of initial concentration for the sand and the sand/soil mixture columns, respectively. The NaCl concentration reached to the steady state at 1.5-1.7 pore volumes for both sand/soil mixture and sand columns. It had the same transport behavior as bromide.

During all the experiments, the samples were collected at 5 pore volumes interval from the solution used for the column studies to test the inactivation of MS-2. The result showed that the inactivation during these studies is not distinct and can be ignored.

33.4. DISCUSSION

The retention of microorganisms in porous media has been attributed to several mechanisms (Corapcioglu et al., 1984): straining or filtration at pore constrictions, adsorption onto surface of sediments, sedimentation in the pores, interception, systematic and random motion, growth and die-off. Because of its small size, the main retention mechanism for MS-2 is the adsorption onto the surface of the sediments. There are three kinds of adsorptions: physical, chemical and exchange adsorption (Weber, 1972). Usually, physical adsorption applies to that resulting from Van der Waals forces. Chemical adsorption is formed by the chemical interaction with the adsorbent. Exchange adsorption is caused by electrostatic attraction to charged sediments. Herzig et al. (1970) indicated that Van der Waals forces can overcome the random movement energy of Brownian motion when the particle diameter is greater than 0.1 μm. As for MS-2, because of its small size (26-26.6 nm), physical adsorption can be ignored. Chemical adsorption can also be ignored when the porous media is soil or sand and non-chemical reactive (Weber, 1972). Thus electrostatic attraction becomes the key adsorption mechanism for MS-2.

The surfaces of sand and soil generally carry negative electrostatic charges. These charges may be permanent (not pH dependent) or may depend on the pH of the surrounding electrolyte solution (Gannon, 1991). In the neutral solution (pH = 7) the pH_{iep} of MS-2 is 3.9, which causes MS-2 to be negatively charged. As a consequence, the MS-2 adsorption is expected to be low. The result of the sand/soil mixture column was in agreement with this expectation. Negligible sorption of MS-2 in soils has also been reported by other studies (Jin, 1997; Bales, 1993). As for the sand column, MS-2 showed relatively high adsorption effect, which may indicate the presence of oxides of iron, aluminium, calcium and manganese in the sand. These chemicals are usually present in the sand and provide patches of positively charged sites on the particle surfaces, which will enhance the adsorption of MS-2.

The difference of breakthrough curve of MS-2 with and without SOM indicated the effect of SOM existence to the adsorption of MS-2. In general, the SOM is negatively charged, which compete with virus for the same binding site (Gerba, 1984). Once the adsorption capacity of the sediments was filled, the MS-2 can not be adsorbed by sediments. When the SOM increase, it will further detach the attached virus from sediments (Sobsey et al., 1980). Also, with the existence of MS-2, the adsorption of SOM decreased. The adsorption competition between MS-2 and SOM were observed in the experiment 3.

The ionic strength is another important factor that can affect the adsorption of MS-2 onto the sediment. According to Derjaguin-Landau-Verwey-Overbeek (DLVO) theory, a higher ionic strength compressed double layers, thereby increasing attachment rates (Schijven et al, 2000). Many studies have observed the ionic strength effect in the batch and column studies. (Grant et al., 1993; Penrod et al., 1996). In this research, we also observed that the saltwater (NaCl) dramatically enhanced the adsorption of MS-2. In experiment 4, after 10 pore volumes, very low percentage (around 12%) of MS-2 concentration was transported to the effluent in both sand and sand/soil mixture columns, which indicate high retardation factor in these sediments.

33.5. CONCLUSIONS

Column experiments were conducted to investigate the impacts of soluble organic matter and salinity on the adsorption of MS-2 onto the sand and the sand/soil mixture. The following conclusion may be made:

(1) Adsorption of MS-2 is sediments dependent. In general, the effect of MS-2 adsorption is not distinct because of the same negative charges for both MS-2 and sediments.

(2) The existence of soluble organic matter decreases the adsorption effect of MS-2 due to its competition with MS-2 for the same binding site of the sediments.

(3) Salinity (ionic strength) can compress double layer and hence enhance the adsorption effect of MS-2 onto sediments.

ACKNOWLEDGEMENTS

The research was supported in part from the Louisiana Department of Environmental Quality (DEQ), through the grant of C9 996102-08 and from the Sea Grant Omnibus program 2006-2008. We also acknowledge the generous support from Ms. Sarah Jones, Tye Fitzgerald and Magdalena Angelova for assistance in carrying out the experiments and performing the analysis.

REFERENCES

Adams, M. H., 1959. Bacteriophages, Wiley-interscience, New York.

Bales, R. C., Hinkle, S. R., Kroeger, T. W. and Stocking, K., 1991. Bacteriophage Adsorption during Transport through Porous Media: Chemical Perturbations and Reversibility, Environmental Science and Technology 25:2088-2095

Bales, R. C. and Li, S., 1993. MS-2 and Poliovirus Transport in Porous Media: Hydrophobic Effects and Chemical Pertubations. Water Resources Research 29(4):957-963

Chu, Y., Jin, T., Baumann, T., and Yates, M. V., 2003, Effect of Soil Properties on Saturated and Unsaturated Virus Transport through Columns. Journal of Environmental Quality 32:2017-2025

Corapcioglu, M. Y. and Haridas, A., 1984. Transport and Fate of Microorganisms in Porous Media: A Theoretical Investigation. Journal of Hydrology 72:149-169

Corbin, A. L., 2004. Recovery of F+ RNA Sepcific Bacteriophage for the Evaluation of a Marshalnd Upwelling System in Low Saline Waters. MS thesis, Department of Biological Sciences, Nicholls State University, Thibodaux, LA.

Gannon J., Tan Y., Baveye, P. and Alexander, M., 1991. Effect of Sodium Chloride on Transport of Bacteria in a Saturated Aquifer Material, Applied and Environmental Microbiology 57(9):2497-2501

Gerba, C. P., 1984. Applied and Theoretical Aspect of Virus Adsorption to Surfaces, Advanced Applied Microbiology 30:133-168.

Grant, S. B., List, E. J., and Lidstrom, M. E., 1993. Kinetic Analysis of Virus Adsorption and Inactivation in Batch Experiments. Water Resource Research 29:2067-2085

Guan, H., Schulze-Makuch, D., Schaffer, S. and Pillai, S. D., 2003, The Effect of Critical pH on Virus Fate and Transport in Saturated Porous Medium, Ground Water 41(5): 701-708

Herzig, J. P., Leclerc, D. M. and Le Golf, P., 1970. Flow of Suspensions Through Porous media-Application to Deep Filtration. In: Flow through Porous Media, American Chemistry Society, Washington, D. C., 129-157

Jin, Y., Yates, M. Y. and Jury, W. A., 1997. Sorption of Viruses during Flow through Saturated Sand Columns, Environmental Science and Technology 31:548-555

Kramer, M.N., Herwaldt, B. L., Graun, G. F., Calderone, R. L., and Juranek, D. D., 1996. Waterborne Disease: 1993 and 1994, Journal of the American Water Association 88:66-80

Levenspiel, O., 1999. Chemical Reaction Engineering, 3rd edition, New York, John Wiley & Sons.

National Oceanic and Atmospheric Administration (NOAA), 1997. The 1995 National Shellfish Register of Classified Growing Waters, Silver Spring, MD

Penrod, S. L., Olson, T. M., and Grant, S. B., 1996. Deposition Kinetics of Two Viruses in Packed beds of Quartz Granular Media, Langmuir 12:5576-5587

Richardson, S. D. and Rusch, K.A., 2005. Fecal Coliform Removal within a Marshland Upwelling system Consisting of Scatlake Soils. Journal of Environmental Engineering 131(1):60-70

Richardson, S. D. and Rusch, K.A., 2004. Use of Rhodamine Water Tracer in the Marshland Upwelling System, Ground water 42(4):678-688

Schijven, J. F. and Hassanizadeh, S. M., 2000. Removal of Viruses by Soil passage: Overview of Modeling, Processes, and parameters, Critical reviews in environmental Science and Technology 30(1):49-127

Sobsey, M.D., Dean, C. H., Knuckles, M. E., and Wagner, R. A., 1980. Interaction and Survival of Enteric Viruses in Soil Materials. Applied Environmental Microbiology 40:92-101

Sundram, A., Donnelly, L., Ehlers, M.M., Very, A., Grabow, W. and Bailey, I.W, 2002. Evaluation of F-RNA Coliphages as Indicators of Viruses and the Source of Fecal Pollution, Water SA special Edition: WISA Proceedings 2002, 86-91

U.S. Environmental Protection Agency, 2002. Onsite Wastewater Treatment Systems Manual, Office of Water EPA, 625-R-00-008

Watson Jr., R. E. and Rusch, K. A., 2002. Performance Evaluation of a Marshland Upwelling System for the Removal of Fecal Coliform Bacteria from Domestic Wastewater. Water Environment Research 73(3):339-350

Weber, W. J., 1972. Physicochemical Processes for Water Quality Control, Wiley-Interscience, New York

34

SALTWATER INTRUSION BARRIER DEVELOPMENT AND MANAGEMENT IN COASTAL AQUIFER SYSTEMS

Mohamed H. Elrawady and Frank T-C. Tsai
Department of Civil and Environmental Engineering, Louisiana State University
3507 CEBA Building, Baton Rouge, LA 70803, USA

ABSTRACT

Saltwater intrusion in freshwater aquifers due to increasing ground water withdrawal in coastal areas renders water supply shortage and drinking water contamination, which has detrimentally impacted coastal residential quality and economic development. One method to manage the saltwater intrusion without affecting needed water quantity is the development of a hydraulic barrier (saltwater intrusion barrier) via surface water injection. Injection of quality surface water into aquifers intuitively leads to several advantages and has minimum to no impact on natural environments. However, the barrier efficacy and economy rely on a well-planned design over a set of injection wells. This study develops a saltwater intrusion barrier (SIB) management model to optimize the design by finding the least number of injection wells and minimum injected surface water amount such that human-built saltwater intrusion is efficiently mitigated and ground water availability is enhanced. The SIB management model involves optimization of injection well selection, injection rate scheduling, and injection rate optimization, which leads to a mixed integer nonlinear programming (MINLP) problem. The objective of the SIB model is to minimize the overall pumping rates while the desired ground water heads and chloride levels in the intruded areas are met.

34.1. INTRODUCTION

Saltwater intrusion in coastal aquifers is one of the significant issues in water resources over the last decades. Its negative impacts include drinking water shortage and aquifer contamination. Moreover, it also leads to land subsidence and estuary ecosystem destruction. Mitigation of saltwater intrusion requires an appropriate ground water management strategy, which involves optimization techniques to deal with the conjunctive use of surface water and ground water. Many research efforts have been focused on optimization of ground water management. Shamir et. al. (1984) developed an annual operation plan for a coastal aquifer based on four decision variables, these variables were; a desired ground water surface map, a desired freshwater-saltwater interface toe, a desired contaminant concentration map and finally the minimization of energy consumed in the pumping or recharge processes. Hallaji and Yazicigil (1996) developed seven ground-water management models to determine the optimal operating policies to prevent saltwater intrusion at a coastal aquifer in southern Turkey. In their research, simulation models were linked to linear and quadratic optimization models. They had three management objectives; maximizing water withdrawal to enhance agriculture, minimizing drawdown and minimizing pumping costs. They reached an optimum policy that featured the use of excess pumping water from highly productive wells to meet shortage in demands in less productive wells through overland transport. Emch and Yeh (1998) developed a nonlinear multiobjective management model for a hypothetical case study inspired by the Waialae aquifer in Hawaii. The study focused on the two traditional ground water objectives of cost-effective allocation of surface and ground water supplies while minimizing saltwater intrusion. Their study featured a flow simulation model using a quasi-three-dimensional finite-difference model based on the sharp interface assumption. The simulation model was linked directly to the optimization procedure. The constraint method was used to solve the multiobjective aspect of the problem. Recently, Rao et.al. (2004) combined a simulation-optimization model that was solved as a nonlinear, nonconvex combinatorial problem using a simulated annealing and an existing sharp interface model. The authors were seeking the optimal conjunctive use of surface and ground water for a coastal delta in the east of India where surface water availability fluctuates in time and ground water availability fluctuates in space.

34.2. SALTWATER INTURSION IN EAST BATON ROUGE, LOUISIANA

The East Baton Rouge Parish is located on the Southeastern Louisiana aquifer system which is bounded by the Mississippi River, Pearl River and Lake Pontchartrain. In the East Baton Rouge Parish, the aquifer system is comprised mainly of 10 aquifers that reach as deep as 3,000 feet. The Baton Rouge Fault is located in the system and restricts the southward flow into it. Historical evidence show that the fault has prevented the saltwater coming from the Gulf of Mexico from moving northward, but due to the economic growth of the southeastern Louisiana community, the ground water demand has drastically increased and enormous ground water withdrawals has occurred in the Baton Rouge area without an appropriate management plan. Excessive withdrawal of ground water has led to a continuous decline in ground water levels and as a result, undesirable chloride concentrations were detected in some monitoring wells.

Many research efforts has been devoted to the southeastern Louisiana area, The United States Geological Survey (USGS) had a large contribution in such research efforts. In 1965, USGS conducted a feasibility study on using Scavenger–well system as a solution to the problem of vertical saltwater encroachment. The methodology proved to be beneficial for a system of two separate wells where freshwater overlies saline water, where it is possible to pump fresh water from the upper section and salty water from the lower section. In another study conducted by Louisiana Department of Transportation and Development (DOTD) with the cooperation of USGS, Whiteman, Jr. (1979) pointed out the possible future threats that could result from the saltwater encroachment at the "600-foot" and "1500-foot" sand aquifers of Baton Rouge and he called for immediate attention. He recommended several possible control strategies including discharge wells to pump salty water from the aquifers, and also injection wells to dispose the salty water. In another DOTD and USGS joint effort, Torak and Whiteman, Jr. (1982) developed a finite-difference, digital-computer model simulating ground water flow in the "2000-foot" sand aquifer of Baton Rouge as it responds to variations in pumping. A similar study for the "1500-foot" and "1700-foot" sand aquifers of Baton Rouge was conducted by Huntzinger et.al. (1985).

34.3. SALTWATER INTRUSION MODELING

Modeling of ground water flow and saltwater concentration transport in aquifers is a complex problem due to density variation in saline ground water. Darcy's velocity that accounts for the variation in density is (Guo and Langevin, 2002):

$$\mathbf{V} = -\frac{K_f}{n}\left(\nabla h_f + \frac{\rho - \rho_f}{\rho_f}\nabla z\right) \tag{1}$$

where $\mathbf{V}$ is the seepage ground water flow velocity in the pore space in vector form; K_f is the freshwater hydraulic conductivity; h_f is the freshwater ground water head ; n is the porosity; ρ and ρ_f are the saltwater and freshwater densities, respectively. Accordingly, the density-dependent ground water flow equation is formulated as the following (Guo and Langevin, 2002; Langevin 2003)

$$\rho S_f \frac{\partial h_f}{\partial t} + n\frac{\partial \rho}{\partial C_{salt}}\frac{\partial C_{salt}}{\partial t} = \nabla.\left[\rho K_f\left(\nabla h_f + \frac{\rho - \rho_f}{\rho_f}\nabla z\right)\right] + \rho_s q_s \tag{2}$$

where S_f is the freshwater storage coefficient; C_{salt} is the saline concentration; q_s is the flow rate at sinks or sources and ρ_s is the fluid density at sinks or sources. The corresponding saltwater transport equation is

$$\frac{\partial C_{salt}}{\partial t} = \nabla.(\mathbf{D}\nabla C_{salt}) - \nabla.(\mathbf{V}C_{salt}) - \frac{q_s}{n}C_{salt,S} \tag{3}$$

where $\mathbf{D}$ is the hydrodynamic dispersion coefficient; $C_{salt,S}$ is the saline concentration at sinks or sources. A numerical model, SEAWAT (Guo and Langevin, 2002), has been developed to solve Eqs.(2)-(3), which is based on the MODFLOW (McDonald and Harbaugh, 1988) and MT3DMS (Zheng and Wang, 1998). Usually, the freshwater has the density $\rho_f = 1000\,Kg/m^3$ and seawater has the density $\rho_{sea} = 1025\,Kg/m^3$ around the concentration 35 ppt (parts per thousand) of dissolved salts. When the saltwater concentration is low, the terms $(\rho - \rho_f)/\rho_f$ and $\partial\rho/\partial C_{salt}$ become insignificant.

34.4. SEMI REAL-CASE STUDY

This paper presents a semi real-case study at the East Baton Rouge Parish, Louisiana, where a saltwater intrusion model was developed based on available data (e.g., pumpage data and electrical resistivity log data). The study area was determined according to a recent USGS technical report (Griffith 2003), where the saltwater presence is indicated in the aquifers north of the Baton Rouge fault. As shown in Fig. 1, the western boundary of the study area is the Mississippi River and the eastern boundary is the Amite River. The time-varying water levels of the two rivers were used to determine the left and right groundwater head boundary conditions in the model. We collected the river stage data over a fifteen-year period from the NOAA website (www.riverwatch.noaa.gov). The southern boundary was located around 230 meters south of the Baton Rouge fault. The Baton Rouge fault could be a natural impermeable boundary condition due to its historical flow restriction northward; however, we consider the fault inside the domain in order to investigate the fault impact on the on-going saltwater intrusion. We assumed that the ground water head at the southern boundary is 1 foot above the sea level. The northern head boundary was determined by the averaged ground water head (33 feet) of three observation wells which are close to the designated boundary.

The Southeastern Louisiana aquifer system in East Baton Rouge Parish consists of at least 10 aquifers. Electrical resistivity log data of 44 sites in the study are were obtained from the USGS, Water Resources Division in Louisiana District to construct the subsurface stratigraphy. We interpreted the resistivity data into sand aquifers and clay layers based on the erratic change in the resistivity distribution where the sand aquifers have much higher resistivity value than that for the clay layers. We used the USGS technical report (Griffith, 2003) to verify the correctness of the developed stratigraphy. The resistivity data showed that some aquifers and clay layers are missing at several locations in the study area. As for the modeling purpose, an artificial layer with a thickness 0.1 feet and corresponding hydraulic conductivity value were assigned to the disappeared sand and clay layers. A natural neighbor interpolation method (Tsai et al., 2005) was used to interpolate elevations for each layer. In this study, only the "1500-foot" and "1700-foot" sand aquifers were considered due to their heavy pumping history. It was assume that there is no significant leakage at the top of the "1500-foot" sand and at the bottom of the "1700-foot" sand.

Detailed pumping data of production wells from year 1990 to year 2004 were collected from Louisiana Capital Area Ground Water Conservation Commission (CAGWCC) for 5 parishes and 10 aquifers. Again, only the pumping data relating to the "1500-foot" and "1700-foot" sand aquifers within the study area are considered. There are 13 pumping wells shown in Fig. 1. The monthly pumpage data were compiled for the saltwater intrusion model which simulated over a 15-year time period from year 1990 to year 2004. A total number of 180 stress periods (180 months) was considered.

The hydraulic conductivity values for the "1500-foot" and "1700-foot" sands are assigned 120 m/day; and the clay layer between the two sands has hydraulic conductivity 40 m/day. Using the average thickness, the transmissivity values are close to those in the USGS technical report No.34 (Huntzinger, 1985). The storage coefficient is 0.000951 m^{-1} for the sands and 0.000003 m^{-1} for clay. The study area was discretized into a grid of 200 rows and 106 columns. The grid cell dimension is of 123 meter by 110 meter. The ground water simulation is based on MODFLOW (McDonald and Harbaugh, 1988).

Specifically, the Baton Rouge fault was characterized by the Horizontal Flow Barrier (HFB) (Hsieh and Freckleton 1993), where the fault has a relative low conductivity value to restrict the horizontal flow of ground water. The HFB allows additional thin cells with low permeability and zero storage between two regular computation cells to function as a fault (horizontal barrier). For the Baton Rouge fault, we assigned 161 horizontal flow barriers (thin cells) for each layer shown in Fig.1. The physical features of the fault were characterized by the parameter termed Hydraulic Characteristic (HC). HC is defined as the barrier hydraulic conductivity per barrier width. We assume that the Baton Rouge fault has an average width of 10 meters and a low conductivity value 20 m/day. Therefore, The HC value is 2 for this study.

The saltwater transport was simulated by MT3DMS (Zheng and Wang, 1998), which used the flow velocity data from the results of MODFLOW. In this study, we assume that the porosity is 0.3, longitudinal dispersivity is 50 m, ratio of transverse dispersivity to

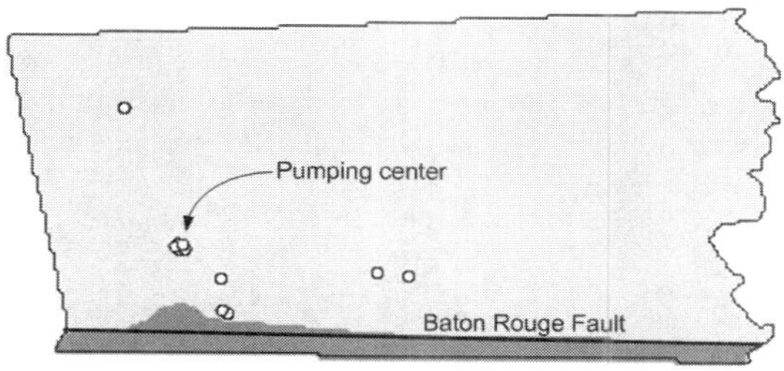

Fig. 1. Initial chloride concentration distribution at the beginning of 1990.

longitudinal dispersivity is 0.1, and the ratio of vertical dispersivity to longitudinal dispersivity is 0.01. Again, this paper conducted a semi real-case study. In the MT3DMS, we exaggerated the saltwater intrusion over a 15-year period by introducing large dispersivity value and the fault permeability value. The starting concentration plume is delineated to be similar to that in year 1992 published in the USGS website (water.usgs.gov/wid/html/la.html). A uniform concentration (C_0) was assumed. For a semi real-case study, the model southern boundary was also assumed to have a constant concentration boundary condition with C_0, which was use to normalize the simulated concentrations. Other three boundaries have zero concentration condition. Fig. 1 shows the initial saltwater encroachment before simulation. Six check point (C_1 to C_6) were selected in the study area shown in Figure 2(a)-(c). The check points C_1 and C_2 monitor the saltwater intrusion from the right side of the fault. C_1 check point is between the fault and the proposed saltwater intrusion barrier (SIB). SIB will be discussed in the next section. C_3 check point monitors the saltwater intrusion for the center of the fault. C_4, C_5 and C_6 check points monitor the saltwater intrusion for the left side of the fault. All of the six check points are in the "1500-foot" sand aquifer.

34.5. SALTWATER INTRUSION BARRIER (SIB) DEVELOPMENT

The goal of the SIB is to develop a hydraulic ridge using a series of injection wells to prevent saltwater intrusion. It is understood that finding the best SIB alignment is challenging due to a great number of possible injection locations. In this study, the proposed SIB alignment was selected approximately 500 meters north of the fault and parallel to it. The optimal SIB should consider the number of injection wells, well spacing, injection rates, and injection schedule to construct an appropriate management strategy. This will make the saltwater intrusion management model a very complex mixed integer nonlinear programming (MINLP) problem. In this study, we specify a set of injection wells with a uniform spacing to reduce the problem complexity. And we also consider a constant injection rate for all injection wells. The remaining problem is to find the optimal injection rate. Before implementing the saltwater intrusion barrier (SIB), the concentration simulations were made to understand the saltwater encroachment during a 15-year period. The results were used as a baseline to evaluate the SIB performance. Also, the baseline results help setup the constraints in the saltwater intrusion management model.

First, we conducted an extreme case (Case I) that featured 161 injection wells with a constant injection rate to all time periods. A design criterion was decided such that concentration values at the three check points (C_1, C_2 and C_3) should be at most 10% of the initial concentration at the ending period. After several trials, beginning from the low constant injection rate of 100 m^3/day and increasing the injection rate in increments of 100 m^3/day, an injection rate of 1000 m^3/day was found to produce chloride concentrations all satisfying the designed criterion.

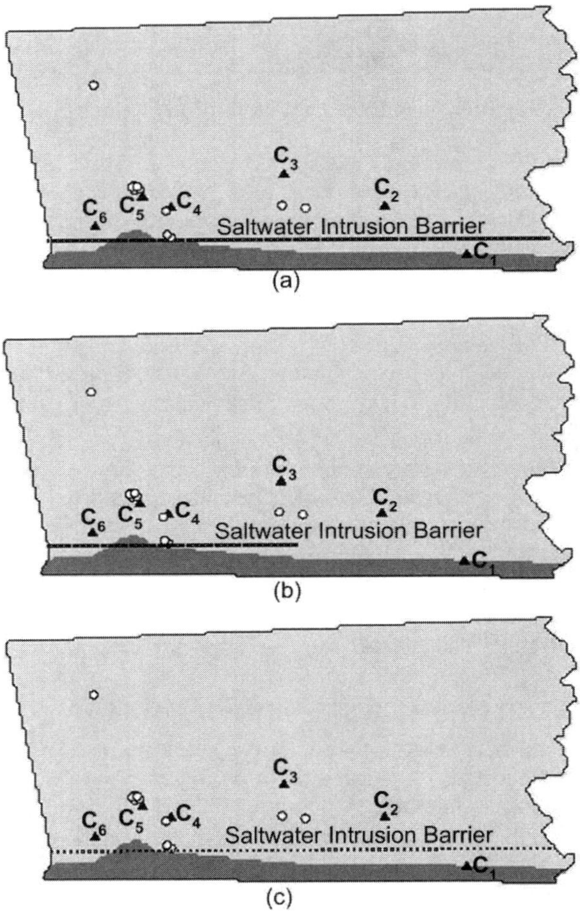

Fig. 2. (a) Full saltwater intrusion barrier (Case I), (b) Segment saltwater intrusion barrier (Case II), and (c) Double-spaced saltwater intrusion barrier (Case III).

The next step is to consider reducing the number of injection wells in the SIB in order to lower the costs of installation, operation and maintenance of wells. However, the designed criterion in Case (I) still has to be satisfied. This will certainly require increasing the injection flow rate.

In Case II, we only consider a half segment of injection wells left side of SIB in Case I, where the pumping centers are present. The number of injection wells was reduced to 80.

Figure 2(b) shows the segment of the barrier used in Case II. The same injection rate of 1000 m^3/day was applied to all wells through the 15-year period. However, the designed criterion at the C_1 and C_2 check points can not be fulfilled as they are located away from the injection barrier. The concentrations at the C_1 and C_2 locations are over 90% of the initial concentrations at the ending period. We don't anticipate the designed criterion at these two locations will be satisfied while increasing the injection rate within a reasonable magnitude. It indicated that a saltwater intrusion barrier that also covers the right side of the study area is needed. Case II was not further considered.

Case I is regarded as the extreme solution which has the minimum injection rate but requires a great number of injection wells and which may not be economically feasible. An optimization scheme will be needed to achieve results similar to Case I, but with less injection wells.

Next, Case III was conducted, with 80 injection wells that cover the whole region. Figure 2(c) shows the saltwater intrusion barrier alignment for Case (III). The spacing between two consecutive wells in Case III was double the space of that in Case I where the well spacing was one grid cell in the horizontal direction. Again, the injection rate in Case III is uniform in time for all wells. The chloride concentration distribution is close to that of Case (I), but with higher magnitudes as expected.

Table 1 shows a comparison between Case I and Case III. The comparison is based on the ratio C_i / C_0 at the check points, where C_i is the final concentration after 15 years of simulation. Based on the results of Cases (I) and (III), the concentration ratios that resulted from Case (I) will be regarded as the target ratios that should be reached by Case (III). To achieve that, the injection rate in Case (III) has to be increased gradually until the resulting concentration ratios resemble those of Case (I), which can be formulated as an optimization problem. The uniform injection rate will be considered.

Table 1. C_i/C_0 ratios for cases (I) and (III) after the 15 years simulation period.

Check Point	Case (I)	Case (III)
C_1	0	0
C_2	0	0
C_3	0.0015	0.004
C_4	0.73	0.87
C_5	0.61	0.80
C_6	0.31	0.58

34.6. THE SALTWATER INTRUSION MANAGEMENT MODEL

In this section, we wanted to predict the barrier performance in Case III against Case I for the period from 2005 to 2008. The initial groundwater head and saline concentration distributions are obtained at the end of 2004 from the baseline simulation without SIB. To optimize the injection strategy, the decision variable will be the injection rate applied uniformly in both time and space. The management objective function is to minimize the injection rate subject to conditions dictated by the concentration ratio C_i / C_0 at the six check points and a range of injection rates.

The simulation period for optimization will be different than the 15-year simulation period in the previous section. The purpose of using the 15-year simulation period for Cases (I) and (III) was to increase the accuracy of results by using real pumping data that is available for those 15 years. In other words, the 15-years simulation results show what the saltwater encroachment would have looked like if a certain injection scenario was implemented during that period. The main purpose of the saltwater intrusion barrier management model is to plan a future injection scenario. Hence, the simulation period for the simulation-optimization model is a 4 year period starting at the beginning of 2005 and ends at the end of 2008. The pumping rates in this period are assumed to follow the same trend of the last 4 years of the 15 year period.

The base simulation is Case (I) with the new future simulation period. Its resulting C_i / C_0 ratios at the same six check points will be used as the target ratios that the optimizer will look for when different injection rates are applied to Case (III) using the same future simulation period. The objective function can be written as follows:

$$\min \quad Q \tag{4}$$

subject to

$$1000 \le Q \le 2000 \tag{5}$$

$$(C_i / C_O)_{II} \le (C_i / C_O)_I \tag{6}$$

where Q is the injection rate; (C_i / C_O) is the ratio of final to initial concentration for cases (I) and (III). Eq.(5) dictates the range of injection rates that the optimizer should look at while Eq.(6) sets the concentration ratios that the optimum injection rate has to satisfy at every check point. However, at check points C_4, C_5 and C_6 where high chloride concentrations are expected, the ratios of Case (III) were allowed to exceed those of Case (I) within a certain limit, a factor of 1.2 was multiplied by the ratios of Case (I) in the inequality as follows:

$$(C_i / C_O)_{II} \le 1.2(C_i / C_O)_I \tag{7}$$

Based on all the previous criteria, the optimum injection rate was found to be 1500 m^3/day. Table 2 shows the C_i / C_0 ratios for Case (I) and the optimized solution for Case (III). Figures 3(a) and 3(b) show the saltwater concentration distribution at the end of 4-year prediction period without SIB against that with the optimized SIB in case (III).

Table 2. C_i/C_O ratios for Case (I) and Case (III) with the optimum injection rate after the 4 years simulation period.

Check Point	Case (I)	Case (III)
C_1	0	0
C_2	0	0
C_3	0	0
C_4	0.75	0.80
C_5	0.68	0.75
C_6	0.36	0.43

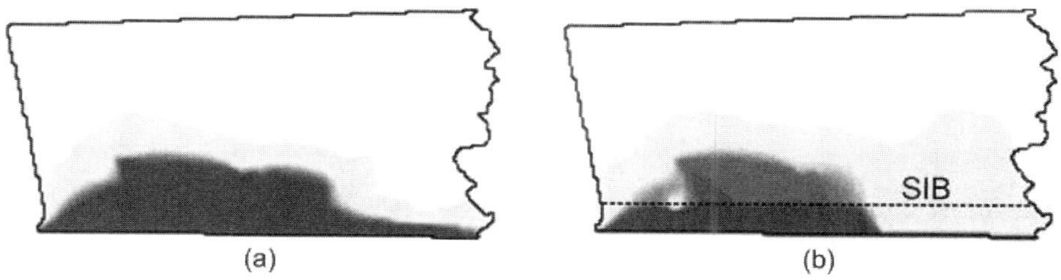

Fig. 3. (a) Saltwater concentration distribution at the end of 2008 without SIB, and (b) Saltwater concentration distribution at the end of 2008 with the optimized SIB of case III.

34.7. CONCLUSIONS

A flow-transport simulation model was developed for the East Baton Rouge Parish; the saltwater intrusion barrier was then developed and tested on that model. The preliminary results showed that developing such a barrier has a great significance in terms of reducing the saltwater encroachment. Three options were tested for the barrier's alignment. The option where the barrier covers the whole model domain with an increased spacing between every two consecutive injection wells proved to be the most beneficial when the injection rate is optimized. An optimization model was formulated; the constraints of that model were influenced by the preliminary results. An optimum injection rate for a 4-year period was obtained. The whole study is a good start for more complex simulation-optimization research where the injection rate is not uniform in time or space; also more complexity could be introduced by optimizing the barrier's location and spacing between injection wells.

ACKNOWLEDGEMENTS

This research was supported in part by Louisiana Board of Regents under award LEQSF(2005-08)-RD-A-12 and Department of the Interior, U.S. Geological Survey under Grant No. 05HQGR0142. This manuscript is submitted for publication with the understanding that the United State Government is authorized to reproduce and distribute reprints for governmental purposes. The views and conclusions contained in the document are those of the author and should not be interpreted as necessarily representing the official policies, either expressed or implied, of the U.S. Government. The thorough and constructive reviews provided by three anonymous reviewers as well as the Associate Editor are greatly appreciated. Special thank you goes to Louisiana Capital Area Ground Water Conservation Commission (CAGWCC) and USGS, Water Resources Division in Louisiana for providing data and invaluable discussions.

REFERENCES

Emch, P. G., and Yeh, W. W-G., 1998. Management model for conjunctive use of coastal surface water and ground water. Journal of Water Resources Planning and Management - ASCE 124 (3): 129-139.

Griffith, J. M, 2003. Hydrogeologic Framework of Southeastern Louisiana, USGS, Water Resources Technical Report NO. 72

Guo, W., and Langevin, C. D., 2002. User's Guide to SEAWAT: A Computer Program for Simulation of Three-Dimensional Variable-Density Ground-Water Flow. U.S. Geological Survey, Open-File Report 01-434.

Hallaji, K., and Yazicigil, H., 1996. Optimal Management of a Coastal Aquifer in Southern Turkey, Journal of Water Resources Planning and Management - ASCE 122 (4): 233-244.

Harbaugh, A.W., Banta, E. R., Hill M.C., and McDonald, M. G., 2000. MODFLOW-2000, The U.S Geological Survey Modular ground water model- user guide to modularization concepts and the ground-water flow process, USGS, Open-File Report 00-92.

Hsieh, P. A., and Freckleton, J.R., 1993. Documentation of a computer program to simulate horizontal-flow barriers using the U.S. Geological survey's modular three-dimensional finite difference ground-water flow model, USGS, Open-File Report 92-477.

Huntzinger, T. H, Whiteman, Jr. C.D., and Knochenmus, D.D., 1985. Simulation of ground-water movement in the " 1,500- and 1,700-foot" aquifer of the Baton Rouge area, Louisiana, USGS, Water Resources Technical Report NO. 34.

Langevin, C.D. 2003. Simulation of submarine ground water discharge to a marine estuary: Biscayne Bay, Florida.Ground Water 41, no. 6: 758-771.

Long, R.A, 1965. Feasibility of a scavenger-well system as a solution to the problem of vertical salt-water encroachment, USGS Water Resources Pamphlet NO. 15.

McDonald, M.G., and Harbaugh, A.W., 1988. A modular three-dimensional finite-difference ground-water flow model: U.S. Geological Survey Techniques of Water Resources Investigations, book 6, chapter, A1, 586 p.

Rao, S.V.N, Bhallamudi, S. M, Thandaveswara, B.S., and Mishra, G.C., 2004. Conjunctive use of surface and groundwater for coastal and deltaic systems. Journal of Water Resources Planning and Management - ASCE 130(3): 255-267.

Shamir, U., Bear, J., and Gamliel, A., 1984. Optimal annual Operation of a coastal aquifer, Water Resources Research 20(4): 435-444.

Torak, L.J., and Whiteman, Jr. C.D., 1982. Applications of Digital Modeling for evaluating the ground-water resources of the"2,000-Foot" sand of the Baton Rouge area, Louisiana, USGS, Water Resources Technical Report NO. 27.

Tsai F. T-C., Sun, N-Z., and Yeh W. W-G., 2005. Geophysical parameterization and parameter structure identification using natural neighbors in groundwater inverse problems, Journal of Hydrology 308 (1-4): 269-283.

Whiteman, Jr. C.D., 1979. Saltwater encroachment in the "600-Foot" and "1,500-Foot" sands of the Baton Rouge Area, Louisiana, USGS, Water Resources Technical Report NO. 19.

Zheng, C., and, Wang, P.P., 1998. MT3DMS, A modular three-dimensional multispecies transportmodel for simulation of advection, dispersion and chemical reactions of contaminants in groundwater systems: Vicksburg, Miss., Waterways Experiment Station, U.S. Army Corps of Engineers.

35

DATA COLLECTION NETWORK TO SUPPORT ECOSYSTEM FORECASTING FOR THE BARATARIA BASIN – MISSISSIPPI RIVER DOMAIN

Gregory D. Steyer
U.S. Geological Survey National Wetlands Research Center, Coastal Restoration Field Station
P.O. Box 25098, Baton Rouge, LA 70894, USA

Alaina B. Owens
Coastal Louisiana Ecosystem Assessment and Restoration (CLEAR) Office, Louisiana State University, 3201 Energy, Coast & Environment Bldg., Baton Rouge, LA 70803, USA

Brady R. Couvillion
IAP World Services, U.S. Geological Survey National Wetlands Research Center, Coastal Restoration Field Station, P.O. Box 25098, Baton Rouge, LA 70894, USA

ABSTRACT

Ecosystem forecasting is limited by a number of uncertainties including inadequate initialization information, unknown boundary conditions, inaccurate model physics and atmospheric forcing functions, and inadequate algorithm development of geomorphic and ecological responses to hydrodynamic and geophysical processes. Monitoring can help reduce these uncertainties by providing numerical information on those variables that can improve simulation capabilities. A data collection network was designed for the Mississippi River, south of Tarbert Landing, Ms. to Venice, La., and for the inland and nearshore waters of Barataria Basin, La. The network considered existing monitoring efforts that effectively link oceanographic and water quality observations to system drivers to support ecosystem hindcasting, nowcasting, and forecasting capabilities. The design was driven by the diverse needs of the modeling community and utilized questionnaire surveys, workshops, and web-based applications to inventory past and present modeling applications, domains, and attributes. The use of existing monitoring stations, including variables measured and frequency of data collection, were also considered, as well as recommendations on where new monitoring stations should be located, what types of data should be collected, and at what temporal frequency. Compiled databases are presented as a series of GIS layers to represent overlap between existing and requested stations, and priorities are established for either maintaining or augmenting existing stations or deploying new stations. An optimization plan will be prepared from the results of this study that will provide a rationale for the ecosystem forecasting data collection network that includes a description of existing monitoring programs to support ecosystem forecasting and a set of justifiable locations for "megastations" that meet the needs of the modeling and assessment communities.

35.1. INTRODUCTION

The Louisiana Coastal Area (LCA) Ecosystem Restoration Study (USACE-MVN, 2004) Science and Technology Program recognized the need for the development of a system-wide assessment and monitoring plan (SWAMP) as a critical component in moving forward with a science-based coastal restoration planning and implementation strategy. A properly designed monitoring approach can support not only the assessment of restoration efforts, but also modeling programs for forecasting ecosystem response. Ecosystem forecasting is limited by a number of uncertainties including inadequate initialization information, unknown boundary conditions, inaccurate model physics and atmospheric forcing functions, and inadequate algorithm development of geomorphic and ecological responses to hydrodynamic and geophysical processes. Monitoring can help reduce these uncertainties by providing data at sufficient spatial and temporal scales to describe system drivers and the state of the system that models strive to match (Steyer et al., 2005).

The process of developing an integrated monitoring program to support numerical models capable of forecasting at ecosystem scales requires collaboration among multiple research scientists with complementary modeling skills and forecasting objectives. Collaboration with the monitoring and assessment communities that maintain existing monitoring stations and observation networks is also necessary. Additionally, the development of a holistic monitoring network that extends from the upland/coastal boundary of Louisiana to the nearshore/offshore waters of the Gulf of Mexico must take into account, and preferably be nested within, existing and proposed regional, national, and global observing systems (Table 1). Finally, to optimize an observation system over time and space, data collection locations, variables measured, and sampling frequencies should be based on both qualitative (i.e., knowledge-based) and quantitative (i.e., statistics-based) approaches.

The SWAMP framework is designed to monitor biological, chemical, physical and climatological variables in four modules: wetlands, barrier islands, inland waters and rivers, and nearshore waters. The wetlands module, known as the Coastwide Reference Monitoring System (Steyer et al., 2003) is currently being implemented, but the barrier island module is under development. This paper will describe the inland waters and rivers module and the nearshore coastal waters module within an integrated coastal waters data collection plan to support an ecosystem forecasting system for the Barataria Basin and Mississippi River (BA-MR domain, Figure 1). The Mississippi River component of the domain extends from Tarbert Landing, Ms. (River Mile 307) to Venice, La. The plan is built off of the Coastal Louisiana Ecosystem Assessment and Restoration (CLEAR) Open Water Monitoring framework document developed by Swenson and Swarzenski (2004). It broadly describes data collection needs of the assessment community. The data collection network is being designed at a programmatic scale and will concentrate on the selection of critical locations to gather data on oceanographic (hydrodynamic and geomorphic) and water quality (fate and effects) variables that will provide initialization and input data to support and formulate numerical models capable of predicting coastal ecosystem dynamics with a minimum of a 1-km resolution throughout the domain. The data collection network described herein is not designed to support project-specific modeling efforts; however, project-specific efforts could be nested within this integrated framework.

Our objective is to effectively link oceanographic and water quality observations to system drivers to support ecosystem hindcasting, nowcasting and forecasting capabilities,

Table 1. Observation and ecosystem forecasting systems identified within the Barataria Basin and northern Gulf of Mexico that were considered in developing an optimized data collection network.

	Program Title	Program Acronym	Web Site
1	nowCOAST web mapping portal	NOAA	http://nowcoast.noaa.gov/
2	U.S. Coastal Observing Systems Western Gulf of Mexico	NOAA	http://www.csc.noaa.gov/coos/texas_gulf.html
3	USGS Water Resources Louisiana Hydrowatch	Hydrowatch	http://la.water.usgs.gov/hydrowatch.htm
4	USGS Water Resources	NWIS	http://waterdata.usgs.gov/nwis
5	National Data Buoy Center	NDBC	http://www.ndbc.noaa.gov/
6	National Water Level Observation Network	NWLON	http://www.co-ops.nos.noaa.gov/d_nwlop.html
7	ERDC Wave Data Sites	ERDC	http://sandbar.wes.army.mil/public_html/pmab2web/htdocs/SouthEast.html
8	USACE Water Control Data	USACE	http://www.mvn.usace.army.mil/eng/edhd/watercon.htm
9	USACE Navigation Data Center	USACE	http://www.iwr.usace.army.mil/ndc/data/data1.htm
10	U.S. Environmental Protection Agency Environmental Monitoring and Assessment	EMAP	http://www.epa.gov/emap/html/data.html
11	Louisiana Universities Marine Consortium	LUMCON	http://www.lumcon.edu/
12	LSU Earth Scan Lab	LSU	http://www.esl.lsu.edu/
13	Louisiana Agriclimatic Information	LSU	http://www.agctr.lsu.edu/weather/
14	UNO Coastal Research Laboratory	UNO	http://coastal.uno.edu/
15	Louisiana Department of Natural Resources Monitoring Program	LDNR	http://www.dnr.state.la.us/crm/coastres/monitoring.asp
16	Louisiana Department of Environmental Quality	LDEQ	http://www.deq.state.la.us/technology/tmdl/index.htm
17	Wave-Current-Surge Information System	WAVCIS	http://wavcis.csi.lsu.edu/
	Long-term Estuary Assessment Group	LEAG	http://leag.tulane.edu/
18	Louisiana Hydrometeorological Networks	JOSS	http://www.joss.ucar.edu/gapp/networks/louisiana/
19	Ocean.US was created by the National Oceanographic Partnership Program - IOOS	IOOS	http://www.ocean.us/
20	Gulf of Mexico Coastal Ocean Observing System	GCOOS	http://www-ocean.tamu.edu/GCOOS/gcoos.html
21	The Ocean Research Interactive Observatory Networks	ORION	http://www.orionocean.org/
22	Southeastern Universities Research Association Coastal Research Initiative	SCOOP	http://www.sura.org/programs/coastal.html
23	Coastal Observation Technology System (COTS)	COTS	http://www.csc.noaa.gov/cots/
24	Tributary Forecasts in the Lower Mississippi River Forecast Center	LMRFC	http://www.srh.noaa.gov/lmrfc/forecast/tributaries/index.shtml
25	Terrestrial Observation & Prediction System	TOPS	http://ecocast.arc.nasa.gov/research/tops.html
26	NASA Ames, Ecological Forecasting Program	NASA	http://ecocast.arc.nasa.gov
27	NOAA National Ocean Service Ecological Forecasting	NOAA	http://www.oceanservice.noaa.gov/topics/coasts/ecoforecasting/welcome.html
28	Naval Research Laboratory Gulf of Mexico Monitoring and Forecast Systems	NRL	http://www7320.nrlssc.navy.mil/GOM_NFS/observations/observations.htm#_Aircraft_Salinity_Mapper
29	Applied Sciences Directorate – Stennis Space Center Ecological Forecasting	NASA	http://www.asd.ssc.nasa.gov/application.aspx?app=ecological
30	Geochemical and Environmental Research Group	GERG	http://www-gerg.tamu.edu/menu_research/gerg_cur_res.htm
31	Central Gulf Ocean Observing System	CenGOOS	http://www.cengoos.org/
32	Global Climate Observing System	GCOS	http://www.wmo.ch/web/gcos/gcoshome.html
33	Global Terrestrial Observing System	GTOS	http://www.fao.org/GTOS/

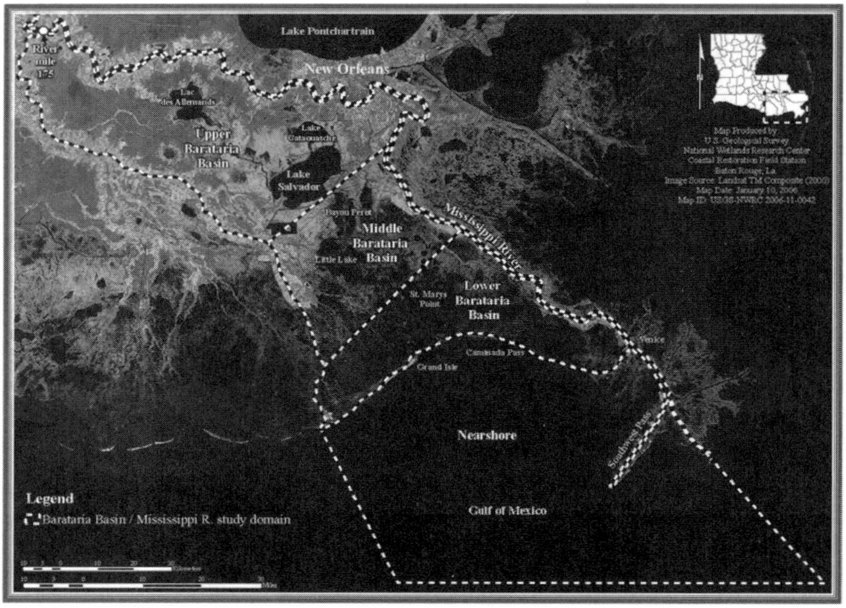

Fig. 1. Map of the Barataria Basin and Mississippi River study domain.

taking into account existing monitoring efforts to support an optimized data collection network. Additionally, this effort involves developing a comprehensive monitoring database and disseminating information regarding the locations of monitoring stations and the availability of data to the modeling and assessment communities. Full utilization of existing monitoring data may afford greater accuracy in assessing the skills of different modeling tools and may enhance the efficiency of modeling efforts.

35.2. METHODS

The basis for the development of an optimized data collection network is to assemble a comprehensive database of existing monitoring activities. This database was assembled in ArcGIS with the most current and accurate data available from existing monitoring activities. The consolidated database includes existing monitoring databases from the U.S. Army Corps of Engineers New Orleans District (USACE-MVN) and Engineer Research and Development Center (USACE-ERDC); the Louisiana Departments of Natural Resources (DNR), Environmental Quality (DEQ), Health and Hospitals (DHH), and Wildlife and Fisheries (DWF); the Louisiana Oil Spill Coordinator's Office (LOSC); Louisiana Offshore Oil Port, Inc. (LOOP); Long-term Estuary Assessment Group (LEAG); Louisiana State University Coastal Ecology Institute (LSU-CEI); Louisiana Universities Marine Consortium (LUMCON); Wave-Current-Surge Information System (WAVCIS); National Oceanic & Atmospheric Administration (NOAA); U.S. Fish and Wildlife Service (USFWS); and the U.S. Geological Survey (USGS). As part of this study, the following information was extracted from the consolidated database: geographic location, variables measured, and sampling frequency. The consolidated monitoring database supports a globally unique identification for each record so that future changes to each agency's database can be rapidly incorporated.

To determine which monitoring stations modelers are utilizing and to assess needs, modelers from Federal and State agencies, universities, and private industry that have conducted or are conducting oceanographic, hydrodynamic, geomorphic, climate, and water quality modeling within the BA-MR domain (Figure 1) were queried through personal, telephone, and e-mail inquiries. The modelers involved in these inquiries provided the knowledge-based approach for determining data needs to support development of the data collection network. The modelers were asked to provide information regarding their modeling efforts in a standardized questionnaire survey that included: (1) Investigator(s) and Collaborator(s); (2) Model Application; (3) Modeling Domain; (4) Modeling Objectives/Questions Asked; (5) Validation of Model Attributes; (6) Data Transfer Format; (7) Metadata Availability; (8) Monitoring Stations Currently Utilized; (9) Critical Monitoring Station Locations; and (10) Variables Collected and Frequency Sampled.

From the aforementioned survey, modeler responses were consolidated into a tabular format and made available to the modeling community through an internet FTP site. Participants were given the opportunity to review and comment on the consolidated responses via e-mail and in a workshop setting. The workshop served to integrate responses from the diverse group of modelers with respect to the currently used monitoring stations and with respect to the positioning of new stations to better understand system dynamics. Each participant was given the opportunity to identify essential data collection needs to support modeling efforts within the BA-MR domain, and these stations were consolidated into GIS layers. Draft maps were prepared and distributed to all task participants for review, and comments and recommendations were incorporated into final maps. The data layers representing monitoring stations in current use and requested future stations were overlaid with the consolidated monitoring station layer using a 5-km buffer to assess overlaps in station locations, variables measured, and frequency of sampling. The overlays could then be used as a qualitative assessment to support recommendations with regards to maintaining existing critical stations, fulfilling data gaps, and possibly augmenting existing stations to support a "megastation" (i.e., for measurement of 3 or more categories of variables). Quantitative assessments of salinity and precipitation monitoring within the BA-MR domain is currently being conducted using uncertainty analysis (Habib et al., 2004). Our study provides a tool for assessing correlations among stations, whereas the results of the uncertainty analysis will refine the temporal resolution and spatial distribution of monitoring stations. This study, together with the uncertainty analysis, forms the rationale of an optimized data collection plan to support ecosystem forecasting within the BA-MR domain.

35.3. RESULTS

35.3.1. Modeler Participation

Surveys were sent to 74 individuals from 19 affiliations comprising the Federal, State, academic, and private consulting community that were identified as having modeling expertise within the BA-MR domain. Respondents provided information on 35 modeling efforts conducted by over 55 investigators and collaborators. The workshop was attended by 22 individuals that included investigators and/or collaborators on 11 of the 35 modeling efforts. Workshop participants predominantly represented hydrodynamic and water quality model applications, while wave and sediment transport applications were under-represented. The workshop recommendations were refined during the review process, and ultimately, investigators representing 32 of the 35 modeling efforts participated in determining essential data needs.

35.3.2. Modeling Efforts

Modeling applications within the BA-MR domain were primarily related to hydrodynamics, sediment transport, water quality, and wave dynamics, with differences between applications arising from the scope of variables modeled, dimensionality, and how the governing equations were solved. Within the hydrodynamic applications, associated sediment and water quality models included TABS-MD (RMA2, RMA4, RMA10, and SED2); HEC-RAS; CH3D-SED; CH3D-WES; POM; FVCOM; MIKE11; ADCIRC; CELSS; BLESS; CE-QUAL-ICM; LAQUAL; AUTOQUAL; and WASP, together with their modified versions. Box models were also used to assess salinity and nutrients. The survey identified that 71.4% of the modeling efforts had domains that included the near shore waters, indicating the importance of extending the geographic boundary conditions of the models well offshore. Wave modeling applications tended to focus on wave forecasting (average conditions and surges associated with large storm events) and sediment transport (sediment budget, sand erosion and deposition, and longshore sediment movement), using applications or modified versions of applications such as STWAVE, GENESIS, SBEACH, WISWAVE, and FEMA Surge.

Efforts varied widely in spatial coverage. Model domains targeted waterbodies (e.g., Lac des Allemands), restoration project areas (e.g., Myrtle Grove), geographic regions (e.g., Mississippi River Delta), and entire coastal basins (e.g., Barataria Basin). The scope and objectives of the modeling efforts were also influenced by scale. For example, the targeted water bodies were modeled to assess pollutant load capacities and assimilations; whereas, for the purpose of making planning and operational decisions, the restoration project areas were commonly modeled to assess the effects of freshwater diversions on salinity, water level, flow, sediment and/or nutrient patterns. The larger-scale geographic regions and basins commonly addressed larger scientific uncertainties, such as the influence of circulation dynamics with respect to the distribution of particular dissolved chemical and biological compounds, the implications of global climate change on northern Gulf of Mexico estuaries and coastal waters, and characterizations of wave climate along the U.S. Gulf of Mexico coast.

35.3.3. Data Collection

The consolidated monitoring station database, compiled as of January 1, 2005, identified 1,376 stations within inland waters and rivers and nearshore waters for the BA-MR domain. An overwhelming majority of these stations (91.9%) were discrete stations with no fixed instrumentation, and nearly 60% of these stations were sampled less frequently than monthly. There were 571 stations sampled at least monthly, of which 112 stations were sampled at an hourly frequency or better.

The survey also identified that 32 stations in the BA-MR domain were being utilized by multiple modelers for their specific applications (Figure 2). The characteristics of these stations showed that 73% have at least an hourly sampling; 18% are sampled monthly, and

9% are sampled less frequently than monthly. The optimal transmission of data collection needed to capture ecosystem dynamics and for integration into ecosystem forecasts is real-time. The stations most commonly used in Barataria Basin are Davis Pond, Lake Cataouatche, Lake Salvador, Barataria Bay Waterway, Little Lake, Mud Lake, St. Mary's Point, Hackberry Bay, central Barataria Bay, Barataria Bay pass, Grand Isle, and Grand Terre. Stations most commonly used in the Mississippi River domain are Tarbert Landing, Audubon Park, Venice, and Southwest Pass. Many respondents utilized stations from the National Data Buoy Center offshore of Southwest Pass and Mobile Bay, because of limited amounts of data from the nearshore.

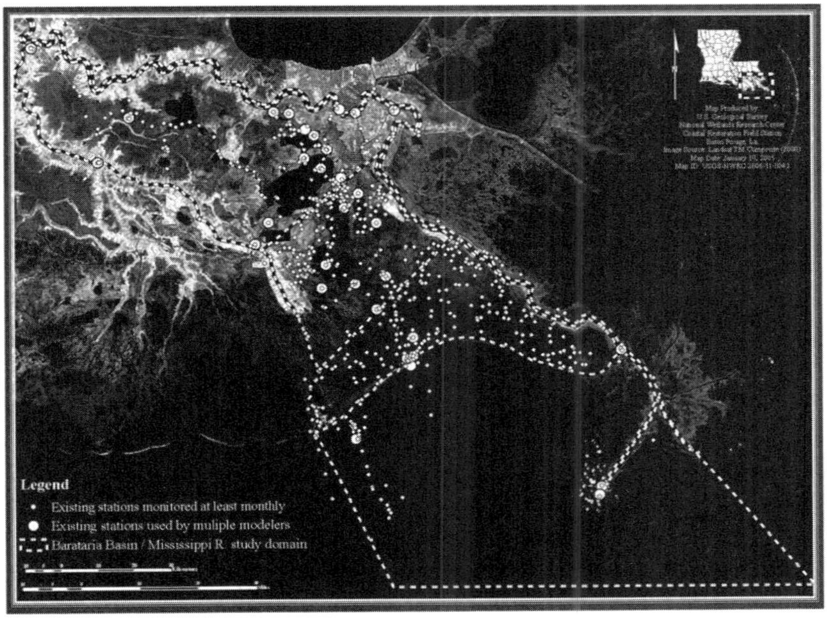

Fig. 2. Existing open-water monitoring stations with a minimum sampling frequency of monthly. Stations most commonly used in modeling efforts are identified.

35.3.4. Data Needs

The variables that were identified by the modelers as most critical in capturing the system drivers that influence ecosystem dynamics are identified in Table 2. The variables were categorized to enhance clarity in Figure 3. All variables are recommended to be incorporated into permanent gauging stations, except for discharge, which should be conducted synoptically. Synoptic water quality surveys were also recommended in addition to water quality measurements at permanent gauging stations. There was no consensus on frequency of collection, other than all variables should be sampled at least on an hourly basis. The most frequent sampling period recommended for each category is described in Table 2.

Table 2. List of essential variables to be measured by category, as identified by respondents.

Category	Variables	Sampling Frequency
Water Condition	Salinity, Conductivity, Temperature, Water Level, Turbidity	15 minute inland; 2 minute nearshore
Water Quality	Dissolved Oxygen, Chlorophyll, Fluorescence	15 minute inland; 2 minute nearshore
Suspended Sediment	Suspended Sediments	15 minute inland; 2 minute nearshore
Wave	Current Velocity, Current Profile, Wave Height, Wave Period, Wave Direction, Wave Spectrum	2048 samples per hour
Current Velocity	Current Speed, Current Direction	6 minute inland; 2 minute nearshore
Discharge	Discharge transects	Monthly
Meteorologic Conditions	Precipitation, Wind Speed, Wind Direction, Evaporation, Air Temperature, Relative Humidity, Barometric Pressure, Solar Radiation, Quantum Radiation (PAR)	Hourly

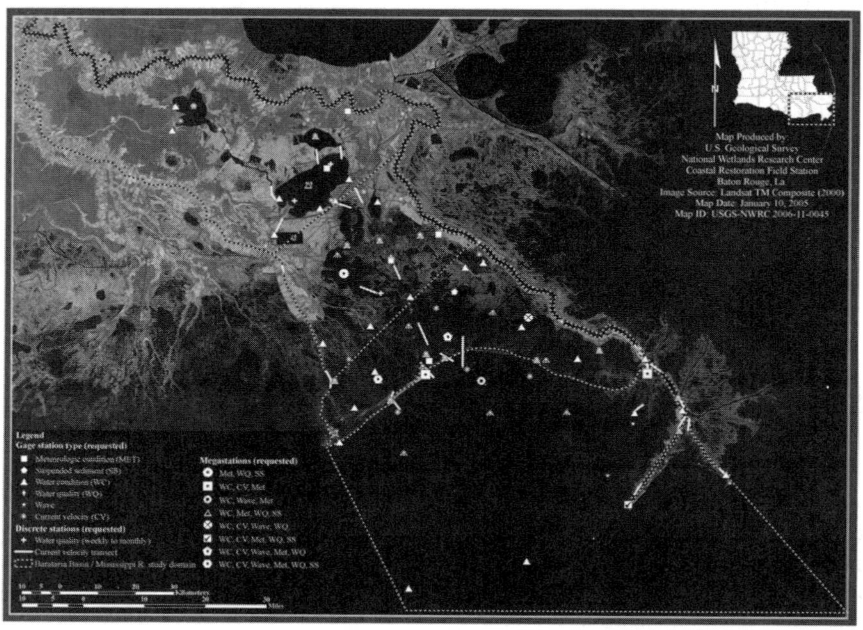

Fig. 3. Requested future station locations and variables measured, as identified by the modelers surveyed.

The requested station locations and variables to be measured, as identified by the modeling community, are illustrated in Figure 3. The requested stations in the upper Barataria Basin (north of Bayou Perot and Rigolets) are spatially distributed in the freshwater sources to Lac des Allemands, Lake Cataouatche, and Lake Salvador, as well as within these larger water bodies. In the upper Barataria Basin, the categoric variables identified were primarily water condition, current velocity, water quality, and meteorologic conditions. In the middle Barataria Basin (Bayou Perot and Rigolets to St. Mary's Point), stations were requested along three east-west transects in the vicinity of Bayou Perot and Rigolets, Little Lake, and St. Mary's Point. The following variables are recommended for measurements along transects in the following locations: (1) water condition and current velocity: Bayou Perot and Rigolets; (2) water condition, current velocity, suspended sediment, and meteorologic conditions: Little Lake; and (3) water condition: St. Mary's Point. The lower Barataria Basin had the greatest needs, covering all variables. Critical stations were also identified in all the major passes into the Barataria Basin, for example, Belle Pass, Caminada Pass, Barataria Pass, Pass Abel, and Grand Bayou Pass. In addition to synoptic discharge measurements, water condition, current velocity, and water quality were also recommended to be measured in these passes. For the nearshore waters, essential stations were requested along two east-west transects, in depths of approximately 8-10 m and 20 m following the plume of the Mississippi River. These stations should measure water condition, water quality, and waves. Locations for the Mississippi River stations were requested in reaches near St. Francisville, Baton Rouge, New Orleans, Venice, and Southwest Pass. These stations are recommended to have measurements for a full suite of variables: water condition, current velocity, water quality, and suspended sediment.

There were a total of 116 future stations requested within the BA-MR domain (Figure 3). The modeler queries identified the need for an hourly sampling frequency or better at permanent gauging stations. We found 78 existing stations equipped with hourly or better recorders located within 5 km of requested stations (Figure 4). The existing stations may provide opportunities for augmentation (e.g., the addition of variables or future modification of collection frequencies) to support ecosystem forecasting applications. Of all the stations requested, 9 locations were identified as "megastations," with one in the upper basin, one in the middle basin, four in the lower basin along an east-west transect, two in the nearshore, and one in the Mississippi River. Ideally, these stations would measure a complete suite of variables. Data collection gaps are illustrated in Figure 4. These gaps exist in areas where the 5 km buffers of existing stations do not overlap. Specifically, they are found in Lac des Allemands, the east and west flanks of the lower Barataria Basin, and throughout the nearshore waters. In order to address current data collection deficiencies, stations requested in these areas need new deployments.

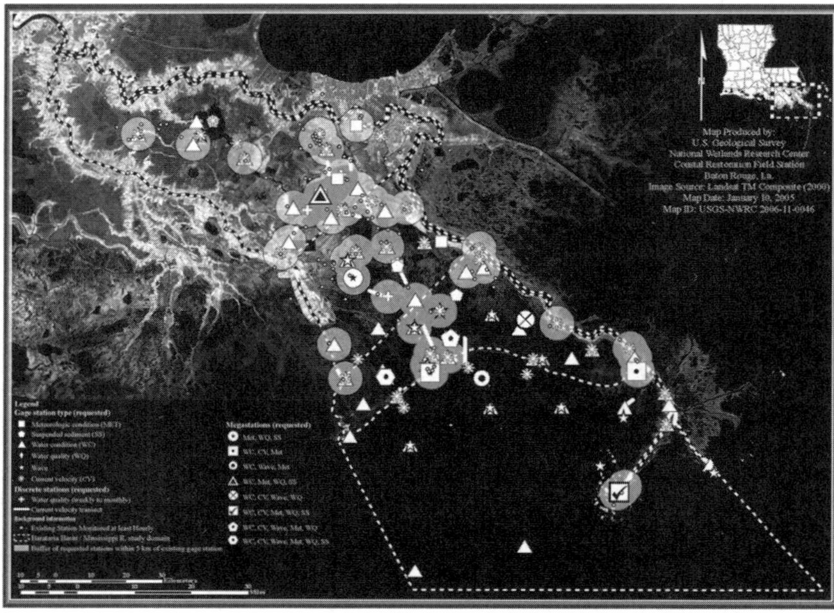

Fig. 4. Requested future station locations and existing open-water stations within 5 km of those requested and with a minimum sampling frequency of hourly.

35.4. DISCUSSION

A review of modeling efforts in the northern Gulf of Mexico by Day et al. (2000) found that, even though, current simulation models have great capabilities to project cumulative impacts at different temporal and spatial scales, they have been utilized in a piece-meal manner, without a consistent spatial framework, and without an integrated field data approach to calibrate and assess skill. The 35 modeling efforts that have been conducted within the BA-MR domain revealed the importance of developing an integrated, programmatic, data collection network in which project-specific efforts could also be nested. Additionally, this integrated approach could serve as a backbone for initiatives within the larger northern Gulf of Mexico region. Although this study did not address modeling efforts and associated monitoring needs in offshore Gulf of Mexico waters, the majority of the surveyed modeling efforts had model domains that extended into the offshore waters. For example, the wave and circulation modeling efforts covered the largest domains, extending into the northern Gulf of Mexico to 27° North and between 85° and 95° West. Within this domain, the modeling community must establish the offshore and east-west boundary conditions. The selection of locations to assess boundary conditions should be made considering other regional observation systems, such as those being taken into account under the Gulf of Mexico Coastal Ocean Observing System (GCOOS, Table 1). It may be possible to pursue these additional stations, as well as nearshore stations, through the GCOOS because they are currently drafting priorities for enhancing their national backbone of stations in the Gulf of Mexico.

Within the BA-MR domain, the assessment community is currently collecting an impressive amount of data. Unfortunately, the extent of variables collected and the frequency of sampling does not match well with the needs of the modeling community. The deficiencies in existing monitoring efforts within the Barataria Basin and nearshore are primarily related to variables within the categories of water quality, meteorologic conditions, current velocity, and waves. Within the Mississippi River, the deficiencies are primarily water quality, suspended sediment, and current velocity. The modeling community requested that the spatial distribution of future deployments should adequately assess (1) freshwater inputs into the BA-MR domain, including all of the major passes; (2) along-shore variations in wave conditions; and (3) impacts on hydrodynamics, morphology, and sediment and water quality of the Mississippi River because of river diversions. The variables in Table 2 were identified as being most critical to the modeling community for describing the current system state, and for detecting and predicting change in the BA-MR inland and nearshore coastal waters. These variables are consistent with the subset of common variables identified for measurement under the Global Ocean Observing System (UNESCO, 2003), and with recommended data collection improvements identified in Day et al. (2000).

Based on the needs of the modeling community and the lack of data collection in critical locations, we recommend that the consolidated monitoring station database, as well as uncertainty analysis, be used to identify new locations for future station deployment and to identify existing stations as potential candidates for augmentation. This study identified data collection gaps within Lac des Allemands, lower Barataria Basin, and throughout the nearshore waters. Therefore, initial deployment of new stations should be made in these areas. Additionally, it is recommended that the 32 existing stations that are commonly utilized in modeling efforts should be maintained with long-term funding commitments as part of this open-water network. We also propose the construction of "megastations" when addressing the augmentation of existing stations or deployment of new stations. The benefits of "megastations" include decreases in the cost per unit of monitoring efforts as well as reduced reliance on extrapolation between temporally or spatially disparate stations. Over 67% of all requested stations have existing stations within 5 km, providing substantial opportunities to augment stations that are spatially distributed throughout the domain.

With improved coordination between assessment agencies and the modeling community when citing stations, selecting variables, and determining sampling frequencies, as well as when removing stations, an optimized data collection plan can be developed that addresses the needs of both communities. The development of the optimized data collection plan is a two-step process. The first step, described in this report, requires that the modeling community identify critical existing stations and any modifications to be considered, and identify the placement of new stations to close spatial data gaps. The second step is integrating the results of this identification phase into existing monitoring and assessment programs. Once the second step is complete, an optimization plan can be prepared that provides a rationale for a holistic ecosystem forecasting data collection network, a description of existing monitoring programs that support ecosystem forecasting, and a set of justifiable locations for "megastations" (including the monitoring variables and temporal frequencies of data acquisition). The linkage to existing monitoring programs will incorporate discussions regarding data management requirements, computational infrastructure, and skill assessment to ensure compatibilities with preliminarily findings from our questionnaire surveys. We suggest the development of a Data Management and Communications (DMAC) plan that describes metadata standards and levels of quality that should be uniformly accepted by all entities that operate stations within the BA-MR domain. The standards and protocols should conform to the greatest extent possible with those described in the Integrated Ocean Observing System DMAC plan (Hankin, S., and the DMAC Steering Committee, 2005).

Once an optimization plan is developed and a foundation of data is available for all modeling efforts, model development and enhancement will rely more on intellectual input, and less on the amount of data used for model initialization and validation. The collected data should support both operational needs (e.g., nowcasts and forecasts), and with the help of hindcasts and nowcasts, the data should aid skill assessments of a variety of oceanographic (hydrodynamic and geomorphic) and water quality (fate and effects) models. Benefits from coordinating the development of this data collection network with pre-existing regional, national, and global observation networks include reduction in potential overlap and duplication of effort; enhancement in the efficiency of modeling efforts; and increase in the likelihood of leveraging financial support for future implementation, operation, and maintenance of monitoring stations.

REFERENCES

Day, J.W., Inoue, M., and Reyes, E., 2000. Establishment of an input database and modeling framework for the Northern Gulf of Mexico at the watershed level. Louisiana Sea Grant contract number 169-12-5187.

Habib, E., Rivera-Monroy, V.H., Visser, J.M., Steyer, G.D., Twilley, R.R., Swenson, E., Rose, K.A., Justic, D., Nuttle, W., and Hyfield, E., 2004. Uncertainty analysis of the CLEAR ecosystem model: issues, needs, and future directions. Chapter 17 *in* R.R. Twilley (ed.), Coastal Louisiana Ecosystem Assessment and Restoration (CLEAR) Model of Louisiana Coastal Area (LCA) Comprehensive Ecosystem Restoration Plan, Volume II: Tasks 9-15, Final Report to Department of Natural Resources, Coastal Restoration Division, Baton Rouge, LA, Contract No. 2511-02-24, 355 pp.

Hankin, S., and the DMAC Steering Committee, 2005. Data management and communications plan for research and operational integrated ocean observing systems: I. Interoperable data discovery, access, and archive. Ocean, US, Arlington, VA 304 pp. http://dmac.ocean.us/dacsc/imp_plan.jsp.

Steyer, G.D., Twilley, R.R., and Raynie, R.C., 2005. An integrated monitoring approach using multiple reference sites to assess sustainable restoration in coastal Louisiana, Proceedings of the Monitoring Science and Technology Symposium, RMRS-P-000, Rocky Mountain Research Station. (In Press).

Steyer, G.D., Sasser, C.E., Visser, J.M., Swenson, E.M., Nyman, J.A., and Raynie, R.C., 2003. A proposed coast-wide reference monitoring system for evaluating wetland restoration trajectories in Louisiana. Environmental Monitoring and Assessment 81:107-117.

Swenson, E.M., and Swarzenski, C.M., 2004. Open water monitoring to support coastal restoration. Chapter 16 *in* R.R. Twilley (ed.), Coastal Louisiana Ecosystem Assessment and Restoration (CLEAR) Model of Louisiana Coastal Area (LCA) Comprehensive Ecosystem Restoration Plan, Volume II: Tasks 9-15, Final Report to Department of Natural Resources, Coastal Restoration Division, Baton Rouge, LA, Contract No. 2511-02-24, 355 pp.

UNESCO, 2003. The Integrated Strategic Design Plan for the Coastal Ocean Observations Module of the Global Ocean Observing System, GOOS Report No. 125: IOC Information Document Series N°1183.

U.S. Army Corps of Engineers, New Orleans District, 2004. Louisiana Coastal Area (LCA), Louisiana Ecosystem Restoration Study, Volume 1-4.

36

REHABILITATION OF WETLANDS IN SRI LANKA FOLLOWING THE 2004 INDIAN OCEAN TSUNAMI

Rusyan Jill E. Mamiit
Joint Institute for Marine and Atmospheric Research
University of Hawai`i – Manoa / NOAA-NMFS-PIFSC
2570 Dole St., Honolulu, HI 96822, USA

ABSTRACT

Mangroves are considered as special types of wetland ecosystems, which provide a wide array of economic goods and ecological services. Mangroves worldwide have been exposed to anthropogenic- and naturally-induced interventions. Of late, mangroves in Sri Lanka were subject to severe natural damage caused by the tsunami that hit the Indian Ocean on December 26, 2004. The impacts of tsunami magnified the significance and vulnerability of mangroves at the event of natural disasters. Anecdotal evidences and observations following the tsunami indicate that the existence of healthy mangroves reduced the impacts of the tsunami in two ways: (i) dense mangrove forests offered resistance from speeding water and tidal waves to reach inland areas and (ii) the natural water funnels in the mangrove ecosystem distributed the volume of tidal water, which in turn lowered the impact in inland settled areas (IEMA, 2005). In recognition of the protective function of mangroves during the tsunami, this paper reports the findings of household surveys in the villages of Kapuhenwala and Wanduruppa, mangrove sites characterized by functioning and degraded ecosystems, respectively. Results demonstrate better public recognition and awareness of the protective value of mangroves after the tsunami. There is also an increased interest among members of coastal communities to lobby for the rehabilitation of mangrove ecosystem to support local, regional and global environmental concerns. Community accounts on the role of mangroves during the tsunami suggest the environmental and socioeconomic viability of including mangrove rehabilitation efforts in the post-tsunami reconstruction and rebuilding plans.

36.1. INTRODUCTION

Sri Lanka is an island with 1,585 km of coastline (UNEP, 2001). As in many tropical island regions with long stretch of coastal waters, the country is abundant with coastal and marine vegetation. Wetlands, specifically mangroves, are among the most common and important coastal vegetations in Sri Lanka. Although mangrove areas in the country are limited to 12,000 hectares in size (Pemadasa, 1984) and 23 mangrove species (WRI, 2003), it has great environmental, commercial, recreational, aesthetic and economic value.

Mangrove ecosystems are located in the brackish water margin between land and sea in most tropical and sub-tropical regions (Spaninks and van Beukering, 1997). They are among the highly productive and taxonomically diverse ecosystems on earth. Mangroves are also biologically and economically the most dynamic among coastal resources (Lal, 1990). Because of the wide array of goods and services it offers, mangroves are dubbed as 'biological and economic supermarkets' (Mitsch and Gosselink, 1993). For people in coastal communities, mangroves are significant sources of fish, which in turn supports the protein needs of coastal populace. Mangrove ecosystems also provide essential timber and non-timber products for domestic and commercial uses. Apart from these goods, mangroves provide ecological, environmental and biological services, of which nursery grounds for marine life, storm and shoreline protection and stabilization, control of soil erosion and flooding are included.

In Sri Lanka, mangroves have been exposed to both anthropogenic- and naturally-induced interventions. The increased migration to coastal regions since Sri Lanka's independence in 1948 resulted in degradation of the country's coastal vegetation, particularly mangroves (Global Oceans, 2005). The World Rainforest Movement (1999) disclosed that over the last ten years, many of the lagoons and estuaries in Sri Lanka have been subjected to rapid destruction of its mangrove vegetation due to tourism and commercial aquaculture developments (Liyanage, 1995). The intensification and rapid expansion of prawn farming in the west coast of Sri Lanka contributed to the conversion and subsequent decline of mangroves. The inadequacy and lack of implementation of governing mechanisms to regulate the use of the coastal and marine resources exacerbated the conditions of mangroves in the southern Sri Lankan coast. Of late, mangroves in the southeastern coastal belt of Sri Lanka were subjected to severe natural damage caused by the tsunami that hit the Indian Ocean on December 26, 2004. The impacts of tsunami magnified the significance and vulnerability of mangroves at the event of natural disasters.

Anecdotal evidences and initial observations following the aftermath of the tsunami indicate that the existence of healthy mangroves helped reduce the destructive impacts of the tsunami (MAP, 2005). In the process, however, these resources suffered most of the damage. According to reports, mangrove forests reduced the impact of the tsunami in two ways: (i) dense mangrove forests offered resistance from speeding water and tidal waves to reach inland areas and (ii) the existence of natural water funnels in the mangrove ecosystem distributed the volume of tidal water, which in turn lowered the impact in inland settled areas (IEMA, 2005).

In recognition of the protective role of mangrove ecosystems in mitigating the impacts of the tsunami and in response to the results of the rapid environmental assessment conducted by UNEP, IUCN and other collaborating agencies, the need to include environmental rehabilitation efforts in the tsunami reconstruction and rebuilding plans is paramount. Owing to this exigency, it is important, therefore, to present a justification for the inclusion of mangrove rehabilitation and restoration efforts in the current tsunami reconstruction projects in Sri Lanka. Accordingly, a socioeconomic perception survey of the shoreline protection value of mangroves is imminent to provide a sound basis for future plans of action.

In consideration of the preponderance of assessing the values attached by coastal residents to wetlands *i.e.* mangrove ecosystems; this study reports the findings of household surveys in Kapuhenwala and Wanduruppa, areas characterized by broad and narrow stretch of mangroves, respectively. These areas were severely damaged by the tsunami and the survey demonstrates the awareness of coastal residents on the protective function of mangrove ecosystems during the disaster. This study also provides insights and economic basis for more informed decisions and efficient allocation of limited resources for mangrove management as well as effective incorporation of human and environmental concerns in the implementation of reconstruction and rehabilitation plans in the post-tsunami development process. The following section provides a brief background on the study areas. This is followed by a discussion on the survey methodology. Results of the data analysis and its implications are presented in the fourth section. Finally, the last section draws conclusions and provides recommendations on post-tsunami mangrove management.

36.2. STUDY AREA

With the intention of assessing the role of mangrove ecosystems in reducing the impacts of the tsunami, the areas of Kapuhenwala and Wanduruppa were selected as the study sites. The suitability of the sites was investigated through a physical assessment of the areas using the approach developed by the IUCN. It was determined in the reconnaissance survey that the two sites were apposite since they exhibit the characteristics of sites with both functioning and degraded mangrove ecosystems. Despite the tourism development in Kapuhenwala, a broad stretch of mangrove still exists in the area. The village of Wanduruppa, on the other hand, has a narrower and patchy mangrove ecosystem, which many residents attributed to population encroachment in the area in the last decade.

Kapuhenwala and Wanduruppa villages are located in the Hambantota district, 200 km south of Colombo. The district is one of the most dynamic coastal areas in Sri Lanka. It has 27 km of coastline extending from Rekawa to Ussangoda in the southernmost coast. The sites demonstrate a diversity of marine and coastal resources. The dominant natural vegetation includes mangroves, scrublands, salt marshes, reed beds and grasslands.

36.2.1. Kapuhenwala

Kapuhenwala is one of the villages in the larger division of Tangalle. The village is characterized with estuarine and deltaic coastline. The major resources in Kapuhenwala upon which its 59 registered households depend for livelihood include farming and lagoon and offshore fisheries. Tourism has been an important part of the village economy for the last decade. There are three major hotels in the area.

The mangrove areas within the village represent 14 species. These include *Lumnitzera racemosa*, *Ceriops tagal*, *Avicennia officinalis*, *Sonneratia*, *Excoecaria agallocha*, *Bruguiera gymnorhiza*, *Rhizophora mucronata* and *Aegiceras corniculata* (Wijayaweera, 2005). The National Aquatic Resources Research and Development Agency and the Rekawa Development Foundation have successfully reintroduced some mangrove species in the immediate periphery of the village lagoon. The Sri Lanka Forest Department also initiated a process to demarcate the mangrove area of the lagoon in order to implement community-based restoration activities (IUCN Sri Lanka, 2004).

Severe damage was noted in the Kapuhenwala area following the tsunami on December 26, 2004. Substantial damage to coastal vegetation, particularly mangroves, was observed around the village lagoon. The impact of the tidal waves was observed to have gone beyond 350 m from the shoreline. This explains the massive destruction in most of the mangrove areas and human settlements next to the lagoon. Fifty out (50) of the 59 houses were either partially or totally destroyed by the tidal waves (Jeewananda, 2005). The village, however, was fortunate to have the lowest number of fatalities with only two casualties recorded from the tsunami. Livelihood, specifically, lagoon and offshore fisheries were largely affected by the tsunami, as most of the fishing boats and gears were destroyed. Beach areas with tourism infrastructure such as the Hotel Lagoon Paradise and the Lagoon Wadiya Beach Cabanas were reduced into rubble during the tsunami.

36.2.2. Wanduruppa

The village of Wanduruppa is a coastal and farming area, which belongs to the Ambalantota Divisional Secretariat in the southern province of Sri Lanka. The area is one of the progressively growing coastal villages with approximately 512 households. Similar to other coastal villages, farming and fishing are the major sources of income in Wanduruppa. Compared to other villages in Ambalantota, Wanduruppa has a higher living standard with 41% of the families receiving an average monthly income of SLR 7,500 (US$ 75). This, in part, is due to diversity of livelihood opportunities in the area, which include cottage industries.

Despite the economic development in the area, the village of Wanduruppa possesses a number of coastal and marine resources. Fishery is the most dominant followed by 34 hectares of agricultural lands. These resources, which support the livelihood activities of both residents and non-residents, were severely affected by the December 2004 tsunami. Before the tsunami, there were 26 beach seine nets that were in operations in the village. After the disaster only one net suitable for operation remained. The rapid environmental impact assessment in the area indicated seawater has run through a distance of 1 km from the coast. A total of 348 out of 512 houses were damaged (De Silva, 2005). Road infrastructure was also impaired by the tsunami.

The crown-shaped patches of mangroves in Wanduruppa, which prior to the tsunami have been exposed to considerable human intervention and land conversion, were also severely impacted by the disaster. The mangrove area starts from Walipatanwila and extends to Godawaya, covering an average area of 76 hectares. *Sonneratia*, *Acrosticum aureum*, *Aegiceras corniculata*, *Acanthus illicifolius*, *Avicennia officinalis* are found as the dominant stands in the area affected by the tsunami (Wijayaweera, 2005).

36.3. SURVEY METHODOLOGY

Apart from the comprehensive implementation of customary research practices and extensive secondary literature reviews, primary data were also obtained from household surveys. A purposive sampling approach was developed to determine whom to target for the household interviews. This tactic, although not as statistically representative of the population as other approaches, was deemed the most practical approach for the purposes of the study. The number of samples was determined using the recommended sample size developed by Bunce and Pomeroy (2003) for household interviews in coastal areas.

The primary data collection was extensively carried out separately for the two study sites. Household surveys were primarily implemented in Kapuhenwala over a period of 10 days in April 2005. With equal number of days, the same survey was completed for Wanduruppa in April 2005 as well. The questionnaires were conducted in both English and Sinhala with 44 respondents (out of 59 households) in Kapuhenwala and 107 (out of 512 registered households) in Wanduruppa.

One major aspect of the survey was the assessment of the conditions and roles of mangroves as perceived by the residents before and after the tsunami. In order to ascertain the ranking pattern of the different functions attached to the mangrove ecosystem before and after the tsunami, a series of questions on resource conditions were sought from the residents.

Mangroves act as a natural barrier to waves coming inland. They help reduce the effects of storm surges and flooding. While this particular function of mangroves is widely acknowledged, the scientific information on its particular role during the tsunami is still limited. In the light of better understanding the protective role of mangroves, the villagers were asked to give anecdotal experiences on the role(s) of mangroves at the event of a natural disaster. Questions on property damages and loss economic activities were solicited.

Despite the many goods and services provided by the mangroves in Kapuhenwala and Wanduruppa, they are under intense pressure from anthropogenic and natural interventions. It was crucial to elicit perceived problems on resource governance, management options and inclusion of mangrove rehabilitation efforts in the post-tsunami reconstruction and rebuilding process to properly evaluate the needs, problems and concerns faced by the communities in their struggle to sustain an intact mangrove resource.

In order to capture the willingness to participate of the community to post-tsunami mangrove rehabilitation efforts, residents were asked to express their willingness to contribute in terms of time and labor. Relevant socioeconomic and demographic information, both in the community and household levels, were also solicited.

36.4. HOUSEHOLD SURVEY

A consensus among coastal communities indicates the significance of intact mangroves during the recent tsunami in the Indian Ocean. In several coastal areas in Sri Lanka, anecdotal evidence shows that the broad stretch of mangroves in the Hambantota district served as protective barrier from tidal waves coming inland. So as to validate the contention on the actual role of the mangroves during the tsunami and to assess the economic importance of sustaining healthy mangroves, the villages of Kapuhenwala and Wanduruppa in Sri Lanka, areas characterized by intact and degraded mangrove ecosystems, respectively, were investigated.

Personal household surveys were conducted to ascertain the shoreline protection function of mangroves during the tsunami. A total of 151 completed questionnaires were analyzed for the study. The relatively small size and compactness of the Kapuhenwala village allowed the survey completion of 75% of the entire village households. Forty-four out of 59 households were interviewed. Correspondingly, the large number of households in Wanduruppa, which totals to 512, did not permit for the same degree of completion. Completing 107 questionnaires in Wanduruppa, in any case, was a close representation of the population.

36.4.1. Respondent Profile

The residents in the study sites, generally, belong to nuclear family households. The residents of Kapuhenwala and Wanduruppa are middle-aged as reflected by 67% and 61% of the villagers belonging to the following age brackets: 31-40 years old and 41-50 years old, respectively. Apparently, there are a high proportion of males in both villages – 86% in Kapuhenwala and 71% in Wanduruppa. It should be noted that the study survey targeted head of households, who in Sri Lanka are typically males. The proximity of Wanduruppa to the Ambalantota town and the accessibility of private and public educational institutions in the area as well as the relatively higher monthly cash income of the households explain the high proportion (71%) of residents with an average formal education of 8 years compared to 47% in Kapuhenwala. The high level of literacy in Wanduruppa is manifested in a high degree of participation and membership of 62% of the population in environmental and coastal-based organizations.

36.4.2. Resource Condition

One of the objectives of the survey was to establish the level of awareness of coastal communities on the different functions of the mangrove ecosystems. In order to ascertain the cognizance of the communities to mangroves and its associated goods and services, a series of questions were designed to capture the residents' degree of familiarity.

Residents in Kapuhenwala and Wanduruppa have a reasonable good understanding of the coastal resources in the area as shown by almost 100% familiarity of the residents to mangroves in both sites. Before the tsunami, 82% of the residents in Kapuhenwala agree that mangroves in the area were intact and in healthy conditions. Remarkably, only 79% of the villagers in Wanduruppa have the same opinion. The low degree of recognition for healthy mangroves in Wanduruppa was expected since prior to the tsunami, the mangroves in the area have been constantly threatened by human activities. This is further explained by 72% of the households in Wanduruppa who do not perceive any difference between the pre- and post-tsunami conditions of the mangroves compared to 90% of the residents in Kapuhenwala who recognized the substantial devastation on the resource following the tsunami.

There is a high level of awareness on the different goods and services offered by the mangrove ecosystems in the study sites. In Kapuhenwala, it is interesting to note that prior to the tsunami disaster, the functions associated to mangroves are ranked as follows: 1) source of fish, 2) breeding grounds for fish, 3) tourism, 4) source of timber and non-timber forest products, 5) research area and 6) shoreline protection (Table 1a). The ranking was largely influenced by the presence of a large fishing community in the village as demonstrated by 52% of the residents engaged in both lagoon (mangrove) and offshore fisheries. The residents of Wanduruppa, on the other hand, do not share a similar ranking.

The regular flooding caused by the re-channeling of the river mouth in Wanduruppa has influenced the hierarchical assignments of functions to mangroves. The ranking of residents in Wanduruppa manifests that the shoreline protection and flood control functions of mangroves are of highest importance, before and after the tsunami (Table 1b). The dominance of the protective role of mangroves was realized in Kapuhenwala only after the tsunami. Despite the high recognition of the protective function afforded by the mangrove ecosystems in Wanduruppa prior to the tsunami disaster, community members remained unresponsive to the incessant degradation of the mangroves in the area. Village residents in Wanduruppa never imagined that anything worst than the frequent inundation that devastates the village can occur. Prior to the tsunami, it was felt that the existing strip of mangroves was sufficient to protect the village from flooding. Hence, residents continued with their unsustainable use of mangroves in the area. After the tsunami, the residents gained a better understanding that mangroves can protect them from disasters worst than floods.

Table 1a. Ranking of the different functions of mangroves in Kapuhenwala.

BEFORE THE TSUNAMI	FUNCTION	AFTER THE TSUNAMI
4	Source of timber and non-timber products (excluding fish)	6
1	Source of fish (lagoon)	2
3	Recreational / tourist area	3
6	Protection from natural disaster (storms, tsunami, cyclones)	1
5	Area for educational and environmental research	5
2	Nursery grounds for lagoon and offshore fish species	4

Table 1b. Ranking of the different functions of mangroves in Wanduruppa.

BEFORE THE TSUNAMI	FUNCTION	AFTER THE TSUNAMI
6	Source of timber and non-timber products (excluding fish)	6
3	Source of fish (lagoon)	2
4	Recreational / tourist area	3
1	Protection from natural disaster (storms, tsunami, cyclones)	1
2	Area for educational and environmental research	5
5	Nursery grounds for lagoon and offshore fish species	4

The ratio of mangrove area to the population is lower in Wanduruppa. Wanduruppa has a narrower stretch of mangroves, which supports 512 households; whereas, Kapuhenwala has a broader and continuous strip of mangroves, which supports only 59 households. The mangrove to population ratio, partly, explains why mangrove resources in Wanduruppa are more degraded despite the high recognition of the community of the different goods and services that can be derived from it.

In terms of involvement in mangrove conservation activities, the residents of Kapuhenwala showed a higher level of recognition (84%) of the different organizations (*e.g.,* Rekawa Development Foundation, National Aquatic Resources Research and Development Agency) involved in mangrove research and conservation activities in the village. On the contrary, only 71% of the residents in Wanduruppa are aware of the mangrove-related organizations in the area (*e.g.,* IUCN-The World Conservation Union, Biodiversity People's Organization). The familiarity, awareness and involvement of the residents in coastal-based environmental organizations in both sites underscore the relevance of mangroves as natural indicators of environmental quality for the communities.

36.4.3. Shoreline Protection

The provision of indirect support and protection to economic activity and property by the natural functions or regulatory 'environmental' services is the primary attribute of indirect use values associated with mangrove ecosystems (Spaninks and van Beukering, 1997). Shoreline protection and flood control functions are two of the most important indirect services provided by mangroves. The full value of these services, however, is not fully recognized. The recent tsunami demonstrated these functions and generated interest among scholars on assessing the actual protective role of mangroves as buffer from storm damage, wave action and alluvial materials, at the event of natural disasters.

Owing to the significance of conducting an actual assessment of the protective function of mangroves, the residents were asked to give brief anecdotal experiences of the tsunami and their perceptions on the actual role(s) of the mangroves during the disaster. Residents from Kapuhenwala and Wanduruppa articulated that mangroves were able to reduce the impacts of waves coming inland. Ninety-five percent (95%) of the households in Kapuhenwala recognized the shoreline protection function of the mangroves during the tsunami; whereas 85% of the households in Wanduruppa acknowledged the service. It was felt that patches of narrow stretch of mangroves in Wanduruppa were not able to offer as much protection compared to broader and continuous stretch of mangroves in Kapuhenwala. Infrastructure and property damages as well as casualties were perceived to have been more substantial if there were no mangroves surrounding the villages according to 85% of the residents in Wanduruppa and 98% of the villagers in Kapuhenwala.

The damages induced in both sites are presented in Figure 1. Structures such as residential buildings that were fully damaged, partially damaged (unusable) and partially damaged (usable) were all considered 'damaged'. Initial village survey reveals that 68% of the houses in Wanduruppa were damaged (De Silva, 2005). The results of the study survey, however, showed that 90% of the houses were damaged in Wanduruppa. It can be noted that in all categories of damages, the village of Wanduruppa registered greater percentage of damages compared to Kapuhenwala. This confirms the initial claims and immediate observations after the tsunami that areas with broader mangroves parallel to the coasts experienced relatively lesser damages than areas with more disturbed mangrove ecosystems. Interestingly, damages on boats and fishing gears were greater in Kapuhenwala despite the fact that broader mangroves characterize the site. This was the case since most fishermen in Kapuhenwala leave their boats and gears along the coast after their daily fishing activities.

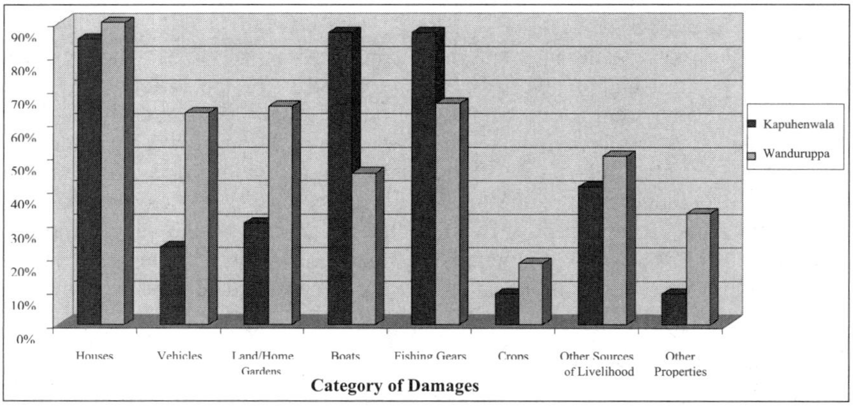

Fig. 1. Percentage of households that experienced damages on their properties and livelihood during the tsunami.

Previous studies, which investigated the shoreline protection function of mangroves, have assessed the function based on expected property and livelihood damages or expenditures on preventing such damage from taking place. This study, however, is using actual damage estimates on private properties, livelihood and public infrastructure from two sites severely affected by the tsunami. Costs of damages from the tsunami were estimated for properties, which include houses, vehicles, hotels and residential lands. The assessment of damage to houses was based on the proportion of houses damaged in both sites. The actual damage costs provided by the three registered hotel operators in Kapuhenwala were also used. The damage to livelihood includes destruction of fishing gears, boats and crops. The estimates were adjusted based on the number of households involved in fishing and farming. The costs of damage to community infrastructure include damage costs on public roads, schools, causeways and public parks. The information provided by the Grama Niladari (village leader) and Divisional Secretariat officers was used in the assessment. Table 2 summarizes the damages incurred in the study sites.

Table 2. Costs of tsunami damages to property and livelihood.

DAMAGES	KAPUHENWALA Total Weighted Costs (in SLR)	WANDURUPPA Total Weighted Costs (in SLR)	DAMAGE COSTS AVOIDED[B] (in SLR/household)
Property	SLR 6,666,750 (US$ 66,668)	SLR 116,136,395 (US$ 1,161,364)	SLR 191,716 (US$ 1,917)
Livelihood	SLR 8,676,092 (US$ 86,761)	SLR 19,952,467 (US$ 199,525)	SLR 19,748 (US$ 197)
Community Infrastructure	SLR 1,980,000 (US$ 19,800)	SLR 1,700,000 (US$ 17,000)	(SLR 490) (-US$ 5)
Others (*e.g.* Appliances, Clothes, etc.)	SLR 32,625 (US$ 326)	SLR 8,594 (US$ 86)	(SLR 42) (-US$ .50)
Total	SLR 17,355,467 (US$ 173,555)	SLR 137,797,456 (US$ 1,377,975)	SLR 210,932 (US$ 2,109)

Note:

a. Exchange rate as of May 2005: US$ 1 = SLR 100.

b. These estimates are costs that could have been avoided if the area in Wanduruppa has intact and broader stretch of mangroves. The estimates per household were determined by calculating the difference between the monetary values of the damages in both sites. The monetary values were adjusted based on the number of households affected in each site.

The severity of the damages in Wanduruppa and the relatively disturbed mangrove ecosystem in the village was expected to generate a higher damage cost value since it was postulated that the degree of destruction is greater in areas where mangrove ecosystem is more degraded. Using the damage costs avoided approach, the findings revealed that threatened mangroves reduce the protection afforded to inland properties, community infrastructures and livelihood by SLR 210,932 per household (US$ 2,109/household). It should be noted that the damage estimates upon which the costs were derived was influenced by the intensity of the tsunami and the number of waves that hit the coastal villages. Based on the anecdotal accounts, residents from Kapuhenwala observed two 14-ft waves during the tsunami, while in Wanduruppa an average of 3 waves with an approximate height of 13-ft were observed on December 26, 2004.

The damage cost avoided estimates as shown in Table 2 underscore the significant role of intact and healthy mangroves at the event of disasters such as tsunami. These estimates can be translated as the total monetary value of the shoreline protection function of intact mangrove ecosystems.

36.5. RESOURCE GOVERNANCE AND MANGROVE REHABILITATION

The recognition of the shoreline protection function offered by mangroves prior to the tsunami is highly remarkable among residents in Wanduruppa. The villagers, however, remain unresponsive to mangrove resource management efforts in the area as indicated by 79% of the households who are familiar but not involved with the different mangrove programs and projects in the village. This attitude towards mangroves is supported by 40% of the residents in Wanduruppa who do not think that the mangrove management options in the area are effectively implemented. The residents of Kapuhenwala, on the other hand, have better understanding and more involvement as suggested by 90% of the households with good recognition of the mangrove management activities taking place. Seventy-seven percent (77%) of the residents in Kapuhenwala agree that mangrove management options in the area are working efficiently for the protection and enhancement of the resource. Despite the perception that the current management options are working well, 91% of the residents in Kapuhenwala believe that there is a need to change mangrove management strategies in the area especially after the tsunami. Sixty-seven percent (67%) of the households in Wanduruppa, however, expressed indifference in changing the management options.

Notwithstanding the difference in opinions in both sites, there is a consensus among the residents that mangrove rehabilitation efforts should be an integral part of the tsunami reconstruction and rebuilding plans. Ninety-five percent (95%) of the households in both sites expressed their willingness to participate on ongoing and future mangrove rehabilitation efforts, mostly by volunteering their time and labor. Some residents, however, are willing to cooperate and participate but are not certain if what kind of participation they are willing to render. There is a high unwillingness to relocate among residents to relent to mangrove rehabilitation and restoration efforts. This was expressed by 73% of the households in Kapuhenwala and 54% in Wanduruppa.

36.6. CONCLUSIONS

The recent tsunami that hit the island state of Sri Lanka damaged many properties and infrastructure. Compounding the destruction was the devastation it brought to coastal vegetation upon which most coastal residents depend for food and livelihood. The protective role of mangrove ecosystems has been substantially and tragically demonstrated by the recent tsunami. Theoretical claims on the role of healthy and intact mangroves as important tidal barrier against tsunamis were corroborated by actual observations and valued in economic terms as equal to SLR 210,932/household (US$ 2,109/household).

The findings demonstrate that damage costs for sites with intact mangroves are significantly lower compared to areas with degraded ecosystems. The results of the damage costs avoided assessment for the shoreline protection function of mangroves provide a sound basis for the inclusion of mangrove rehabilitation efforts in the current post-tsunami reconstruction and rebuilding programs. The estimate suggests that good mangroves, if managed sustainably, represent a much higher overall livelihood value.

It is apparent that there is a high degree of dependence on mangroves among coastal households. The resource is important for both subsistence living and cash income sources of the residents. The absence of incentive tools to effectively manage the resource results in an alarming rate of decline. This calls for a strengthened institutional framework for livelihood and environmental management. Effective tools such as land-use planning, environmental impact statements and economic incentives to promote sustainable use of the mangrove resources should be implemented. Local level participatory disaster mitigation planning is also important to incorporate in future plans as expressed by the communities.

ACKNOWLEDGEMENTS

This study would have not been possible if not for the generous financial and professional support provided by the members of the IUCN Regional Ecosystems and Livelihood Groups and the IUCN Sri Lanka. The author is especially grateful to Lucy Emerton, Mikkel Kallesoe, Usman Iftikhar and Shamen Vidanage for initially developing the topic for the study. Equal gratitude is given to Kaushalya Wijayaweera, Kumi Ekarathne and Diana Alwis for the field assistance. The author is also grateful to Sandun Perera, Gayathri Sriskanthan and Mattias Rust for the assistance in the reconnaissance survey.

REFERENCES

Bunce, L., and Pomeroy, R., 2003. Socioeconomic Monitoring Guidelines for Coastal Managers in Southeast Asia: SocMon SEA. World Commission on Protected Areas and Global Coral Reef Monitoring Network/Australian Institute of Marine Science, Townsville, Australia.

De Silva, J. 2005. President, Pradeshiya Sabha of Ambalantota and GN of Wanduruppa North. Personal Communication. 20 April. Ambalantota, Hambantota, Sri Lanka.

Global Oceans. 2005. Sri Lanka – ICM Country Profile. http://www.globaloceans.org/country/srilanka/sri lanka.html [Accessed: 14 March 2005]

IEMA. 2005. IEMA News. http://www.iema.net/article.php?sid=3609 [Accessed: 14 March 2005]

IUCN Sri Lanka. 2004. The Environmental Profile of the Rekawa, Ussangoda and Kalametiya (RUK) Coastal Ecosystems. Coast Conservation Department, Colombo. VIII + 148pp.

Jeewananda, R., 2005. Grama Niladari of Medilla. Personal Communication. 5 April. Medilla, Tangalle, Sri Lanka.

Lal, P., 1990. Conservation or Conversion of Mangroves in Fiji: An Ecological Economic Analysis. Environment and Policy Institute East-West Center Occasional Paper No. 11. Honolulu: East-West Center.

Liyanage, S., 1995. Pilot Project on Participatory Management of Seguwanthive Mangrove Habitat in Puttalam District of Sri Lanka. Paper presented to the International Conference on Wetlands and Development, 8-14 October 1995, Selangor, Malaysia.

MAP. 2005. Mangrove Action Project. http://www.earthisland.org/map/tsunami.htm#1 [Accessed: 14 March 2005]

Mitsch, W., and Gosselink, J., 1993. Wetlands. Van Nostrand Reinhold: New York. 2nd Edition.

Pemadasa, M., 1984 Grasslands. In Ecology and Biogeography in Sri Lanka, edited by C.H. Fernando. The Hague, The Netherlands, W Junk Publishers, pp. 99–131.

Spaninks, F., and van Beukering, P., 1997. Economic Valuation of Mangroves: A Review of Techniques and Case Studies with Special Attention to the Production Function Approach. CREED Working Paper Series. IIED: London.

UNEP, 2001. Sri Lanka: State of the Environment 2001. UNEP-RRCAP: Pathumthani, Thailand.

Wijayaweera, K., 2005. IUCN Staff. Personal Communication. 5 April. Colombo, Sri Lanka.

WRI, 2003. Earth Trends: The Environmental Information Portal. http://earthtrends.wri.org [Accessed: 15 March 2005]

WRM, 1999. Local Fisherfolk Protect the Mangroves in Sri Lanka. http://www.wrm.org.uy/bulletin/20/SriLanka.html [Accessed: 15 March 2005]

37

DEVELOPMENT OF BIOLOGICALLY DOMINATED ENGINEERED STRUCTURES FOR EROSION REDUCTION AND ENVIRONMENTAL RESTORATION

Steven Hall[1], Robert Beine[2], Matthew Campbell[3], Tyler Ortego[1]
[1]Biological and Agricultural Engineering, Louisiana State University AgCenter, EB Doran, LSU, Baton Rouge, LA 70803, USA
[2]Department of Agricultural Chemistry, LSU AgCenter, AgChem Bldg, and LDAF, Baton Rouge, LA 70803, USA
[3]ORA Technologies, LLC, Baton Rouge, LA 70810, USA

ABSTRACT

Controlled density biologically dominated engineered reef structures were developed to reduce coastal erosion in areas like the Gulf Coast. Louisiana has been losing coastal wetlands at an alarming rate measured in square kilometers per year. While subsidence is a significant component of these losses, these losses have been exacerbated by coastal erosion due to wave and storm action on low lying and minimally protected coastal islands and wetland areas. Traditional rock breakwaters have been emplaced offshore and onshore in critical areas, but can be cost prohibitive. A second weakness of this technique in Louisiana is that the heavy rock breakwaters tend to sink through poorly compacted mud sediments. Development of relatively low density, biologically dominated erosion reduction structures can help address these issues. Aquatic sessile reef building animals such as oysters can build mass on structures which encourage such growth. In this process, waterborne material is converted to biomass and thus added to the coastal area, while simultaneously, the reduction of wave energy behind the structure tends to encourage deposition of suspended sediments. The relatively light weight of the structure reduces the rate at which sinking occurs, and the vertical growth of organisms may offset sinkage as well.

37.1. INTRODUCTION

Louisiana's coast has been experiencing rapid land loss. Louisiana is losing coastal wetlands at a rate of 65-90 square kilometers per year (Coast 2050). The average short term rate of shoreline erosion is 9.4m/yr, up from a long term average of 6.1m/yr (Penland, Conner et al., 2005). Rapid subsidence, eustatic sea level rise and drastic alteration to the natural Mississippi River building processes are the main reasons for this accelerated land loss. Storm events, such as the hurricanes of 2005, create high energy waves and washover events that breach beaches and barrier islands. These features recover during fair weather conditions, but not to their original conditions (Penland, Conner et al., 2005). Similar problems (though not necessarily of the same magnitude) occur along the coasts of Texas (Rodriguez, Fassell et al., 2001) and other gulf states.

37.1.1. Breakwaters

Breakwaters have been used to stop or reduce wave action reaching a shoreline, thus reducing or even reversing erosional losses. The reduction in wave energy slows littoral drift, induces sediment deposition and creates a shoreline bulge in the lee of the structure (CEM V-3). Breakwaters can be either emerged or submerged. Emerged breakwaters are designed to completely stop wave energy in the lee. Submerged structures are used where total wave energy reduction is either not practical (read too expensive) or not desirable (Ahrens, 1987). The USACE provides guidelines for predicting shoreline response based relationships between the structure and the beach (CEM VI-5). Other studies suggest methods for estimating the structure's effects on incoming waves (Ahrens, 1987).

37.1.2. The Oysterbreak

Campbell (2004) suggested a method to grow an oyster reef into the shape of a submerged breakwater. The method consists of placing a light weight support structure on which the oysters can grow. The structure has been designed with materials to stimulate oyster growth, and a shape to allow the oysters to dissipate wave energy. As the oysters grow, the structure will become completely dominated by the oyster reef. The structure has been termed the "oyster break". A physical model study (Campbell, 2004) concluded that the oysterbreak effectively reduced wave action in the lee of the structure. As growth occurred on the structures, performance approached that of Ahrens's predictive model (Ahrens, 1987). Finally, Campbell (2004) developed a model for predicting wave attenuation with time based on structure geometry and rates of oyster growth.

37.1.3. Oyster Biology

The eastern oyster, *Crassostrea virginica,* is a sessile invertebrate bivalve most commonly found in estuarine waters with a salinity above 5 to 10 parts per thousand (ppt), and is most abundant where salinity ranges between 10 and 20 ppt. Oysters thrive at salinities above 20 ppt. However predation is heaviest in these areas (Shumway, 1996). The oyster is very temperature tolerant and can be found from the Western Gulf of Mexico to Canada (Shumway, 1996).

37.1.4. Life Cycle

The oyster begins life as a free swimming, planktotrophic larva. Larvae are distributed primarily by water currents, but are capable of swimming vertically at speeds up to 2mm s^{-1}. Oysters at this stage suffer a high mortality rate due to predation, disease and possibly poor food supply. Larvae that survive the pelagic stage eventually obtain a benthic stage. Now called pediveliger larvae, these oysters crawl around on a hard substrate typically called "cultch". If conditions are right, pediveliger larvae will cement to the cultch and metamorphoses into juvenile oysters or "spat" (Kennedy, 1996) in a process known as "setting". Once set, an oyster can reach market size (>9 cm) in 2-5 years, depending on temperature (Shumway, 1996). It has been shown that some oysters in Louisiana can reach this size in less than 9 months (Menzel, 1951). In southern waters, oysters can grow throughout the year. In colder areas, growth is limited to 7 or 8 months of the year (Shumway, 1996).

The eastern oyster is a dioecious alternate hermaphrodite. It is sexually undifferentiated for part of the year, and then develops gonads as spawning season approaches. After spawning, the gonad reenters the undifferentiated phase. Spat typically develop gonads 8 to 12 weeks after settlement. Spat are usually male in the first year, though factors such as temperature, health and male/female ratio can influence this (Eble and Scro, 1996). The spawning period of the eastern oyster varies with location. In the Gulf of Mexico, spawning lasts from May until late October. In the Chesapeake Bay, spawning lasts from early June until October, and farther north, only lasts from June until August. The geographical and seasonal nature of oyster spawning indicates that temperature has a significant effect on spawning (Thompson, Newell et al., 1996). In fact, spawning can be triggered in hatcheries by increasing the temperature of the water (Castagna, Gibbons et al., 1996). After spawning, the oyster begins a period of glycogen storage, commonly called fattening. Depending on location, the oyster will either continue to build glycogen stores throughout the winter, or metabolize its glycogen stores when it is too cold to feed. Glycogen stores reach a peak immediately before gametogenesis, and are practically depleted by the end of the spawning season (Thompson, Newell et al., 1996).

37.1.5. Factors Affecting Setting

A variety of physical factors affect setting. Increased temperature has been shown to stimulate setting, but not salinity. Larvae tend to settle on undersurfaces, and have been shown to prefer shade in the lab. They have also been shown to prefer highly irregular or pitted surfaces. Oysters are very gregarious. It has been shown that they will almost inevitably select a surface near other spat, probably due to a waterborne pheromone (Kennedy, 1996).

Oyster larvae are also highly influenced by the presence of a biofilm (Anderson, 1995). It has been shown that certain bacteria in naturally occurring biofilms produce chemical messengers such as L-3-4-dihydroxyphenylalanine (L-DOPA) and melanin that stimulate setting (Kennedy, 1996). It has also been shown that ammonia (NH_3) induces settlement behavior (Kennedy, 1996).

37.1.6. Nutrition

Oysters feed primarily on phytoplankton, but can also feed on smaller zooplankton, detritus, bacteria and suspended particulates (Langdon and Newell, 1996). Oysters have also been shown to absorb nutrients directly from the water (Langdon and Newell, 1996). Studies have shown that increasing carbohydrates or supplementing diets with omega-3 fatty acids can increase oyster growth (Langdon and Newell, 1996). Furthermore, it has been shown that in an abundance of food, both adult and juvenile oysters will select more nutritious food particles, and reject others (Newell and Langdon, 1996).

37.1.7. Bio-enhanced Concrete

Concrete consists of aggregates such as sand or stone bound together in a cement matrix. The two main types of cements are hydraulic cements which harden due to hydration reactions and air-set cements which harden through drying. The most commonly used hydraulic cement is Portland cement. A mixture of only cement and water is known as neat cement. Neat cement combined with fine and coarse aggregates produces concrete. Coarse aggregates come from gravel, crushed stone, blast furnace slag and recycled concrete. Coarse aggregates are those larger than 5mm, fine aggregates are smaller than 5mm. Aggregates occupy most of the volume of the concrete. Typically, aggregates are stronger than the cement matrix (Kett, 2000). The strength of the concrete is based on the strength of the cement matrix, the strength of the aggregates and the strength of the matrix-aggregate interface. Besides water, cement and aggregates various chemicals known as admixtures can be added to the concrete. These include accelerators, air entraining agents, water reducers and plasticizers.

Anderson (Anderson, 1995) hypothesized that a pH increase due to calcium hydroxide in concrete would increase oyster settlement, and found an increase in oyster settlement beyond that of an increase in pH alone. It is proposed here that combining concrete with a nutrient source will serve to both attract more oysters and provide more food for their increased growth. The proposed nutrient source is cottonseed because of its relatively high levels of proteins and free fatty acids (Salunkhe, Chavani et al., 1992). It is generally accepted that the presence of any organic material is deleterious to the strength of concrete, however, Mannan and Ganapathy (2002) have recently made structurally sound concrete with oil palm shell as a coarse aggregate. It is also proposed that the addition of crushed oyster shell will stimulate more setting.

37.1.8. Mechanics of materials

If we are to design a structure with reduced strength concrete, it is necessary to know the relevant engineering properties of the concrete. The necessary properties are density, ultimate compressive strength, modulus of elasticity and modulus of rupture, which is a measure of flexural strength. Because the flexural strength of concrete is typically only 10 to 20% of the compressive strength, flexural strength tends to be a limiting factor in designing beams of un-reinforced concrete.

As long as sufficient moisture is present, a concrete's strength increases throughout its life, but can be assumed to reach maximum strength at 28 days (Washa, 1998).

37.2. OBJECTIVES

The main objectives of this experiment are 1) to determine the ability of various mixes of regular and biologically enhanced concrete to attract and grow oysters, 2) determine the material properties of said concrete and 3) select a material for oysterbreak construction.

37.3. MATERIALS AND METHODS

37.3.1. Field tests

Nine trapezoidal cross section concrete blocks were hung in the water column in Grand Isle, LA in June of 2005. Some blocks had no additives, some had cottonseed, and some had crushed oyster shell. Each block was divided into three segments. An arbitrary end of the block was selected and measurements taken at 13 cm, 36 cm and 56 cm from that end. An approximate cross sectional perimeter was determined by wrapping the block with a piece of string and measuring the length of string. The blocks were measured before placement, at 5.5 months, and at 7 months. Because the blocks were not all the same size, the original values were subtracted from the values at 5.5 months and 7 months to determine the changes in perimeter. The bases and heights of the trapezoidal cross section were also measured using calipers before placement and after seven months. An approximate cross sectional area of the concrete bars with oyster growth was calculated using the equation for the area of a trapezoid. The thickness of each bar, from upper surface to lower surface as oriented in water, was also measured. The original measurements were, again, subtracted from the measurements at 7 months to determine the changes in area and thickness.

The following parameters were tested: total increase in perimeter, increase in cross sectional area and increase in thickness. A Microsoft Excel spreadsheet was created to tabulate the values and compare groups of samples with one tailed students' T test. The following groups were considered: samples with cottonseed, without cottonseed, with oyster shell, without oyster shell, with cottonseed and oyster shell and with neither cottonseed nor oyster shell.

37.3.2. Compression Tests

An array of molds was created by hot gluing 2" (5.1 cm) nominal diameter by 10 cm long schedule 40 pvc pipes to a sheet of sheet steel. This was laid over a ¾" (1.9 cm) nominal thickness plywood board of the same dimensions and placed into a plastic bin with a sealable lid. Before assembly, the sheet metal and pvc molds were cleaned thoroughly and inspected for dirt and old concrete. Concrete samples were mixed to a proportion of 1 part Type III Portland cement to 3 parts sand to 2.5 parts gravel by mass. Cottonseed was mixed was added to 0%, 0.5%, 1% 5% and 10% m/m of cement. Three water contents were tested: 0.7, 0.6, 0.5 m/m of cement. Six to eight samples were made for each mix (2 tests, 7 days and 28 days, 3-4 samples per test). Samples were hand mixed by first placing the mortar, cottonseed and water into a clean five gallon bucket. The ingredients were stirred until fully incorporated. The sand was added and incorporated, then the gravel. The concrete mixture was then scooped and hand guided into the appropriate pvc molds. After filling the molds, the mixture was rodded with a half inch diameter steel rod to settle the concrete and remove air pockets. Additional concrete was then added to each mold to fill. After all of the mixes were made and all molds filled, a neat cement cap was made by mixing 3 parts cement to 1 part water (by mass), applying to the top of each sample, and tapping the pvc molds until the surface of the cap was smooth and level. The filled mold array was then covered with plastic wrap, covered with the plastic bin lid and allowed to cure in an air conditioned lab. After 24 hours, the pvc molds were pulled off of the sheet metal, and the concrete samples removed with a press. The caps were ground to remove irregularities. Each sample was then labeled, wrapped in plastic wrap and placed vertically in the air conditioned lab. At seven days, each sample was carefully measured and weighed.

Two strength tests were conducted for each mix, one at 7 days ± 6 hours and another at 28 days ± 20 hours. The tests were conducted with a Quotium Qtest load testing apparatus. The head speed was set to compress at a rate of.13 cm per minute. Because the software only allows a load of up to 44.5 KN, if the load exceeded 44.5 KN, the load and displacement were zeroed, and a second test run without moving the heads. The data sets are then combined in processing. The load vs. displacement curves were exported to tab delimited text files. Stress and strain were computed by dividing the load and displacement by the cross sectional area and length of the samples, respectively. Ultimate stress is defined as the peak stress sustained by the sample. After correcting the strain, a correction factor was applied according to ASTM C 39 Standard Test Method for Compressive

Strength of Cylindrical Concrete Specimens. Modulus of elasticity is defined as the slope of the linear portion of the stress-strain curve.

A second group of tests was performed on a series of proprietary low fines (little to no sand), cottonseed enhanced concrete mixes provided by ORA Technologies, LLC. Compression tests were performed as described above. Also, mixes were formed in the shape of long trapezoidal beams and subjected to center point load flexure tests. Testing setup and calculation of the modulus of rupture were done according to ASTM C 293-02 Standard Test Method for Flexural Strength of Concrete (Using Simple Beam With Center Point Loading).

37.4. RESULTS

37.4.1. Field Tests

Table 1 shows the mean increase in perimeter and the rate of increase for all samples at 5.5 months and 7 months. Growth is statistically higher at 7 months than at 5.5 months

Table 1. Increase in perimeter and rates of increase at 5.5 months and 7 months.

Date	Mean increase (cm)	P	Mean rate of increase (cm/month)	P
5.5 months	0.33	0.00856408	0.05	0.026653
7 months	1.96		0.28	

(p=.027). Table 2 compares mean percent increase in perimeter, total increase in perimeter and rates of increase between different groups. We can see that the addition of either cottonseed or oyster shell had a statistically significant effect ($p<.05$) on oyster growth. There appears to be a negative growth at 5.5 months for the oyster shell enhanced concrete compared to regular concrete. This is because on of the bars did not set properly and was slightly tapered at the ends. This bar was originally measured around the middle. This was not taken into account during the measurements. Comparison between samples with additives to those without show a significant increase in growth rates only at 7 months. This is again probably due to tapering in the bars containing oyster shell. Comparison between cottonseed enhanced concrete is significantly greater ($p<.003$) than in oyster shell enhanced concrete at 5.5 months, but at 7 months is less, though not significantly (p=.37). Again, errors due in measurement due to the tapering likely affected this. Similar results were found for area and thickness tests.

Table 2. Comparison of changes in perimeter for various groups.

Increase in perimeter (cm)					Increase in perimeter (cm)				
	5.5 months		7 months			5.5 months		7 months	
Bio additives vs no additives					cottonseed vs no bio-additive				
with	mean	0.64	mean	0.75	cs	mean	1.54	mean	2.15
	cm/month	0.12	cm/month	0.11		cm/month	0.28	cm/month	0.31
none	mean	0.92	mean	2.21	none	mean	0.64	mean	0.75
	cm/month	0.17	cm/month	0.32		cm/month	0.12	cm/month	0.11
	t test: p=	0.221	t test: p=	5.27E-05		t test: p=	0.012	t test: p=	0.001
cotton seed vs oyster shell					oyster shell vs no bio-additve				
cs:	mean	1.54	mean	2.15	oyster	mean	-0.32	mean	2.34
	cm/month	0.28	cm/month	0.12		cm/month	-0.06	cm/month	0.33
oyster:	mean	-0.32	mean	0.92	none	mean	0.64	mean	0.75
	cm/month	-0.06	cm/month	0.13		cm/month	0.12	cm/month	0.11
	t test: p=	0.002	t test: p=	0.370		t test: p=	0.034	t test: p=	0.008

37.4.2. Compression tests

Figure 1 is a graph showing the density of concrete as cottonseed is added. Figure 2 shows ultimate compressive strength as concrete is added. We can see that our concrete ranges from 2.25 grams per cubic centimeter (g/cc) for no cottonseed to 2.05 g/cc at high levels of cottonseed. Strength ranges from 27 megapascals (MPa) to 7 MPa for high levels of cottonseed. An exponential curve was fit to the curves in Figure 2 with a relatively good fit ($.75<R^2<.99$). Above about 10% cottonseed to cement, the concrete often failed to harden and, therefore, was not included in the data. Table 3 summarizes the compressive strength, flexural strength and the ratio between flexural strength and compressive strength for samples provided by ORA Technologies. Compressive strength ranged from about 22.5 MPa for OC4 to about 11.3 MPa for NF. Flexural strength ranged from about 6.2 MPa for OC4 to about 3.7 MPa for NF. The ratios between the two ranged from a remarkable high of.45 for OC1 to a more expected value of.26 for OC4 with a mean of about.33. OC2f showed a relatively high flexural strength to compressive strength ratio (.34).

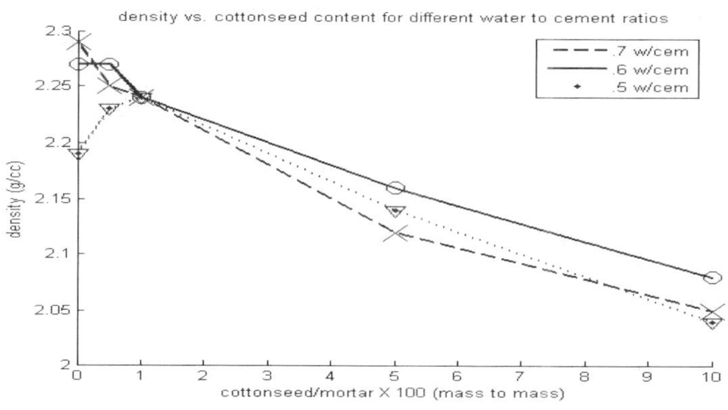

Fig. 1. Density vs. cottonseed content for different water to cement ratios.

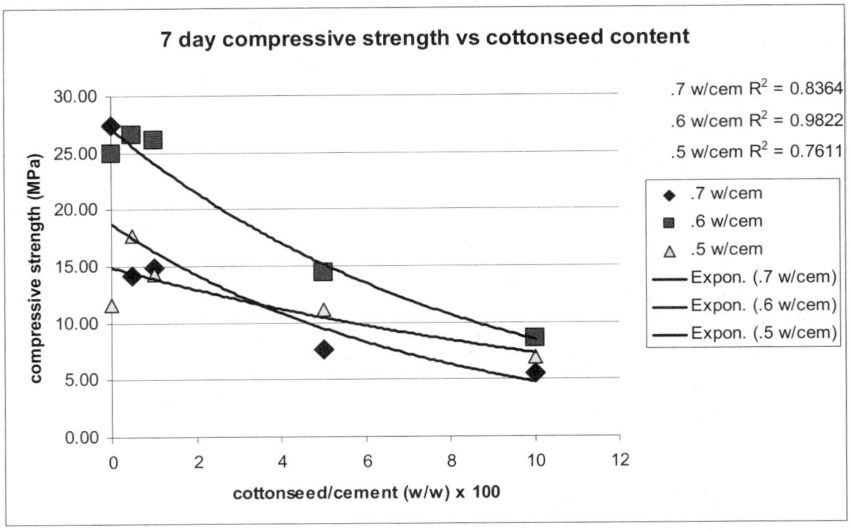

Fig. 2. Seven day compressive strength vs. cottonseed content for different water to cement ratios.

Table 3. Summary of compressive strength, modulus of rupture and the ratio between the two for samples provided by ORA Technologies LLC.

Sample		24 hr Compressive strength (MPa)	7 day Compressive strength (MPa)	Modulus of Rupture (MPa)	Ratio
OC1	**mean**	5.44	11.16	5.05	0.45
	stdev	0.30	2.70	1.30	
OC2	**mean**	11.51	22.54	6.63	0.29
	stdev	2.57	2.13	0.49	
OC2f	**mean**	8.53	18.35	6.19	0.34
	stdev	0.69	1.90	0.22	
OC4	**mean**		26.09	6.90	0.26
	stdev		1.36	0.99	
NF	**mean**		11.33	3.66	0.32
	stdev		1.28	0.45	

37.5. DISCUSSION

The increase in growth from 5.5 months to 7 months probably reflects seasonal variations in oyster growth rates, though it may also be due to errors associated with tapered concrete bars. It is evident from the 7 month measurements that the addition of either cottonseed or oyster shell will have a positive effect on oyster growth, though it is not clear which has the greater effect. Growth on bars with cottonseed is higher at 5.5 months ($p=.002$) but less at seven months, though not significantly so ($p=.37$). This is probably again due to the tapered bar or bars containing oyster shell. However, the fact that growth on the oyster shell enhanced concrete went from less than on cottonseed enhanced concrete to greater than on cottonseed enhanced concrete (despite tapered ends) might indicate higher growth rates.

The concrete followed a predictable pattern in terms of density and strength. Increasing cottonseed occupied space that would have other wise been occupied by cement or aggregates, thus reducing density and effective load bearing cross sectional area. Even with high levels of cottonseed, the concrete maintained a compressive strength of 7 MPa, which should allow for effective structure design. The concrete from ORA Technologies had a relatively high flexural strength to compressive strength ratio. This indicates that using a conservative estimation (20-25%) we can predict, for design purposes, the flexural strength of the concrete based on compression tests. Based on this estimation, we can predict flexural strengths of 5.5-6.9 MPa for regular concrete to 1.4-1.7 MPa for concrete with very high levels of cottonseed.

37.6. CONCLUSION

It is concluded that cottonseed enhanced concrete will produce a structurally sound concrete that enhances oyster growth. Furthermore, using the model developed by Campbell (Campbell, 2004), we can predict and compare the performance of oysterbreaks made form from standard concrete and cottonseed enhanced concrete. Further studies will be needed to determine the mechanical properties of oyster shell enhanced concrete. Studies to determine oyster growth on concrete enhanced with both oyster shell and cottonseed are also recommended. Finally, a study is being planned to determine oyster affinity for various mixes of concrete based on surface roughness.

ACKNOWLEDGEMENTS

The authors wish to thank faculty, staff and students of the department of Biological and Agricultural Engineering for their assistance, Wayfarer Nursery and Supply (http://www.wayfarernursery.com/) and the LSU AgCenter for funding. This manuscript has been approved by the LSU AgCenter as manuscript 06-22-0044.

REFERENCES

Ahrens, J. P., 1987, Technical Report CERC-87-17 Characteristics of Reef Breakwaters. W. E. S. Department of the Army, Corps of Engineers. Vicksburg, MS.

Anderson, M. J., 1995. A Chemical Cue Induces Settlement of Sydney Rock Oysters, Saccostrea commercialis, in the Laboratory and in the Field. Biology Bulletin 190: 350-358.

Campbell, M. D., 2004. Analysis and Evaluation of a Bioengineered Submerged Breakwater. Biological and Agricultural Engineering. Baton Rouge, LA, Louisiana State University. Master of Science in Biological and Agricultural Engineering.

Castagna, M., Gibbons, M. C., et al., 1996. Culture: Applications. in The Eastern Oyster Crassostrea virginica. V. S. Kennedy, R. I. E. Newell and A. F. Eble. College Park, MD, Maryland Sea Grant College, University of Maryland.

Eble, A. F., and Scro, R., 1996. General Anatomy. in The Eastern Oyster Crassostrea virginica. V. S. Kennedy, R. I. E. Newell and A. F. Eble. College Park, MD, Maryland Sea Grant College, University of Maryland.

Kennedy, V. S., 1996. Biology of Larvae and Spat. in The Eastern Oyster Crassostrea virginica. V. S. Kennedy, R. I. E. Newell and A. F. Eble. College Park, MD, Maryland Sea Grant College, University of Maryland.

Kett, I., 2000. Engineered Concrete: Mix Design and Test Methods. Boca Raton, FL, CRC Press, LLC.

Langdon, C. J., and Newell, R. I. E., 1996. Digestion and Nutrition in Larvae and Adults. The Eastern Oyster Crassostrea virginica. V. S. Kennedy, R. I. E. Newell and A. F. Eble. College Park, MD, Maryland Sea Grant College, University of Maryland.

Mannan, M. A., and Ganapathy, C., 2002. "Engineering Properties of Concrete with Oil Palm Shell as Coarse Aggregate." Construction and Building Materials 16: 29-34.

Menzel, R. W., 1951. Early Sexual Development and Growth of the American Oyster in Louisiana Waters. Science: 719-721.

Newell, R. I. E., and Langdon, C. J., 1996. Mechanisms and Physiology of Larval and Adult Feeding. in The Eastern Oyster Crassostrea virginica. V. S. Kennedy, R. I. E. Newell and A. F. Eble. College Park, MD, Maryland Sea Grant College, University of Maryland.

Penland, S., Conner, P. F., et al., 2005. Changes in Louisiana's Shoreline: 1855-2002. Journal of Coastal Research (Special Issue no. 44): 7-39.

Rodriguez, A. B., Fassell, M. L., et al., 2001. Variations in the shoreface progradation and ravinement along the Texas coast, Gulf of Mexico. Sedimentology 48: 837-853.

Salunkhe, D. K., Chavani, J. K., et al., 1992. Chapter 7, World Oilseeds, Chemistry, Technology and Utilization. Cotton. New York, Van Nostrand Reinhold: 251-2.

Shumway, S. E., 1996. Natural and Environmental Factors. in The Eastern Oyster Crassostrea virginica. V. S. Kennedy, R. I. E. Newell and A. F. Eble. College Park, MD, Maryland Sea Grant College, University of Maryland.

Thompson, R. J., Newell, R. I. E., et al., 1996. Reproductive Processes and Early Development. in The Eastern Oyster Crassostrea virginica. V. S. Kennedy, R. I. E. Newell and A. F. Eble. College Park, MD, Maryland Sea Grant College, University of Maryland.

Washa, G., W., 1998. Strength and Elastic Properties. Concrete Construction Handbook. J. A. Dobrowolski, McGraw-Hill.

Section 7:

HURRICANE KATRINA AND RITA IMPACTS

38

USING LANDSAT IMAGERY TO MODEL SALINITY CHANGE IN LAKE PONTCHARTRAIN DUE TO HURRICANES KATRINA AND RITA

Fugui Wang[1], Y. Jun Xu[1], and Andrea Bourgeois-Calvin[2]

[1] School of Renewable Natural Resources
Louisiana State University Agricultural Center, Baton Rouge, LA 70803, USA

[2] Lake Pontchartrain Basin Foundation
Three Lakeway, Ste. 2070, 3838 N. Causeway Blvd., Metairie, LA 70002, USA

ABSTRACT

Satellite remote sensing has been identified as a powerful tool for assessing water quality over large open water bodies. Satellite image technology offers the advantages of overcoming space and time constraints and providing relatively inexpensive, yet reliable information on surface water characteristics. This study carried out a multi-temporal analysis of satellite images to evaluate the immediate and after effects of hurricanes Katrina and Rita on salinity changes in a large estuarine system, Lake Pontchartrain, located in the Northern Gulf of Mexico, United States of America. Historical water quality data was gathered from eight monitoring locations in the lake. Surface water reflectance values were derived from four scenes of Landsat 5 TM and Landsat 7 ETM imagery acquired on 3 August 2001, 27 February 2002, 15 September 2002, and 22 August 2005. Two regression models were developed based on observed salinity measurements and the derived water reflectance. These models were then applied to estimate salinity changes from four image dates after Katrina and Rita. The results showed an immediate, powerful impact on the magnitude and distribution of the salinity levels across Lake Pontchartrain. On August 22, a week before Katrina, the estuary showed an average salinity of 4.8 ppt. On August 30, one day after Katrina made its landfall, the average salinity level of the estuary nearly doubled (8.1 ppt). The salinity levels decreased rapidly during the following week to an average of 6.1 ppt, however they increased once again directly after Rita made its landfall on 24 September 2005, and remained high (7.6 ppt) through the end of October 2005. Under normal circumstances, Lake Pontchartrain has typically shown a salinity distribution with relatively low levels (about 2.7 ppt) along the north shore, northeast shore and northwest shore, but higher levels (about 4.4 ppt) around the south and southwestern shores of the lake. The spatial distribution has mainly been influenced by freshwater river inflow from the drainage basins located north and west of the lake estuary. After Katrina and Rita, indications of saltwater intrusion were evidenced across the estuary, up to the river mouths. The results highlight the importance of further research into interrelationships between the hydrologic regimes of freshwater inflow and salinity intrusion in Lake Pontchartrain. Beyond assessment of Katrina and Rita effects on salinity, this study yielded insights into the applicability of combining multi-date satellite measurements for predicting large-scale water quality changes caused by extreme weather events.

38.1. INTRODUCTION

On the morning of August 29, 2005, the storm surge of Hurricane Katrina pushed elevated levels of saltwater from the Gulf of Mexico into Lake Pontchartrain, a 1,619 square kilometer oligohaline estuary located north of New Orleans, Louisiana, USA. The next day, water from the lake poured into the city through breaches in flood walls and levees. Only three weeks later, a second hurricane, Rita, compounded the region's devastation and introduced more saltwater into the lake. While there exists a need for a comprehensive assessment of the hurricanes' impacts on salinity levels in the large estuary, there are a limited number of field measurements available from the end of August to the end of October, 2005 due to damage or temporal shutdowns of water quality monitoring stations in the area.

Remote sensing with multispectral technology, such as the Thematic Mapper (TM), may offer an alternative means for assessing the surface water salinity changes over this period of time because of the distinctive spectral reflectance characteristics of water constituents (Fraser 1998; Ostlund et al., 2001; Dekker et al., 2002). The unique spectral reflectance of lake water primarily depends on components and concentrations of optically active substances such as phytoplankton, clays and silts from shoreline erosion, resuspended bottom sediments, and organic detritus from stream and/or wastewater discharges (Ostlund et al. 2001). In general, there is a direct correlation in reflectance of water with suspended sediment concentration, turbidity, and algae biomass (positive correlation) and an indirect correlation with salinity (negative correlation) (Bhargava and Mariam 1992; Wang 2004). Based on these relationships, a series of parameters, including salinity, can be assessed by employing remote sensing techniques. Lavery et al. (1993) developed multitemporal, empirical algorithms for predicting salinity in southwestern Australian estuaries with Landsat TM imagery. Braga et al. (1993) demonstrated that salinity at Guanabara Bay, Rio-De-Janeiro, Brazil was very well correlated with the simultaneous Landsat 5 TM data, especially during high tides. Baban (1993) showed a close relationship between salinity in the Norfolk Broads and the water reflectance values derived from Landsat TM imagery. Baban (1997) demonstrated that the spatial pattern of salinity in Breydon Water Estuary, derived from Landsat TM imagery, is thought to be due to the influence of suspended solids and turbidity on the optical characteristics of water in this environment. Dewidar and Khedr (2001) mapped the salinity distribution of Manzala lagoon with a regression equation using band 3 of the Landsat TM with a band 2 to band 4 ratio as predictors.

One of the essential steps in assessing water quality using remote sensing techniques is the development of numerical relationships between spectral radiance measured by sensors, and "ground truth" data of water quality parameters collected simultaneously in the field (Lathrop 1992; Ostlund et al. 2001). When Katrina made its landfall on 29 August 2005, the hurricane storm surge pushed the gulf waters into Lake Pontchartrain, which may have substantially affected the chemical, physical and biological properties of the lake water. Only three weeks after Katrina, Hurricane Rita came ashore, ravaging the remainder of the fragile Louisiana coastline. The devastating hurricanes made water sampling impossible. Therefore, this study was carried out to (1) develop algorithms for predicting surface water salinity distribution across Lake Pontchartrain after Katrina and Rita by using historical *in-situ* salinity measurements and multi-date satellite imagery, and (2) evaluate the effects of the two extreme weather events on spatiotemporal variations of salinity levels in the lake.

38.2. LAKE PONTCHARTRAIN BASIN ENVIRONMENT

Draining a total land area of 12,173 square kilometers, the Lake Pontchartrain Basin encompasses 16 parishes in southeast Louisiana and four counties in Mississippi. The surface water area comprises two large lakes, Pontchartrain to the east and Maurepas to the west, and several smaller sister lakes (Figure 1). The entire estuary extends about 82 km from west to east at the longest point, and 39 km from north to south at the widest point, with an average water depth of 3.7 - 4.3 m and a total surface water area of 1,619 square kilometers.

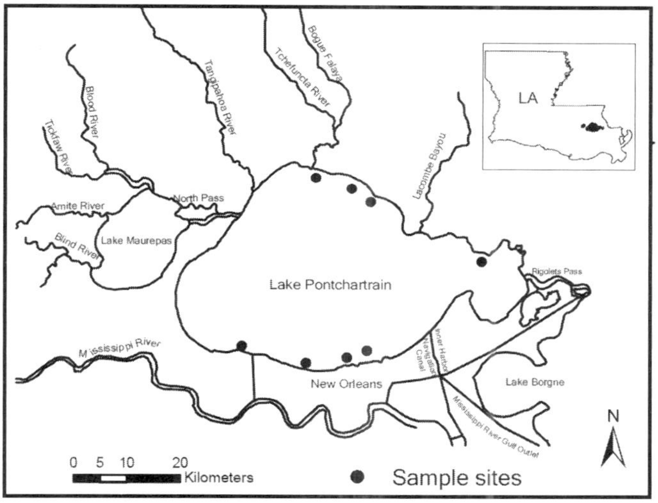

Fig. 1. The Lake Pontchartrain-Maurepas System and water quality sampling locations.

Connecting to the Gulf of Mexico via the Rigolets Strait, Chef Menteur Pass, and the Mississippi River Gulf Outlet (MRGO) - a 76-mile long man-made navigation channel - Lake Pontchartrain and its sister lakes are a precious natural resource containing numerous fish and wildlife habitat while providing a variety of benefits to the surrounding communities. The waterbodies and the surrounding wetlands provide protection and food for many estuarine animals such as ospreys, bald eagles, dolphins, manatees, sea turtles, sharks, crabs, crawfish, shrimp, oysters, least killifish, speckled trout, alligator, and red snapper among others. The silty bottom of the estuary supports many invertebrate animals, that in turn are a principal food source for fish.

Nearly 1.5 million people lived directly adjacent to Lake Ponchartrain before Katrina and Rita. Historically, this large estuary has been subject to numerous anthropogenic impacts including urban and agricultural runoff, shell dredging, shoreline alteration, and industrial discharge (Howarth et al., 1991; Fraincie and Poirrier, 1999; Penland et al., 2002). Over the past 15 years, however, pollution to the estuary has decreased largely and Lake Pontchartrain has been considered an environmental success story.

Lake Pontchartrain receives freshwater from several major rivers in the northern watersheds including (from east to west): Bayou Lacombe, Tchefuncte, Tangipahoa,

Tickfaw, and Amite Rivers, the latter 3 of which have been reported (Xu and Wu, in press) to contribute a significant portion of the total freshwater inflow. The freshwater flows directly over the saltwater entering the lake from the Gulf of Mexico through Rigolets Pass, Chef Menteur Pass and MRGO. Therefore, the magnitude, timing, and duration of the freshwater sources may greatly affect fluctuations in estuarine chemical and physical properties including, among others, salinity, temperature, turbidity, and concentrations of nutrients, sediment, and dissolved oxygen.

38.3. SALINITY MEASUREMENTS

Salinity was measured *in-situ* utilizing a YSI 85 meter at eight sample sites along the shore of Lake Pontchartrain. The measurements have been taken weekly since 2001, as part of the Lake Pontchartrain Basin Foundation's Basin-wide Water Quality Monitoring Program. All monitoring activities of the program are conducted under an EPA-approved Quality Assurance Project Plan (Bourgeois-Calvin, 2000-2005). These measurements were used to establish a pre-hurricane salinity baseline, and to develop regression models for predicting salinity changes following Katrina and Rita.

38.4. SATELLITE IMAGERY PROCESSING

The remotely sensed data used for this study included five scenes of Landsat 5 TM and three scenes of Landsat 7 ETM (path 22, row 39). The information on image sources and dates is summarized in Table 1.

Table 1. Sources, dates and uses (SMD = Salinity Model Development; SPI = Salinity Pattern Identification) of the Landsat 5 TM and 7 ETM imagery.

Sensor	ID	Image Date	Source	Use	Date of Salinity Measurements
ETM	7022039000121550	08/03/2001	Courtesy University of Maryland The Global Land Cover Facility (GLCF) (http://glcfapp.umiacs.umd.edu)	SMD	08/07/2001
ETM	7022039000205850	02/27/ 2002	Courtesy Louisiana State University	SMD	02/26/2002
TM	5022039000225810	09/15/2002	Ordered from USGS through http://glovis.usgs.gov/	SMD	09/17/2002
TM	5022039000523410	08/22/2005	Ordered from USGS through http://glovis.usgs.gov/	SMD	08/23/2005
ETM	71022039_03920050830	08/30/2005	Courtesy University of Louisiana at Lafayette	SPI	No measurements
TM	5022039000525010	09/07/2005	Courtesy Louisiana State University CADGIS Lab	SPI	No measurements
TM	5022039000528210	10/09/2005	Courtesy Louisiana State University CADGIS Lab	SPI	No measurements
TM	5022039000529810	10/25/2005	Courtesy Louisiana State University CADGIS Lab	SPI	No measurements

In order to extract the spectral reflectance of each monitoring site from the corresponding pixels on the multi-date images, the first step in image processing was to geometrically correct positional errors using the image acquired on August 22, 2005 as a reference map. Over 10 ground control points (GCP) were selected from the distorted images and the reference image. A first order polynomial equation and nearest neighbor resampling method were employed in the geometric image correction. The corrected imagery indicates a high location accuracy.

The second step in the image processing was to correct the imagery radiometrically, whereby the effects of atmosphere, sun illumination geometry and instrument calibration on reflectance leaving the water body were minimized and normalized. Many studies have been done on radiometric correction, and various correction algorithms have been established. This study adopted the image-based COST (cosine of the solar zenith angle) model proposed by Chavez (1996). On the basis of the COST model, a tool has been developed by RS/GIS Laboratory in Utah State University to create an ERDAS Imagine TM spatial model (.gmd format). The COST model incorporated in the tool is given below:

$$\rho_{bandN} = \frac{\pi((L_{bandN} * Gain_{bandN} + Bias_{bandN}) - (H_{bandN} * Gain_{bandN} + Bias_{bandN})) * D^2}{E_{bandN} * (COS((90-\theta) * \pi / 180))^2}$$

where

ρ_{bandN}	=	Reflectance for Band N
L_{bandN}	=	Digital Number for Band N
$Gain_{bandN}$	=	slope of response function (channel gain)
$Bias_{bandN}$	=	intercept of response function (channel offset)
H_{bandN}	=	Digital Number representing Dark Object for Band N
D	=	Normalized Earth-Sun Distance
E_{bandN}	=	Solar Irradiance for Band N

Utilizing this tool, digital numbers (DN) of each image were converted into reflectance leaving the water body, a ratio of the upwelling radiance over downwelling irradiance.

38.5. DEVELOPMENT OF SALINITY REFLECTANCE MODELS

Water reflectance values from the monitoring sites were extracted from the corresponding pixels of multi-date Landsat imagery, band 1 through band 7 (except for band 6). Ratios of band 3 to band 2, band 4 to band 3, and band 2 to band 1 were also extracted from a 3 by 3 window with a cell size of 28.5m x 28.5m. Stepwise multivariate regression was used to develop the relationship between the reflectance values and the field salinity measurements. It was assumed that there was no significant difference in reflectance between the dates when the Landsat images were acquired and the dates when field water salinity samples were collected. To minimize errors, salinities measured on the dates that were closest to the image acquisition date were used (Table 1).

Two regression models were developed individually for the Landsat 5 TM sensor and the Landsat 7 ETM sensor. Stepwise multivariate regression was carried out to reduce the number of model predictors. The regression coefficient and a 0.05 level of significance were used to justify the goodness of fit in the regression models. All statistical analyses were performed using the SAS software program (SAS Institute Inc., 1996). All satellite imagery and map processing were carried out in ERDAS Imagine 7.8 (Leica) and ArcGIS 9.0 (ESRI).

38.6. SALINITY PREDICTION AND SPATIAL ANALYSIS

Salinity levels across Lake Pontchartrain for two post-Katrina dates, August 30 and September 7, and two post-Rita dates, October 9 and October 25, were estimated for each cell using the developed algorithms. To quantify the changes pre- and post-hurricanes, estimates from 10 points on a latitudinal transect (30.18° N) and 8 points on a longitudinal transect (90.16° W) over Lake Pontchartrain were extracted as shown in Figure 2. Salinity levels of four pre-Katrina and four post-Katrina images were averaged to quantify the hurricane effect.

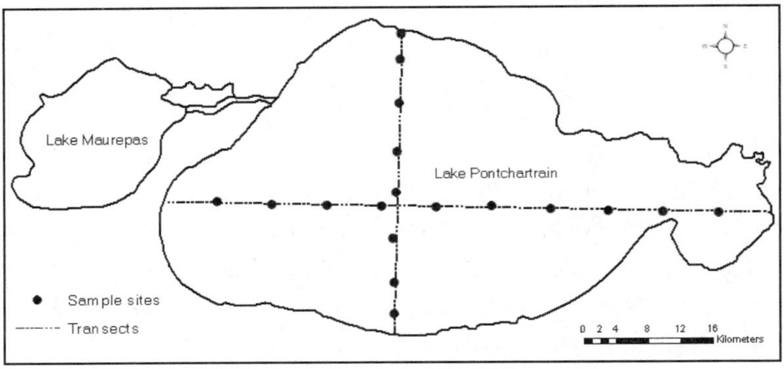

Fig. 2. Transects and Locations of sampling predicted salinity.

Each point represents the average salinity of a 200m x 200m area. The average salinity levels at all 18 points, pre- and post-hurricane, were compared to determine the level of salinity change across the Lake Pontchartrain estuary, both from east to west and north to south.

38.7. RESULTS AND DISCUSSION

38.7.1. Relationships between Multispectral Reflectance and Salinity Levels

Results from the multivariate regression analysis showed that for both Landsat 5 TM and Landsat 7 ETM, the reflectance values from two bands explained over 50% of the variation in salinity levels over time (Table 2). Landsat 5 TM band 3 and band 4 showed the closest relationships with the salinity levels on 15 September 2002 and 22 August 2005. The two bands accounted for 24.3% (band 3) and 28.4% (band 4) of the variation in salinity changes. Landsat 7 ETM bands 2 and 3 also showed relatively similar relationships with salinity in the lake on 3 August 2001 and 27 February 2002. Together, bands 2 and 3 explained 61% of the variations in salinity, with band 2 contributing 35.6% alone.

Table 2. Multivariate Regression Analysis of Salinity and Spectral Reflectance

Sensor	Regression Equations	R^2	$P <$
TM	Salinity=3.26+172.56*band 3-157.80*band 4	0.5271	0.0077
ETM	Salinity=1.60+220.01*band 2 -114.53* band 3	0.6122	0.0021

The relatively low regression coefficients were mainly caused by three factors. First, the salinity measurements used in developing relationships of reflectance and salinity were collected near the shore of the lake. The derived reflectance values from some of those measurement locations may not have represented the actual reflectance leaving water bodies well, because of the shallow water depth. Secondly, the difference in dates and times between the satellite imagery acquisition and the field measurements may have affected the determined relationship between reflectance and salinity. Third, the sampling locations near the shore may not have accurately represented the salinity variations in the center portion of the lake.

Dewidar and Khedr (2001) found a correlation coefficient of 0.47 and of 0.50 between salinity and reflectance of Landsat 5 TM band 2 and band 3, respectively, for a large shallow brackish lake in Egypt. In another study on water quality of Guanabara Bay in Brazil with Landsat 5 TM data, Braga et al. (1993) found a very poor correlation between salinity and reflectance. In contrast, Lavery and others (1993) were able to develop an algorithm using Landsat 5 TM band 4 to estimate salinity levels in southwestern Australian estuaries with a relatively strong correlation (R^2=0.75). The difference in salinity-reflectance relationships found in those studies may have been a result of the three factors discussed previously. Compared with these findings, the regression models developed in this study achieved satisfactory goodness of fit to the multi-date sets from the shallow Lake Pontchartrain estuary.

38.8. PRE-KATRINA SALINITY BASELINE

Salinity data collected from the eight monitoring locations showed that prior to Hurricane Katrina, from 2001 to 2004, salinity levels in Lake Pontchartrain averaged 3 ppt, varying from 0 to 11 ppt. These field measurements generally showed an increasing trend of salinity from the lake's north shore towards the south shore.

Spatially detailed maps of salinity for the pre-Katrina period were generated using the developed regression models. The maps (Figures 3 through 10) clearly exhibited salinity variability over time and across the entire Lake Pontchartrain estuary. Before Hurricane Katrina, the spatial patterns on 3 August 2001 and 15 September 2002 were very similar with low salinity along the northeast shore and high salinity along the southwest shore (Figures 3 and 4). Salinities on 22 August 2005 and 27 February 2002 were low in the western part and high in the eastern part of the lake. The extent of low salinity regions from the west to the center divisions of the lake pushed the high salinity region to the east (22 August 2005) and the southeast (27 February 2002) portions of the lake (Figures 5 and 6). On these two dates salinity levels were relatively low across most parts of the lake.

Under normal circumstances, salinity of Lake Pontchartrain is influenced by freshwater discharge from the Upper Lake Pontchartrain Basins, saltwater intrusion by the Rigolets and Chef Menteur Passes, as well as MRGO, and water currents caused by winds and tidal levels. On 3 August 2001 and 15 September 2002, the extent of saltwater intrusion through

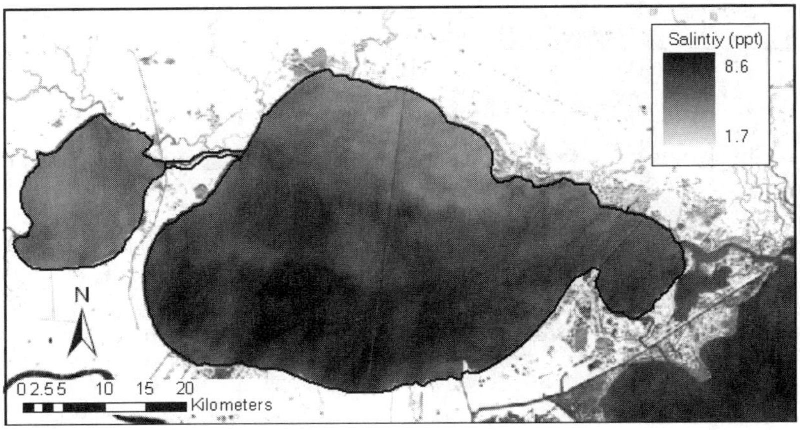

Fig. 3. Salinity distribution on August 3, 2001.

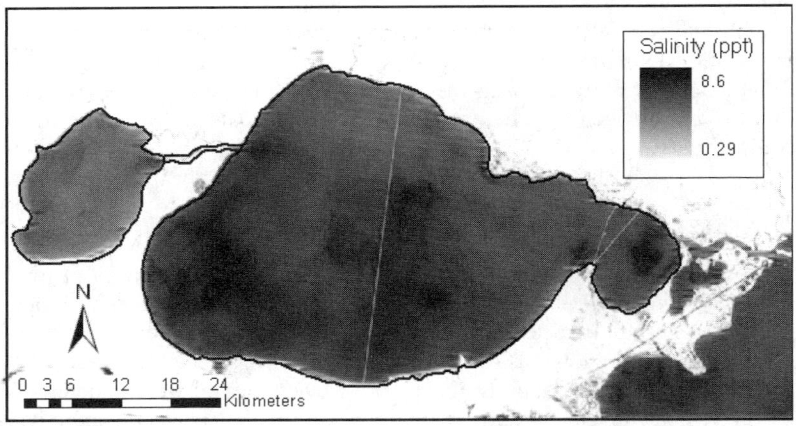

Fig. 4. Salinity distribution on September 15, 2002.

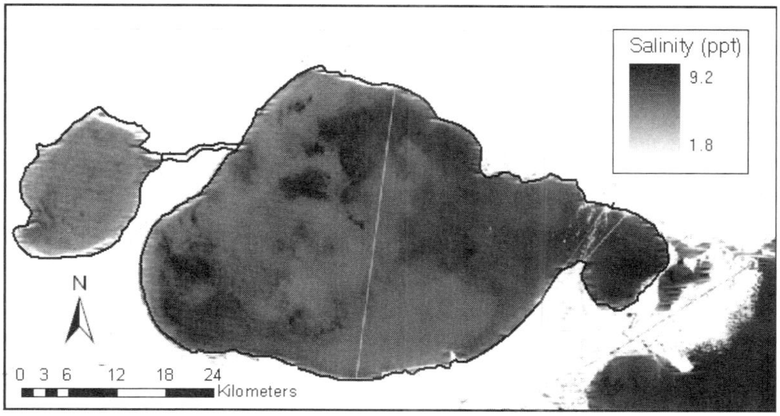

Fig. 5. Salinity distribution on August 22, 2005.

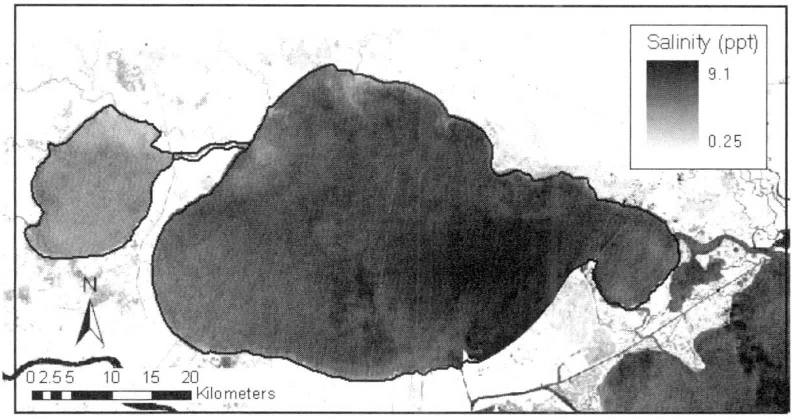

Fig. 6. Salinity distribution on February 27, 2002.

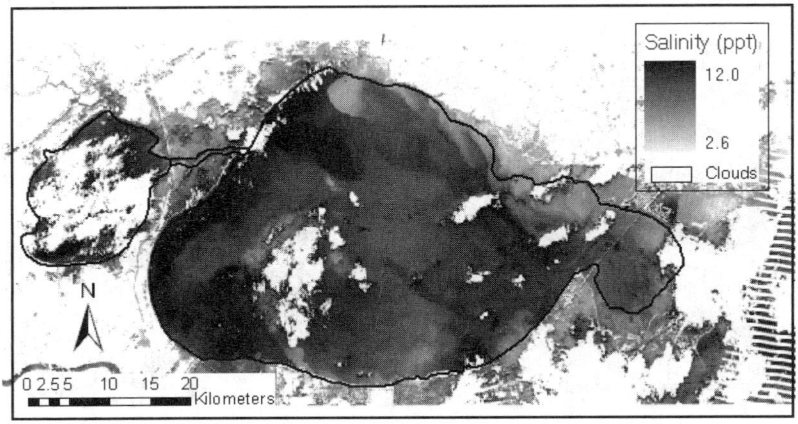

Fig. 7. Salinity distribution on August 30, 2005.

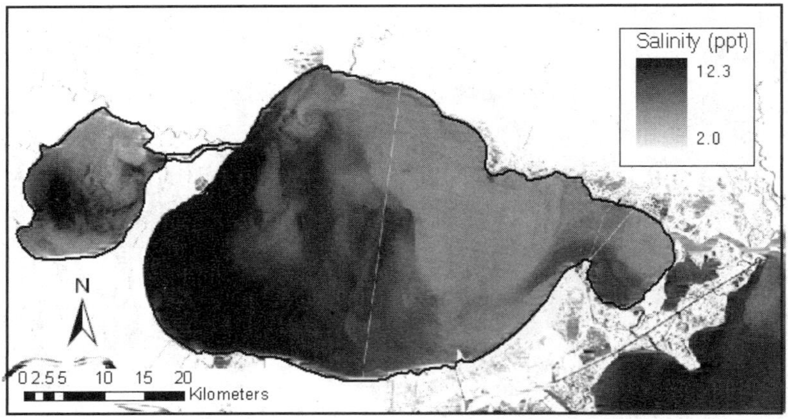

Fig. 8. Salinity distribution on September 7, 2005.

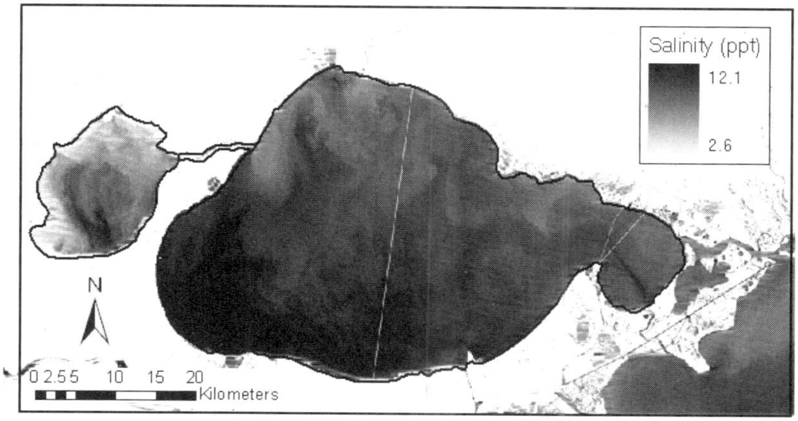

Fig. 9. Salinity distribution on October 9, 2005

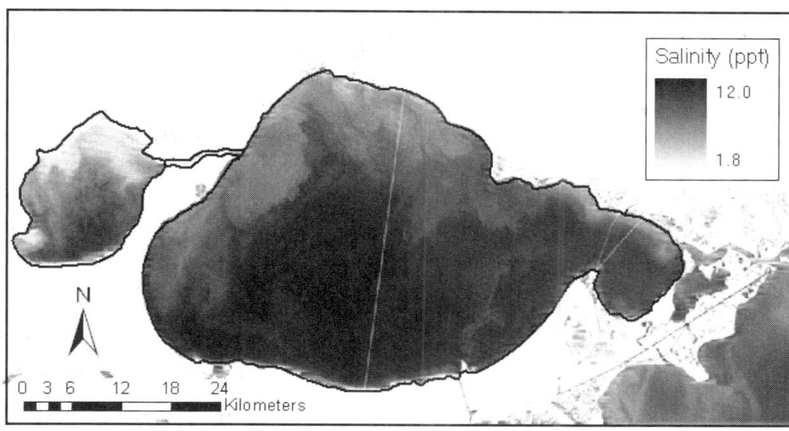

Fig. 10. Salinity distribution on October 25, 2005.

Rigolets and Chef Menteur Passes may have been low because of a lack in wind speed and the direction of tidal levels, when compared to 22 August 2005. Therefore, low salinity regions appeared in the eastern portion of the lake on 3 August 2001 and 15 September 2002. On 27 February 2002, during a typical wet season for the region, freshwater discharge from the Upper Lake Pontchartrain Basins was high, which may have contributed to the overall low salinity distribution across the lake. The only exception was the lake's southeast portion which showed high salinity levels.

38.9. CHANGES IN SALINITY DISTRIBUTION AFTER KATRINA AND RITA

The two developed regression models were applied to four images acquired after Hurricane Katrina (Table 1). The modeled results provided insights into immediate effects of the hurricane on salinity distribution across Lake Pontchartrain.

These results showed that on August 30, 2005, one day after Katrina, the salinity levels were high across the lake except for one small area at the mouth of the Tchefuncte River where lower salinity (~2.6 ppt) appeared to be maintained due to freshwater inflow. From this area, throughout the center of the lake and down to the south shore, there was a region of moderate salinity (~7.5 ppt). Along the west shore and at the southwest corner of the lake, the salinity is very high (~10.0 ppt). The high salinity regions apparently formed a narrow strip along the west shore of Lake Pontchartrain. High salinity also occurred at the lower northwest corner of the lake, where salinity was supposed to be low because of the freshwater discharge from the Tangipahoa River (Figure 7). This spatial distribution was significantly different from that prior to the hurricane. The storm surge, resulting from the hurricane, apparently constricted the amount of freshwater entering into the lake. Overall, the entire lake area appeared to be well mixed with the saltwater from the Gulf of Mexico.

The modeling results for September 7, a week after Katrina, showed a retreat of the saltwater intrusion back towards the south shoreline (Figure 8). However, after Rita's landfall on September 24, salinity encroached northwards again (Figure 9) and remained high through the end of October (Figure 10).

Based on the salinity levels modeled for the eight dates, averages before and after Katrina were quantified (Table 2). The results showed a strong hurricane impact on salinity across the entire Lake Pontchartrain estuary. Before Katrina, the lake had an average salinity between 4.4 and 5.5 ppt, with a maximum range of 8.6 - 9.2 ppt and a minimum range of 0.3 - 1.8 ppt. After Katrina, the average salinity increased to 6.1 - 8.1 ppt, with a maximum range of 12.0 - 12.3 ppt and a minimum range of 1.8 - 2.6 ppt.

Table 2. Modeled salinity levels in Lake Pontchartrain.

Date	Maximum	Minimum	Average in the Entire Lake	Standard Deviation	Skewness
August 3, 2001	8.6	1.7	5.4	1.5	-0.4
September 15, 2002	8.6	0.3	5.5	1.3	-0.4
August 22, 2005	9.2	1.8	4.8	1.0	-0.4
February 27, 2002	9.1	0.3	4.4	1.4	-0.1
August 30, 2005	12.0	2.6	8.1	2.1	-1.7
September 7, 2005	12.3	2.0	6.1	2.2	0.6
October 9, 2005	12.1	2.6	7.1	1.8	0.4
October 25, 2005	12.0	1.8	7.6	2.4	0.1

Salinity levels of four pre-Katrina and four post-Katrina (Table 1) images, taken from the 18 locations along the longitudinal and latitudinal transects (Figure 2), were averaged. The results showed a clear increase of salinity levels after the hurricane. The magnitude of the increase was more distinctive on the north-south transect than on the west-east transect (Figures 11 and 12). The salinity levels at two north shore locations appeared to be little

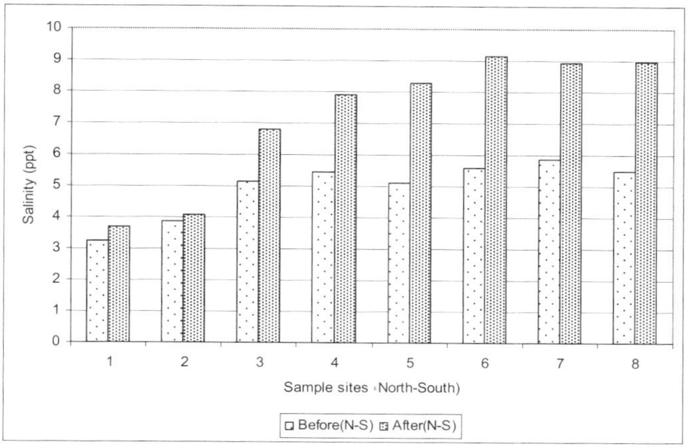

Fig. 11. Salinity changes along the north-south transect.

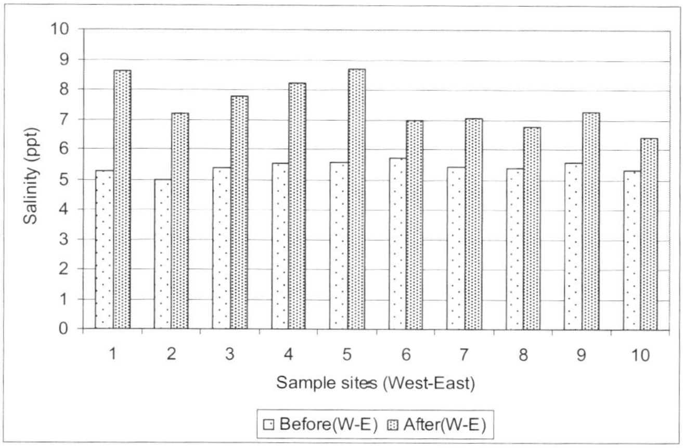

Fig. 12. Salinity changes along the west-east transect.

affected by the hurricanes. However, a drastic change in salinity could be seen at a third location off the north shore (about 10 kilometers; Figure 11). Along the latitudinal transect, the hurricane caused a higher increase of salinity in the western part of the lake (Figure 12).

Some researchers (Georgiou, 2001; Tate et al., 2002) conducted modeling studies on salinity intrusion to Lake Pontchartrain from the Mississippi River Gulf Outlet (MRGO). The MRGO was completed in 1968 to connect the Gulf of Mexico directly to the port of New Orleans' inner harbor (Figure 1). The studies suggested a profound saltwater flux from the MRGO which contributed to increasing the salinity concentration of the southern portion of Lake Pontchartrain. The higher salinity near the Inner Harbor Navigation Canal of New Orleans found in this study not only supports the previous findings, but also indicates increased impacts of the MRGO during the storm surges.

38.10. CONCLUSIONS

This study provides the first simulation of Hurricane Katrina and Rita impacts on salinity levels in Lake Pontchartrain. It is also the first attempt to use satellite imagery for predicting water quality parameters of this large lake estuary in the northern Gulf of Mexico. The results not only revealed the powerful, immediate effects of the two hurricanes on saltwater intrusion, but also demonstrated the great potential of satellite measurements for time-sensitive assessment of environmental conditions during extreme weather events.

REFERENCES

Baban, S. M. J. 1993. Detecting water-quality parameters in the Norfolk Broads, UK, using Landsat imagery. International Journal of Remote Sensing 14:1247-1267.

Baban, S. M. J. 1997. Environmental monitoring of estuaries estimating and mapping various environmental indicators in Breydon Water Estuary, UK, using Landsat TM imagery. Estuarine Coastal and Shelf Science 44:589-598.

Bhargava, D. S., and D. W. Mariam. 1992. Cumulative effects of salinity and sediment concentration on reflectance measurements. International Journal of Remote Sensing 13:2151-2159.

Braga, C. Z. F., A. W. Setzer, and L. D. Delacerda. 1993. Water-quality assessment with simultaneous Landsat-5 TM data at Guanabara Bay, Rio-De-Janeiro, Brazil. Remote Sensing of Environment 45:95-106.

Chavez, P. S. 1996. Image-based atmospheric corrections revisited and improved. Photogrammetric Engineering and Remote Sensing 62:1025-1036.

Dekker, A. G., R. J. Vos, and S. W. M. Peters. 2002. Analytical algorithms for lake water TSM estimation for retrospective analyses of TM and SPOT sensor data. International Journal of Remote Sensing 23:15-35.

Dewidar, K., and A. Khedr. 2001. Water quality assessment with simultaneous Landsat-5 TM at Manzala Lagoon, Egypt. Hydrobiologia 457:49-58.

Fraser, R. N. 1998. Multispectral remote sensing of turbidity among Nebraska Sand Hills lakes. International Journal of Remote Sensing 19:3011-3016.

Georgiou, I. 2001. Plot of the Month-Lake Pontchartrain Estuarine System New Orleans, LA. Contours.

Lathrop, R. G. 1992. Landsat Thematic Mapper monitoring of turbid inland water-quality. Photogrammetric Engineering and Remote Sensing 58:465-470.

Lavery, P., C. Pattiaratchi, A. Wyllie, and P. Hick. 1993. Water-quality monitoring in estuarine waters using the Landsat Thematic Mapper. Remote Sensing of Environment 46:268-280.

Ostlund, C., P. Flink, N. Strombeck, D. Pierson, and T. Lindell. 2001. Mapping of the water quality of Lake Erken, Sweden, from Imaging Spectrometry and Landsat Thematic Mapper. Science of the Total Environment 268:139-154.

Tate, J. N., Carrillo, A. R., Berger, R. C., and Thibodeaux, B. J., 2002. Salinity changes in Pontchartrain Basin Estuary, Louisiana, resulting from Mississippi River-Gulf Outlet Partial Closure Plans with width reduction - Final Report. Army Engineer Research and Development Center, Coastal and Hydraulics Lab., Vicksburg, MS USA. 42p.

Wang, F. 2004. Applications of hyperspectral reflectance and Landsat-5 imagery in assessing and mapping water quality in Reelfoot Lake, Tennessee. Thesis.

Xu, Y.J., and Wu, K., in press. Seasonality and interannual variability of freshwater inflow to a large oligohaline estuary in the Northern Gulf of Mexico. Estuarine, Coastal and Shelf Science.

EXPERIMENTAL STORM SURGE SIMULATIONS FOR HURRICANE KATRINA

Hassan S. Mashriqui, G. Paul Kemp, and Ivor van Heerden
LSU Hurricane Center, Suite 3221 CEBA Bldg.
Louisiana State University, Baton Rouge, LA 70803, USA

Brian D. Ropers-Huilman
Center for Computation & Technology
Louisiana State University, Baton Rouge, LA 70803, USA

Emily Hyfield
Coastal Ecology Institute
Louisiana State University, Baton Rouge, LA 70803, USA

Young Yang, Kate Streva, and Ahmet Binselam
LSU Hurricane Center, Suite 3221 CEBA Bldg.
Louisiana State University, Baton Rouge, LA 70803, USA

ABSTRACT

Louisiana is vulnerable to hurricanes and major flooding as 70 percent of the population lives on 36 percent of land at or below sea level. That includes cities as large as New Orleans. Operational meteorological forecasts for hurricanes have been available for decades and are now quite reliable 24 hours prior to landfall. Comparable oceanographic predictions of storm surge elevation, water velocity and levee overtopping are only now becoming feasible. Early, detailed and reliable forecasts of storm effects on population centers, levees and evacuation routes can save lives, and did, we believe, during Hurricanes Katrina and Rita. The ADvanced CIRCulation model (ADCIRC) began to be used experimentally to forecast surge for approaching hurricanes beginning in 2002. Since then, a research team at the Louisiana State University Center for the Study of Public Health Impacts of Hurricanes and the Hurricane Research Center has submitted 33 forecasts to emergency management personnel for 10 hurricanes making landfall in the Gulf of Mexico. The operational potential was well demonstrated during the approach of Hurricane Katrina when 7 forecasts were issued prior to the landfall on August 29, 2005. Very good agreement between simulated surge elevation and high water marks was achieved in ADCIRC runs based on National Hurricane Center advisories issued up to 2 days before landfall. Here we focus on other information produced by ADCIRC, particularly (surge) current velocities, that can be useful in predicting levee overtopping and failure, in guiding pre-landfall flood-fighting efforts, and in staging post-landfall breach closure equipment and supplies.

39.1. INTRODUCTION

Low lying coastal Louisiana is susceptible to damage caused by surge, currents and waves as tropical cyclones make landfall. It has been known for at least a decade that surge associated with a Saffir-Simpson Category 3 hurricane, approaching at a range of speeds from a number of directions could overtop New Orleans flood defenses on both the east and west banks of the Mississippi (IEM Inc., 2004; Brouwer, 2003). Hurricane Katrina made landfall in Louisiana on August 29, 2005, as a fast-moving Category 3 storm, flooded 80 percent of the City of New Orleans before it smashed into Mississippi and Alabama (Knabb et al. 2005). A fully parallel, 2-dimensional (depth averaged) ADvanced CIRCulation model (ADCIRC) run at Louisiana State University (LSU) was used to provide operational storm surge forecasting support over the last 51 hours prior to landfall (Kemp et al. 2006). ADCIRC predicts water surface elevation changes and current velocities generated by an approaching hurricane (Westerink et al. 1994; Luettich and Westerink, 1995). The S08 version of ADCIRC grid used by LSU has been run experimentally since 2002 for storm operations support in the Gulf of Mexico by the LSU Center for the Study of Public Health Impacts of Hurricanes and Hurricane Research Center (Kemp et al. 2006).

LSU modeled 33 National Hurricane Center (NHC) advisory forecasts associated with 10 tropical cyclones that entered the Gulf of Mexico during the 2004 and 2005 hurricane seasons, in addition to the mythical Hurricane Pam. Hurricane Pam was a Category 3 storm created by the National Hurricane Center (NHC) for a preparedness exercise in July 2004 that was forecast to cause flooding eerily similar to Katrina one year later (Collins, 2006; IEM Inc., 2004). This information was posted for emergency managers and the public as each run was completed (www.hurricane.lsu.edu/floodprediction). We report on the potential for multiple, near real-time forecast simulations to improve preparedness and public safety response. ADCIRC predicted storm surge elevations during hurricane Katrina to within 15 percent of observed values more than 24 hours prior to landfall (Kemp et al. 2006). Here, we focus on the spatial pattern of water velocity, in addition to elevation, in an effort to better explain the dynamics that led to widespread levee failure on the east side of the city.

39.2. METHODS

The LSU team used the ADCIRC S08 bathymetry grid developed by Westerink et al. (2004) and Feyen et al. (2002). It consists of 314,442 finite element nodes in a domain large enough to reduce boundary effects that includes the western half of the Atlantic Ocean, the Caribbean Sea and the Gulf of Mexico (Figure 1). Computational mesh elements were large in deep water and very small and refined in the area around Louisiana. About 85 percent of these nodes are concentrated within Louisiana's wetlands and estuaries to reproduce the complex geometry of natural and manmade channels and levee systems (Figure 1). Dynamic hurricane wind stress and pressure fields are generated with a U.S. Army Corps of Engineers (USACE) version of the Planetary Boundary Layer model (Thompson and Cardone, 1996) using interpolated track and storm information derived from advisories issued by the NHC.

For hurricane Katrina simulations, the LSU Center for Computation & Technology provided up to 254 computer processors on a Beowulf cluster for each run. This reduced

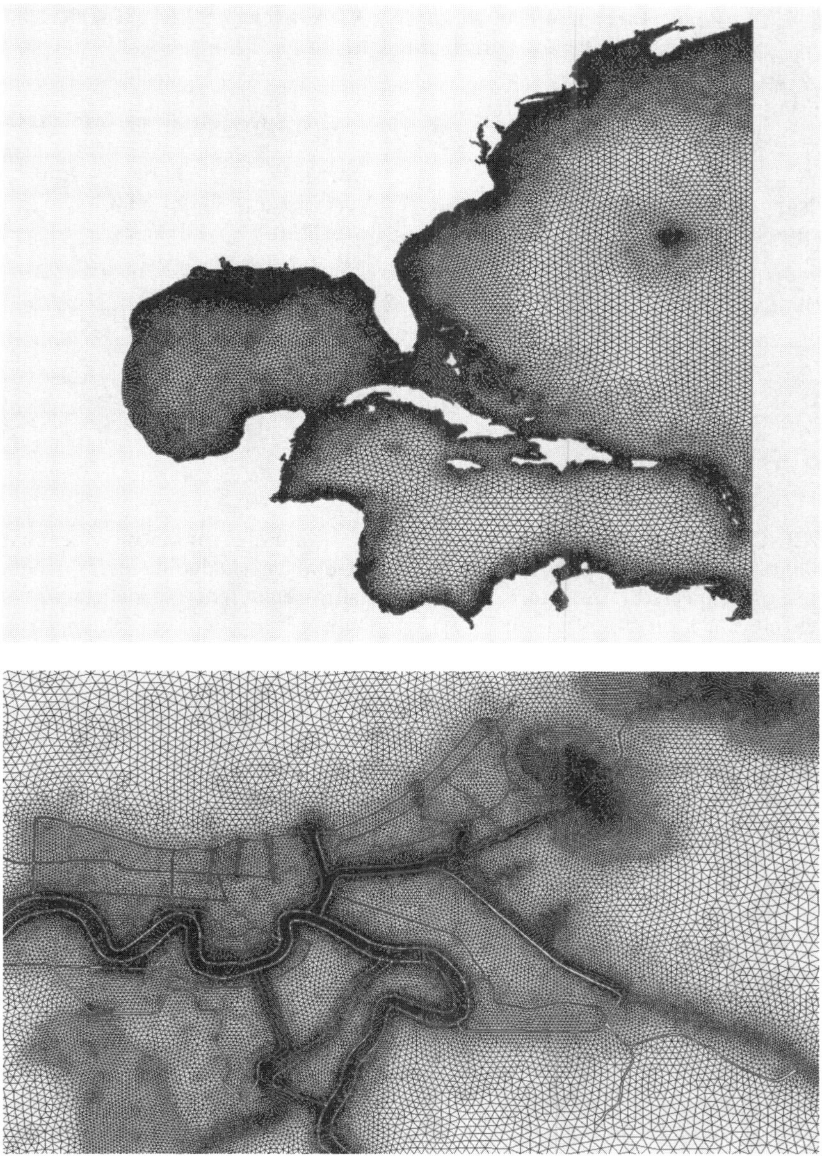

Fig. 1. ADCIRC "S08" computational model domain (top) and detailed mesh outlines around New Orleans area (for S08 grid reference see Feyen et al. 2002).

ADCIRC simulation time for a typical run to 2.5 hours for a total 8-days of storm simulation. The pre- and post-processing steps required another 2 to 3 hours. Most runs required 6 to 8 hours after the NHC advisory to produce a surge map (Table 1).

Table 1. National Hurricane Center (NHC) Advisories and LSU Hurricane Center Surge Forecasting for Hurricane Katrina.

National Hurricane Center (NHC) Advisory Information			LSU Surge Analysis Information		
Advisory Number	**NHC Advisory Date:Time (UTC)**	**Time to Landfall (h)**	**Surge Forecast Date:Time (UTC)**	**Time From NHC Advisory (h)**	**New Orleans Flooding Predicted**
16	8/27/05 0900	51	8/27/05 1930	10.5	NO
17	8/27/05 1500	45	8/27/05 2000	5	NO
18	8/27/05 2100	39	8/27/05 2700	6	YES
22	8/28/05 1200	24	8/28/05 2000	8	YES
25	8/29/05 0300	9	8/29/05 0930	6.5	YES
31	8/30/05 1500	-27	Post Storm		YES

ADCIRC produces a surge map showing the maximum elevation of water reached at every point in the relevant portion of the model domain, including areas where levee overtopping is expected. This information is crucial for emergency managers who want to know when evacuation routes will be flooded and where levees will be challenged. In some cases, movie animations are also produced to study surge propagation in more detail. Surge elevations have received the most attention to date from various groups investigating the levee failures that occurred on the eastern side of the city, but ADCIRC can also produce snapshots of predicted circulation and current velocities, which may also have played a role. LSU issued surge forecasts based on NHC Advisories 16, 17, 18, 22 and 25, and a postcast based on Advisory 31, issued 27 hours after landfall (Table 1). An LSU surge forecast was printed in the New Orleans Times-Picayune newspaper on the day before the storm struck and has been credited with saving lives by encouraging many to leave before it was too late.

39.3. RESULTS AND DISCUSSION

The NHC track prediction for Katrina was exceptionally consistent and accurate for more than 2 days prior to landfall (Knabb et al. 2005). The center of the eye was forecast to pass between 13 and 59 km east of downtown New Orleans Superdome for all ADCIRC simulations discussed in this paper (Figure 2). The forecast track shifted gradually to the east over this simulation period within this narrow corridor. Areas of surge buildup (Figure 3) and patterns of high-velocity circulation (Figure 4) were apparent in all simulation results initially along the south side of St. Bernard polder and in the triangular "funnel" that opens to the east between the Gulf Intracoastal Waterway (GIWW) and the Mississippi River Gulf Outlet (MRGO). The MRGO is an artificial ship channel that provides an avenue for surge that builds up in Lake Borgne to move into the Inner Harbor Navigation Canal (IHNC) in the heart of New Orleans (Figure 2). Levees on the south (west side) bank of the MRGO failed more completely than any others during Katrina, while floodwalls and levees lining the IHNC also experienced extensive overtopping and breaching.

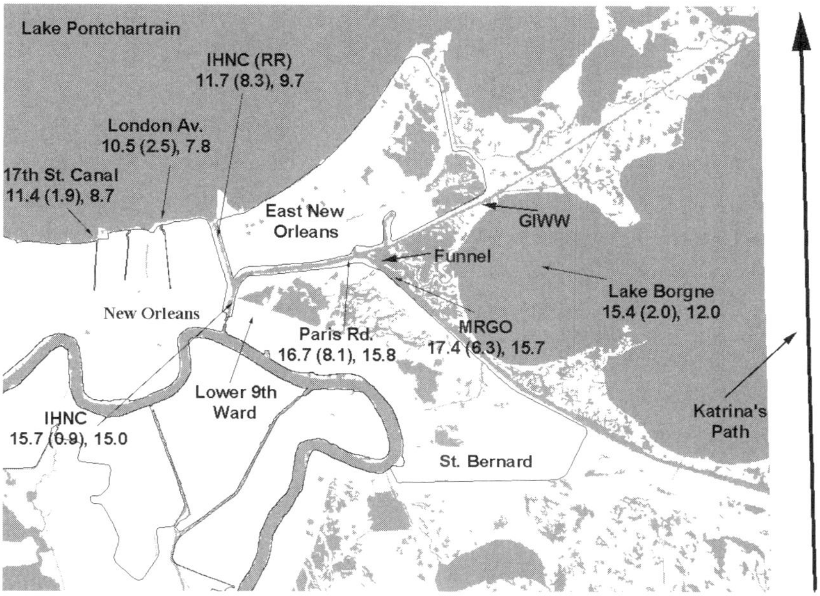

Fig. 2. Predicted ADCIRC maximum surge elevation and current velocity based on NHC Advisories 31 and 18. The arrow indicates the point at which the surge and velocity is reported. The first number under the name indicates the surge elevation (ft, NGVD29) predicted from the Advisory 31 postcast. The second number in parentheses is the maximum current velocity (ft/s). The final number is the surge elevation (ft) forecast 39 hours prior to landfall.

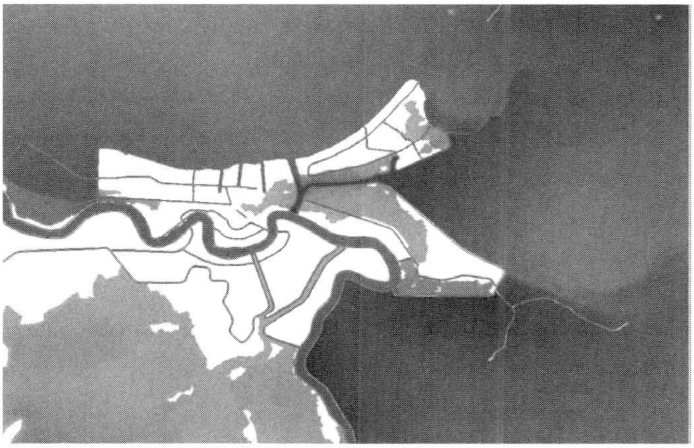

Fig. 3. Flood map showing forecast flooding of New Orleans posted 39 hours prior to hurricane Katrina's landfall based on NHC Advisory 18. Darker shades indicate higher surge elevations.

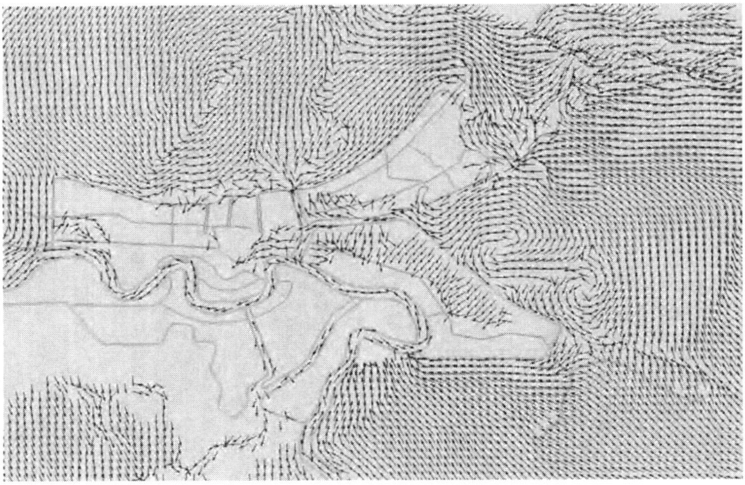

Fig. 4. Flows generated by Katrina toward and into New Orleans over the tops of levees near the onset of flooding. Note that circulation in the triangular "funnel" that opens to the east of the IHNC includes flows both normal and parallel to the MRGO levee.

Table 2. Maximum predicted surge for the advisory 18 and 31. Surge numbers were compared with IPET simulated hind cast. Surge velocities based on the advisory 31 were also shown.

Location (Figure 1)	Surge Forecast (ft) Adv. 18	Surge Postcast (ft) Adv. 31	Surge Hindcast (ft) IPET	Velocity Postcast (ft/s) Adv. 31
IHNC south near the Lower 9th Ward	15.0	15.7	15.0	0.9
IHNC north near the Railroad	9.7	11.7	-	8.3
Lake Borgne	12.0	15.4	-	2.0
MRGO at Bayou Bienvenue	15.7	17.4	18.1	6.3
GIWW /MRGO near Paris Rd.	15.8	16.7	17.2	8.1
Lake Pontchartrain near the **17th Street Canal**	8.7	11.4	12.4	1.9
Lake Pontchartrain near the **London Avenue Canal**	7.8	10.5	11.4	2.5

The first simulation to predict significant overtopping of New Orleans flood defenses is based on Advisory 18, released by the NHC 39 hours before landfall, and by LSU 6 hours later (Table 1). Overtopping is depicted in this simulation along the south levee of the St. Bernard polder, on both sides of the GIWW west of where it joins the MRGO, and into New Orleans Metro along the southern half of the IHNC (Figure 3). The most extensive flooding is shown as spilling out of the GIWW/MRGO into a large impounded marsh (St. Bernard Wetlands) that is bounded on the south by a secondary levee. The only overtopping predicted in this simulation that did not take place later is that into the western side of the city (Figure 3). Surge elevation forecasts generally increased through Advisory 31 as the NHC ramped up its predictions for storm intensity (Figure 2) but the expected locations of highest surge and levee overtopping remained consistent through landfall (Table 2). ADCIRC results produced by the LSU forecast model were also very similar to a hindcast reported by the ADCIRC modelers working with the USACE Interagency Performance Evaluation Team (IPET) and to high water mark elevations measured by LSU and others (IPET 2006; Kemp et al. 2006).

The east side of the city of New Orleans that faces Lake Borgne experienced a higher level of storm surge than the metro center south of Lake Pontchartrain (Figure 2). The surge from Lake Borgne propagated through the MRGO and GIWW to the IHNC causing water levels in that canal to rise above 15 ft (NGVD), while surge along the lakefront to the west was 4 to 5 ft lower. The highest surges predicted and observed for the greater New Orleans area were in the throat of the funnel just east of Paris Road (Figure 2). Among the highest velocities anywhere were indicated for the MRGO near a gate structure that pierces the hurricane protection levee on the south bank of the channel (Table 2). The wingwall on that structure (Bayou Bievenue) failed and resulted in scour 50 ft deep on the protected side of the levee. If the predicted time-series of velocity and surge in the MRGO channel at that point are compared (Figure 5), it can be seen that velocity increases proportionally with surge. A mean current velocity in excess of 3 ft/s is predicted as the surge rises above 8 ft.

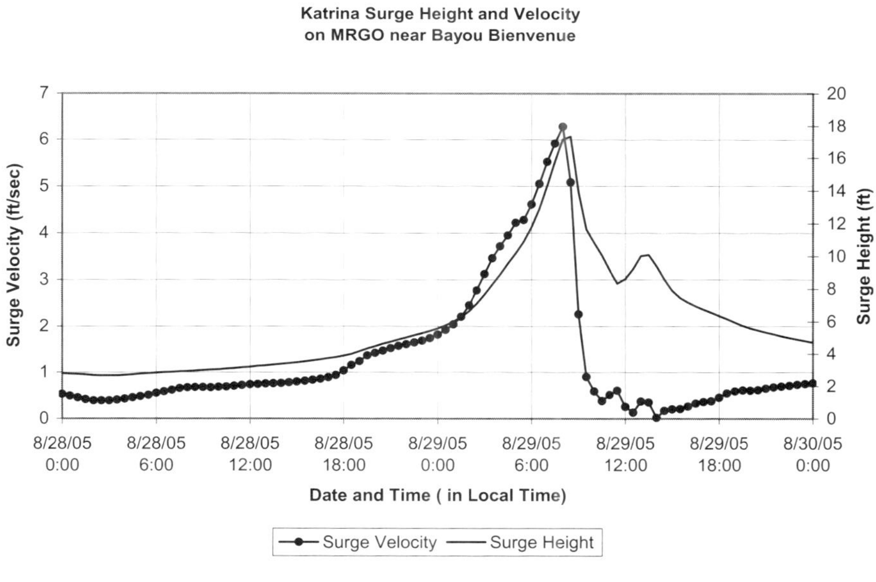

Fig. 5. ADCIRC simulated surge height and velocity on MRGO near Bayou Bienvenue for hurricane Katrina based on NHC Adv. 31 postcast. Note that ADCIRC predicts that scour velocity for non-cohesive material (> 3.0 ft/sec) was reached when surge was below 9.0 ft (NGVD27), several hours before the peak surge arrived.

This is the point where Chow (1983) would predict a serious potential for erosion of the largely non-cohesive soils that made up the adjacent levee. ADCIRC predicts that mean velocities remained above this traditional threshold for more than five hours. It is interesting to see how quickly the velocities drop as the surge begins to fall.

Maximum current velocities to the west were greater than 8 ft/s at Paris Road in the combined GIWW/MRGO channel that connects to the IHNC (Table 2). While concern about erosion of earthen levee faces is also warranted at this location, this jet to the west also suggests a potential to produce dynamic loads on the I-walls and other levee structures that line the IHNC, a number of which failed during Katrina.

39.4. CONCLUSIONS

Louisiana is very susceptible to inundation by hurricane surges. Storm surge builds up over a large area but reaches its highest elevations when it encounters a steep shoreface as along the Mississippi and Alabama coasts or is trapped against levees and floodwalls as it was around New Orleans. Man-made canals such as MRGO, GIWW and IHNC allowed the surge to propagate into the heart of the city and appear to have increased current velocities in the vicinity of earthen hurricane protection levees. We have found that the ADCIRC S08 model experiences a linear speed-up that is proportional to the number of processors to at least 256 thus making operational emergency management support feasible. Accurate modeling of storm surge propagation as Katrina approached allowed for adjustment of short-term plans, and helped ensure a more complete evacuation. In future, it can be expected that storm surge forecasting will become an important addition to the excellent meteorological forecasting services provided today by the NHC that may encourage managers to undertake more flood-fighting efforts to temporarily bolster levees and other structures likely to be challenged, and to preposition supplies to repair anticipated breaches.

REFERENCES

Brouwer, G., 2003. The Creeping Storm. Civil Engineering-ASCE 73(6): 46-55.

Collins, S.M., 2006. Preparing for a Catastrophe: the Hurricane Pam Exercise. Opening statement of Senator Susan M. Collins, Chair, Senate Committee on Homeland Security and Governmental Affairs, January 24, 2006, Washington, D.C. (hsgac.senate.gov /_files/0124065MCOpen.pdf).

Chow, V.T. 1983. Open Channel Hydraulics, 20th Printing, 1983, McGraw-Hill Book Company, New York, NY. 680 pp.

Feyen, J.C., J.H. Atkinson and J.J. Westerink, 2002: GWCE-Based Shallow Water Equation Simulations of the Lake Pontchartrain - Lake Borgne Tidal System. Computational Methods in Water Resources XIV, S.M. Hassanizadeh et al. Eds., Volume 2, Developments in Water Science, 47, Elsevier, Amsterdam, 1581-1588.

IEM Inc., 2004. Southeast Louisiana Catastrophic Hurricane Functional Plan, August 06, 2004, Innovative Emergency Management (IEM), Baton Rouge, Louisiana. 121pp.

IPET (Interagency Performance Evaluation Task Force), 2006. Performance Evaluation Plan and Interim Status, Report 1 of a Series: Performance Evaluation of the New Orleans and Southeast Louisiana Hurricane Protection System. U.S. Army Engineers MMTF 00038-06, January 10, 2006. 165 pp + Apps.

Kemp, G.P., Mashriqui, H.S., van Heerden, I.Ll., Levitan, M., Braud, D., Ropers-Huilman, B.D., Cunningham, R., Evers, E., Hyfield, E., and D. Dartez, Submitted 2006. Forecasting surge flooding of New Orleans during Hurricane Katrina. EOS -weekly newspaper of the geophysical sciences, the American Geophysical Union (AGU), 2000 Florida Avenue N.W. Washington, DC 20009, USA.

Knabb, R.D., Rhome, J.R., and Brown, D.P., 2005. Tropical Cyclone Report: Hurricane Katrina. NOAA National Hurricane Center. 20 Dec., 2005. 42pp.

Luettich, R.A., Jr., and Westerink, J.J., 1995. Continental Shelf Scale Convergence Studies with a Barotropic Tidal Model, Quantitative Skill Assessment for Coastal Ocean Models, D. Lynch and A. Davies [eds.], Coastal and Estuarine Studies series, vol. 48, pp. 349-371, American Geophysical Union press, Washington, D.C.

Thompson, E. F., and Cardone, V. J., 1996. Practical modeling of hurricane surface wind fields. Journal of Waterway Port Coastal and Ocean Engineering - ASCE 122(4): 195-205.

Westerink, J.J., Luettich, R.A., Jr. and Muccino, J., 1994. Modeling Tides in the Western North Atlantic Using Unstructured Graded Grids. Tellus 46a(2):178-199.

Westerink, J.J., Feyen, J.C., Atkinson, J.H., Richard A. Luettich, R.A., Dawson, C.N., Powell, M.D., Dunion, J.P., Roberts, H.J., Kubatko, E.J., Pourtaheri, H., 2004. A New Generation Hurricane Storm Surge Model for Southern Louisiana, University of Notre Dame, Notre Dame, IN 46556, http://www.nd.edu/~adcirc/index.htm.

IN THE WAKE OF THE GULF COAST HURRICANES: A COASTAL FLOOD FREQUENCY ANALYSIS

Marlene Tomaszkiewicz, Wilbert O. Thomas, and Nicole Metzger
Michael Baker, Jr., Inc.
3601 Eisenhower Avenue, Alexandria, VA 22304, USA

ABSTRACT

Extreme storm surges from the 2005 hurricane season caused extensive flooding in the coastal areas along the Gulf Coast. The objective of this presentation is to characterize the magnitude and frequency of storm surges from these tropical storms, to describe damages resulting from this extreme event, to describe frequency analyses at long-term tidal stations (greater than 30 years of record) in the impacted area, and to compare frequency analyses at these long-term tidal stations to elevations from the effective Flood Insurance Studies.

Tidal elevation data for the long-term stations is available from the National Oceanic and Atmospheric Administration, the U.S. Army Corps of Engineers, and the U.S. Geological Survey. Data sets will be assembled indicating the annual peak tidal elevations at various stations along the Gulf Coast. This data can then be used to estimate events of varying frequencies.

The frequency analyses will be performed using four methods for the annual maximum and peak elevation above a threshold data series. The four methods include fitting the two data series to the Pearson Type III and log-Pearson Type III frequency distributions, and using both the Weibull and Cunnane plotting positions. The best approach will be chosen through comparison to the observed data.

40.1. INTRODUCTION

The Atlantic Basin Hurricane Season has been quite active in recent years. Names such as Ivan, Charley, Dennis, Katrina, Rita, and Wilma still resonate loudly with the residents along the Gulf Coast of the United States. Hurricanes Katrina, Charley, Ivan, and Wilma all rank within the top five of the costliest U.S. Atlantic Hurricanes, causing damages that total nearly $120 billion. Despite these statistics, the Gulf Coast remains an alluring place to call home. The Federal Emergency Management Agency (FEMA) has published coastal stillwater elevations for the 10-, 50-, 100-, and 500-year events in these areas in Flood Insurance Studies (FIS). However, perhaps future damage potential can be mitigated if a coastal flood frequency analysis is re-examined along the Gulf Coast.

First, let us focus on the background of each of these tropical events. The landfall locations can all be seen in Figure 1. Hurricane Charley made landfall at Punta Gorda,

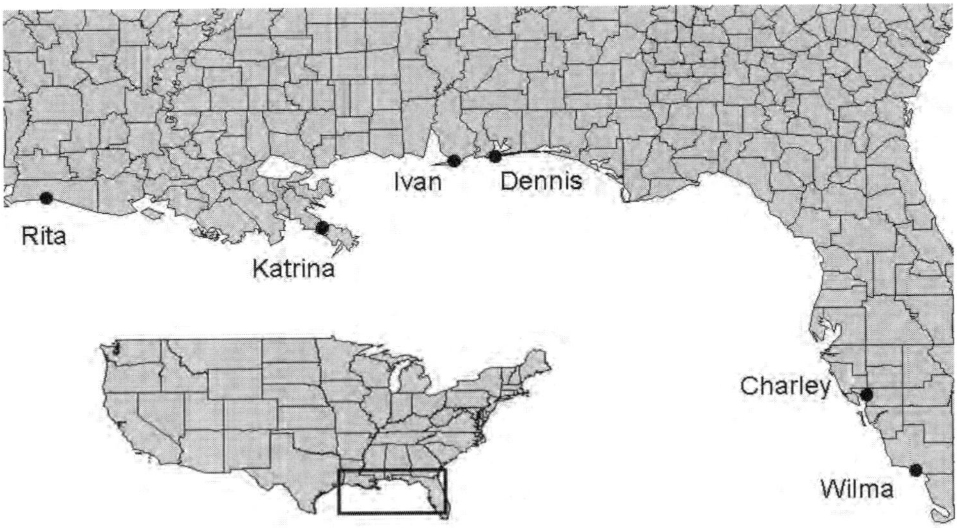

Fig. 1. Landfall locations for major hurricanes in 2004 and 2005.

Florida on August 13, 2004 as a Category 4 storm. Property damages from the storm placed it as the second costliest storm in the United States at the time (and it is now in third place as a result of Hurricane Katrina). Just over a month later, Hurricane Ivan struck the United States. Ivan came ashore at Gulf Shores, Alabama on September 16, 2004. Ivan broke several hydrological records, and is credited with potentially the highest wave height ever recorded, at 91 feet. The storm surge and winds from the storm brought extensive damage to the Florida Panhandle. Further inland, the storm caused rivers and streams to reach approximate 100-year levels. The first major hurricane of 2005 was Dennis. It made landfall in nearly the same location as Hurricane Ivan did, less than a year earlier, on July 10, 2005 at Santa Rosa Island, Florida. Like Ivan, Dennis made landfall as a Category 3 hurricane.

Next was Hurricane Katrina, which forced a re-evaluation of hurricane preparations in its aftermath. Hurricane Katrina made landfall at Grand Isle, Louisiana as a strong Category 3 storm on August 29, 2005. Its storm surge devastated the Mississippi Coastline and caused levee breaches in New Orleans. The sister storms of Rita and Wilma followed soon thereafter. Hurricane Rita made landfall at Johnson's Bayou, Louisiana on September 24, 2005. Towns along the southwestern Louisiana coast were essentially destroyed as a result of the storm. The active storm season of 2005 closed with Hurricane Wilma, which once again struck the Florida Gulf Coast, at Cape Romano, Florida on October 24, 2005 as a Category 3 hurricane.

40.2. DATA

Annual maximum elevation data was obtained from the New Orleans and Mobile Districts of U.S. Army Corps of Engineers (USACE) web site, from the U.S. Geological Survey

(USGS), and from the National Oceanic and Atmospheric Administration (NOAA) Center for Operational Oceanic Products and Services (CO-OPS). Stations were selected based on the proximity to the Gulf Coast and years of record. Stations with more than 30 years of record were used in the analysis. Supplemental elevation data for Hurricane Katrina was obtained from the NOAA *Preliminary Report – Hurricane Katrina Storm Tide Summary* (NOAA, 2005a) and for Hurricane Rita from the NOAA *Preliminary Report – Hurricane Rita Storm Tide Summary* (NOAA, 2005b). Selected USACE and NOAA stations and their locations are shown in Figure 2 and Table 1.

The frequency analyses were performed on the annual maximum elevations (one value per year) at each station. For the USACE stations, once-daily readings at 8:00 am were used in addition to high water marks for the extreme events. High water marks were also used at the NOAA stations, and this is documented later in the report. For the USACE stations in eastern Louisiana, the USACE provided instantaneous maximum elevations and historical data for selected events from 1812 to the beginning of systematic data collection.

Fig. 2. Locations of gages used in analysis.

Table 1. Selected gages used in analysis.

No.	Source	Gage	Location	Latitude	Longitude
1	NOAA	8771450	Galveston Pier 21, TX	29.31 N	94.79 W
2	NOAA	8771510	Galveston Pleasure Pier, TX	29.29 N	94.79 W
3	NOAA	8770570	Sabine Pass North, TX	29.73 N	93.87 W
4	USACE	73650	Calcasieu River and Pass Near Cameron, LA	29.78 N	93.35 W
5	USACE	73472	Calcasieu River Salt Water Barrier Channel, Lake Charles, LA	30.25 N	93.22 W
6	USACE	70900	Mermentau River at Grand Chenier, LA	29.77 N	92.99 W
7	USACE	70600	Lacassine Wildlife Refuge Headquarters, LA	30.00 N	92.78 W
8	USACE	76592	Freshwater Canal at Freshwater Bayou Lock, LA	29.55 N	92.31 W
9	USACE	76600	Schooner Bayou, LA	29.76 N	92.26 W
10	USACE	76720	Leland Bowman Lock, LA	29.78 N	92.20 W
11	USACE	88550	Atchafalaya Bay NE of Eugene Island, LA	29.46 N	91.34 W
12	USACE	03780	Lower Atchafalaya River at Morgan City, LA	29.70 N	91.21 W
13	USACE	85550	Lake Pontchartrain at Frenier, LA	30.11 N	90.40 W
14	USACE	85625	Lake Pontchartrain at West End, LA	30.02 N	90.12 W
15	USACE	82875	Barataria Waterway at Lafitte, LA	29.67 N	90.11 W
16	NOAA	8761724	Grand Isle, LA	29.26 N	89.96 W
17	USACE	85575	Lake Pontchartrain at Mandeville, LA	30.36 N	90.10 W
18	USACE	85800	Bayou Yscloskey at Shell Beach, LA	29.85 N	89.68 W
19	USACE	85750	Chef Menteur Pass, Lake Borgne, LA	30.07 N	89.81 W
20	USACE	85725	Lake Borgne at Rigolets, LA	30.14 N	89.63 W
21	NOAA	8747766	Waveland, MS	30.28 N	89.37 W
22	USACE	-	Back Bay Biloxi at Biloxi, MS	30.40 N	88.84 W
23	USACE	-	Pascagoula River at Pascagoula, MS	30.37 N	88.56 W
24	NOAA	8735180	Dauphin Island, Mobile Bay, AL	30.25 N	88.08 W
25	NOAA	8729840	Pensacola, Pensacola Bay, FL	30.40 N	87.21 W
26	NOAA	8729108	Panama City, St. Andrew Bay, FL	30.15 N	85.67 W
27	NOAA	8575512	Apalachicola, FL	29.73 N	84.98 W
28	NOAA	8725520	Fort Myers, FL	26.65 N	81.87 W
29	NOAA	8725110	Naples, FL	26.13 N	81.81 W
30	NOAA	8724580	Key West, FL	24.55 N	81.81 W
31	NOAA	8723970	Vaca Key, FL	24.71 N	81.11 W

40.3. ANALYSIS APPROACH

The historical data was analyzed using four different methods to estimate elevations for various storm recurrences. Methods of analysis include the Weibull and Cunnane plotting positions, and the Pearson Type III and log-Pearson Type III distributions. The Pearson Type III distribution is a flexible three-parameter distribution commonly used in flood frequency analysis. Procedures for applying the Pearson Type III distribution to the logarithms of annual peak flow are outlined in *Bulletin 17B* (Interagency Advisory Committee on Water Data (IACWD, 1982).

Plotting position formulas can be used to determine the recurrence intervals of climatic and hydrologic data independent of using a theoretical frequency distribution. The basic plotting position formula for determining the recurrence interval in years is

$$PP = \frac{N + 1 - 2a}{m - a}$$

where N is the number of years of record, m is the rank with $m = 1$ for the largest value, and a is a constant unique to a given plotting position. The two plotting positions selected for this study were the Weibull ($a = 0$) and Cunnane ($a = 0.4$) plotting positions. The Weibull plotting position tends to be conservative as the largest flood in 100 years is assigned a recurrence interval of 101 years, whereas the Cunnane plotting position is an unbiased plotting position. Additionally, the Cunnane method is a relatively distribution-free plotting position, implying it is appropriate when the underlying distribution of the data is not known (Cunnane, 1978). The largest event in 100 years is assigned a recurrence interval of 167 years for the Cunnane plotting position.

A sample Pearson Type III frequency curve is shown in Figure 3. The frequency curve is compared to the Cunnane plotting positions.

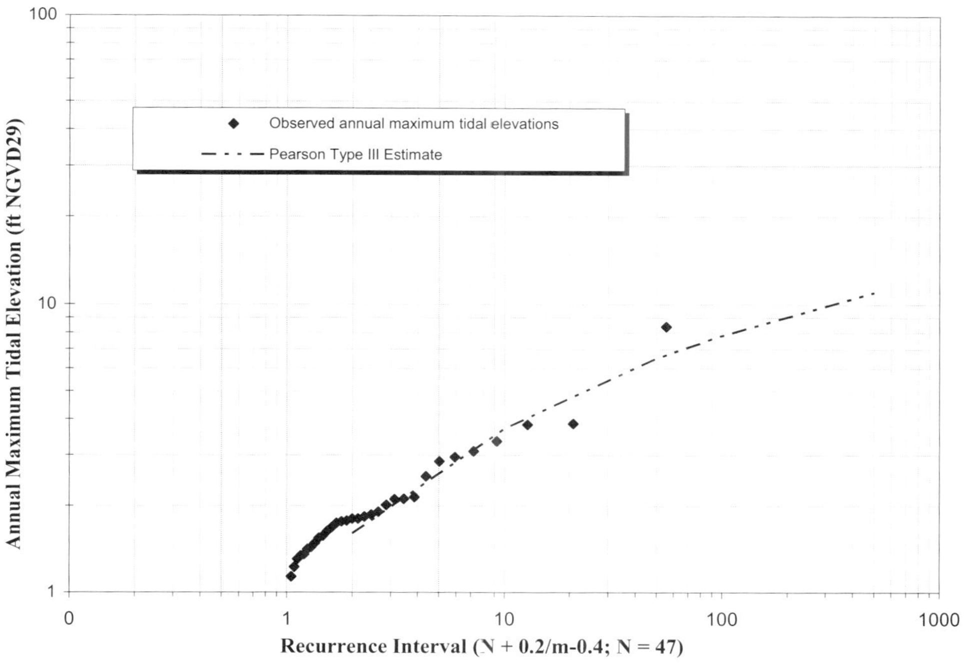

Fig. 3. Pearson Type III frequency curve compared to Cunnane plotting position for Grand Isle, LA.

40.4. RESULTS

A summary comparing the estimated 100-year elevations, the peak elevations from the hurricanes, the FEMA effective 100-year stillwater elevation (SWEL), and the FEMA 100-year or Base Flood Elevation (BFE) is shown in Table 2 for the various methods. The

Table 2. Comparison of estimated 100-year elevations (in feet NGVD 29) to hurricanes (C: Charley, I:Ivan, K: Katrina, R: Rita, W: Wilma).

Location	Hurricane Elev	Cunnane	Log Pearson Type III	Pearson Type III	FEMA SWEL	FEMA BFE
Galveston Pier 21, TX	2.98 (R)	10	5.2	7.0	11.1	AE 11
Galveston Pleasure Pier, TX	3.58 (R)	11	5.5	6.5	13.1	VE 19
Sabine Pass North, TX	9.06 (R)	9.0	7.4	7.9	11.7	VE 13
Calcasieu River and Pass Near Cameron, LA	11.5 (R)	9.9	8.9	9.3	12.3	VE 13
Calcasieu River Salt Water Barrier Channel, Lake Charles, LA	8.65 (R)	8.9	8.1	8.0	8.1	AE 9
Mermentau River at Grand Chenier, LA	10.5 (R)	10.0	9.0	9.4	10.0	AE 11
Lacassine Wildlife Refuge Headquarters, LA	6.44 (R)*	6.6	6.7	6.5	6.3	VE 11
Freshwater Canal at Freshwater Bayou Lock, LA	12.1 (R)	6.5	9.9	10.5	11.9	V 14
Schooner Bayou, LA	7.50 (R)	7.0	6.5	6.7	9.3	A 11
Leland Bowman Lock, LA	9.65 (R)	8.0	8.2	8.6	9.3	A 11
Atchafalaya Bay NE of Eugene Island, LA	10.09 (R)	13.0	12.3	10.6	12.6	VE 19
Lower Atchafalaya River at Morgan City, LA	7.02 (R)	8.1	8.7	8.3	10.0	AE 12
Lake Pontchartrain at Frenier, LA	8.1 (K)	12	10.8	11.3	12.5	---
Lake Pontchartrain at West End, LA	10.9 (K)	11	9.3	10.2	11.6	---
Barataria Waterway at Lafitte, LA	5.91 (R)	6.8	5.9	5.8	7.8	VE 12
Grand Isle, LA	8.38 (K)	9.2	7.8	7.8	9.4	VE 13
Lake Pontchartrain at Mandeville, LA	8.82 (K)	10.1	9.9	10.4	11.6	---
Bayou Yscloskey at Shell Beach, LA	19.65 (K)	16	12.1	13.9	12.2	V 16
Chef Menteur Pass, Lake Borgne, LA	16.13 (K)	18	11.2	13.0	12.2	V 15
Lake Borgne at Rigolets, LA	19.65 (K)	20	13.5	14.0	12.2	V 18
Waveland, MS	23 (K)	23	25.8	22.6	9.2	VE 17
Back Bay Biloxi at Biloxi, MS	24 (K)	17.2	15.7	16.8	10.6	VE 16
Pascagoula River at Pascagoula, MS	13 (K)	13.3	11.9	11.3	8.3	A 6
Dauphin Island, Mobile Bay, AL	6.58 (I)	8.5	7.5	6.7	7.8	VE 10
Pensacola, Pensacola Bay, FL	6.15 (I)	8.5	7.3	7.0	8.0	VE 11
Panama City, St. Andrew Bay, FL	4.07 (I)	7.5	6.6	6.2	8.0	VE 11
Apalachicola, FL	4.33 (I)	9.5	7.5	6.8	8.7	---
Fort Myers, FL	3.53 (C)	7.2	4.7	4.4	8.2	AE 9
Naples, FL	3.21 (C)*	4.5	4.9	4.7	10.1	VE 17
Key West, FL	4.05 (W)	4.0	3.0	3.2	8.4	VE 10
Vaca Key, FL	6.26 (W)	3.0	5.2	5.5	9.2	---

FEMA stillwater elevations were identified in the effective Flood Insurance Study (FIS) reports for the various counties and parishes of the affected areas. The FEMA BFEs were taken from the effective Flood Insurance Rate Maps (FIRMS). The Pearson Type III distribution appears to provide the most reasonable estimates across all recurrence intervals. The log-Pearson Type III estimates for the 500-year events appeared to be unreasonably high for some stations. In some instances, the log-Pearson Type III results appeared to be low for the 100-year event. Based on the qualitative assessment, the Pearson Type III distribution appears to provide the most reasonable statistical estimates for the stations analyzed and will be used to estimate the frequency of the hurricanes.

A map showing the approximate 100-year elevations is shown in Figure 4. This data can be used in future community planning to potentially mitigate damages from future tropical storms.

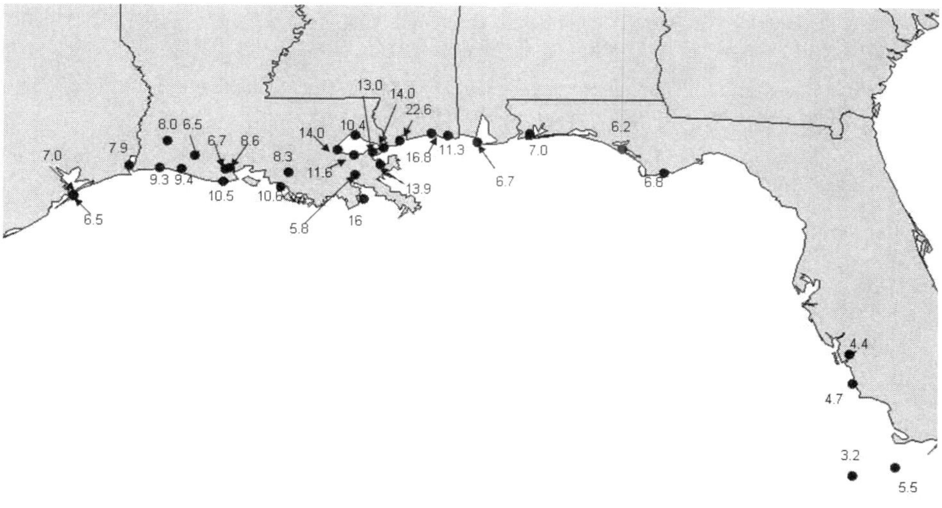

Fig. 4. Approximate 100-year elevations at each study location.

REFERENCES

Cunnane, C., 1978, Unbiased Plotting Positions – A Review: Journal of Hydrology, Vol. 37, pp. 205-222.

Interagency Advisory Committee on Water Data, 1982, Guidelines For Determining Flood Flow Frequency: Bulletin 17B of the Hydrology Subcommittee, Office of Water Data Coordination, U.S. Geological Survey, Reston, Virginia, 183 pages.

National Oceanic and Atmospheric Administration. Center for Operational Oceanographic Products and Services. 15 Sep 2005. <http://140.90.121.76/index.html>

National Oceanic and Atmospheric Administration, 2005a, Preliminary Report – Hurricane Katrina Storm Tide Summary, NOAA Ocean Service, Silver Spring, Maryland, 20 pages.

National Oceanic and Atmospheric Administration, 2005b, Preliminary Report – Hurricane Rita Storm Tide Summary, NOAA Ocean Service, Silver Spring, Maryland, 22 pages.

U.S Army Corps of Engineers, Mobile District. Gulf Coast Tidal Gages. 15 Sep 2005. *<http://water.sam.usace.army.mil/tides.htm>*

U.S. Army Corps of Engineers, New Orleans District. Water Control Data. 15 Sep 2005. <http://www.mvn.usace.army.mil/eng/edhd/watercon.htm>

Wikipedia: The Free Encyclopedia. 14 Jan. 2006. Wikimedia Foundation. 14 Jan. 2006 <http://en.wikipedia.org>.

REBUILDING A CITY ON UNCERTAIN TERRAIN: 300 Years of Human Intervention Altering Louisiana's Deltaic Plain Require Creative Approaches to Flood Hazard Mitigation and Rebuilding of New Orleans

Bruce G. Sharky
Department of Landscape Architecture
Louisiana State University
Baton Rouge, LA 70803, USA

ABSTRACT

This paper presents some urban design strategies for rebuilding more flood-resilient neighborhoods in New Orleans, Louisiana. 80 percent of the city flooded with as much as 10 to 12 feet of water during Hurricane Katrina in 2005. Prior to Hurricane Katrina the city's population was approximately 450,000. It is estimated that six months later the population is about 150,000 with 300,000 people living in diaspora throughout the United States. An historical context leading up to what has been called America's worst natural disaster is presented. The author argues that hurricanes Katrina and Rita should referred not as the nation's worst natural disasters but rather the nation's worst human disaster. The impacts from the two storms are the sum results of nearly 300 years of man's intervention in the highly fragile and complex coastal environment: the Mississippi River deltaic coastal plain. The author presents alternative strategies for rebuilding neighborhoods in flood devastated New Orleans based on non-structural, land use allocations for resettlement. The much larger question is there the will of government leaders to develop and implement a comprehensive regional storm flood protection and hazard mitigation plans that will provide storm protection for category 5 hurricanes. Seasoning the debate are the very real concerns of protecting the rich cultural heritage that together with its natural coastal environment define New Orleans as America's unique city.

41.1. HURRICANES KATRINA AND RITA

The intent of this paper is to suggest a series of land use, urban design strategies for rebuilding more storm resilient neighborhoods in New Orleans. To set the rationale for these strategies, a abbreviated overview of the historical events leading to the hurricane catastrophe of 2005 is presented.

The causes of tropical storm related damages to New Orleans have been attributed to three principal sources in addition to wind related damage:

1. Rain induced flooding.

2. Overtopping of levees and flood walls from storm surge.

3. Levee and floodwall breaches attributed to inadequate design, construction, and maintenance.

The conditions that exacerbated the damages have historical dimensions that go far beyond the actual impacts from Katrina and Rita tropical storms. These conditions include:

1. Loss of the natural defenses against tropical storms such as diminishing coastal marshlands and barrier islands. Also the wholesale destruction of cypress swamps and marshlands in the vicinity of New Orleans, especially the so called back swamp along the shore of Lake Pontchartrain that existed prior to urbanization in Jefferson, Orleans and St. Bernard parishes.
2. Sinking or settling of ground in Jefferson, Orleans and St. Bernard parishes as land was cleared for urban development, reducing moisture of organic soils causing their compaction and settling.
3. Relaxing of building code enforcement especially in areas developed after World War II. Subdivisions were built, in many cases below the 100-year flood level. The ground elevations of many of these areas were below sea level and sank further from the de-watering process of the soils. (See Figure 1 for relative elevations above and below sea level in New Orleans)
4. Rising ocean levels contributing to increasing flooding from tropical storms in urban and coastal areas.

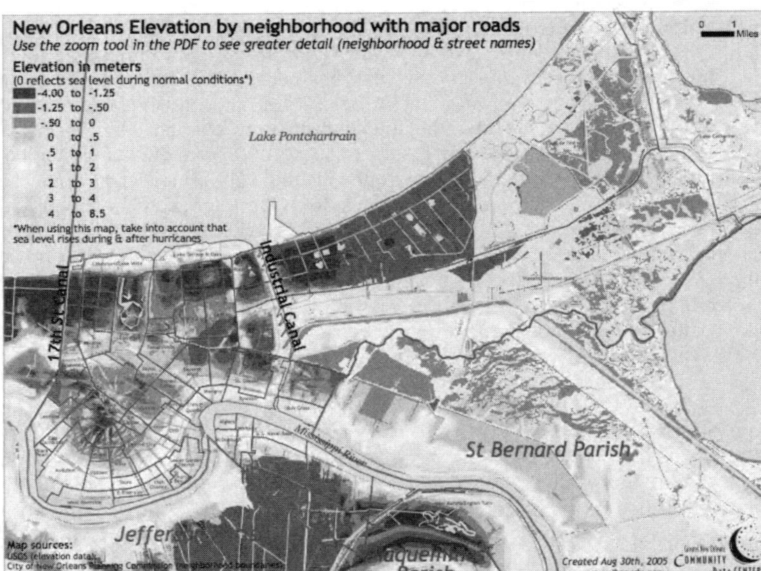

Fig. 1. New Orleans Elevation in Meters. Source: Times Picayune.

The hurricane storm of 2005 and the epic human tragedies surrounding hurricanes Katrina and Rita that devastated coastal Louisiana have been vividly presented to the world by the national and international media. The devastation caused by hurricanes Katrina and Rita has been called the most destructive natural disaster in America's history. Whole communities along the Gulf of Mexico coast from Mississippi to Texas were washed away by the seven-meter plus storm surges that crossed over the coastal marshlands and bays. 400,000 residents of New Orleans evacuated leaving some 50,000 individuals stranded for days and weeks before rescue operations were carried out. The levee and canal system designed for a category 3 hurricane managed initially to perform their flood-protection function as Hurricane Katrina swept in off the coast. However, 80 percent of the city flooded to depths of 3 to 5 meters in areas within the city at elevations well below sea level. The floodwater that inundated the vast majority of the city did not come from storm water that overtopped the levee and canal system. This was what was reported after the initial impact of the hurricane. However, as communications improved the floodwaters were confirmed to have come not from water overtopping the canals but rather from three separate canal breeches. These breeches were caused by floodwall failures, the result of a combination of inadequate design, construction, and maintenance. Large expanses of the city stood in 2 to 4 meters of standing flood water for two to three weeks before pumping operations were completed.

The scale of destruction caused by flooding and hurricane-force winds must be experienced on the ground in order to comprehend the historic extent of property damage, effects on human lives, and disruption of infrastructure. Estimates of the economic losses range from \$100 to \$200 billion in physical losses of property with still yet to be accounted for losses to businesses and other economic activities. Government agencies and various environmental, social, and economic scientists are still appraising the scale of economic productivity losses, and property damage as well as the impact on coastal marshlands and other natural systems. It is estimated that as much as 100 to 125 square miles (160-200 square kms.) of coastal marshes and barrier islands were lost during the two storms. The media has also reported in vivid detail the slow and ineffectual response by federal and state government to the disaster. A combination of poor disaster response planning, delayed response, mismanaged action, and lack of communication further exacerbated the tragedies that unfolded in the days, weeks, and months in the storm's aftermath.

41.2. BRIEF HISTORY

The Frenchman Bienville was charged with finding the optimum location for locating a new city that would afford strategic military advantages in protecting the future port city of New Orleans while providing advantageous river access to the vast North American interior. See Figure 2, of equal importance the location had to be adequate in facilitating maritime commerce by providing safe and convenient transport of goods to and from Europe by way of the Gulf of Mexico via Lake Pontchartrain and the Mississippi River. The natives had been living in the rich coastal marshes and intricate web of wetlands that a future city of New Orleans would be situated. They advised Bienville to locate and build his new capital city of Louisiana on the high ground of the natural levee formed by the Mississippi River. The natural levee and areas of high ground were made higher than the nearby surrounding low swamp lands by the annual deposit of silt from the Mississippi River. This meandering high ground was a natural levee, built up over eons of annual flooding and deposition of silt and debris. Bienville located the new city in what is now called the French Quarter on the natural levee. As the town prospered and grew, successive

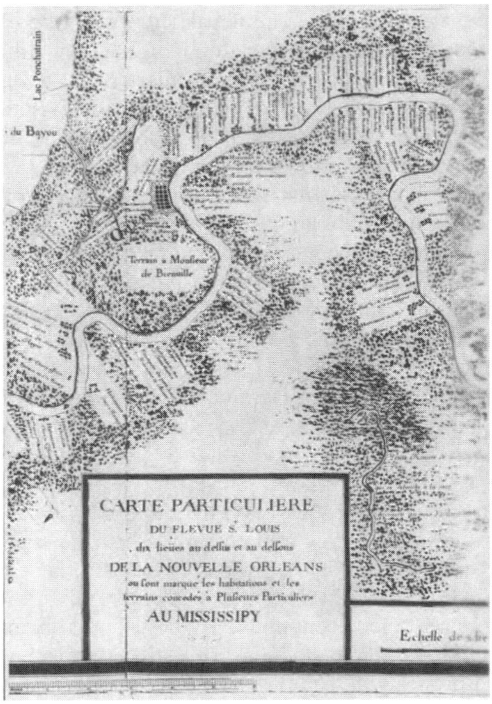

Fig. 2. New Orleans Environs 1723. Note location of French Quarter, extent of swamp areas, and the arpent plantation parcels. Source: LSU Hill Memorial Library.

new development continued to follow the natural levee or high ground, approximately representing the 20 percent of New Orleans that sustained minor or no flooding during Hurricane Katrina. As The Crescent City continued to grow and prosper urbanization gradually creped into the wetlands and marshes which were also on lower lying terrain that often experienced seasonal flooding during heavy rains and the periodic hurricane storms that annually buffet the Mexican Gulf Coast.

Over the years as Louisiana's sovereignty was transferred (France to Spain then back to France), ultimately purchased in 1803 by the US government, New Orleans grew, urbanization replaced the valuable wetlands, and the annual hurricane season came to dominate and influence daily life. Figure 3 delineates urbanization in New Orleans along the Mississippi River's natural levee high ground with low-lying cypress swamps between the Metairie and Gentilly ridges and Lake Pontchartrain. The certainty of a hurricane with the strength of a category 3 or 4 was anticipated. Although a category 5 hurricane never hit, during recorded history, the event was always seen as inevitable. Even under this cloud of the possible one-day demise of New Orleans, the city attained a population of 1.3 million by the 1970s then to decline to 450,000 by end of the 20th century. The explanation for the

Fig. 3. New Orleans 1879 Source: Times Picayune

population decline does not fall in the scope of this paper. During the time, beginning with French rule, structural solutions to the annual storm flooding were designed and implemented.

There are three basic components that make up the infrastructure defense system constructed to protect against storm flooding in New Orleans and along the Mississippi River. The first was the construction of earthen levees on the property lines facing the Mississippi River. The height and location of the levees were generally established after each major storm. The levee system that contains the Mississippi River proved adequate for Katrina with little flooding attributable from the Mississippi River. The second and third structural components put in place to handle flooding and storm surges, particularly in the low-lying areas built up between the Mississippi River and Lake Pontchartrain after World War I consisted of a system of canals later supplemented with pumps to hoist the waters from old wetland bottomlands up over a series of other canals and natural levees into Lake Pontchartrain or the Mississippi River. It was the failure of the canals and pump systems that led to the flooding of nearly 80 percent of New Orleans in the disastrous events that begun with Hurricane Katrina. Figure 4

Other issues effecting urban rebuilding include the economic viability of the city and its capacity to attract investment and sustain viable economic growth since Katrina. Will there be sufficient employment opportunities to support the return of a large portion of the pre-Katrina population? There is currently an insufficient labor force to meet the demand of employers. The inadequacy of the labor force can be explained in part by the lack of sufficient, affordable housing. This is a "what came first the chicken or the egg" issue not so easily solved in the America without governmental intervention Many businesses have moved out of New Orleans and do not plan to come back, at least until the safety of the area from flooding and the social and economic future is made more clear. A healthy recovery will require a vision to attract new types of economic engines to invest in New Orleans with a more general plan of improving economic opportunities in the state.

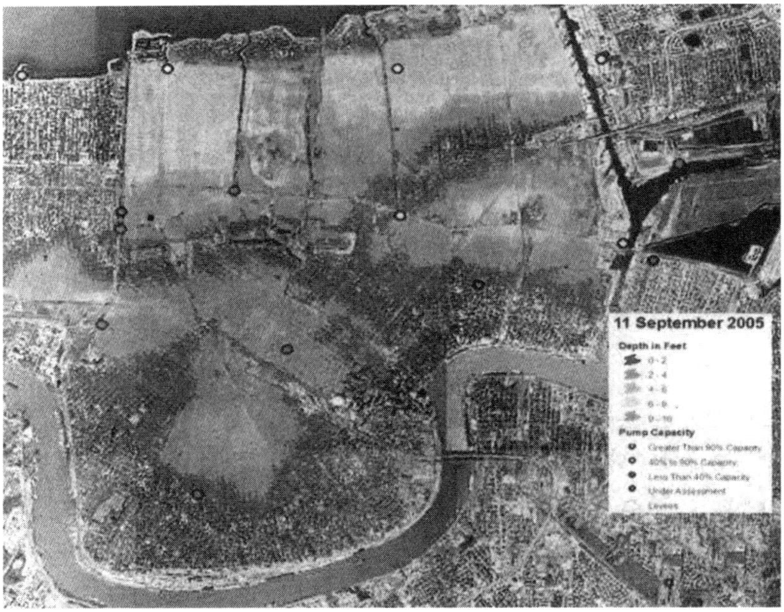

Fig. 4. Depth of Flooding after Hurricane Katrina: New Orleans. Source *Times-Picayune*, New Orleans.

41.3. HUMAN INTERVENTION HAS DIMINISHED THE STORM-FLOOD PROTECTION CAPABILITY OF THE LOUISIANA COASTAL ENVIRONMENT

The underlying tragedy of the 2005 storm events is that much of what was experienced during and after hurricanes Katrina and Rita in terms of property damage, and human losses, and economic disruption cannot be blamed entirely by natural phenomenon. Rather, the damage and losses are the results of a long history of human intervention in the region beginning with the initial French settlement of New Orleans and coastal Louisiana in early 1700s.. A history of government policies beginning with the arpent settlement pattern along river-ways, construction and maintenance of levee systems to control seasonal flooding, finally evolving in the 20th century with the construction of more elaborate structural flood control and hurricane defense systems that included canals, levees, and pumping systems that essentially compartmentalized the city into a 13 bowl-like configurations. Through man's intervention much of the city has subsided well below sea level. It is estimated that ground subsidence of 15 feet (4.5 meter) has occurred in New Orleans since 1788.

As the importance of the Mississippi River for commerce grew a combination of dredging to maintain river transportation routes and the expanding and heightening of the river levees served to cut off the sediment replenishment chain vital for maintaining the coastal marshes and barrier islands. Sediment that used to flow from the Mississippi and Atchafalaya rivers to rebuild the coastal marshes and barrier islands was effectively cut off. Instead the silt-laden floodwaters of the two rivers were channeled by a system of levees that deposited the silt-laden river into the deeper waters of the Gulf of Mexico Continental Shelf. Current estimates indicate that 25 square miles (40 square kilometers) of Louisiana marshes and coastline are lost each year, starved from the lack of seasonal sediment deposition together with a subsiding coastline, and the rise of ocean levels attributed to global warming. Further acts of man have weaken and diminished the natural hurricane defenses of the coast with the construction of canals, dikes, and other modifications of the marshland as part of oil and gas development on and offshore of the coast. Human intervention has weakened the coastal marshes, reducing their ability to dissipate storm surge energy as well as diminish their capacity to recover from inundation after storm-driven flooding.

A fragmented governmental framework further complicates the application of comprehensive, rationale flood management and urban growth activities. Although the levee, canal, and pumping infrastructure are for the most part designed and paid for by the national government, maintenance of large sections of the levee system are divided amongst a large number of local levee boards. A serious lack of coordination exists amongst national, state, and local government frustrated and limited the design and implementation of a comprehensive approach to hurricane flood protection. Well-known issues involving graft and corruption—seemingly a characteristic of the southern Louisiana culture—further exacerbate the plethora of inadequacies of the region's flood control management response.

41.4. WHAT HAVE WE LEARNED AND WHAT IS POSSIBLE TO DO

So what to do? Move New Orleans north to the neighboring state of Arkansas? That may not be that extreme of an idea considering that there is a precedent where communities in the United States and in Louisiana have been relocated to safer ground after serious flooding has re-occurred. Also, predictions of continuing land losses suggest that New Orleans and much of the coast south will no longer exist within the next 50 to 100 years. But the option of relocating New Orleans is not a politically acceptable option at this time. Rationale thinking (considering lessons learned from history and how flooding rarely occurred on areas of natural higher ground in New Orleans) would suggest that, any rebuilding should focus on building on the higher ground that either did not flood or had minimal flooding during Hurricane Katrina. This would mean infilling along or near the natural levee system of the Mississippi River or on the Gentilly-Metarie Ridge, a natural levee formed when the Mississippi River passed further east of its present location. This natural ridge is above sea leave, is located in the middle of the city, and does not traditionally experience flooding including in 2005. This would require increasing building density, constructing 3-5 story or more living units, building a first-rate public transit system, and building in a mix-use configuration. This would follow the model of successful urban places such as Greenwich Village in New York City with shops and stores on the ground floor, with living units on the upper floors. This has proven to be a good model, supporting a high quality of life potential. For New Orleans, employing this approach would mean that people would have to give up living in single-family homes. This is not such a bad alternative for residents who evacuated but want to return to live in New Orleans. It may also be more affordable for people and a city government whose finances are severely stressed.. By consolidating the city, this would make it more affordable to rebuild the infrastructure: including: power, water, sewer, schools, public transportation, etc. The city is bankrupt and really can't afford to rebuild the infrastructure for the entire city now or in the foreseeable future. And this means that about 1/3 to 1/2 of the area of the city could be converted to wetlands and green open space. In areas of the city that that are below sea level, this strategy has several benefits and would of course mean that people would not be rebuilding on terrain that is most likely going to flood again during the next major hurricane. We are in a hurricane cycle that will see several more years of Hurricane Katrina strength events. This cycle suggests an even greater threat of catastrophic impact given continuing ground subsidence coupled with rising ocean levels.

41.5. DISASTER MITIGATION AND REBUILDING STRATEGIES

The goal of this paper is to expand the current discussion of rebuilding strategies for New Orleans to include strategies that are sustainable and increase the resiliency of the rebuilt neighborhoods against future storm flooding.. The most likely strategy that government agencies, principally The US Army Corps of Engineers (USACE) is to rebuild the levee and canal protection systems to tropical storm category 3 standards. Additional measures being considered include enforcement of building restrictions within the 100-year flood plain and enforcement of design standards of residential, commercial, and other structures. This paper sets forth an argument that a more comprehensive plan for flood control protection should include non-structural measures. The non-structural measures include establishing a redundant flood detention system 200-300 feet (60 to 110 meters) on either side of the canals and into the lowest lying, flood-prone areas of the city (including Lake View, Gentilly, and Orleans East). Neighborhoods where structural damage exceeds 50 percent or more would be managed as a combination of reconstructed marsh and wetlands, flood detention floodplain, and green space to serve various park, recreation and trail functions as well as wildlife habitat. The non-structural approach assumes that a large, but still undetermined percentage of residents do not intend to return and that it is in the interest of the larger community that land reuse decisions be made for whole segments of a neighborhood rather than be left up to the piecemeal decisions of individuals to return and rebuild. Finally, it is assumed that there is a once in a lifetime opportunity to rebuild New Orleans as a better functioning, compact, and livable city than it was before Hurricane Katrina. To be successful, this opportunity should insure the positive attributes of diversity of cultures, architecture, and neighborhood-based establishments are made an integral to the rebuilding blueprint.

It is too early to say in what manner or degree the city will be rebuilt to its pre-hurricane condition. Rebuilding plans will most certainly include the upgrading and strengthening of the structural systems. This might include rebuilding some canals, installing locks, and perhaps the redesign of the pumping system to include raising the pumps and installing a self-contained, power generating backup. There are viable arguments supporting a non-structural approach to rebuilding the residential and commercial neighborhoods. This approach is reminiscent of the advice the Native Americans gave Bienville three hundred years ago. They purportedly recommended to Bienville to build on the higher ground on the natural levees of the Mississippi River and not in the swamp and marshlands that lay between the river and Lake Pontchartrain to the north. The marsh and cypress swamps served vital protective purposes of detaining storm waters and dissipating of storm surge energy.

Reconstructing wetlands as part of a comprehensive rebuilding strategy for New Orleans would also provide valuable open space and wildlife habitat for the enjoyment of the city's populations. An overlay system of broad open lands located along the canal system would serve as am effective and important redundant backup system for further detaining and carrying away water during times of flooding.

41.6. PRECEDENTS TO CONSIDER

The creation of an inter-connected system park and greenway system has precedents in many cities in the United States and overseas. Boston, Massachusetts carved out of its dense urban fabric what has become that city's beloved system of parks and trails called the Emerald Necklace. The original intent of the city was to create a single-purpose storm flood system in the 1880s. Designed by Fredrick Law Olmsted, the Emerald Necklace consisting of the Back Bay Fens, Arnold Arboretum, Franklin Park, and the Boston Commons was built flood control purposes at the same time serve the park and recreation needs of the city. Areas of the system have also come to serve as habitat for wildlife a residual benefit contributing much to the quality of urban life. See Figure 5 depicts a comparison of Boston and New Orleans with an overlay highlighting the Emerald Necklace over both cities. The

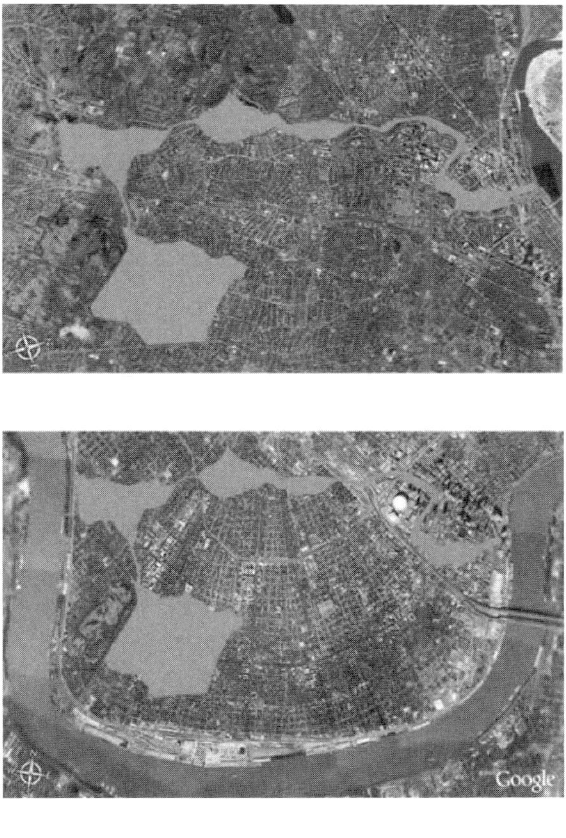

Fig. 5. Boston's Emerald Necklace overlaid on New Orleans. Source: (LSU Landscape Architecture student Tanner Robinson, November 2005).

intent of showing this figure is two-fold: to show, by example, one way of developing a system of parks and floodways designed to serve and knit together many diverse neighborhoods. The figure provides a visual picture of the scale of open space-greenway system that will provide a meaningful degree of flood detention. In the case of New Orleans, the proposed open space-greenway would serve as a redundant flood handling system to the primary levee and canal system. The greenway system would also offer trail, park, and recreation opportunities for residences, in some cases for residences in neighborhoods with limited park and recreation facilities.

Other cities have created similar multi-purpose park and flood management systems. Houston, Texas has recently—since Hurricane Allison in 2002—converted large parcels of land adjoining Buffalo Bayou and Brays Creek into an extensive system of park and trail corridors. See Figures 6A and 6B. Portions of established residential neighborhoods were razed creating storm water detention areas to protect the Texas Medical Campus from flooding. The impetus for both projects was flood control from tropical storms and heavy seasonal rains.

(a)

(b)

Fig. 6. (a) Buffalo Bayou-before, Artist's rendering. Source Dodson Associates, Ltd.; (b). Buffalo Bayou-after, Artist's rendering. Source Dodson Associates, Ltd.

The project provided continuous trails, riparian plantings, and corridor lighting while providing for the central task of flood management. Project designers proposed green fingers extending into develop areas of the city and neighborhoods.

Brays Bayou is also located in Houston, Texas. See Figure 7. Planning fro the project was by the SWA Group of Houston, TX. Project plans call for twenty-one miles of channel improvements and riparian restoration. The plans include provisions for greenways, habitat improvements, and capabilities of managing floods with several storm water detention basins. The project required acquisition of flood-prone land. The client was the Harris County Flood Control District.

Fig. 7. Brays Bayou, Houston, TX. Harrison County Flood Control District.

41.7. PROPOSED NEW ORLEANS GREENWAY SYSTEM

The success of the greenway flood control systems implemented in Boston and Houston, provide a valuable lessons for potential application in New Orleans. New Orleans has in place a web of canals and levee systems that could serve as the foundation of establishing an expanded non-structural flood management system. The planners of the canals created narrow rights-of-way for new canals to be built in existing neighborhoods. The rationale for this approach was to minimize the costly taking of broad cross-section of existing neighborhoods such as Lake View, Gentilly, and Orleans East. Now that the margin of safety is in question with regards to building narrow canals adjacent to residents, considering the extensive damage to homes and property within one to two blocks of failed sections of canals (17th Street, London Avenue, and the Industrial canals) consideration would suggest that at least 200 feet of right-of-way to either side of existing or new canals would provide a reasonable margin of safety. This dimension falls within the range of properties heavily damaged that were in the path of canal breeches experienced during Hurricane Katrina. This right-of-way cross section already exists along a mile or so of several existing canals: Orleans Street, London Avenue, and St. John Bayou. What is being suggested is that existing properties be purchased, structures demolished or moved, and that the remaining land be converted into a combination lineal park and trail system, designed to serve as a redundant system of flood water detention and containment in the event of future tropical storms or heavy rains. See Figures 8a and 8b for an illustration of what is being proposed.

Overall System

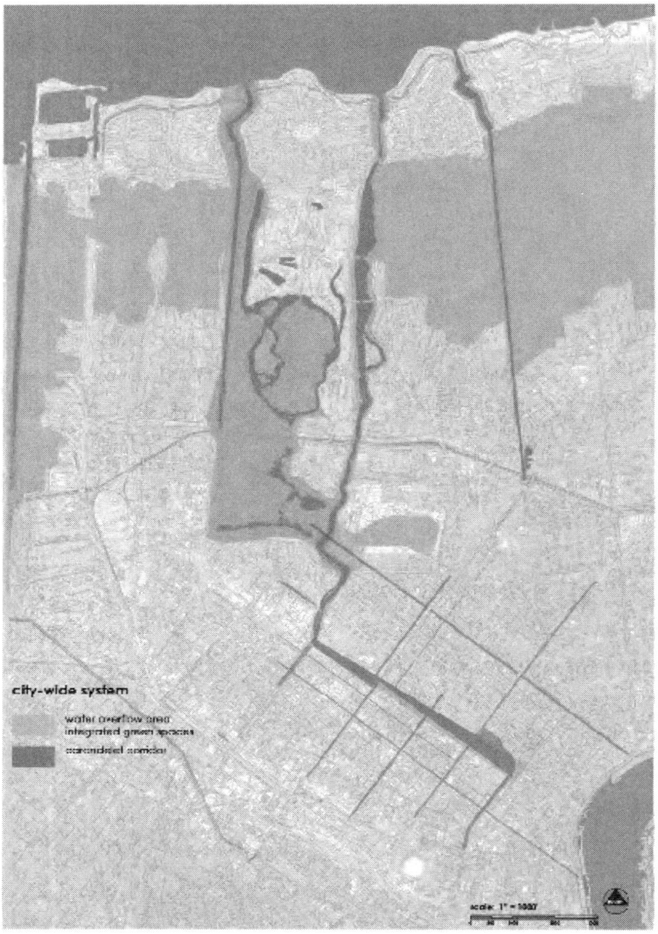

Fig. 8a. Proposal for comprehensive park and open space system with reconstructed wetlands and storm water detention. (LSU Landscape Architecture student Tanner Robinson, November 2005).

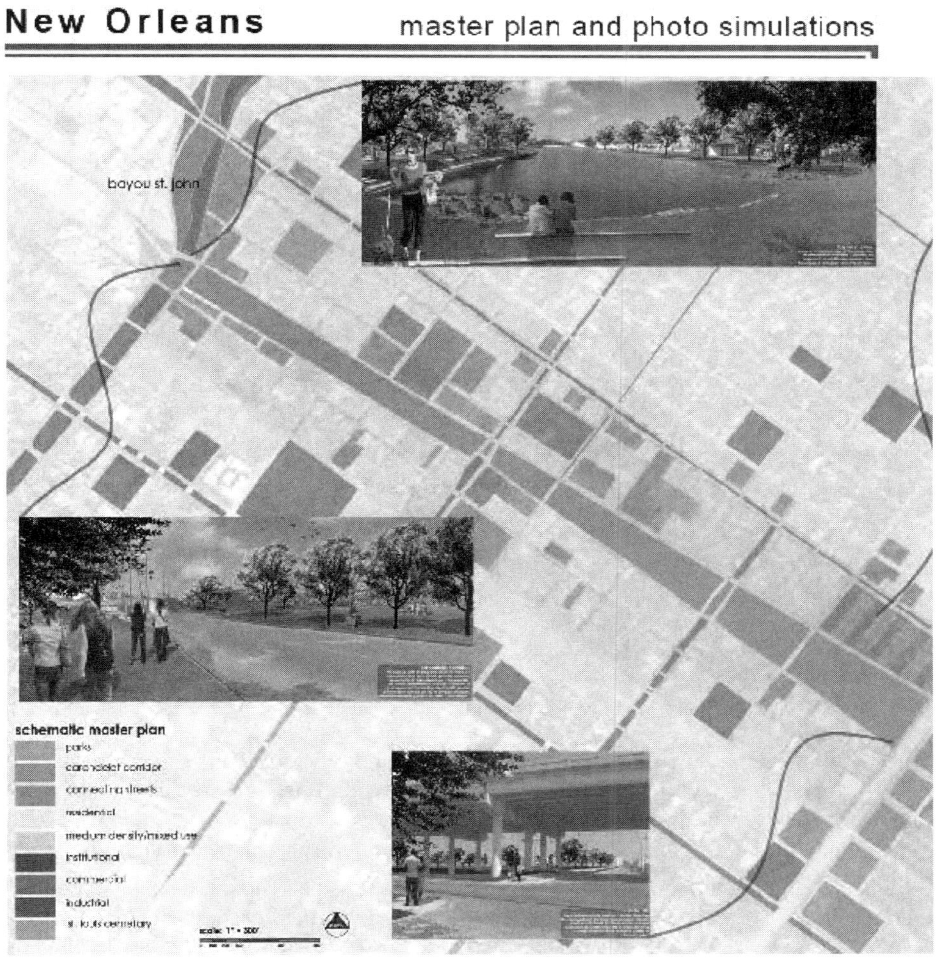

Fig. 8b. Photo Simulations proposal for rebuilding by infilling with open space lands for recreation and storm water retention. (LSU Landscape Architecture student Tanner Robinson, November 2005).

As a means of adding additional storm water storage capacity, neighborhoods adjoining the proposed canal greenway system that are situated well below established 100-year flood elevations and that experienced the greatest percent of building structural damage from to flooding should be purchased. Questions of who should purchase condemned properties with what financial resources is the subject outside the scope of this paper. Current discussion, however, includes the creation of a New Orleans re-development district that would coordinate property buyouts while encouraging residents wishing to return with incentives to rebuild or purchase in designated areas. The low-lying lands would then be converted to park and open space, ultimately serving as additional flood storage and detention function. Some of the land could be converted into wetland and wildlife habitat, in effect allowed to build up with periodic flooding with silt-laden water. (See Figure 9) The goal would be to recreate a naturalized band of coastal vegetation serving as a redundant storm surge barrier (or dissipate storm-surge energy) of the levees and floodwalls along Lake Pontchartrain. The amount of land that may fall within the criteria of high-risk flood susceptibility may be extensive. However, the commitment of this land to open space is reversible should conditions in the future merit reconsideration. The low-lying, flood-prone lands—considering current thinking of providing the city with storm level 3 protection—could be thought of as land placed in a bank in order to give future generations options and for making choices that better serve their interests and needs.

Fig. 9. Proposed reconstruction of wetlands for storm retention, park and recreation for Lake View to Gentilly Areas. (LSU Landscape Architecture student Tien Vo, November 2005).

There are architectural issues involving such topics as maintaining the architectural styles so distinctive to the New Orleans character. Beyond the debate of style are the fears of gentrifying the rebuilt areas of the city and the potential outcome of creating new neighborhoods unaffordable to those most traumatized and who lost the most from Hurricane Katrina. Affordable housing must be provided together with an updated infrastructure. This infrastructure must include better schools and other public faculties and services, adequate public transportation (a system designed considering the needs of evacuating large numbers of people who do not own a vehicle.

41.8. REBUILDING QUALITY NEIGHBORHOODS

It is in large part a quality of life and economic efficiency issue to avoid having individual homeowners decide whether or not to move back and rebuild neighborhoods. A neighborhood scattered with 20, 30 or 50 percent of boarded up homes and cleared lots does create a viable, healthy living environment. What will the value of their invest be with that scenario? Rather, alternate approach is for government to decide what areas to rebuild and what areas to bulldoze.

Bulldoze the low-lying areas susceptible to flooding (hard rains or storm). Bank this land for future consideration, in the meantime convert it into parkway, greenway, trails, and other non-built public uses including flood detention, back up to the canal and other levee flood control systems. Banked greenway could be held in reserve, so that in 30 or more years as the city population grows (if it does) and as better, more effective flood control systems are put in place, some of the banked lands or greenway system could be developed under set of strict and specialize design guidelines. The guidelines would have to take into account potential flooding.

41.9. CONCLUSION

Great uncertainty exists as to whether or not governmental institutions have the vision or courage to implement a comprehensive, flood hazard mitigation plan based on sound science and lessons learned from the events themselves and from what we could learn from the precedents of others such as the Netherlands. William Shakespeare popularized the questioning of whether or not it is wise to continue a tenuous enterprise or set of conditions. Whether it is wise or morally defensible position to continue building a major metropolitan city that is without question in harm's way, is not the question that is being listened to at this time, with respect to fate of New Orleans. Her fate is to rise out of the ashes and the vision is that the city achieve a renaissance, that of becoming not only the most culturally engaging city in America but a city that embraces nature to the extent that valuable wetlands and greenways compliment the cultural attraction that bring millions of visitors a year and provides for a high quality of life for its residents. Maintaining the cultural integrity--the soul of the region and city—is a companion concern of social and culturally priceless importance. Not only are the engineers and scientists working in a fragile, natural environment, the cultural environment of the region is equally fragile and perhaps not as resilient. Vision and caution must guide any proposal for mitigation and rebuilding.

REFERENCES

Busquets, Joan and Felipe Correa, 2005. New Orleans: Strategies for City in Soft Land. Harvard University Graduate School of Design, Cambridge, MA.

Colton, Craig E., 2005. An Unnatural Metropolis: Wresting New Orleans from Nature. Louisiana University Press, Baton Rouge, LA.

Mc Quaid , John, 2005. Mystery Surrounds Floodwall Breaches. Times Picayune, New Orleans, LA. September 13, 2005

Urban Land Institute, 2005. Moving Beyond Recovery to Restoration and Rebirth: Urban Land Institute Makes Recommendations on Rebuilding New Orleans, Washington, D.C. November 2005 http://www.fema.gov/fima/planning.shtm.

AUTHOR INDEX

KEYWORD INDEX

This Keyword Index was created based upon the authors' submitted keywords. The editors made changes to better direct readers to the papers of interest for each topic.